Direct-Write Technologies for Rapid Prototyping Applications

Direct-Write Technologies for Rapid Prototyping Applications:

Sensors, Electronics,

and Integrated Power Sources

ALBERTO PIQUÉ
DOUGLAS B. CHRISEY
Naval Research Laboratory
Materials Science and Technology Division
Washington, D.C.

ACADEMIC PRESS

A Division of Harcourt, Inc.

San Diego   San Francisco   New York   Boston
London   Sydney   Tokyo

Academic Press
A Division of Harcourt, Inc.
525 B Street, Suite 1900, San Diego, CA 92101-4495, USA
http://www.academicpress.com

Academic Press
Harcourt Place, 32 Jamestown Road, London NW1 7DX, UK
http://www.hbuk.co.uk/ap/

Library of Congress Catalog Card Number 2001092384

Internal Standard Book Number: 0-12-174231-8

Printed in the United States of America

01 02 03 04 05 IP 9 8 7 6 5 4 3 2 1

DEDICATION

To Our Families

My wife Cindy Piqué and children, Alexander and Michael Piqué

and

My wife Linda Ann Chrisey, and children, Ryan Heath and Kevin Robert Chrisey

To acknowledge their constant love, humor, enthusiasm, and understanding throughout the preparation of this book

CONTENTS

Part II: Materials

Karel Vanheusden, Paolina Atanassova, James Caruso, Hugh Denham, Mark Hampden-Smith, Klaus Kunze, Toivo Kodas, Allen Schult, and Aaron Stump

Part III: Direct-Write Techniques

Part IV: Comparison to Other Approaches to Pattern Material

PREFACE

Direct-write technologies are the most recent and novel approaches to the fabrication of electronic and sensor devices, as well as integrated power sources, whose sizes range from the meso- to the nanoscales. The term direct write refers to any technique or process capable of depositing, dispensing, or processing different types of materials over various surfaces following a preset pattern or layout. The ability to accomplish both pattern and material transfer processes simultaneously represents a paradigm shift away from the traditional approach for device manufacturing based on lithographic techniques. However, the fundamental concept of direct writing is not new. Every piece of handwriting, for instance, is the result of a direct-write process whereby ink or lead is transferred from a pen, or pencil onto paper in a pattern directed by our hands. The immense power and potential of direct writing lies in its ability to transfer and/or process any type of material over any surface with extreme precision resulting in a functional structure or working device.

Direct-write technologies are a subset of the larger area of rapid prototyping and deal with coatings or structures considered to be two-dimensional in nature. With the tremendous breakthroughs in materials and the methods used to apply them, many of which are discussed in this book, direct-write technologies are poised to be far-reaching and influential well into the future. The industry's push toward these technologies and the pull from applications—rapidly changing circuits, designs, and commercial markets—are documented for the first time here. Although direct-write technologies are serial in nature, they are capable of generating patterns, of high-quality electronic, sensor, and biological materials—among others—at unparalleled speeds, rendering these technologies capable of satisfying growing commercial demands.

The value of commercialization in promoting new technologies and direct write's potential in commercial markets both deserve further attention. For instance, the extent to which the use of integrated circuit processes continues to grow amazes even those active in the electronics field. Indeed, as with all things, the physical aspects of such processes can make their use restrictive, thus designers are looking elsewhere to fulfill customer demand for their overall miniaturization. One area, conformal passives, has already been recognized as having great potential in this regard. Meanwhile, the design, testing, and commercial production of miniaturized electronics are rapidly changing.

Many of the direct-write techniques covered in this book can be used to provide a working prototype in a mere matter of minutes. This is remarkable when compared to conventional approaches whose mask design and patterning can take days or weeks. While most direct-write techniques are in the mesoscale regime in size (from millimeters to 10 μm)—in other words, not as small as photolithography—they are agile in the sense that one prototype can be totally different from the next with almost no effort. Furthermore, the speed of these direct-write techniques makes them viable for the final production, especially if it is a small lot.

With such powerful technologies on the brink of widespread commercialization, the appearance of this text is very timely. *Direct-Write Technologies for Rapid Prototyping Applications* is the first and only book on this topic that discusses the wide range of existing direct-write technologies, including both those that are already poised for commercialization and others still in the development stage that are creating physical structures pushing the limits of nanoscale fabrication. This book is the result of the vision of Dr. Bill Warren and the Mesoscopic Integrated Conformal Electronics (MICE) program, which Dr. Warren lead while at DARPA. It was his interest and motivation that resulted in various teams forming across the U.S. to further the development of new materials and deposition techniques, and this text is a natural consequence of their work. But this book goes far beyond the scope of the MICE program's work to include issues and approaches that one program alone could not. The goal of this text is to cover a representative cross-section of these topics, thereby providing a foundation for further work and commercial development. It can be used as a tool as well as a reference by both commercial production engineers and bench-level research scientists.

This book is organized into four parts: Applications, Materials, Direct-Write Techniques, and Overview of Technologies for Pattern and Material Transfer. We felt it was best to begin our discussion of direct-write technologies with the area driving their developmental research. This first part consists of four chapters covering applications in passive and active electronic devices, micro-electronic manufacturing, electrical power sources, and sensor systems.

The second part of the book focuses on the key to making most of those applications become a reality—the starting materials.

The third part of the book is dedicated to an examination of the numerous direct-write techniques either currently available or in development. The cross-section of different techniques addressed by these eleven chapters is unique because the scale with which each process works varies from centimeters to nanometers. The cross-section is thus meant to represent the incredibly broad impact direct-write technologies can have in research and industry.

The first four chapters of this part describe dispensing-type approaches to direct writing such as Ink Jet, Micropen^{TM}, thermal spraying, and dip-pen nanolithography. The next two chapters cover direct-write techniques based on electron and focused ion beams, while the last five chapters deal with laser-based direct-write techniques.

The book concludes with a technology overview chapter. Putting direct-write technologies in perspective for the electronics and sensor communities at large was a daunting task. This chapter's contributor deserves special mention for the monumental and timely effort he put forth to be comprehensive and clear about the value of these new technologies against the backdrop of decades of lithographic and other patterning processes.

As with all books covering a wide-ranging area, it is impossible to be fully comprehensive addressing all techniques and issues. Given the space limitation, we chose those topics that we thought were the most important representations of this growing field. Still, we know there are novel and exciting topics that we were not able to include or of which we were unaware. In addition, as with any text in a rapidly changing field, by the time this book goes to print, some new development will already have been made. But we feel that this text will have a significant life as it is the first book on this topic as well as the first to group comprehensive overviews of the numerous apects of direct-write processes in a single volume.

Perhaps more importantly, we have asked each contributor to discuss the potential of that chapter's technique in the growing field of direct-write technologies.

While this book serves as the most current overview of direct writing, conference proceedings on direct-write technologies and related topics are other sources for the latest information in the field. For the reader desiring more information on any given topic we would suggest a look at the extensive references provided with each chapter as well as a literature search on that particular contributor.

Lastly, the editors would like to thank all the contributors to this text for the hard work, dedication, and timeliness that it takes to produce such a valuable contribution to the field. The chapters' authors are uniquely qualified to have the best perspective on their respective areas and they have tried to convey that in their work. Furthermore, the authors made a special effort to address their area's shortcomings and potential for future work.

This book would not have been possible without the support of Graham Hubler and Donald Gubser at then Naval Research Laboratory. We would like to thank Bill Warren at DARPA for all his encouragement. Finally, we would like to thank Gregory Franklin and Marsha Fillion at Academic Press for all their help througout the preparation of this book.

Alberto Piqué
Douglas B. Chrisey

CONTRIBUTORS

PAOLINA ATANASSOVA (55, 123), Superior MicroPowders, Albuquerque, New Mexico 87019

PLAMEN ATANASSOV (55), Superior MicroPowders, Albuquerque, New Mexico 87109

RAY AUYEUNG (517), Naval Research Laboratory, Materials Science and Technology Division, Washington, D.C. 20375-5345

R. BASS (313), Naval Research Laboratory, Electronics Sciences and Technology Division, Washington, D.C. 20375

NELSON S. BELL (229), Sandia National Laboratories, Albuquerque, New Mexico 87185-1411

RIMPLE BHATIA (55), Superior MicroPowders, Albuquerque, New Mexico 87109

GEOFF L. BRENNECKA (229), Sandia National Laboratories, Albuquerque, New Mexico 87185-1411

JAMES CARUSO (123), Superior MicroPowders, Albuquerque, New Mexico 87109

DOUGLAS B. CHRISEY (1, 385), Naval Research Laboratory, Materials Science and Technology Division, Washington, D.C. 20375-5345

C. PAUL CHRISTENSEN (385), Potomac Photonics, Inc., Lanham, Maryland 20706

PAUL G. CLEM (229), Sandia National Laboratories, Albuquerque, New Mexico 87185-1411

W. ROYALL COX (177), MicroFab Technologies, Inc., Plano, Texas 75074

LINETTE M. DEMERS (303), Institute for Nanotechnology and Center for Nanofabrication and Molecular Self-Assembly, Nortwestern University, Evanston, Illinois

HUGH DENHAM (123), Superior MicroPowders, Albuquerque, New Mexico 87109

DUANE B. DIMOS (229), Sandia National Laboratories, Albuquerque, New Mexico 87185-1411

KLAUS EDINGER (347), Institute for Research in Electronics and Applied Physics, University of Maryland, College Park, Maryland 20742

MARCELINO ESSIEN (475), Optomec, Inc., Albuquerque, New Mexico 87109

JAMES M. FITZ-GERALD (517), Department of Materials Science and Engineering, University of Virginia, Charlottesville, Virginia 22904-4745

C. FOTAKIS (493), Foundation for Research and Technology-Hellas (FORTH), 71 110 Herklion, Crete, Greece

RICHARD GAMBINO (261), Center for Thermal Spray Research, State University of New York, Stony Brook, New York 11794-2275

DANIEL GAMOTA (33), Motorola Advanced Technology Center, Schaumburg, Illinois 60196-1078

ROBERT GREENLAW (261), Integrated Coating Solutions, Inc., Huntington Beach, California 92646

MARK HAMPDEN-SMITH (55, 123), Superior MicroPowders, Albuquerque, New Mexico 87109

DONALD J. HAYES (177), MicroFab Technologies, Inc., Plano, Texas 75074

HENRY HELVAJIAN (415), Center for Microtechnology, The Aerospace Corporation, Los Angeles, California 90009-295

HERBERT HERMAN (261), Center for Thermal Spray Research, State University of New York, Stony Brook, New York 11794-2275

SEUNGHUN HONG (303), Institute for Nanotechnology and Center for Nanofabrication and Molecular Self-Assembly, Northwestern University, Evanston, Illinois 60208

BRUCE H. KING (229, 475), Optomec, Inc., Albuquerque, New Mexico 87109

TOIVO KODAS (55, 123), Superior MicroPowders, Albuquerque, New Mexico 87109

G. KOUNDOURAKIS (493), Foundation for Research and Technology-Hellas (FORTH), 71 110 Herklion, Crete, Greece

KLAUS KUNZE (123), Superior MicroPowders, Albuquerque, New Mexico 87109

JON LONGTIN (261), Center for Thermal Spray Research, State University of New York, Stony Brook, New York 11794-2275

GREG MARQUEZ (475), Optomec, Inc., Albuquerque, New Mexico 87109

C.R.K. MARRIAN (313), Defense Advanced Research Projects Agency, Electronics Technology Office, Arlington, Virginia 22203-171

R. ANDREW McGILL (93), Naval Research Laboratory, Materials Sciences and Technology Division, Washington, D.C. 20375-5345

W. DOYLE MILLER (475), Optomec, Inc., Albuquerque, New Mexico 87109

CHAD A. MIRKIN (303), Institute for Nanotechnology and Center for Nanofabrication and Molecular Self-Assembly, Northwestern University, Evanston, Illinois 60208

ROHIT MODI (517), Department of Mechanical and Aerospace Engineering, George Washington University, Washington, D.C. 20052

DAVID NAGEL (557), Department of Electrical and Chemical Engineering, George Washington University, Washington, D.C. 20052

PAUL NAPOLITANO (55), Superior MicroPowders, Albuquerque, New Mexico 87109

M.C. PECKERAR (313), Naval Research Laboratory, Electronics Sciences and Technology Division, Washington, D.C. 20375-5345

ALBERTO PIQUÉ (1, 385), Naval Research Laboratory, Materials Science and Technology Division, Washington, D.C. 20375-5345

PHILIP D. RACK (517), Department of Materials Science and Engineering, The University of Tennessee, Knoxville, Tennessee 37996-2200

MICHAEL J. RENN (475), Optomec, Inc., Albuquerque, New Mexico 87109

K.-W. RHEE (313), Naval Research Laboratory, Electronics Sciences and Technology Division, Washington, D.C. 20375-5345

BRADLEY RINGEISEN (93, 517), Naval Research Laboratory, Materials Science and Technology Division, Washington, D.C. 20375-5345

SANJAY SAMPATH (261), Center for Thermal Spray Research, State University of New York, Stony Brook, New York 11794-2275

ALEN SCHULT (123), Superior MicroPowders, Albuquerque, New Mexico 87109

JAMES SKINNER (33), Motorola Advanced Technology Center, Schaumburg, Illinois 60196-1078

AARON STUMP (123), Superior MicroPowders, Albuquerque, New Mexico 87109

JOHN SZCZECH (33), Motorola Advanced Technology Center, Schaumburg, Illinois 60196-1078

ELLEN TORMEY (261), Sarnoff Corporation, Princeton, New Jersey 08543

N.A. VAINOS (493), Foundation for Research and Technology-Hellas (FORTH), 71 110 Herklion, Crete, Greece

KAREL VANHEUSDEN (123), Superior MicroPowders, Albuquerque, New Mexico 87109

DAVID B. WALLACE (177), MicroFab Technologies, Inc., Plano, Texas 75074

WILLIAM L. WARREN (17), Defense Advanced Research Projects Agency, Arlington, Virginia 22203-1714

HUEY-DAW WU (517), SFA, Inc., Largo, Maryland 20774

PETER K. WU (93), Southern Oregon University, Ashland, Oregon

DANIEL YOUNG (517), Naval Research Laboratory, Materials Science and Technology Division, Washington, D.C. 20375-5320

IOANNA ZERGIOTI (493), Foundation for Research and Technology-Hellas (FORTH), 71 110 Herklion, Crete, Greece

JIE ZHANG (33), Motorola Advanced Technology Center, Schaumburg, Illinois 60196-1078

CHAPTER 1

Introduction to Direct-Write Technologies for Rapid Prototyping

DOUGLAS B. CHRISEY AND ALBERTO PIQUÉ

Naval Research Laboratory, Washington, D.C.

1. Direct-Write Technologies
2. Electronics
3. Biomaterials
4. Miscellaneous Application Areas
5. Conclusions

1. DIRECT-WRITE TECHNOLOGIES

The ability to deposit and pattern different thin-film materials is inherent to the fabrication of components and systems such as those found in electronic devices, sensors, MEMS, etc. The trend toward miniaturization has been led by developments in lithography techniques, equipment, and resists materials. But with increased capabilities comes limited flexibility as well as increased complexity, time, and cost. Today there remains, and will remain in the future, applications for rapid prototyping thin film material patterns with CAD/CAM capabilities. A wide range of different direct-write technologies is being developed to satisfy this need. They differ in resolution, writing speed, 3-D and multimaterial capabilities, operational environment (gas, pressure, and temperature), and basically what kinds of final structures can be built. Direct-write technologies do not compete with photolithography for size and scale, but rather complement it for specific applications requiring rapid turnaround and/or pattern iteration, for minimizing environmental impact, for conformal patterning, or for prototyping and modeling difficult components, circuits, subassemblies, etc. (1).

Direct-Write Technologies for Rapid Prototyping Applications
Copyright © 2002 by Academic Press. All rights of reproduction in any form reserved.

The area of electronic components might be the greatest driver for the development of new direct-write technologies and materials with special emphasis on electronic material quality, writing speeds, and processing temperatures. In particular, there is a strong need in industry for rapid prototyping and "just in time methods" (JITM), materials, and tools to direct write passive circuit elements on various substrates, especially in the mesoscopic regime, that is, electronic devices that straddle the size range between conventional microelectronics (sub-micron range) and traditional surface-mount components (10-mm range). Integral passives and high-density interconnects are important and listed on the National Electronics Manufacturing Initiative (NEMI) roadmap. The need is based, in part, on the desire to: (1) rapidly fabricate prototype circuits without iterations in photolithographic mask design, in an effort to iterate the performance on circuits too difficult to accurately model; (2) reduce the size of PCBs and other structures ($\sim$30–50% or more) by conformally incorporating passive circuit elements into the structure; and (3) fabricate parts of electronic circuits by methods that occupy a smaller production scale footprint, that are CAD/CAM compatible, and that can be operated by unskilled personnel or totally controlled from the designer's computer to the working prototype. The savings in time is especially critical to the quick-changing electronics market of today. The novel direct-write approaches described in this book will contribute to new capabilities satisfying next-generation applications in the mesoscopic regime.

There is no single book that summarizes the different direct-write technologies available and emerging today. This is due, in part, to the advances made only recently in the different approaches in this field, such as ink jet printing, laser forward transfer techniques, laser chemical vapor deposition (LCVD), matrix-assisted pulsed-laser evaporation direct write (MAPLE-DW), and Micropen (2). As an example of the relevance of these technologies, it is useful to consider the size of a few markets on which direct-write tools are expected to have an impact. For example, the development of the next generation of sensors for medical applications hinges on the ability to further miniaturize their size as well as tailor their response to controllably varying analytes. The market for biomedical sensors exceeds several billion dollars. In the area of electronic devices and systems, direct-write technologies could play a pivotal role in such markets as Smart Cards, with a global sales exceeding a billion dollars. The development of new multilayer circuit board assemblies, where current technologies are reaching global sales of 30 billion dollars, will also be improved. These examples do not take into consideration the potential for new opportunities and further growth that a new technology, such as direct write, would offer, in which case the market size will grow faster. The goal of this book is to provide a platform to introduce ideas and approaches that are fundamental to direct-write technologies, to present in a similar format a cross-section of some of the conventional and exciting new direct-write techniques, and lastly to objectively discuss the spectrum approaches for patterning materials where direct writing is a subunit.

Electronics and biomaterials are two areas that offer the greatest opportunity and at the same time the greatest challenges for the development, implementation, commercialization, and impact of direct-write processes. The remainder of this chapter uses these two areas to introduce and discuss different direct-write techniques, material issues, and potential for future applications.

2. ELECTRONICS

There are numerous direct-write technologies in existence today and they are of increasing importance in materials processing—enabling, for example, the simplification of printed circuit-board manufacture at reduced costs (3). Other exciting areas for direct-write applications include chemical and biological sensors, integrated power sources, and 3-D artificial tissue engineering. Figure 1 gives a representative sample of some of the different direct-write techniques that exist today. Each of the techniques in Fig. 1 are different in the way they direct write material and each has advantages and disadvantages with respect to a given metric of direct-write ability. There exists no universal tool with which to direct-write all materials, so it may be necessary to combine two or more disparate approaches in order to achieve a direct-write tool capable of a wide range of materials and structures. With a direct-write approach, 3-D structures can be built directly without the use of masks, allowing rapid prototyping, product development, and cost-effective small-lot manufacturing. As materials and processing challenges are being met with increasing success— low processing temperatures and high write speeds, for instance—direct-write techniques move toward an even wider range of possible applications. Passive electronic components (e.g., capacitors, resistors, ferrite-core devices) and conducting interconnects with properties comparable to conventional thick-film approaches (e.g., screen printing and tape casting) have been made by direct-write techniques using a variety of materials. In a parallel development demonstrating the new capabilities of different approaches, direct writing of heterogeneous biomaterials (proteins, DNA, living cells) is being used for tissue engineering and array-based biosensors.

To optimize different direct-write techniques for electronic device fabrication, high-quality electronic starting materials must be developed and tailored for each processing method, transfer method, and required final electronic (or other) device performance. Many different approaches exist to direct-write or transfer-patterned materials as shown in Fig. 1 and each technique has

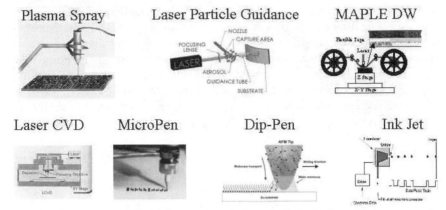

FIGURE 1   To compare different direct-write techniques, a schematic picture of some approaches that are commonly used in commercial or research institutions today is shown. Note the wide range of approaches used to transfer the material and the subsequent effect on material properties, the flexibility to adapt to other material systems, as well as feature resolution.

strengths and shortcomings in terms of ultimate capabilities. The techniques include plasma spray, laser particle guidance, matrix-assisted pulsed-laser evaporation (MAPLE), laser chemical vapor deposition (CVD), Micropen, ink jet, e-beam, focused ion beam, and several novel liquid or droplet microdispensing approaches (4–11). These techniques differ in the way that they transfer, deposit, or dispense materials and they are compared principally in terms of cost, speed, resolution, flexibility to work with different materials, final material properties, and processing temperature.

While the direct-write techniques of Fig. 1 are wide ranging, they are a subset of a larger research-and-development area called rapid prototyping. The classic definition of rapid prototyping is "a special class of machine technology that quickly produces models and prototype parts from 3-D data using an additive approach to form the physical models" (12). The length scales of common rapid prototyping techniques such as solid freeform fabrication (SFF), stereolithography (SLA), selective laser sintering (SLS), fused deposition modeling (FDM), and computer numerical controlled (CNC) milling are centimeters to meters—macroscopic—and extend visibly in three dimensions. Typically they are only to-scale models and are structural as opposed to functional. When viewing the evolution, current status, and future potential for direct-write techniques, it is useful to consider the lessons learned from rapid prototyping. The key aspect of solid freeform and additive fabrication is the ability to deposit or build up material only where it is required to produce finished parts. Novel materials combined with the additive nature of these processes leads to tremendous flexibility in the shape and complexity of parts that can be fabricated. The choice of starting materials and the specific rapid prototyping technique will produce unique microstructures that impact final performance, especially of macroscopic parts. The ability to do point-wise deposition of one or more materials provides the opportunity for fabricating materials with novel microstructural and macrostructural features such as micro-engineered porosity, graded interfaces, and complex multimaterial constructions.

There is a thread of continuity between all direct-write prototyping techniques: their dependence on high-quality starting materials, typically with specially tailored chemistries and/or rheological properties (e.g., viscosity, density, and surface tension). The starting materials, sometimes termed "pastes" or "inks," may consist of combinations of powders, nanopowders, flakes, surface coatings, organic precursors, binders, vehicles, solvents, dispersants, and surfactants. These materials have applications as conductors, resistors, and dielectrics and are currently being developed specifically for very low-temperature deposition (< 200 to $400\,^{\circ}$C). Low-temperature processing will allow fabrication of passive electronic components and radio-frequency devices with the performance of conventional thick-film materials, but on low-temperature flexible substrates, such as plastics, paper, and fabrics. The desired final electronic materials may be silver, gold, palladium, and copper conductors or alloy conductors; polymer thick film and ruthenium oxide-based (cermet) resistors; and silicate glass and metal titanate-based dielectrics.

Figure 2 shows a schematic diagram outlining the evolution and future of rapid prototyping technologies and materials development as it applies to direct writing. Today, most rapid prototyping techniques involve depositing single materials and they are typically not electronic materials with the exception of some limited metals. To expand this capability to multiple materials will principally involve developing the rapid prototyping tool. New techniques or approaches are needed to address the issues with handling different materials such as the compatibility of multimaterial, multifunctional components and the co-firing of multifunctional, multimaterial systems. Subsequent to the rapid prototyping of multiple materials will be the expansion of applications to flexible polymer or other low-processing temperature substrates. Research in this area is ongoing and is centered on novel approaches to processing materials at lower temperatures. One popular approach is the use of laser surface sintering, because under the right conditions (e.g., wavelength, fluence, pulse width) it allows the surface of the direct-written material to achieve temperatures far beyond the decomposition temperature of common polymer substrates. That is, the heat-affected zone does not penetrate significantly beyond the penetration depth of the laser.

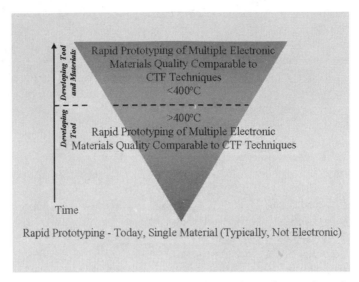

FIGURE 2 A schematic diagram is shown outlining the evolution of rapid prototyping technologies and materials development compared to conventional thick film (CTF) techniques. The dotted line indicates the approximate maximum processing temperature for the polymer substrate Kapton.

Figure 3 shows a schematic diagram of the approximate temperature scales associated with direct-write processing. In order for direct-write techniques to increase their area of applications and compatibility with almost any type of substrate, most all direct-write processing needs to occur in an ambient atmosphere at room temperature. Fabrics can only withstand temperatures of about 200 °C and Kapton about 400 °C. Chemical precursors react over a large temperature range with minimum temperatures of about 200 °C. For high-k ceramic dielectrics, the thermal sintering temperature and atomic-diffusion-distances-per-unit time increases with temperature and with noticeable changes starting at about 500 °C. The ambient increase in the surface temperature by laser sintering can be estimated for a given set of conditions, but a conservative estimate of best case sintering is approximated at about 200 °C. In Figure 3, we show this increase against two possible ambient laser-sintering temperatures: one at which proteins denature and another that is the maximum temperature at which Kapton can be processed. Ceramic materials used in conventional thick-film approaches typically require sintering at temperatures around 900 °C for 1 hour that would be catastrophic for temperature-sensitive plastic substrates. For infrared laser wavelengths (1 to 10 μm), the penetration depth is very dependent on laser and material

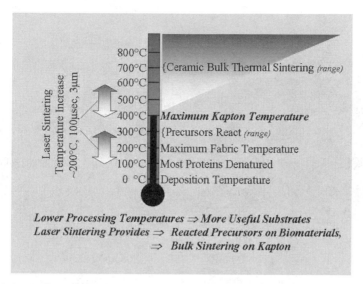

FIGURE 3   A schematic diagram is shown of the approximate temperature scales associated with direct-write processing as well as the approximate improvement due to laser sintering.

properties, but for most combinations good, localized surface absorption is achieved between 1 and 10 micrometers. Annealing just the surface brings with it the benefits of higher temperature processing on thermally sensitive, but technologically important, plastic or paper substrates.

Serious technological problems occur for the fabrication of high-quality crystalline materials, required for state-of-the-art electronic performance of the final device, because it is nearly impossible at these processing temperatures ($\leq 400\ ^\circ C$) even with laser sintering. That is, high-quality crystalline ceramic processing of refractory materials is synonymous with high temperatures for long times as required for atoms to diffuse and defects to be annealed (see Fig. 3). Consider, for example, dielectrics for high-k, low-loss capacitors applications. The highest possible packing density for spherical powders is $\sim 74\%$ for the face-centered cubic structure and it is even lower ($\sim 64\%$) for random close packing (13,14). This means that there is at least 26% air in the structure. According to the logarithmic mixing rule for dielectrics, 26% air reduces the effective dielectric constant by almost an order of magnitude, highlighting the importance of reducing the porosity in transferred materials (15). Figure 4A illustrates the difficulty involved in forming a high-density coating with spherical powders alone. Significant particle–particle bonding does not occur at low processing temperatures. One approach to overcoming the porosity problem is with a poly-dispersed powder, that is, spherical powders of different

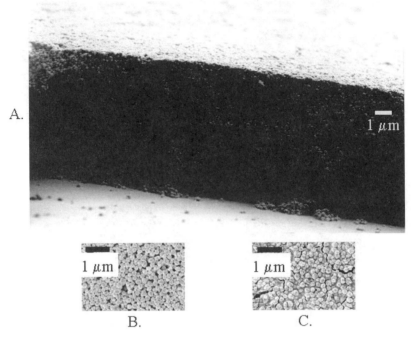

FIGURE 4   Scanning electron micrographs of a fracture cross-section (A), and surface images (B and C) demonstrate the extremely uniform and optimized packing of $BaTiO_3$ nanopowders.

particle sizes. Furthermore, porosity is a source of dangling bonds and adsorbed moisture, neither of which lends it to either low loss in the microwave or to the millimeter frequency regime.

Another way to overcome this liability, which most of the aforementioned techniques can use to some extent, is to begin with a high-density packed powder of differing particle sizes combined with chemical precursors that form low-melting-point nanoparticles *in situ*, welding the powder together chemically (16). The precursor chemistries used are diverse and include various thermal, photochemical and vapor, liquid, and/or gas coreactants. The precursor chemistries carefully avoid incorporation of carbon and hydroxide (which would cause high losses at microwave frequencies) and chemistries that are incompatible with other fabrication line processing steps. Part of optimizing the materials for different direct-write processes will be to find the ideal ratios of powders, nanopowders, and chemical precursors to produce the most dense film with the desired electronic properties. Figure 4B,C demonstrate two cases in this optimization. In Fig. 4B there is a low precursor/powder ratio as is clear from the porosity in the micrograph. In Fig. 4C the ratio is closer to ideal since

FIGURE 5   An example of a loaded dipole antenna and log-periodic fractal antenna fabricated by direct-write methods.

the voids observed in Fig. 4B are filled with reacted precursor. Also in Fig. 4C, cracking is present due, in part, to the thermal reaction of the precursor. Figure 4 shows examples for conventional thermal processing at low temperatures. To further improve the electronic properties for low-temperature processing, especially of oxide ceramics, laser surface sintering is used to enhance particle–particle bonding and reduce porosity.

In most cases, individual direct-write techniques make trade-offs between increasing particle bonding to help the transfer process and optimizing direct-write properties such as resolution or speed. The resolution of direct-write lines can be on the micrometer scale, speeds can be greater than 200 mm/s, and the electronic material properties are comparable to those of conventional screen-printed materials. The use of electronic materials that have been optimized for direct-write technologies results in deposition of finer features, minimal process variation, lower prototyping and production costs, higher manufacturing yields, decreased prototyping and production time, greater manufacturing flexibility, and reduced capital investments.

An excellent example of the application of direct writing to low-cost product development is the rapid prototyping of multiband antennae. Examples of two of these antennae are shown in Fig. 5. The direct writing of these antennae is accomplished in minutes with the commercial, purchased software that produces the design file, whereas conventional lithographic mask fabrication, patterning, and process development would take days. Furthermore, selective dielectric loading to enhance particular band performance can be done without resorting to complicated photolithographic lift-off techniques.

3. BIOMATERIALS

Materials advances have driven advances in the direct writing of electronic materials. In contrast, recent advances in the direct writing of biomaterials have been driven by advances in the transfer technology, in particular, the recognition that in some cases the transfer process can be extremely gentle and selective. Two laser transfer techniques in particular have been successful with novel biomaterial depositions. First, laser-guided particle deposition has the ability to deposit almost any material on any substrate with micrometer-scale accuracy. This technique has been applied to the deposition of individual "living" neural cells (see Fig. 6). The deposited cells remained viable in spite of being exposed to the laser transfer process (17,18). Micrometer-scale patterns of viable cells and the methods to manufacture them are required for next-generation tissue engineering, fabrication of cell-based microfluidic biosensor arrays, and selective separation and culturing of micro-organisms. (No current technology is capable of writing adjacent patterns of different viable cells.) Second, at the Naval Research Laboratory, patterns of viable *Escherichia coli* bacteria, as well as of various types of eukaryotic cells, proteins, enzymes, and antibodies have been transferred onto various substrates with a laser-based forward transfer technique (19). These techniques are covered in depth later in the text.

With these new computer-controlled tools, we are now in a position to physically manipulate living and structural life-supporting biomaterials from their natural environment. In particular, we can create three-dimensional mesoscopically engineered structures of living cells, biocompatible cell scaf-folding, proteins, DNA strands, and antibodies as well as cofabricate electronic devices on the same substrate to rapidly generate cell-based biosensors and

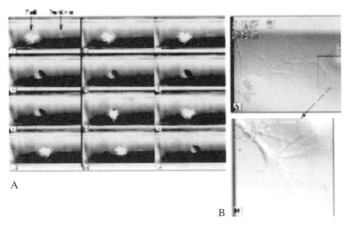

A

B

FIGURE 6 An individual neuron cell (embryonic chicken spinal cord) was deposited by laser guidance in a hollow-core fiber. Inner diameter of the fiber: 30 mm; cell diameter: ~9 mm. Time lapse between images: 0.3 s. (A) and (B) illustrate cell viability after guidance with normal adhesion and neurite growth. Image width in (B): ~30 mm [Adapted from (17)].

bioelectronic interfaces. This will, for example, allow us to probe intercellular signaling in cancer cells or to construct tissue-based biosensors. An exciting application of direct writing biomaterials is a computer-controlled (i.e., bioCAD/CAM) system that will be able to help identify a human wound or tissue abnormality, prompt human input in order to three-dimensionally map the heterogeneous biomaterial profile needed for repair (i.e., cell/biomolecule structure), and form the complementary tissue or organ structure designed from the input. These systems could potentially be used in battlefield, hospital trauma, and general surgical applications to speed diagnosis, the repair process, and recovery. Furthermore, these techniques have the unique ability to construct tissue cell by cell. Different cell types can easily be placed adjacently or in multilayers by changing the culture from which we perform the transfer. This ability will enable us to create miniature organs that are engineered to mimic natural tissues. These methods represent an important advance in biomaterial processing and the manipulation of natural systems.

4. MISCELLANEOUS APPLICATION AREAS

Direct-write technologies also have potential in the medical device markets with the fabrication of micromechanical components, labeling, and biocompatible and active coatings. UV micromachining, a subtractive direct-write process, has already shown the ability to reduce the size and improve the performance of a number of medical devices (20).

Various direct-write methods could be utilized for efficient assembly of components in the display and interconnect industry. For example, the "drop on demand" ink-jet approach has shown itself to be an outstanding path for the microdeposition of organic light-emitting polymers and phosphors, solder bumps, spacer balls, electrical interconnects, and adhesive sealant/bond lines in the manufacture of display panels. Some of the advantages of this route to fabrication include low cost, high speed, noncontact, and environmentally friendly processing.

5. CONCLUSIONS

In this chapter, we introduced the direct writing of electronic materials of biomaterials, and of other miscellaneous materials, and discussed some of the issues surrounding their use in future applications. Direct writing is a subset of the larger field of rapid prototyping and has experienced significant growth recently through the development of new transfer methods and improved materials that react at lower temperatures. The impact these advances should have on the field of direct writing will make many of the approaches commercially viable for rapid prototyping and small-lot-size manufacturing. For electronics, these technologies offer opportunities in both manufacturing improved discrete electronic devices on flexible substrates and rapid prototyping machines with increased flexibility. Furthermore, as a research tool, direct writing is already breaking new ground in the computer-controlled three-dimensional construction of biological materials. Direct writing breaks the paradigm outlined in the first paragraph in this chapter: the assumption that increased capabilities must come with limited flexibility. Instead, there are advantages of lower cost with increased flexibility of manufacturing, enhanced process integration, and reduction in environmental impact (21,22).

Because this is the first edited textbook on this subject, we organized the chapters in such a way that we hope will provide a strong foundation for understanding the many issues common to most direct-write approaches. Care was taken to also include information not only about conventional approaches to direct writing, but also about novel ones. It was our goal to have each author (or team of authors) be as objective as possible in comparing techniques and to list both a technique's strengths and its weaknesses. The value of this text to the general reader will be in its comprehensive overview of the field and the stimulation of ideas for new application areas of continued research.

REFERENCES

1. D. B. Chrisey, *Science* 289, 879 (2000).
2. D. B. Chrisey, D. R. Gamota, H. Helvajian, D. P. Taylor, *Materials Development for Direct Write Technologies*, Materials Research Society Symposium Proceedings Vol. 624 (2001).
3. A. R. Ehsani and M. Kesler, *IEEE Spectrum* (May 2000), p. 40.
4. H. Esrom, J.-Y. Zhang, U. Kogelschatz, A. J. Pedraza, *Appl. Surf. Sci.* 86, 202 (1995).
5. M. J. Renn, R. Pastel, H. J. Lewandowski, *Phys. Rev. Lett.* 82, 1574 (1999).
6. P. Blazdell and S. Kurdoda, *Surf. Coat. Tech.* 123, 239 (2000).
7. M. K. Herndon, R. T. Collins, R. E. Hollingsworth, P. R. Larson, M. B. Johnson, *Appl. Phys. Lett.* 74, 141 (1999).
8. H. Herman and S. Sampath, *Industr. Ceram.* 18, 29 (1999).
9. B. H. King, D. Dimos, P. Yang, S. L. Morissette, *J. Electroceram.* 3, 173 (1999).
10. D. B. Chrisey et al., *Appl. Surf. Sci.* 154, 593 (2000).
11. J. M. Fitz-Gerald et al., *Appl. Phys. Lett.* 76, 1386 (2000).
12. T. Grimm and T. Wohlers, Time-Compression Technologies May (2001), p. 40.
13. S. Torquato, T. M. Truskett, P. G. Debenedetti, *Phys. Rev. Lett.* 84, 2064 (2000).
14. O. Pouliquen, M. Nicolas, P. D. Weidman, *Phys. Rev. Lett.* 79, 3640 (1997).
15. M. P. McNeal, S.-J. Jang, R. E. Newnham, *Proc. IEEE Intl. Symp. Appl. Ferro.* 2, 837 (1996).
16. P. Atanassova, K. Kunze, T. Kodas, M. Hampden-Smith, Superior Micropowders, unpublished results.
17. D. Odde and M. J. Renn, *Biotechnol. Bioeng.* 67, 312 (2000).
18. D. Odde and M. J. Renn, *Trends Biotechnol.* 17, 385 (1999).

19. B. R. Ringeisen, D. B. Chrisey, A. Piqué, H. D. Young, R. Modi, M. Bucaro, J. Jones-Meehan, B. J. Spargo, *J. Biomaterials* in press.
20. C. P. Christensen, Medical Device and Diagnostic Industry, January (1995).
21. X. Yan and P. Gu, *Computer-Aided Design*, 28, 307 (1996).
22. R. Maddox and J. Knesek, *Aerospace Am.* 31, 28 (1993).

Applications

In many cases, applications tend to drive the development of new technologies, and direct writing is one such technology. The need for direct writing electronic and sensor materials is founded in exciting and often revolutionary applications, numerous examples of which will be given here. The specific applications presented individually in each chapter are representative of some areas where direct-write technologies could have an impact. As successful applications are commercialized—demonstrating the inherent flexibility of direct-write techniques—the potential for using direct-write products in other areas grows. Part I is devoted to applications of direct-write material deposition, in particular, applications to defense electronics, chemical and biological sensors, industrial applications, and small-scale power-management applications. Other exciting applications are on the horizon for use in medicine, tissue engineering, wireless and other communications, optoelectronics, and semiconductors.

CHAPTER 2

Overview of Commercial and Military Application Areas in Passive and Active Electronic Devices

WILLIAM L. WARREN

Defense Advanced Research Project Agency, Arlington, Virginia

1. Introduction
2. Direct-Write Electronic Component Manufacturing
3. Making Direct-Write Processes a Reality
4. Applications of Direct-Write Manufacturing
5. Conclusions

1. INTRODUCTION

Passive components (capacitors, resistors, and inductors) are commonly soldered to a printed wiring board using surface mount technologies, or are being embedded into low-temperature cofireable ceramic (LTCC) packages. The usual technique for making these LTCC packages is based on cofiring of laminates made from layers of green tapes with screen-printed features at temperatures around 850 °C. The relatively high processing temperatures employed make it impossible to use these multilayer ceramic packages on low-temperature/conformal substrates. Dozens of process steps can be required to fabricate a thick film or cofired ceramic substrate; in its entirety, weeks of effort can be expended on these processes for a new design. Another limitation is that screen printing has limited feature tolerances.

An additional complexity is that simulations are not currently precise enough to accurately predict the performance of the passive components fabricated in a dense multilayered fashion (three dimensions). This being the case, new designs must be iterated several times by an ad hoc methodology to arrive at the required functionality. Consequently, commercial development of

Direct-Write Technologies for Rapid Prototyping Applications

sophisticated integrated ceramics has emphasized a highly empirical approach, making the technology impractical for low-volume, specialty components or for rapid prototyping. This trend toward higher-level integration, which is the passive component analog of an integrated circuit (IC), places increasing demand on fabrication processes and manufacturers. Indeed, we are in an electronic age of sensing, analyzing, actuating, controlling, and monitoring just about everything. Development of these intertwined functionalities requires many iterations of prototyping and testing. These long iterations not only cost precious research and development dollars, but also can cost a company a competitive edge at the marketplace.

2. DIRECT-WRITE ELECTRONIC COMPONENT MANUFACTURING

To overcome the aforementioned limitations regarding the fabrication of electronic components (resistors, capacitors, inductors, interconnects, batteries, antennas, etc.), direct-fabrication approaches are being actively considered to simplify the processing and provide greater flexibility than would otherwise be possible using conventional approaches (tape casting, screen printing, lamination, surface mount approaches, etc.). Direct-write technologies will have an impact in all size ranges, but perhaps the greatest will be in the mesoscale arena. Mesoscale electronic devices are defined as electronic components/devices that straddle the size range between conventional microelectronics (submicron-range) and traditional surface mount components (10 mm-range). This is an all-but-forgotten size regime in the electronics community, but is important to the military, wireless communications, and medical communities to name a few. There is still much to do outside of the integrated circuit.

It is anticipated that the direct-write technologies will provide a rapid prototyping and agile manufacturing approach to integrating mesoscale devices and modules that will revolutionize the design process (1). These technologies will also allow the direct printing of metals, polymers, ceramics, etc. for fabricating multilayer-multicomponent structures of batteries, antennae, and other functional mesoscale devices. For instance, a direct-write technology, which is capable of dispensing high dielectric constant materials and ferrite-based materials, can be used to reduce the manufacturing cost of a miniature antenna by depositing the backing material of a patch antenna only at the specified location.

The direct-write computer aided design (CAD) and computer aided machines (CAM) can be used to control the application of material layers, allowing excellent control of the device design and a highly flexible system design. The ultimate advantage of the direct-write approach is that it integrates many functions into one conformal electronic system including advanced batteries, advanced electronics, and functional devices. The direct-write of mesoscale electronic devices will also allow for front-end inventiveness (due largely to rapid prototype agility—prototype time from weeks to hours), low-cost small-batch processing, and back-end processing for modular upgrades to commercial and Department of Defense (DoD) components. Indeed, the National Electronics Manufacturing Initiative (NEMI) roadmap argues that direct-tools "technology allows for very rapid time to market."

One of the major promises of direct-write technologies is to the potential ability for rapid prototyping and manufacturing of miniaturized and rugged mesoscale electronics on any surface through the three-dimensional integration of passive components (resistors, capacitors, inductors, high-gain antennae, and interconnects) and active components (batteries, etc.). An ultimate direct-write machine goal would be to develop a single system capable of rapid (hours) prototyping/manufacturing CAD/CAM that can deposit a wide variety of materials (conductors, insulators, ferrites, ruthenates, metals, ferroelectrics, glasses, polymers, etc.) for customized, robust, electronic components at low-substrate temperatures in a conformal manner on virtually any substrate (paper, plastic, ceramics, metals, etc.).

The following is a list of some of the needs, impacts, or advantages of a direct-write process.

- The ability to combine many materials, devices, and power for complete electronic module functionality. Figure 1 shows an example of this using a micro-air vehicle.

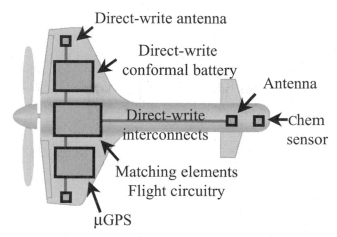

FIGURE 1   Example of the ability to combine many materials, devices, power for complete electronic module functionality using direct-write approaches for a micro-air vehicle.

- Integration with integrated circuits. The integrated passive components will be rugged, light weight, and will save space. There are 125 passive components associated with the 486 chip, while the Pentium II has 345 for power conditioning, filtering noise, decoupling, etc.
- Rapid prototyping and manufacturing. The turnaround time on new designs will occur in hours, not weeks or months.
- Satisfies DoD and commercial products for small-lot manufacturing or for specialized one-of-a-kind parts.
- Machines and electronics become one—these direct-write mesoscopic electronic devices may be integrated with the physical structure on which the electronic systems will be used; there will be no need for a traditional printed wiring board. The ability to print electronic features on flexible substrates enables unique applications for deployable electronics, such as emplacing electronics in projectiles, use in flexible satellite solar arrays and in rolled-up forms that can be inserted into symmetric or odd shapes, installation on warfighters gear, as well as in various types of surveillance equipment.
- Saves space, reduces weight, has three-dimensional integration.
- Dramatic cost savings by elimination of the majority of passive components both in automated fabrication and minimizing procurement. This is an area that has substantial benefit to all DoD systems.
- Reduced inventories of electronic components or parts.
- Specialty parts "on the fly" can be built without mass production set-up prices for "sunk" capital.
- Increased reliability—rugged electronic components due to the automated assembly process and no solder.

3. MAKING DIRECT-WRITE PROCESSES A REALITY

Building a revolutionary direct-write tool requires new materials, fresh ideas, and an open path to the marketplace. Many features are necessary to make direct-write manufacturing come to fruition on a large scale: (1) machine (tool) capabilities, (2) materials development, (3) end-user pull, and (4) applications.

3.1. MACHINE CAPABILITIES

Direct-write tool characteristics include the ability to direct write excellent feature and edge definition through controlled materials-dispensing, the ability to direct erase materials such as laser micromachining, the rapid deposition of multicomponent precursors/materials and fast write speeds ($\sim$ 100s mm/sec) over conformal surfaces. Very loosely, there are generally two categories of the direct-write tool: (1) those that can deposit functional materials during the deposition process, and (2) those that deposit materials which have to be subsequently laser or oven processed to induce controlled and reproducible functionality. A few examples of tools that fall into the first category are laser chemical vapor deposition (LCVD) (2–8), plasma/thermal/solid-state spray processes (9–12), and laser-guided particle deposition (LGPD) tools (13–15). Similarly, a few examples of tools that fall into the second category are the Micropen (or stylus printing) (16), Matrix Assisted Pulsed Laser Evaporation-Direct Write (MAPLE-DW) (17–19), ink jets (20), electrostatic printing (21). Many direct-write processes involve laser interactions with materials. To this extent, it is important to have a detailed understanding of laser processing parameters.

Laser Process Parameters

Laser scan speed	Laser scan period
Laser scan spacing	Material layer thickness
Laser power	Powder/precursor bed temperature
Air velocity over bed	Chamber temperature
Beam spot size	Beam profile

3.2. MATERIALS DEVELOPMENT

Among the considerations that are critical to the design and fabrication of direct-write electronic components is that of the appropriate material(s)/material precursors for each of the components that make up the device (22–27). Selection must be based on factors such as solvent and binder removal, the coefficient of expansion, reactivities, chemical compatibility, stress development, and the direct-write tool. Precursor development will be an important part of the direct-write effort, especially if deposition is to occur at low temperatures. Therefore, development of material property database for a specific fabrication method may be crucial to the successful design and operation of the direct-deposition tool and resulting electronic components. Indeed, a wide variety of material precursors will be needed (metals, ferrites, complex oxides, dielectrics, etc.). What follows is a list of a few of the material parameters that need to be considered during the direct-write process.

Material Parameters

Viscosity, rheology	Melting temperature
Mean particle size	Surface tension/wetting properties
Particle size distribution	Particle shape/geometry
Specific heat	Thermal conductivity
Density	Emissivity, diffusivity, reflectivity
Solids loading	Substrate material
Sintering rate parameter	Porosity

As discussed in the preceding two sections, a wide range of parameters influence laser sintering of solids. These parameters may be classified as properties of the direct-write tool (laser) or properties of the materials. At first blush, the large number of operating parameters, coupled with the even larger number of material-specific parameters seems intangible. To make the process less daunting, advances in computational modeling will likely play a large role in direct-write processes (28–30). For instance, as illustrated in Fig. 2, Computational Fluid Dynamics Corp. (30) is developing advanced look-up tables that can enable real-time control of the laser sintering process.

3.3. END-USER PULL

When developing manufacturing or prototype tools, it is imperative to have commercial and military end-users engaged from the inception. This helps ensure that the tools being developed meet the stringent requirements of those who will eventually purchase them. There is a very limited market for a direct-write tool that cannot write at mm/sec speeds at least in the 100s, but the market opportunities blossom when these write speeds are met or exceeded. Needless to say, it is important to get end-user feedback early on so that it can help steer tool development.

In a manufacturing environment, most especially in an electronics manu-facturing facility, both short product life cycles and a large mix of product features put a premium on flexible processes and systems. A few notable examples include cellular phones, digital cameras, laptop computers, personal information devices, and pagers. Short product life cycles require fast devel-opment cycles (rapid prototype) and a flexible production plant to avoid excess inventory. Likewise, the large capital investment in manufacturing tools makes companies more efficient in that these tools are often used in a wide range of products. The electronics industry is driven by products that are smaller, thinner, lighter, faster, and more cost-effective. Direct-write technologies can meet those five criteria—including faster—with some ingenious designs.

23

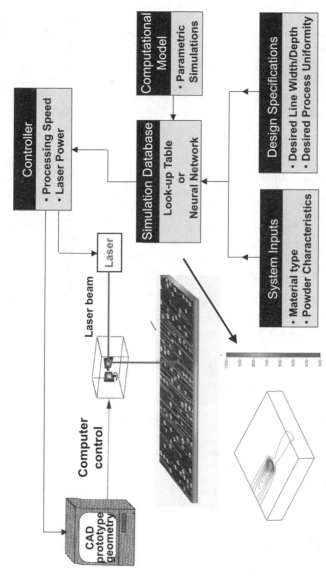

FIGURE 2 Schematic of a virtual direct-write system in which a model is used to create a look-up table to determine laser properties.

For the military, there is also a huge push for flexible and more rugged systems. Rapid prototyping will reduce the development time for new electronics systems. More importantly, the ability to produce components and systems on demand at a low cost will reduce costly inventory, not withstanding its associated maintenance costs. As the military drives toward more expeditionary missions, the long logistics (inventory) tail must be reduced. The DoD is moving increasingly toward "decentralized" mission scenarios, in which there is no major "center" for logistics or supplies, but rather many mobile and flexible units. The direct-write processes help enable this concept for electronics.

The ability to extend direct-write technologies into low-cost, small-lot manufacturing work provides the military with an important cost-conscious capability.

3.4. APPLICATIONS

The last aspect to the development of direct-write tools is that of choosing applications in concert with the end users. The applications should be designed in a fashion that complements the tool-development cycle. For instance, lower hanging fruit will be the direct-write of metals, thus applications in which metal lines are enabling should be chosen first (antennas, interconnects, off-chip connections, etc.).

4. APPLICATIONS OF DIRECT-WRITE MANUFACTURING

4.1. FROM CIRCUITS TO POWER

The ability to directly integrate thin film batteries with electronics will have a high pay-off in many applications. In so doing, it is anticipated that the system weight and volume will be reduced by minimizing the electrical interconnects, reducing packaging, and improving performance. Likewise, the ability to use thinner layers of current collecting material that does not have to withstand the mechanical rigors of conventional fabrication techniques will offer system-level benefits. Thus, it would be advantageous to develop enabling techniques for the fabrication of integrated power sources and electronic devices in the same structure. In addition, it would then be possible to include different batteries in one system, placing each battery close to the target device that needs the power. Both low- and high-power batteries can be included to maximize device performance and/or efficiency (24,31).

There is a biological analogy to the ability to print batteries in a distributed fashion and have them fully integrated with the structure. Every animal cell has its own power plant, the mitochondria. Thus, power is distributed evenly throughout all biological systems as protection against a full-system power failure. Nowadays, most power systems in machines, electronic devices, and gadgets are centralized. There are cases in which there is built-in redundancy for emergencies, but in general power is centrally located (e.g., as in a battery pack on a laptop computer, a cellular phone, a pager, a personal data assistant). The ability to print (direct-write) micro-to-macro batteries in a distributed fashion at low-substrate temperatures opens up a new opportunity for computer architects. What if power were distributed? What systems, performance, and reliability opportunities arise if we have this capability? The direct-write processes will provide the manufacturing forum in which to answer some of these questions.

In any event, the ability to print batteries opens up advances to reduce weight, improve performance (specific energy density), and form rugged, flexible batteries integrated with the physical structure.

Other power devices that may find a niche with direct-write processes include solar cells (8). Being able to direct-write solar cells with the physical structure will minimize weight considerably. Several examples in which weight minimization and energy harvesting go hand-in-hand are space applications, autonomous robots, micro-air vehicles, and/or unattended ground sensors. Monolithic integration of solar cells that can be placed with the physical structure and at any specified location may provide exceptional power delivery that is not currently available with traditional solar cells.

4.2. ANTENNA DESIGN AND INVENTION

Electronic components of particular interest to the electronics community are antennas. For example, the patch antenna is compact and can meet the GPS bandwidth (L-band) requirements. In miniature antennas, the antenna is constructed on a high-dielectric material and/or high-magnetic-permeability material, as is the case for any circuit that uses lower frequency telecommunication such as that of a cellular phone, pager, GPS, or Tracker. A high-dielectric or -permeability substrate for the patch antenna will reduce the effective wavelength in the media, and therefore reduce its dimensions. For patch antennas, dimensions are inversely proportional to the square root of the dielectric constant times the magnetic permeability.

A high-gain antenna can also be constructed using an array of antennas on a high-k and/or a high-magnetic-permeability material. The direct-write integration of a high-performance antenna within a polymer-based interconnect structure would have a significant impact on system performance, size, and weight of mixed-mode circuits with telecommunication requirements. At this particular time it is not possible to selectively deposit conformal high-k or high-magnetic-permeability functional materials on inexpensive polymer substrates because these dielectrics typically require high post-deposition temperatures to crystallize. Like the multilayer passive components discussed in this chapter's Introduction, it is not possible to accurately predict antenna performance (gain, losses, etc.) using high-k/high-magnetic-permeable materials. Thus, antenna design can be largely empirical in nature. Another aspect to antennas is how to fabricate and assemble them over conformal structures, as in helical antennas. This is a formidable challenge to traditional manufacturing processes that require flat substrates. To overcome this obstacle, many antennas are hand-assembled, a slow and expensive process. Direct-write manufacturing over conformal structures has the potential to change this dramatically.

Another opportunity for direct-write manufacturing will be with fractal (or self-similar) antennas. The advantage of such antennas is they offer wide bandwidth due to their self-similar nature. They also offer a significant reduction in size (2–4 times smaller) with similar gains and often no impedance-matching networks are required. The major disadvantage to fractal antennas is the trial-and-error process required for their design, which leads to costly research and development because mask sets have to be designed and fabricated for each prototype. Direct-write technologies easily overcome these obstacles with the promise of fast design iterations for antennas in general, as well as by making conformal and flexible designs manufacturable.

4.3. Wireless Communications

Portable electronic gadgets typically have a combination of analog, digital, and radio frequency (RF) circuitry. In these electronic modules, digital circuits have high silicon-chip content, with only a few passive components. On the other hand, analog and RF circuitry (especially handheld wireless components) are predominantly populated with passive components such as capacitors, resistors, and inductors, as well an antenna and a large battery compartment. These passive components are used for tuning, filtering, matching, terminating, and decoupling. Such wireless components include cellular phones, pagers, and RF identification tags. As discussed, most passive components are surface mount devices soldered to metal pads on a printed circuit board along with the active semiconductor devices. It is amazing that the passive components often consume over 50% of the printed circuit board real estate and at times greater than 70%. For the military the PCB and solder joints make the electronics fragile. The ability to direct write electronics on the physical structure is expected to increase the ruggedness of the electronic components. Furthermore, direct-write technology allows for the fabrication of components that are buried in the layers of the physical structure, with a broader range of values and tighter tolerances than can be obtained today.

For instance, a key parameter of conductors used in RF and microwave circuits is the edge definition of the sidewalls of the conductors. Conventional screen-printed conductors do not have straight edges due to two factors of the screen-printing process: (1) the materials are deposited through the mesh of a screen, producing saw-toothed edges; and (2) the screen-printed materials tend to sag before they are fired at relatively high temperatures (850 °C). The sagging tends to thin the edges of the conductor, which are squeezed into very thin, almost pointed cross-sectional shapes by the lamination process if the conductor is buried within a cofired multilayer ceramic structure. The jagged, thin, and sharp conductor trace edges translate into appreciable signal attenuation or loss for metal lines in high-frequency applications.

Direct-write processes are expected (and have been observed) to produce conductor traces with much better feature and edge definition. Fine feature definition can be accomplished with precise and control feedback systems during materials dispensation. For instance, the controlled feedback can be used to detect the amount of material deposited by techniques such as thermal spray, laser-guided particle deposition, or subtractive laser trimming processes.

One example of the application of wireless communications is the use of RF passive ID tags for remote sensing, inventory control, temperature sensors, humidity sensors, covert identification of friend or foe, as well as identification for hospital and theme park admissions. The passive ID tag is a good application for the direct-write technology because each and every tag is customized. This can be easily accomplished using software via CAD in the direct-write tool to enable rapid and unique manufacturing.

4.4. SENSORS

Sensors are becoming more commonplace in that they are often implemented to improve manufacturing processes, medical procedures, smart materials, interactive displays, and condition-based maintenance applications to name a few. Even though embedded sensors offer much promise they end up being limited by high-quality materials and expensive manufacturing processes. Most manufacturing processes do not integrate the sensors as a part of the physical structure; therefore, they are fragile and, ironically, often fail before the part they are monitoring does.

Sensors for condition-based maintenance are used to provide information regarding stress, corrosion, shock, torque, temperature, humidity, wear, etc. The ability to directly write a robust sensor with the physical structure and perhaps integrate a protective coating will aid in their market acceptance. Direct-write processes for the medical community will embed sensors on surgical instruments to monitor parameters such as pressure, temperature, humidity, pH, etc.

4.5. LIVING MACHINES

Laser-guided particle deposition, a direct-write process being implemented by Optomec Design Corporation, has the ability to deposit virtually any material (inorganic and organic) on any substrate, with micron-scale accuracy (15,32,33). Direct-write work by Renn and Odde (32,33) recently demonstrated that LGDW could be applied to the precise deposition of individual, living neural cells and the deposited cells were shown to remain viable despite being exposed to intense laser fields during deposition. Their observations are making it possible to direct write both inorganic and organic materials with the same tool or perhaps two different types of tissues as shown schematically in Fig. 3. In another example, work done at the Naval Research Laboratory (34) has been able to deposit *E. coli* using MAPLE-DW. After the direct-write deposition process, the *E. coli* is left intact and unharmed.

These examples make possible the fabrication of electronic circuits combined with the cell tissue of numerous varieties to produce living machines or cell-based bio-machines. Such machines have an inorganic shell but have organic sensing and self-healing components. A few potential applications include organ- and tissue-targeted drug delivery or even more futuristic applications such as a micro-sized manufacturing plant.

The tissue engineering and artificially simulated tissue-growth techniques could be of immediate application in clinical areas where the natural growth is either slow or impossible by natural means. Bone, bone marrow, spinal chord, and cardiovascular tissues, among others, are all cases of extreme difficulty in clinical practice. This direct-write technique could facilitate rehabilitation processes for such cases and could be especially useful in trauma care associated with casualties of military combat (35).

Their results (32–34) show that it is possible to deposit a picoliter of material, with the possibility of tissue engineering. Some eye-opening possibilities become realized via direct write of tissues with microelectronic arrays:

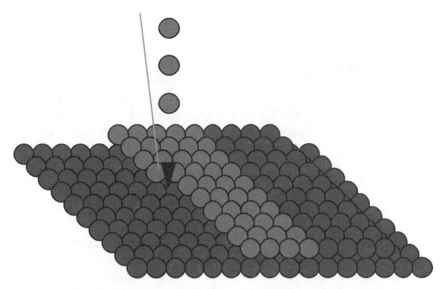

FIGURE 3 Three-dimensional patterning of two cell types. After depositing the cells of one type, an overlay of different cells will be deposited using the laser beam.

1. Recording electrical activity associated with deposited neurons in real time for realization of biosensing and bioelectronic applications.
2. Examining neural network connections and responses to differing environmental responses.
3. Three-dimensional patterning of two or more cell types for artificial tissues, tissue patches for tendon repair, bone marrow, and spinal chord repair or for skin grafts.

There are also many opportunities to combine solid freeform fabrication, direct-write electronics manufacturing, and direct-write tissues for a "complete" machine. The SFF technologies would build the three-dimensional structure with the direct-write electronics and tissues being embedded or combined with the structure during the manufacturing process. This concept brings us a little closer to the *Star Trek* replicator in which a button is pushed and out comes a functional, self-sensing, self-aware machine with "batteries included."

5. CONCLUSIONS

Real-world technologies have always suffered due to the cost of evolutionary and costly prototyping. The direct-write manufacturing vision is to find a way to fabricate an electronic device with optimal functioning in a short period of time at negligible cost. When this is achieved, a new age of innovation will emerge.

Direct-write manufacturing can be described as aiming toward the creation of a robust technology field, which shows good promise of yielding significant advances in a relatively short time. Most importantly the direct-write manufacturing of electronic components promise development of a robust technology field rather than a discrete series of specific product embodiments ("one-off" products) in any single market or industry. Opportunities abound within the military and commercial markets alike: interconnects, rugged electronic components, distributed power, embedded ruggedized sensors, and living machines. The direct-write electronics approach will also significantly reduce design iteration times, earmarking the tool as an "invention" machine with rapid prototyping. Direct-write processes will only find a home in the larger market segment if the tools are amenable to direct-write manufacturing. Often electronic components fabricated via prototype methods are not transferable to a manufacturing environment.

ACKNOWLEDGMENTS

I would like to thank many of the highly ingenious contributors to the DARPA MICE program that are helping make the direct-write electronics manufacturing/prototyping effort a success. Special thanks go to Jeff Bullington, Timothy Schaefer, Lawrence Dubois, Terry Feeley, William Coblenz, Colin Wood, and Michael Goldblatt who were instrumental in helping steer and develop the MICE program.

REFERENCES

1. D. Dimos and P. Yang, Ceramic Industry, 25 (Dec. 1997).
2. A review laser processing including LCVD and LTS methods can be found in Laser Microfabrication: Thin Film Processes and Lithography, D. J. Ehrlich and J. Y. Tsao, eds., pp. 1–536, Academic Press, Boston (1989).
3. D. J. Ehrlich, Revise Inc., Burlington, MA (www.revise.com).
4. T. T. Kodas, T. H. Baum, and P. B. Comita, J. Crystal Growth 87, 378 (1988).
5. N. Nassuphis, R. H. Mathews, S. T. Pamacci, and D. J. Ehrlich, J. Vac. Sci. Technol. B12, 3294 (1994).
6. V. S. Dolat, J. H. C. Sedlacek, and D. J. Ehrlich, Appl. Phys. Lett. 53, 651 (1988).
7. J. Manitt and S. Hull, "Thick Film Copper Conductor Patterning by Laser" US Patent 4,931,323, June 5, 1990.
8. K. H. Church, Sciperio Inc., Stillwater, OK (www.sciperio.com).
9. H. Herman, Plasma Sprayed Coatings, Scientific American, 112 (Sept. 1988).
10. H. Herman and S. Sampath, "Thermal Sprayed Coatings," Metallurgical and Protective Coatings, K. Stern, ed., Chapman and Hall, 261 (1996).

11. A. P. Alkhimov, A. N. Papyrin, V. F. Kosarev, and M. M. Shushpanov, "Gas Dynamic Spraying Method for Applying a Coating" US Patent 5,302,414, April 12, 1994.
12. W. L. Oberkampf and M. Talpallikar, J. Thermal Spray Tech. 5, 53 (1996).
13. M. J. Renn and R. Pastel, J. Vac. Sci. Tech., B16, 3859 (1998).
14. M. J. Renn, R. Pastel, and H. Lewandowski, Phys. Rev. Lett. 82, 1574 (1999).
15. W. D. Miller, Optomec Design Corporation, Albuquerque, NM (www.optomec.com).
16. W. Mathias, Ohmcraft Inc., Pittsford, NY (www.ohmcraft.com).
17. M. Duignan, Potomac Photonics Inc., Landham, MD (www.potomac-laser.com).
18. D. B. Chrisey, A. Piqué, J. Fitzgerald, R. C. Y. Auyeung, R. A. McGill, H. D. Wu, and M. Duignan, Appl. Surface Science 154, 593 (1999).
19. A. Piqué, D. B. Chrisey, R. C. Y. Auyeung, J. Fitzgerald, H. D. Wu, R. A. McGill, S. Lakeou, P. K. Wu, V. Nguyen, and M. Duignan, "A Novel Laser Transfer Process for Direct Writing of Electronic and Sensor Materials," Appl. Phys. A, 69, S279 (2000).
20. D. J. Hayes, Microfab Inc., Plano, TX (www.microfab.com).
21. R. H. Detig, Electrox Inc., Newark, NJ (private communication).
22. Y. Senzaki, J. Caruso, T. T. Kodas, and M. J. Hampden-Smith, J. Am. Cer. Soc. 78, 2973 (1995).
23. M. Nyman, J. Caruso, M. J. Hampden-Smith, and T. T. Kodas, J. Am. Cer. Soc. 80, 1231 (1996).
24. T. T. Kodas and M. H. Hampden-Smith, Superior Micropowders LLC, Albuquerque, NM (www.smpl.com).
25. P. H. Kydd, "Electrical Conductors Formed from Mixtures of Metal Powders and Metallo-Organic Decomposition Compounds," US Patent 6,036,889, March 14, 2000.
26. P. H. Kydd, Parelec LLC, Rocky Hill, NJ (www.parelecusa.com).
27. M. B. Ranade, Particle Technology, Beltsville, MD (www.chemconsultants.org/mbranade.html).
28. J. C. Sheu, M. G. Giridharan, S. Lowry, and A. Krishnan, "Computational Tools to Optimize Direct Write," CFDRC qrtly report 4968/1 (Jan. 1999).
29. M. G. Giridharan, S. Lowry, and A. Krishnan, "A Multi-Block, BFC Radiation Model for Complex Geometries," AIAA-95-2020, 30th AIAA Thermophysics Conference, San Diego, CA (June 1995).
30. "CFD-ACE User Manual," CFD Research Corporation, Version 5 (Oct. 1998).
31. S. Narang and P. Cox, SRI International, Menlo Park, CA (www.sri.com).
32. M. J. Renn and D. J. Odde, Ann. Biomed. Eng. 26, S-141 (1998).
33. D. J. Odde and M. J. Renn, Trends Biotechnol. 17, 385 (1999).
34. B. R. Ringeisen, D. B. Chrisey, A. Piqué, H. D. Young, R. Modi, M. Bucaro, J. Jones-Meehan, and B. J. Spargo, "Generation of mesoscopic patterns of viable Escherichia coli by ambient laser transfer," Biomaterials (in press).
35. P. Atanassov, Superior MicroPowders LLC (personal communication).

CHAPTER 3

Role of Direct-Write Tools and Technologies for Microelectronic Manufacturing

JIE ZHANG, JOHN SZCZECH, JAMES SKINNER, AND DANIEL GAMOTA

Motorola Advanced Technology Center, Schaumburg, Illinois

1. INTRODUCTION

The rapid prototyping, miniaturization, substrate versatility, and three-dimensional (3-D) fabrication capabilities of direct-write (data-driven materials deposition) technologies have gained increased interest from the electronics industry. This growing attention is propelling the development of a number of candidate direct-write systems. These technologies offer the potential for low-cost small-batch manufacturing and unparalleled levels of integration and density that are essential to advanced telecommunications applications. The microelectronics manufacturing industry is evaluating these technologies because they offer the potential for highly compact, ultra-lightweight assemblies. In addition, they can offer rapid product introductions and enhancements, which are becoming increasingly critical to maintain leadership in the accelerating electronics products market. Moreover, it is important to establish an industry standard technology with multiple vendors to ensure a continuous, reliable, long-term source of supply.

Improved component tolerances and unique patterning capabilities have been claimed for many direct-write technologies. Presently, many of these technologies are being evaluated to determine whether each is capable of the requisite reproducibility, robustness, and accuracy. This is necessary for reliable production of complex electronic products containing hundreds of passive and active components and very high densities of interconnects. As the microelectronics industry has learned in the past, a promising and mature research and development technology does not necessarily translate directly to a factory-ready manufacturing process. Several comprehensive evaluation programs are under way within the industry to identify the most promising direct-write equipment and materials systems.

The direct-write technologies under development must be evaluated by fabricating a statistically significant number of test vehicles that are subsequently subjected to a series of factory standard evaluations. This will enable the identification of candidate direct-write equipment, processes, and materials that offer the greatest near-term potential for real manufacturing. The selected solutions must meet microelectronic industry standards for quality, yield, reproducibility, and product reliability in rugged environments. Predictions have been made that direct-write technologies will allow manufacturers to eliminate dedicated chemical vapor deposition systems, sputtering chambers, photolithographic techniques and materials, and enable electronic device and component fabrication in a high-volume factory environment under ambient conditions. The ultimate goals of these development efforts are to build and qualify direct-write technologies and materials systems to manufacture low power, lightweight, robust electronic devices (e.g., flexible Bluetooth modules) at low temperatures on 3-D nonplanar substrates.

The development of materials and direct-write processing technologies will allow the technology developers to gain a competitive advantage within their respective microelectronic infrastructure segments. Initially, these companies will maintain an economic advantage through licensing agreements with competitors and having established the in-house technology capabilities. Due to the competitive environment within the microelectronic industry, the economic benefits to the developers will be short because their competitors will become proficient in the technologies and will begin to implement internally developed technologies enhancements. Irrespective of the total economic benefits achieved by the individual companies, the developed technology will renew growth in the U.S. microelectronics industry and will initiate a paradigm shift in microelectronics manufacturing.

The technology developers must initiate discussions with members of the microelectronics infrastructure in an effort to expose engineers and designers to this new technology. They must provide documentation as well as engineering support to companies that decide to evaluate the developing technology. In addition, the developers must give external presentations to interested suppliers and offer support during the early technology transfer period. If this technology transfer is conducted efficiently, the materials suppliers, equipment developers, and original equipment manufacturers (OEMs) will maintain the momentum and continue to recommend the usage of the advanced direct-write microelectronic manufacturing technology to their customers. Presently, several potential partnerships exist and this list will increase with the introduction of several new suppliers into the microelectronics infrastructure sector.

2. DIRECT-WRITE TECHNOLOGY IN THE MICROELECTRONICS INDUSTRY

Current computer aided design/computer aided manufacturing (CAD/CAM) direct-write processes are limited by the substrate thermo-mechanical material properties. These limitations (e.g., glass transition temperature) reduce the number of potential microelectronic devices and components, which one can fabricate.

Presently, several developing commercial systems offer unique manufacturing advantages. Although each system developer may use a different description or acronym for their technology, the authors decided to group and classify the systems into four general technology categories: (1) microdispensing, (2) laser-assisted transfer, (3) electrostatic assisted transfer, and (4) jetting.

The microelectronics industry has the vision to integrate these systems onto a reel-to-reel platform, which will enable rapid, high-volume manufacturing using flexible plastics as substrates. The compliant nature of the substrate layer(s) will enable the design and fabrication of a truly conformal electronics system. Product assembly line pulse rates will increase and will enable manufacturers to benefit from economies of scale. Furthermore, the substrate type will provide a significant reliability advantage over a nonflexible system. In particular, the strain generated by thermal expansion mismatch of the materials will be more readily dissipated in the flexible system while the strain generated in a "rigid" system will not have a mechanism by which to relax the strain.

2.1. DIRECT-WRITE TECHNOLOGIES

A representative list of the CAD/CAM direct-write tools and the various direct-write materials is given in Table 1. This table identifies materials and tools to fabricate various microelectronic devices and components. Table 2 lists components that can be manufactured by direct-write techniques along with

TABLE 1 Potential Components Fabricated by Direct-Write Technologies and Materials

Components	Direct-write materials	Direct-write technology	Final processing
Capacitors	Metalo-organic materials	Micro-dispensing, jetting	Thermal
	Filled polymer thick films	Micro-dispensing, electrostatic	Thermal, UV, microwave
	Polymer-based dielectric materials	Micro-dispensing, jetting, electrostatic	Thermal, UV
	Ceramics dispersed in an organic binder	Micro-dispensing, jetting	Thermal, microwave, laser
	Ceramic dielectric nano-materials	Laser-assisted transfer	Thermal, laser
	Donor material films	Laser-assisted transfer	Laser
Resistors	Filled polymer thick films	Micro-dispensing	Thermal
	Metalo-organic materials	Micro-dispensing, jetting	Thermal
	Donor material films	Laser-assisted transfer	Laser
Inductors	High μ materials	Micro-dispensing	Thermal, laser
	Donor material films	Laser-assisted transfer	Laser
	Metalo-organic materials	Micro-dispensing, jetting	Thermal, laser
Batteries	Metalo-organic materials	Micro-dispensing, jetting	Thermal, laser
	Electrolyte materials	Micro-dispensing, jetting	Thermal, laser
Antennas	Metallic materials	Laser-assisted transfer	Thermal, laser
	Metalo-organic materials, conductive polymers	Micro-dispensing, jetting	Thermal, laser
Conductors	Metallic nanomaterials	Micro-dispensing, jetting	Thermal, laser
	Donor material films	Laser-assisted transfer	Laser
	Conductive polymers	Jetting, micro-dispensing	Thermal, laser, microwave
	Metalo-organic materials	Jetting, micro-dispensing	Thermal

TABLE 2 Representative Component Values and Tolerances for a Consumer Product

Components	Range of values	Tolerances
Inductors		
Decoupling	$1–10\,\mu H$	±30%
Filter	$10\,nH–0.56\,\mu H$	±10%
Capacitors		
Decoupling	$0.01–100\,nF$	±30%
Filter	$0.4–5,600\,pF$	±10%
Resistors		
Termination	$25–100\,\Omega$	±10%
Logic	$1k–1\,M\Omega$	±20%
Filter	$5–250\,\Omega$	±5%
Antenna	Controlled impedance	±5%

their respective operating range and tolerances. In some cases, a single direct-write tool may have a competitive advantage for a greater variety of electronic devices and components that can be fabricated. However, at present many of these tools have not been subjected to high-volume manufacturing studies that evaluate their applicability toward microelectronics device fabrication. In addition, several of the tools, at this time, have not demonstrated the required attributes necessary to be accepted by manufacturing (e.g., long mean time to failure, short mean time to repair, small footprint, low processing cost environment, and fewer required number of operators, engineers, and technicians). A brief description of the different technologies is presented in the following sections.

2.1.1. Micro-Dispensers

Several different micro-dispensing technologies are currently commercially available: (1) rotary screw, (2) positive displacement piston, and (3) needle valve (1–3). Several companies offer production-ready systems for in-line manufacturing of electronic components and products. Today these systems can cost between \$125 K and \$250 K, require approximately 9.3 m^2 of floor space in a manufacturing microelectronic environment (30 °C/60%RH), and are the most mature of the four tool categories. However, repeatability and precision of the volume of material deposited is related to the dispensing technology. Micro-deposition systems are sensitive to the material rheology formulation because fine features are obtained by using small gauge dispensing tips. In addition, the optimal linear velocity depends on the properties of the material that is being dispensed. Dispensing studies have shown that viscosity, surface tension, needle gauge and length, and flow rate are all critical for dispensing uniform structures. Although the micro-dispensing systems are capable of dispensing materials having a wide range of viscosity, the settling viscosity of thick films must be monitored with an optical imaging system or a feedback weight control loop to observe the morphology of the printed features. The range of dimensions for repeatable traces that these tools have demonstrated are 25 to 300 µm wide and 1.3 to 51 µm thick. Figure 1 is an example of micro-dispensed circuitry. In general, width and thickness (height) are material dependent while "straightness" is machine dependent. In addition, the distance between the dispensing needle orifice and substrate affect the pattern features. For example, too great of a distance will affect the straightness of the line.

Of the commercially available micro-dispensing systems, the displacement piston system is currently used for depositing electronic-grade materials to fabricate integrated, multilayer passive components through a layer-by-layer build-up process that also combines multiple materials within a single layer.

FIGURE 1   Electronic circuitry (40 μm wide and 5 μm thick) fabricated using a micro-dispensing system.

The system is capable of depositing slurries and solutions that cover a wide range of viscosities, from 100 mPa-s to 100,000 mPa-s. It has several attractive features: (1) an accuracy of roughly 2 to 3 μm; (2) capability of writing line widths approaching 25 μm; and (3) dimensional control of better than ±0.5 μm. In addition, the system has demonstrated the ability to fabricate band reject filters with 27 dB suppression at the reject frequency by achieving matched values for the capacitors and resistors. The developers have eliminated the requirements for high processing temperatures by working with polymeric-based conductors, resistors, and dielectrics, which ultimately lead to the capability for printing custom components on low-temperature substrates. Presently, this system is being modified to enable simultaneous multiple material deposition, enhanced adjustment features to fabricate finer-line features, and greater controlled start/stop features in an effort to improve the structure definition at the start/stop points.

A novel linear positive displacement piston pump micro-dispensing system is presently being used in high-volume manufacturing to dispense long lines of material to form resistors and traces on printed wiring boards (PWBs). The typical fluids, which are being used, have viscosities between 1,000 and 500,000 mPa-s and studies have suggested that highly thixotropic materials perform best for continuous line dispensing. The system has demonstrated

PWB line widths on the order of 300 µm using a 30-gauge dispensing needle. Studies suggest that the dispensing envelope is limited by the pressure the system can develop and the viscosity of the material.

A rotary screw equipped micro-dispenser has the ability to deposit materials having viscosities from 100 to 1,000,000 mPa-s using a 30-gauge needle. Resistive pastes can be used to fabricate between 125 and 300 µm wide line resistors. This system has a maximum linear velocity of 13 m/s. A next-generation system is being developed which can operate at velocities approaching 64 m/s over a 152 × 213 cm panel area. Current commercial systems can accelerate/decelerate in 5 milliseconds during negotiation of right angles, which enables the fabrication of structures with dimensional accuracy of ± 0.3 µm. Presently, micro-dispensing developers are designing a device to dispense 10 to 20 µm line widths with a turret subsystem to enable dispensing of multiple materials.

Micro-dispenser technology is the most mature of the four technology groups and several systems have been integrated in high-volume microelectronic manufacturing for low-risk applications. The design of an assembly line based on these systems increases the risk level because they have not demonstrated circuitry and device design that address the microelectronics industry roadmap. The five greatest technical risks are as follows: (1) the ability of these systems to print fine lines requires an exact vertical alignment that is not available on current equipment; (2) a low-cost, reliable, repeatable means of needle-tip fabrication must be identified; (3) acceptable line definition when starting and stopping the needle motion will require improved system control and/or material property modification; (4) multiple pump block modules to improve the speed for simultaneous deposition of multiple materials has not been demonstrated; and (5) minimum demonstrated line width control and writing speed must be improved.

2.1.2. Laser-Assisted Transfer

The laser-assisted transfer technologies are diverse in nature and are currently in different stages of development. The designers of these systems have developed processes to transfer and deposit metals and other materials from donor films to substrates (4–8). Some systems are wafer fab compatible and are presently used by IC manufacturers while others are used in high-volume manufacturing for transferring gold onto pin connectors. Laser scanning techniques are well understood and precision scanning systems for high-volume microelectronics manufacturing are available (Fig. 2). Combinations of laser processes (transfer/sinter/cure/anneal) are being developed by using multiple wavelength lasers. Developers are evaluating the use of higher laser power and/or alternate wavelengths to locally heat regions of the substrate to

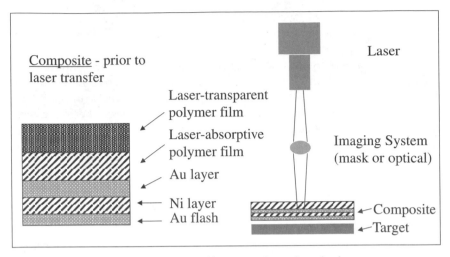

FIGURE 2   Schematic of laser-assisted transfer technology.

high temperatures for sintering and curing. However, care must be taken to ensure the region heated remains small to avoid melting of the substrate.

One laser-assisted transfer technology has demonstrated the ability to fabricate structures having dimensions less than 25 µm. The first studies were conducted in the late 1970s to evaluate the potential for laser-assisted transfer of metals from donor films onto receiving substrates in ambient conditions (5). These studies were performed to develop a low-cost deposition-transfer technique to fabricate metallurgical traces on ceramic and glass substrates using a yttrium aluminum garnet (YAG) laser having an 8-cm × 8-cm stage, a repetition rate of 3,000 Hz, a current of 20 Å, and speed of 20 mm/s. They demonstrated the fabrication of copper and aluminum traces having 100-µm widths on inorganic substrates when processing in ambient conditions. This technology can direct write passive components, conductors, and batteries by supplying materials on a ribbon or donor film. The donor film is comprised of several components—a laser transparent substrate, a thin polymeric film, which has a high optical coefficient of absorption to the wavelength of the laser, and a composite of metals. An example of a composite metal has approximately 120 Å of a gold flash deposited on 1.2 µm of Ni, deposited onto 1.2 µm of gold which is deposited on a 2.5-µm laser absorptive polymer film. This composite structure can be fabricated via evaporation, microwave sputtering, or arc deposition. The thin, high optical coefficient of absorption polymer film is deposited onto a flexible laser-transparent substrate (e.g., polystyrene, polyvinyl acetate, or polyethylene). During exposure to the laser the absorptive layer is vaporized propelling the metal composite toward the substrate and forming a cold weld or pressure bond. Typical laser platforms as those mentioned require approximately 200 sq ft of floor space, and cost between $300 K and $1,000 K.

Another donor film based laser-transfer system uses a 2.2-W infrared (IR) laser source with a spot size of 12 μm, and a drum radius of 25 cm, to cover a 76-cm × 76-cm substrate in 25.4 min. Preliminary studies have suggested that increasing the power of the laser to 9 W will enable the coverage of the same print area in 6 min, thereby meeting one of many manufacturing specifications. The system developer has demonstrated the deposition of single and multiple layers of pure and composite materials. To date, the system is able to consistently fabricate 50- to 200-μm-wide copper lines on an epoxy/E-glass composite and polyimide, and the finest demonstrated features were 25 μm wide and 1 μm thick.

Yet another type of laser system is based on the melting of copper, gold, silver, and ceramic nanomaterials in-flight prior to contact with the substrate. The method of powder delivery impacts the achievable linear velocity. One powder delivery method allows a linear velocity of approximately 50 cm/s with large line dimensions, while the second delivery method is much slower limiting the linear velocity to 0.6 cm/s. Although the developers of the system have not demonstrated multiple material deposition, they are currently building a system that will have this capability. To date, this system has generated lines that were 250 μm wide and approximately 10 μm thick.

In summary, laser-assisted transfer is a promising technology that uses laser energy as either a precision thermal source or as high-energy photons to deposit a variety of electronic-grade materials. The resolution and speed should be more than adequate because of the small spot size and high energy possible with commercial laser systems assuming tool reliability is not an issue. The greatest risk is that the transferred material quality is not well suited for present low-cost characterization techniques.

Similar to micro-dispenser technology, some laser-transfer systems are presently used in the electronics industry. However, the specific applications are not as demanding in the fabrication of circuitry and devices for an electronic product. Further in-depth studies are required before the acceptance of this technology into low-cost, high-volume microelectronics manufacturing.

2.1.3. Electrostatic-Assisted Transfer

Electrostatic-assisted transfer is based on xerographic printing. Two companies offer electrostatic-assisted transfer technologies for the processing of micro-scale and macro-scale microelectronic structure feature dimensions (9,10). Several material properties are critical for an optimal toner, such as particle distribution, volume average particle size, particle charge/unit mass, bath conductivity, dynamic mobility, Zeta potential, and electrophoretic mobility. The micro-scale processing system has demonstrated the ability to consistently deposit 30-μm line widths. However, the imaging portion of the tool has the ability to print features having 10-μm sizes by using a 10-facet polygonal scanner with a 3.5-μm spot size. The 10-cm wide scanner can achieve 500 rev/s and 28,300 spots/scan line with a scan error of ± 1 μm, and imaging speeds of 2.5 cm/s can be obtained. The developers have shown that development, transfer, and fixing takes approximately 16 s for a 10-cm × 10-cm substrate. The footprint for this system is $9 \, m^2$ and during its operation, it requires two system operators.

The commercially available macro-scale electrostatic-assisted transfer technology handles a 340-cm-width web. The technology is capable of reel-to-reel processing and can be used to apply charged organic and inorganic particles to the substrate to fabricate 200-μm features. The charge applied to the particles as well as their size and distribution are critical. The deposition of materials on both organic and metal substrates in high volume has been demonstrated. In addition, deposition onto porous and nonporous substrates that are nonplanar has been shown (see Fig. 3). This system can be miniaturized to accommodate a 30-cm web for typical microelectronic applications and can fabricate sub-100-μm features.

FIGURE 3   Electrostatic-assisted transfer fabricated test vehicle.

Although macro-scale electrostatic technology is well characterized and used in high-volume manufacturing for the fabrication of gross feature structures, a potential risk exists in the scaling of this technology to the micro-scale. Several nonmicroelectronic industries presently use macro-scale electrostatic-assisted transfer technology with no apparent technology related issues. The evaluated micro-scale technology appears promising and efforts were initiated to address the minimal structure feature resolution. This technology holds great promise provided that the developers are able to address the feature scaling related issues.

2.1.4. Jetting

Several jetting tool suppliers offer jetting systems for sale to OEMs and research and development groups (11–14). The suppliers have worked with a variety of commercial sectors (pharmaceutical, electronic, and automotive) to develop the processes for jetting unique materials. Jetting technology is a mature process capable of handling a variety of materials at elevated temperatures (e.g., 220 °C). One ink-jet printer technology consists of a thermally driven process and can only operate within a narrow range of ink viscosities. The material vaporizes readily and obtains sufficient momentum to exit the orifice to form a single droplet that impacts the printing surface. A more robust jetting technology uses a piezoelectric material to rapidly change the volume of the fluid delivery chamber to generate and deliver droplets to the substrate.

Both jetting technologies require a relatively low-viscosity fluid, typically below 100 mPa-s. In addition, there must be sufficient wetting of the jetted fluid on the substrate to promote adequate adhesion. Recently, micro-replication technology has been used for both controlling the flow of the jetted liquids and for increasing the adhesion by modifying the morphology of the substrate material. The overall size of the particles that can be jetted are limited by the nozzle exit orifice diameter and line width/thickness is determined by the jetted liquid properties and the orifice diameter. With a transport mechanism that allows the printhead to reciprocate across the substrate to be printed, it is possible to jet up to four different inks on top of one another. The jetting systems typically require materials having viscosities of 3 to 20 mPa-s, and surface tensions of greater than 35 dynes/cm. Figure 4 shows an example of a jetted pattern with 200-μm trace widths. However, 37-μm-wide traces have been demonstrated. Typical jetting systems require 19 m^2 of floor space and cost between $200 K and $400 K.

A jet printing system is used in manufacturing to deposit palladium oxide (PdO) to fabricate surface conduction electron emitter (SCE) cathodes for flat panel displays. This process uses a piezoelectric-excited device to deposit a dispersion of PdO nanoparticles suspended in a resin and solvent solution.

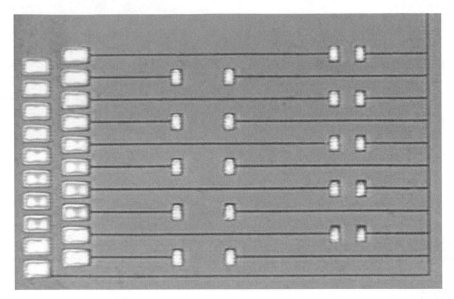

FIGURE 4   Circuitry fabricated with a jetting device: 200 μm line width on a polyimide film substrate.

They are able to achieve displays with high image quality, full motion, and full color pictures that are comparable in quality to those of cathode-ray tube displays. The piezoelectric device is trapezoidal shaped, 125 μm in length, and 62.5 μm × 50 μm and has the capability to deposit 100 to 130 picoliters per pulse. Presently, a higher voltage output system is being developed to enable the dispensing of higher viscosity and higher surface tension materials without heating the device. Alternatively, a heater can be integrated enabling the jetting of materials having high viscosities without requiring elevated excitation voltages.

A second jetting system developer offers a drop-on-demand platform that has a single-nozzle piezoelectric actuated device. The device can be driven at frequencies approaching 10 kHz to create droplets with diameters varying from 25 to 125 μm. Presently, no information has been reported although testing has shown that the operational lifetime of the piezoelectric device is between 24 and 720 hrs. Recently, the jetting device has been integrated on a manufacturing-ready platform to investigate the opportunity to deposit different materials: metalo-organic, filled epoxies, polyimide, fluxes, organic solvents, and phosphors. This system can achieve printing speeds of up to 51 cm/s.

Jetting is an attractive technology because of the perceived simplicity, high speed, and noncontact nature of the process. However, there are significant technical risks in obtaining and formulating suspensions or solutions with optimal rheological and materials properties for jetting (e.g., viscosity, surface tension, density). The issues associated with developing formulations yielding high strengths of adhesion to the substrates present risks as well. Other potential risks concern spatial resolution of the jetted material and the deleterious production of satellites or secondary droplets, and the removal of the carrier or solvent fluid after droplet impact at the substrate to enable the formation of pinhole-free surfaces.

2.2. MATERIALS SYSTEMS

Several commercially available materials systems exist as do a variety of experimental systems that are being developed specifically for the direct-write tools. The commercially available systems are offered in large batch quantities and have demonstrated attributes required for high-volume manu-facturing.

A large supplier and developer of metallic materials offers a variety of nanoparticle metals and mixed metal oxide dielectric systems for laser-assisted transfer for plastic, glass, metal, and ceramic substrates. The tape-like materials are composed of the materials dispersed in a binder and applied to a backing material. The material manufacturer also offers aqueous-based materials having the rheological properties suitable for jetting and micro-dispensing systems.

Another supplier of materials for direct-write systems offers metal-filled organic conductive, carbon-filled resistive, and inorganic-filled dielectric materials. These materials are compatible with polyester, polyimide, paper, and epoxy glass substrates. The typical thermal cure schedule is 5 to 15 minutes at 140 °C to yield a bulk material having optimal electrical and mechanical prop-erties. The materials have a minimum shelf life of 6 months if stored at ambient conditions (23 °C/60%RH) and have a pot life in an ambient electronic manu-facturing environment (30 °C/60%RH) of approximately 8 hours.

Yet another supplier of materials offers metallo-organic solutions, pure metals, metal compounds, and metal-oxide films on high-temperature substrates (e.g., polyimide and ceramic substrates). The thermal-processing temperatures for the materials range from 300 to 850 °C and the processing time at peak temperature is approximately 15 minutes. The materials can be applied to the substrates by micro-dispensing, jetting, or laser-assisted transfer. The typical shelf life of the materials ranges between 3 months to 1 year while their pot life is dependent on factory environment and specified by a change in viscosity.

The composite ribbons used for laser-assisted transfer during high-volume manufacturing are not widely available. Typically, companies have relied on internal efforts for developing proprietary composite ribbons for in-house manufacturing use. Other donor films have been developed although their specific compositions and dimensions are proprietary.

Dispersed ultra-fine copper, gold, and silver nanoparticle suspensions are available in high-volume quantities. The manufacturer of the suspensions offers materials having high viscosities (10 to 20 Pa-s) for micro-dispensing and low viscosities (1–100 mPa-s) for jetting systems. The shelf life of the suspensions is 3 months when the materials are stored at 22 °C and 6 months when they are stored at 0 °C. Meanwhile, the pot life is specified by a maximum allowable change in viscosity.

Several technical risks must be addressed prior to acceptance of the materials systems in high-volume manufacturing. Novel materials systems must be identified having acceptable manufacturing attributes (e.g., viscosity, electrostatic charge, adhesion, ease of handling, low cost, long pot life and shelf life, ease of cleaning, low toxicity, processing in ambient environments).

The direct-write process will be an accepted manufacturing technology provided that materials are developed having the aforementioned attributes. In addition, certificates of compliance containing material specifications for acceptance and the methods for evaluation of the materials to ensure material quality must be generated by the vendors.

3. NEXT GENERATION SYSTEM

The value of direct-write systems to the microelectronics industry depends on many manufacturing-related parameters. The acceptance and long-term market diffusion of direct-write technology are based on several variables (capital costs, materials, manufacturing line pulse rate, etc.). Also, even if these systems demonstrate the capabilities to fabricate structures having the desired electrical performance and dimensions, these systems will not be accepted by the microelectronic infrastructure if cost targets are not achieved.

Product designers were asked to identify an example product and corresponding bill of materials in an effort to perform a series of cost-modeling studies to determine the cost advantages of a direct-write tool. An electronic product was selected and for the cost modeling the corresponding components, physical dimensions, ranges of components, values, and tolerances were taken into consideration. In an effort to compare direct-write systems to conventional manufacturing processes, a cost model was developed based on a cost estimator tool (15). The model was validated and used to calculate the cost involved in the assembly of modules and the manufacture of circuit boards. The model was sensitive to several categories of manufacturing parameters. Table 3 lists the four categories of inputs: (1) general factory

TABLE 3   Direct-Write Cost Model Input Data

General factory data	Equipment	Materials	Processing parameters
Desired throughput	Number	Ag PTF ($/g)	Duration (min)
Hours per month worked	Cost ($/tool)	Carbon ($/g)	Set-up (min)
Labor cost for production	Space (m²)	Substrate ($/cm²)	Set-up quantity (#)
operators	Utilities ($/hr)		Material cost ($/part)
Labor cost for production	MTBF (hr)[a]		Yield (%)
technicians	MTTR (hr)		Operators (#)
Labor cost for production			Technicians (#)
engineers			Engineers (#)
Additional staff costs			
Additional factory costs			
Facilities costs			
Equipment cost depreciation			
cycle			

[a] MTBF, mean time between failure; MTTR, mean time to repair.

data (throughput based on a portable electronic product front-end assembly line), (2) equipment, (3) material, and (4) processing parameters.

Figure 5 shows the direct-write process flow used in the cost model. The data generated by the model was for the fabrication of a 4-layer PWB with embedded passives on a flexible substrate. The data from the cost model was compared to the quoted cost for the product with surface mount technology (SMT) components at medium volume production rates. The key assumptions for processing the PWB are listed in Table 4 (e.g., materials and substrate costs were obtained from vendor quotes based on medium quantity purchasing). The linear direct-write velocity was the speed associated with dispensing of materials while the mean writing speed was calculated by dividing the writing length by the time required for board completion. The completion time was a function of linear writing speed, writing pattern design, and the system motion control optimization scheme.

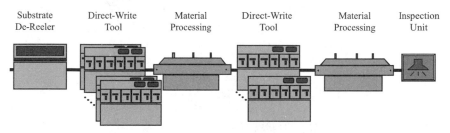

| Substrate De-Reeler | Direct-Write Tool | Material Processing | Direct-Write Tool | Material Processing | Inspection Unit |

FIGURE 5   Direct-write process flow.

TABLE 4   Assumptions Used in the Cost Analysis

Characteristic	Unit of measure
Board size	cm^2
Board density	cm/cm^2
Resistors	#/products unit
Print head	#
Throughput	items/year
Conductive paste	$/g
Resistive paste	$/g
Flexible substrate	$/m^2$
Direct-write linear speed	m/min
Mean writing speed	cm/min

Based on commercially available system-processing parameters, the cost for fabricating a 58-cm^2 populated PWB was 3 times the present cost for the board using mature PWB fabrication technologies. The material cost per item was less than 2% while the cost for the equipment and labor was more than 81% of the total cost. Although there are limitations for further improving writing speed due to the material properties, a multiple-device system is envisioned for the future. This advanced system can provide parallel writing operations thereby increasing manufacturing throughput. The cost reduction associated with using a multiple-device system is shown in Fig. 6 and the cost reduction observed, in this case, is approximately proportional to the number of print heads used.

3.1. DIRECT-WRITE PLATFORM

The next-generation system should have all the attributes of present high-volume microelectronics manufacturing systems. The system would be expected to have a de-reeler that feeds the substrate to the assembly equipment, a direct-write tool, a material processing oven, a test station, and a re-reeler or a packing unit that is based on whether the manufactured item is the final product or is a submodule of another system. Typically, several parameters are used in the development and selection of equipment that will be integrated in the manufacturing system. Parameters such as mean time to failure (MTTF), mean time between failures (MTBF), and mean time to repair (MTTR) are used to determine the productivity of the manufacturing system. The direct-write tool platform should be developed to provide a repeatability of about 0.5 μm in the x and y plane of travel and about 0.001 degrees in theta. These

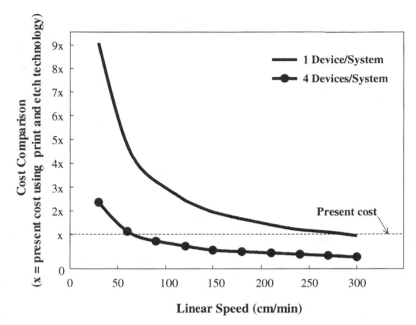

FIGURE 6 Cost reduction as function of device linear velocity and number of devices per system.

characteristics will enable the tool to operate accurately even while writing at high speeds associated with high-volume manufacturing. It is expected that the platform will have multiple devices each operating with a linear speed of at least 200 cm/min. This linear velocity should enable the direct-write platform to achieve a product pulse rate acceptable to high-volume manufacturing. Also, flexibility must be built into the equipment to achieve multitasking through communication with other equipment in the manufacturing system, off-line programming, and upgradeable system hardware and software to facilitate system enhancements.

The development efforts should focus on achieving the lowest possible equipment set-up time and maintenance time which will subsequently improve the productivity of the manufacturing system. Moreover, the "actual" equipment performance will be determined from the data collected through statistical process control (SPC) software, and will be compared with vendor specifications. A manufacturing system that increases productivity through a closed loop failure reporting system complete with a corrective action system must also be developed.

To meet the desired cost targets, the proposed manufacturing system must have the ability to handle substrates in a roll format. This allows for increased use of the substrate material and also results in increased throughput due to the inherent advantage from precluding the use of pallets. The direct-write tools that are available in the market today are not capable of handling substrates in a roll-to-roll form. Equipment must be selected to provide robust hardware and software platforms as well as the flexibility that would allow equipment modifications to accept substrates in a roll format. Also, a mechanism to planarize the substrates must be integrated in the equipment and between processing platforms to maintain planarity of the substrates. In addition, a powerful vision system must be implemented on the platform to compensate for any positional inaccuracies.

In-depth studies are critical to evaluate the efficiency of the equipment in the manufacturing system, processing bottlenecks, and the amount of equipment that is required to maximize productivity. All the equipment developed must comply with factory safety regulations to ensure protection of the personnel. Furthermore, the equipment must have the minimum footprint necessary to maximize the cost-per-unit factory floor space. This will in turn improve the efficiency of the manufacturing system and reduce the cost per manufactured product.

4. TECHNOLOGY DIFFUSION IN MICROELECTRONIC INDUSTRY

A cohesive strategy based on standard microelectronics program-management tools must be created to ensure successful commercialization of the technology. The technology developers must work in tandem with representatives from the microelectronics infrastructure because all will play important roles during the commercialization of these revolutionary technologies.

Once a design has been selected team members should be chosen to integrate all of the elements needed to successfully introduce and transfer direct-write technology to the microelectronic community. By minimizing the number of participants, customer/vendor interactions will be enhanced and meetings with key personnel should be easier to arrange. This will enable the progress, the processing issues, and any potential solutions to be quickly integrated into the program. Companies should plan to have at least quarterly review meetings with the principal program managers from each facility. In addition, direct contact between the appropriate technical members at each facility, face-to-face meetings, and onsite evaluations should be scheduled when appropriate. These meetings and experimental evaluations should be enhanced by the use of video conferencing, e-mail, and voice-mail.

4.1. COMMERCIALIZATION STRATEGY

Focused efforts to identify the potential market segments for the integration of the technology can be helpful for the development of a strategy for commercialization. Three broad segments exist which could benefit by the development of the direct-write technologies: (1) materials developers and suppliers, (2) equipment developers and suppliers, and (3) electronic product assemblers.

The ultimate success is dependent on the commercializing strategies as well as on the technology developers to modify the existing microelectronics manufacturing value chain. The commercialization and licensing of the materials and processing technologies created during the technology evaluation will enable alteration of the existing microelectronics supply chain. The new supply chain structure will shift the value creation to a new segment. However, the commitment made by the technology developers will ultimately determine the level of success. Establishing alliances and teaming relationships with the end users prior to the commercial offering of the developed direct-write system cannot be stressed enough. A teaming approach is typically suggested in microelectronic industry technology transfer models.

Strategies to advertise the technologies and to gain credibility within the microelectronics industry should be developed. Documentation as well as engineering support to those companies willing to evaluate the technology must be made available prior to the sale of the technology. After company "buy-in" is obtained, teams must form technical support organizations, which should both be flexible and have the ability to relocate to the prototype factory to work with its personnel until a desirable level of confidence is established.

The developers can expedite the commercialization of the technologies by establishing relationships with OEMs by offering them quick-turn delivery of fabricated passive-device-populated substrates. The technology developers should begin marketing the technology in the microelectronic market. Entrance into additional markets will be based on revenue and experience gained from the initial efforts. As the systems are transferred to the beta platform, team members will use established procedures to make recommendations to incorporate this technology in future assembly lines and factories. Team members may be met with some resistance, but as the technologies demonstrate their ability to perform in the manufacturing environment, external partners for technology transfer (e.g., material suppliers, equipment developers, and OEMs) will carry the project's momentum forward through recommendation of the technologies' use to their customers.

Past experience suggests that a multistage commercialization strategy can increase the chances for technology commercialization success. The strategy's three general phases are as follows.

4.1.1. Phase 1: Technology Introduction

It is critical to introduce the technology to the microelectronic industry as quickly as possible to enable product design engineers the opportunity to modify existing standard practices. The developers should attend several of the well-attended microelectronic conferences and give presentations discussing the benefits of the technologies such as cost, flexibility, size, etc.

4.1.2. Phase 2: Demonstrate Technology

It is recommended that developers offer beta-level systems to OEMs to generate a database for the direct-write system. While building the beta-level system, frequent update reports should be sent to the industry committee members. The developers should select an OEM site (or sites) for the testing of the system which is representative of a typical manufacturing environment, for instance, 30 °C/60%RH, no clean room facility. This activity will give the developers the opportunity to receive first-hand feedback from the customer. In addition, the developers will receive exposure to the manufacturing environment.

4.1.3. Phase 3: Technology Market Diffusion

During this phase it is suggested that the developers prepare a Design-to-Manufacturing (DTM) Evaluation Questionnaire to help them prepare the platform for the microelectronics manufacturing market. An example questionnaire is given in Fig. 7. It contains questions regarding the six key areas for DTM systems which enhance the rate of diffusion by ensuring that the technology meets the criteria set by manufacturing companies. Also, as technology diffusion will also depend on the economic variables, the technology sales and engineering teams should hold discussions with the OEM purchasing and manufacturing engineering groups to discuss the terms of the agreement for purchase of the system. Standard terms should be used as should customary warranty and technical support terms.

General Use Capabilities

Question
Import/Export Capabilities:
List CAD packages with import capabilities
Import from existing machine programs?
General Setup Capabilities:
Is a graphical product viewer with report generation capability included?
Can product be viewed after CAD translation and prior to production?
Is a line balancer included?
Is there revision control or date/time stamps to identify program versions?
General Work -In-Process Capabilities:
Are colored overlays generated depicting the product and process?
Is real-time data collection and reporting, e.g., yields and alarms, possible ?
What machine data collection protocols are supported?
Are both manual and automated defect tracking supported?
Can defect routes be documented from initial incident to corrective action?
Can program files be downloaded directly to the machines?
Is there a unique security password level for program transmission in addition to program generation?

Optimization Capabilities

Question
Optimization Algorithm Issues:
What attributes are used to optimize fabrication?
What attributes are used to optimize factory performance?
Is it possible to balance with respect to an existing setup?
Is it possible to integrate internal (company-specific) optimizers?

Ease of Use

Question
Is there a window-based, graphical user interface?
How many separate software applications does the user need to run?
Is on-line help provided?
Are there default settings and a default execution procedure?

Architecture

Question
Platform / OS support:
UNIX / AIX
Minimum System Requirements:
UNIX, PC / Windows / Windows NT?
Code Design:
Of how many separate software applications does the complete system consist?

Cost

Question
Licensing:
What is the per-seat cost?
What licenses are available?
Training/ Maintenance Cost:
What is the training cost?
What is the cost of on-site support?
What is the annual fee (maintenance, etc.)?

Support and Training

Question
Product Support Issues:
Is on-site installation / user support available 7X24 hours?
Is support at the same level, regardless of the location, e.g., U.S., Europe, Asia, etc.?
Is a product WWW site available?
Product Enhancement Issues:
What is the procedure for distributing product updates?
What is the procedure for bug submission and tracking?
What is the procedure for user requirements/enhancements submission and tracking?
Training Issues:
Are training classes available for installers/users?

FIGURE 7 A sample of the questions for the design-to-manufacturing evaluation.

5. CONCLUSIONS

The microelectronics manufacturing industry is presently evaluating different direct-write tools for integration into next-generation manufacturing platforms. Several promising tools and complementary materials systems are in different stages of development. The systems developers have demonstrated fine feature structure fabrication as well as lot-to-lot consistency. A few of these tools are enhancements to existing tools presently used in low-volume production of electronic product.

The required system parameters to enable high-volume manufacturing compatibility were discussed. Also, the strategy required for introduction of this technology into the manufacturing industry segment was presented.

REFERENCES

1. Cole, B., "High speed liquid dispensing systems," 2001, *http://www.speedlinetechnologies.com/*
2. Mathias, W., "Micropen direct writing and precision dispensing system," 2001, *http://www.ohmcraft.com/*
3. Hayashi, C., Kashu, S., Oda, M., and Naruse, F., "The use of nanoparticles as coatings," Materials Science and Engineering, A163, 1993.
4. Williams, R., Wrisley, Jr., D., and Wu, J., "Laser transfer deposition," US Patent 4,987,006.
5. Drew, R. F., Schmidt, Jr., F. J., and Vanduynhoven, T. J., "Laser deposition of metal upon transparent materials," T988007, IBM, 1979.
6. Miller, D., "Flow guidance direct write (FGDW) technology," 2001, *http://www.optomec.com/*
7. Owen, M., "Laser system and method for plating vias," US Patent 5,614,114.
8. Chrisey, D. B., and Hubler, G. K., eds., "Pulsed laser deposition of thin films," Wiley Interscience, New York, 1994.
9. Detig, R., "Imaging solutions using conventional and novel liquid toners and inks," 2001, *http://imi.maine.com/completed/chemicals.html*
10. Soszek, P., "How to avoid wet processing," Circuits Manufacturing, 21–25, May 1988.
11. Roussos, G., "Inkjet Technology," 2001, *http://www.inkjet-tech.com/*
12. Bateson, J., "128 Jet MaxRes™ Printhead," 2001, *http://www.leadercorp.com/*
13. Wallace, D., "Microdispensing and precision printing," 2001, *http://www.microfab.com/*
14. Biggs, M., "Spectra piezoelectric technology," 2001, *http://www.spectra-inc.com/prods.html*
15. Hopp, W., Spearman, M., and Zhang, R., "QCOOM 2.0: queuing cost of ownership model," Northwestern University, Evanston, IL, 1994.

CHAPTER 4

Direct-Write Materials and Layers for Electrochemical Power Devices

PAOLINA ATANASSOVA, PLAMEN ATANASSOV, RIMPLE BHATIA,
MARK HAMPDEN-SMITH,* TOIVO KODAS, AND PAUL NAPOLITANO
Superior MicroPowders, Albuquerque, New Mexico

1. INTRODUCTION

In this section, the needs for, and attributes of, direct-write electrochemical power devices are discussed. An electrochemical power device is defined as a device that stores or produces electrical energy. This includes batteries, fuel cells, supercapacitors, and capacitors. The primary focus of this description will be on batteries and fuel cells because capacitors have been described in other sections and supercapacitors are related to fuel cells from a materials viewpoint. The emphasis here will be on the description of the key features of these electrochemical power devices that the authors believe can be significantly improved through direct-write deposition technologies. In this context, direct-write technologies provide various combinations of attributes including conformality, complex patterns, fine features, digital deposition, rapid prototyping, and control over larger thickness and structure. In a complementary sense, the materials enable low-temperature processing ($< 300\,°C$) and realize the combined attributes of the tool. Thus, this section is not intended to be a comprehensive overview of the literature, but rather an illustration of what can be achieved by revolutionary materials systems coupled with direct-write technologies designed specifically to control layer characteristics.

Electrical power devices are extremely important to the U.S. economy and security and are ubiquitous throughout our business and leisure activities. The battery industry alone is estimated to grow to $60b by 2004 (1). Batteries are largely used to power portable electronic devices while nonportable devices are powered by connection to the electrical grid with electrical power supplied by power stations. With the onset of electrical power deregulation and the increasing demand for high-reliability electrical power, fuel cells are gaining momentum as potential sources for distributed power generation from devices as small as cell phones to as large as ships. Distributed power generation can be viewed as a general system design concept to allow "on the spot" and "on demand" power supply. This generalization spans everything from power sources integrated onto a chip to conformal power devices. The suitability of the power output of different electrical power technologies as a function of device size is shown in Fig. 1.

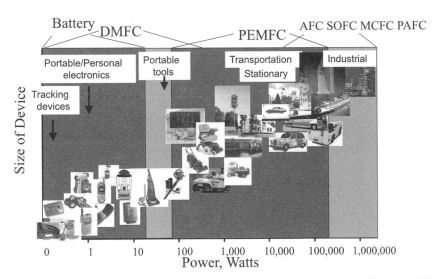

FIGURE 1   Schematic representation of the power requirements for devices as a function of the size of the device (courtesy Superior MicroPowders).

2. BACKGROUND

2.1. CLASSIFICATION OF ELECTRICAL POWER DEVICES

Electrochemical power devices can be broadly classified according to the method by which they convert chemical energy to electrical energy. Fuel cells, in general, are used to convert chemical energy derived from a variety of fuels such as natural gas, hydrogen, gasoline, or methanol into electricity. Batteries and capacitors store chemical energy and convert it directly to electrical energy on demand. Each particular type of electric power device has its own niche in terms of market applications. However, these niches, which exist primarily in the battery industry because fuel cells have not yet achieved broad market adoption, have been driven by the performance characteristics and form factors of the power devices rather than by the needs of the device that it is powering. As a result there is a strong need for application-specific power sources that have electrical performance tailored to the requirements of the device and a form/packaging structure in which the battery dimensions are tailored to the device needs rather than the opposite.

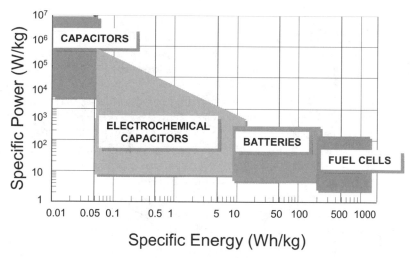

FIGURE 2 Ragone plot for various energy storage and conversion devices. Adapted from R. Kotz and M. Carlen, *Electrochimica Acta*, 45 (2000) 2484.

Specifically, thin form printed batteries are being sought where the power device is integrated into the device packaging.

Fuel cells, batteries, and capacitors all have different performance characteristics that lead to either their individual suitability for a particular application or their use in combination with others. This is an issue particularly when one is considering the power density and energy density requirements of a power source for a particular application. Each different type of device exhibits different power and energy density characteristics as shown in Fig. 2 (2).

For example, a big constraint on the life of a battery is the peak power requirement where the energy storage in the battery has to be compromised for a burst of high power. The energy retention and recharge cycle of a battery could be extended if this compromise did not exist. One solution to this problem would be to provide separate energy storage and power supply units where the battery would handle low power requirements and the electrochemical capacitor could provide short-duration peak-power pulses. A key technological concept enabling such integration is building a set of materials and approaches for integrated and conformal hybrid system manufacturing. This concept of hybrid systems is very important because direct-write technologies might be the only way to build them. A brief overview of the characteristics of fuel cells, batteries, and supercapacitors is given in the next three sections.

2.2. Fuel Cells

Although fuel cells were invented well over 160 years ago by Sir William Grove, they have not previously achieved widespread commercialization due to their high cost to technological issues associated with their efficiency of operation, and manufacturability problems. However, the case for mass-market introduction of fuel cells as an alternative source of electrical power is compelling from many points of view. It has been known for decades that performing the energy conversion process in a fuel cell, instead of in a mechanical engine of any kind, has several advantages. Along with avoiding such pollutants as nitrous oxides, fuel cells usually produce less noise, operate at lower temperatures, and, if configured appropriately, exhibit higher energy efficiencies (especially for combined heat and power applications) (3).

There are five traditional fuel cell technologies: alkaline fuel cells (AFC), phosphoric acid fuel cells (PAFC), the proton exchange membrane fuel cell (PEMFC), molten carbonate fuel cells (MCFC), and solid oxide fuel cells (SOFC) (4). Of these, PEMFC is the most promising candidate for commercialization in electric vehicle, large portable, and stationary applications. The variant, direct methanol fuel cells (DMFC), are most suitable for small portable devices. All of these fuel cell technologies convert a fuel such as hydrogen, gasoline, natural gas, alcohols, or chemical hydrides such as sodium borohydride, into electrical energy. Each fuel cell is based on different technologies to achieve this chemical-energy-to-electrical-energy conversion. As a result, each technology is generally more suited for a specific range of power output. Direct methanol fuel cells are particularly well suited to devices with power requirements in the range of $0.01–\sim 300$ W, while PEMFCs are best suited for devices that require power in the range of $100–\sim 300,000$ W, while most other fuel cell technologies are best suited for devices with powers in the range of $30,000–> 1,000,000$ W (5). However, there is a strong drive to develop fuel cells for all applications and SOFCs are being considered for low-power applications. This corresponds to a huge range of device applications from devices as small as MEMS and RF Tags to industrial power plants, as shown in Fig. 1.

Two of these technologies, MCFC and SOFC, are related and based on high-temperature reactions to achieve electrical power generation. These technologies are not the subject of this discussion. Of interest here are the other fuel cell types including DMFC, PEMFC, PAFC, and AFC—all low-temperature fuel cells that rely on a class of materials called "electrocatalysts." These electrocatalysts facilitate conversion of the chemical energy locked within the fuel to electrical energy through reaction with oxygen in the air to form water. These materials and the structure of the layers formed from these materials are the major determining factors in the cost and performance of these fuel cells. As a result, the methods by which the layers are deposited and those layers' structures are critical to the performance of these devices.

2.3. BATTERIES

In contrast to fuel cells, batteries store chemical energy in a variety of different chemical forms and convert the chemical energy to electrical energy on demand (6). The markets for batteries/portable power can be divided into primary (nonrechargeable, disposable) and secondary (rechargeable). Current battery technologies can in turn be divided into two groups: cylindrical and prismatic. The battery industry is currently undergoing a shift away from standard packaging toward custom-designed application-specific battery performance and form factors for next-generation products. This shift is being driven primarily by two factors in the market: the continuing demand for ever-increasing battery performance and the extremely fast market-adoption rates once a battery capability has been demonstrated.

The consumer demand for ever-increasing battery performance is exacerbated by the semiconductor industry's amazing ability to develop product generations with far superior performance in smaller sizes with extraordinarily short product-development cycles. Not only is the semiconductor industry setting records for short product-development cycles, but also, as quickly as battery companies make relatively marginal improvements in battery performance, the extra energy is eaten up by larger displays, more power-hungry processors, power-hungry wireless communications, and new device features. As Henry Norr, staff writer to the *San Francisco Chronicle* observed in his review of "Power '99" held in Santa Clara, CA, ". . . it's a painful paradox: high-tech manufacturers boast that they can move at warp speed, yet the battery industry still lives, to a remarkable extent, on discoveries made decades ago. While high-tech building blocks like microprocessors and hard drives double in capability every year or two, battery makers count improvements in 10–15%/year as good." The norm is still a 20-year lag between battery-related scientific advances and mainstream products based upon them.

One of the major limitations of the battery industry that has led to this lack of agility is the relatively small number of suitable electrochemistries available. A summary of relevant secondary battery electrochemistries is given in Table 1. However, these electrochemistries are generally not suitable for integration into application-specific designs to achieve thin form structures.

A shift away from traditional battery performance and form factor (shape) was originally predicted and to a large extent is still expected from the move to Li-polymer away from Li-ion electrochemistry. However, while lab tests of Li-polymer variants show good performance characteristics compared to Li-ion, no revolutionary increases in energy density can be expected for Li-polymer because it is based on the same electrochemistry as the current Li-ion cells.

Small, incremental improvements in traditional battery technologies are insufficient and are limiting the development of many of tomorrow's new electronic platforms. Blue tooth technology, for example, will allow virtually all electronic devices to communicate with each other in a wireless manner. This communication will require batteries to provide power to transmit and receive modules, and antennas that carry out the transmission and reception. Examples of these applications include cameras and smart cards, all of which will be able to communicate with each other only if sufficient power is provided from batteries.

There are also other emerging applications that will be enabled by the availability of very thin—in some cases high-power—batteries with small volume. These include active RF tags, toys, powered greetings cards, and magazines. Tomorrow's platforms will require batteries specifically tailored to meet their shape, size, weight, and operating requirements, resulting in the need for a highly flexible battery design and fabrication technology. Size and weight reduction are material and fabrication limited. This area of focus will clearly benefit from the new printable material systems.

TABLE 1 Advantages and Disadvantages of Different Common Battery Electrochemistries[a]

Battery electrochemistry	Typical energy density (Wh/kg)	Advantages	Disadvantages	Common uses
Sealed lead acid	30	Low cost Excellent charge retention Charges fully without deep discharging High drain rates	Lowest energy density Rechargeable only a few hundred times Slow charging Must be stored charged Heavy Lead is an environmental problem	Where weight is not critical Emergency lighting Backup power Non-portable devices
Nickel cadmium	40–60	Relatively inexpensive Fast and simple to charge Can be charged many more times than competing technologies	Trails newer technologies in density Memory effect Cadmium is an environmental problem Heavy	Some portable electronics Tools Video cameras Radios
Nickel metal-hydride	60–80	Higher energy density Some memory effect Relatively environmentally safe	Limited cyclability Needs occasional deep discharge Relatively high self-discharge rate	Low-end laptop computers Cell phones Tools
Li-ion	100–200	High energy density Low rate of self-discharge No memory effect	Expensive due to inclusion of electronics to prevent overcharging and thermal runaway Environmental problems Flammability risk if punctured	Notebook computers Cell phones High-end PDAs Digital cameras

[a]Courtesy of Superior MicroPowders.

This need for application-specific design, high power in a small/thin form is creating a new generation of battery technologies. This move will be enabled by technological breakthroughs in material system performance and in advanced prismatic form fabrication through chemistry and printability. There was an expectation in the battery industry that Li-polymer batteries would fulfill the role of thin form printed batteries. Li-polymer has essentially the same energy density as Li-ion but due to the gelatinous electrolyte, it can be formed into various shapes and sizes for application-specific and custom requirements. However, the markets for Li-polymer batteries in major contrast to Li-ion batteries, have not developed as quickly as expected due to:

- a hefty price premium of about 30% above the already costly Li-ion batteries,
- the long R & D cycle to develop a totally new battery-manufacturing technology,
- the large capital investment required in new battery-manufacturing facilities that can handle flammable materials, and
- the hazardous nature of Li-based battery electrochemistries in the marketplace.

In principle, battery performance can be strongly enhanced by improved materials printed with thinner layer structures; however, many battery electro-chemistries cannot take advantage of direct-write technologies. Metal-air battery electrochemistries have the potential to achieve superior primary and secondary performance compared to other technologies (6). However, break-throughs are required in materials performance and layer structure. This will be the subject of the battery discussion that follows.

2.4. SUPERCAPACITORS

Supercapacitors are a type of capacitor that stores energy within the electro-chemical double layer at the electrode/electrolyte interface (7). They have much higher power density than conventional batteries. Supercapacitors can store much more energy on a weight and volume basis than can most other systems, but can also deliver that energy at a high discharge rate or for longer time periods (see Fig. 2). If used in combination with a battery they can become a highly efficient energy source where high currents are involved. Some of the advantages that supercapacitors have over traditional batteries are:

- they can be charged and discharged almost indefinitely,
- their recharge rate is high, and
- they can provide high discharge currents.

Supercapacitors are essentially electric double layer capacitors (EDLC) and utilize the separation of a charge that occurs when an electrolyte is in contact with a conductor material. Electron accumulation or depletion at the electrode caused by an external power source is counterbalanced by the ionic species in the electrolyte. Because the charge separation in these systems is on the order of molecular dimensions, the resultant capacitance-per-unit area is large. No mass or charge transfer takes place across the interface leading to the benefit of supercapacitors over batteries, they can deliver millions of cycles and maintain high current drains and cycling efficiency. However, a current limitation of the double-layer capacitors is the low cell voltage, practical unit cell voltage of 1 V for aqueous electrolytes and 2–3 V for organic electrolytes.

The structure of a supercapacitor is similar to a membrane electrode assembly (MEA) in a PEMFC and so supercapacitors will not be described further because similar improvements can be made by direct comparison with the PEMFC discussion that follows. High throughput manufacturing solutions for volume MEA production could be successfully employed for new generation supercapacitors.

3. NEED FOR DIRECT-WRITE LAYERS

3.1. INTRODUCTION

It is a well-recognized fact that the performance of both fuel cells and batteries, especially metal-air batteries, can be enhanced through the use of improved materials, thinner, controlled composition layer structures, and in many cases through the deposition of patterned layers with carefully controlled dimensions. Direct-write deposition technologies have the opportunity to make a significant impact on these devices in a number of different ways.

3.1.1. Additive Deposition Processes for Patterned Structures

The additive nature of direct-write deposition processes is valuable in high-volume manufacturing as a result of the more efficient use of the materials, many of which can be extremely expensive (e.g., Pt black used in DMFCs). This is in contrast to potential competing processes that are either subtractive or involve layer deposition by delivering paste through an opening (screen printing, stenciling, slot dye, etc.).

3.1.2. Layer Structure Control

In batteries and fuel cells, control over gas-liquid, gas-solid, and liquid-solid interfaces, and the transport of ions, gases, and electrons between these interfaces is critical to successful operation. The ability of direct-write tools to deliver materials to a surface to provide layer structures that can be controlled over short distances in the lateral or vertical dimension is unique. For example, the ability to tailor the hydrophobicity of a layer by deposition of thin sublayers of varying hydrophobicity (i.e., a layer exhibiting a gradient in hydrophobicity) is extremely valuable in controlling water or liquid electrolyte transport. In this context, a direct-write tool is perhaps best thought of as a tool to deliver a certain volume of material to a specific site on a surface. This is especially true of the ink or paste "dispensing" tools such as micropens and ink-jet print heads that can deliver specified volumes of inks or pastes very reproducibly.

3.1.3. Deposition onto Complex Topography Surfaces

In some cases deposition onto complex topography surfaces is required. It may be necessary to cover the surface in a conformal sense to reproduce the complex topography, or to selectively deposit a layer with varying thickness to provide a smooth coating, or to deposit material into grooves and channels.

3.1.4. Patterned Layers

In many cases patterned layers are required to control functionality. A specific example described in some detail in Section 5.1 is the deposition of a direct-write current collector which has the role of collecting electrons from the active layers in a battery or fuel cell, but in many cases should allow for transport of gases or ions through this layer. As a result of this requirement, and also the need to optimize the electron collection efficiency, a patterned layer is best. The ability to directly write the current collector as a "grid" with certain line width and pitch depending on the application is necessary, and often requires feature sizes that are smaller than can be achieved by alternative patterning processes. It should also be pointed out that the replacement of a traditional current collector, which is typically a nickel mesh structure, already allows for an additional degree of device design and manufacturing flexibility that can be extremely advantageous in terms of cost and/or performance.

As a result of these advantages, there are different reasons to choose direct-write deposition processes over conventional ones and there are cases where a particular direct-write deposition process is more appropriate for a particular design compared to another. For example, for ultraminiature batteries and fuel cells, direct-write deposition methods may be the only viable method to produce a certain structure. As an example, an active RF-Tag battery with an energy requirement to provide 0.003 Wh, using a battery with a volumetric energy density between 400 and 800 Wh/L (higher is better because the battery is smaller), will have a volume of 7.5–3.75 mm^3. Therefore, a printed current collector (and all the other layers) for such a battery would need to be printed on an area of $\sim 500\,\mu m \times 500\,\mu m$. To achieve a high surface area for current collection without blocking the transport characteristics through this layer, lines and spacings of ~ 10 micron features are required. In contrast, larger batteries such as those required to power hand-held portable devices may not require such fine features because the size of the printed area is larger, but they may still require the other advantages of direct-write deposition.

It is likely that direct-write deposition technologies will be used in a manufacturing sense for only portions of battery or for certain specific battery configurations. Further research is required to clarify which types of battery and fuel electrochemistries and configurations are most logical for direct-write manufacture.

A short review of the operation and structure of PEM fuel cells and metal-air batteries is provided. To avoid repetition, an emphasis is placed on the description of PEMFC structure and operation because many features of metal-air battery cathode design are similar.

3.2. PEMFCs

3.2.1. PEMFC System Description

A PEMFC is shown schematically in Fig. 3 and is composed of the following sections: The Fuel Processor or Reformer (3–5), The Power Section of Fuel Cell Stack, and The Power Conditioner and Balance of Plant. These component parts will now be discussed.

3.2.1.1. The Fuel Processor or Reformer

The fuel processor, or reformer, converts natural gas or other fuels into a hydrogen-rich, low-carbon-monoxide-content gas stream. The composition and performance of the electrocatalyst powders in the MEA have a strong influence on the design of this component due to the presence of low concentrations of species (such as carbon monoxide, CO) in the reformed natural gas, which can poison the electrocatalysts. The ideal case is to supply hydrogen to the PEMFC in which case there is no need for a reformer. For residential PEMFC applications in the foreseeable future this is unlikely due to a lack of an existing hydrogen distribution infrastructure. However, for portable/compact applications, the hydrogen infrastructure for refueling is beginning to evolve.

3.2.1.2. The Fuel Cell Stack

The fuel cell stack is composed of a number of component parts as described next and illustrated in Fig. 3.

66

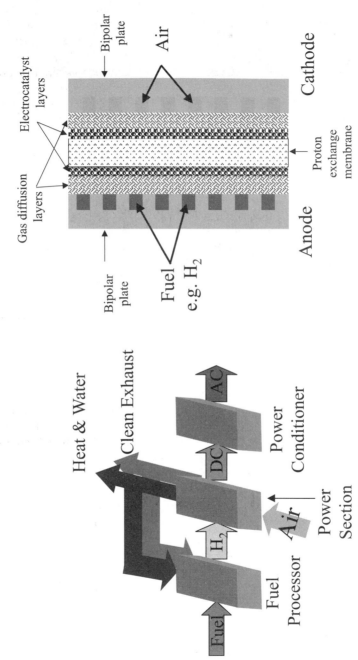

FIGURE 3   Schematic representation of, left, a fuel cell system and right, the construction of an MEA (courtesy of Superior MicroPowders).

Membrane Electrode Assembly (MEA) Each fuel cell stack comprises a number of MEAs. The MEAs are the regions to which the gases (fuel and air) are delivered where the conversion of the chemical to electrical energy takes place as catalyzed by the electrocatalysts. Each MEA will generate a useful voltage of up to ~ 0.8 V. The number of MEAs and their method of connection in the stack dictates the overall voltage of the system. Each MEA in turn, has a number of component parts, which are as follows.

The proton exchange membrane (PEM) The PEM is a proton-conductive electronically insulating membrane that selectively transports protons formed at the anode to the cathode where they react with oxygen ions to form water and electricity. The PEM is typically a sulfonated perfluorohydrocarbon, frequently referred to by its trade name of Nafion, and is produced by various polymer manufacturers, notably DuPont and Gore.

The electrodes The electrodes are mainly composed of electrocatalysts in which the active catalyst is Platinum ("Pt") or Platinum group metals, supported on a conductive support such as carbon, generally written Pt/C. The requirements for the composition of the electrode catalyst are different because a different reaction occurs at each electrode.

$$\text{Anode:} \quad H_2 \rightarrow 2H^+ + 2e^-$$
$$\text{Cathode:} \quad 4H^+ + O_2 + 4e^- \rightarrow 2H_2O$$

The cathode electrocatalyst is generally Pt dispersed on carbon (Pt/C). For the case of pure hydrogen, the anode catalyst is also Pt/C, which dramatically simplifies and lowers the cost of the overall fuel cell. However, due to the presence of CO in the reformed gas mixture that poisons Pt, a mixed catalyst is used containing ruthenium/platinum on carbon, PtRu/C. The performance of these materials and their design to accommodate the reformer performance is critical to cost, reliability, and performance of the fuel cell. The cost and performance of these precious metal-based electrocatalysts is the major contributor to the cost and performance of the fuel cell at the point of broad market introduction.

The gas diffusion layer The gas diffusion layer is a layer of porous hydrophobic material, generally carbon based (carbon cloth on paper), which is provided between the gas delivery channels contained within the bipolar plates and the electrodes to evenly distribute the gas over the surface of the electrodes.

The bipolar plates Sandwiching each MEA is a pair of bipolar plates which generally serve as current collectors to capture the electrons (electricity)

FIGURE 4   Reliant Energy Power Systems 7.5 kW PEM Fuel Cell can operate independently from the electric grid or be connected to it. Market introduction will come only after extensive field testing of this residential unit has been completed (courtesy of Reliant Energy Power Systems).

produced during the conversion of the chemical fuel to electrical power via the gas diffusion layer. The bipolar plates are used to distribute the gas uniformly over the surface of the gas diffusion layers.

3.2.1.3 The power conditioner and balance of plant

The balance of plant entails the remainder of the fuel cell. Operational modules, electrical and fuel interfaces are all part of this group of standard equipment. A picture of an overall system is given in Fig. 4.

3.2.2. MEA Structure-Performance Relationships

The performance of an MEA is primarily judged by reference to the relationship between MEA cell potential and current density, often referred to as a polarization curve. An example of a polarization curve is shown in the figures below and a brief explanation of the influence of the MEA design on the nature of the polarization curve follows.

The polarization curve in Fig. 5 shows the typical shape of the relationship between cell potential and current density. In general it is desired to move the

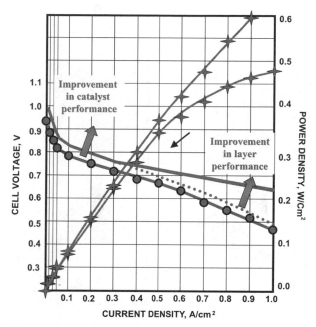

FIGURE 5   Cell voltage and power density versus current density (courtesy of Superior Micro-Powders).

curve vertically (to high voltage) over the entire area of the curve for the following reasons. Operating a PEMFC at a higher voltage leads to higher efficiency of that cell, but it also requires a larger cell (because the power density is lower). This increases capital cost in the construction of the cell and results in a lower operating cost. Operating a PEMFC at a lower voltage generally leads to lower efficiency, but requires a smaller cell (due to the high power density) and therefore smaller capital costs but higher operating costs. The vertical position of the curve is strongly influenced by a number of materials and operating factors including platinum loading (more platinum is better but significantly contributes to the cost), temperature, gas composition, and gas utilization, all of which influence the cost and reliability of the PEMFC. The goal in designing an MEA is to maximize the vertical position of the polarization curve (i.e., performance) while minimizing the cost of the materials components, as well as of capital and operating costs.

The connection between the shape of the polarization curve and the structure of an MEA is well understood and can be divided into different regions as indicated in Fig. 6: the kinetic-, ohmic-, and transport-limited regions of operation of the PEMFC MEA (4).

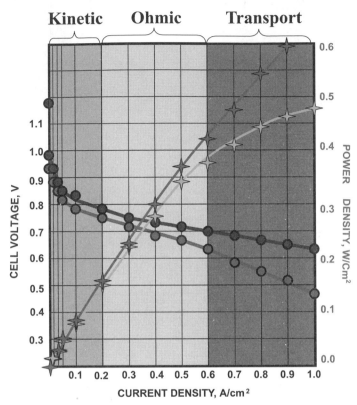

FIGURE 6 Polarization curve showing different loss mechanisms as described in the text (courtesy of Superior MicroPowders).

3.2.2.1. Kinetic Region

In this region, the performance is primarily dictated by the kinetic performance, or reactivity, of the catalyst. The more active the catalyst the higher the cell potential at a given current density. The activity of the catalyst is dictated by its structure, composition, and number of active sites.

3.2.2.2. Ohmic Region

In this region, the performance is primarily dictated by the transport of ions and electrons. Better performance is therefore achieved by good connection between highly conductive carbon particles for electrical conductivity and a good network of proton-conducting polymer connecting the catalytically active sites in the electrocatalyst to the PEM.

3.2.2.3. Transport Region

In this region, good performance is primarily achieved by the diffusion of gaseous species to and from the active site of the electrocatalysis. Better performance is manifested by rapid diffusion through the appropriate pore of the gas from the gas distribution manifold in the bipolar plates through the gas diffusion layer and the electrode.

From this description it is clear that there is a very strong influence of the materials and the structure of the layer comprising these materials on the performance and cost of the PEMFC system. Therefore, direct-write deposition methods can expect to have a strong impact on performance.

3.3. METAL-AIR BATTERIES

Metal-air battery chemistries are well known and describe a number of battery types including Li-, Fe-, Al-, and Zn-air chemistry which have some of the highest power densities of all common battery power sources (5,6,8). In these batteries, the voltage is dictated by the choice of material (metal) at the anode, the capacity/cycle life is limited by the mass of the metal in the anode, and the discharge rate and peak current capacity are limited by the efficiency of the air electrode. However, these technologies have a number of inherent problems that have limited their widespread use as portable power sources. The most active materials, Li, Al, and Fe, have limitations of self-discharge, scaling, and rechargeability (cycle life). For example, the most developed Zn-based technologies have had problems with Zn-dendrite growth that can limit rechargeability. However, all these chemistries rely on the efficient functioning of an air electrode that has a number of historical problems: transport limitations of air (O_2) and OH^- ions, low catalytic activity of the electrocatalyst, recyclability of the electrocatalyst, lack of environmental sensitivity in the drying of the electrolyte, leaking of the air electrode, and reaction with CO_2. Therefore the key to improving the performance of this battery technology is the redesign of the air electrode and improved anode materials.

3.3.1. Construction and Operation of a Metal-Air Battery

The ability to print the layers that compose these batteries, particularly the air electrode, is critical to their improvement in performance for the reasons described in this section. The electrochemistry that occurs during charging and discharging in a Zn-air battery is illustrated in Fig. 7. An illustration of the construction and components of a Zn-air battery is provided in Fig. 8.

The depolarization of the anode requires negligible overvoltage, so this electrode does not limit the kinetics of the electrochemical cell performance. On the other hand, oxygen depolarization at the cathode is an irreversible reaction (significant overvoltage), thus requiring a highly active catalyst for oxygen reduction (8,9). As a result, the oxygen electrode technology is of

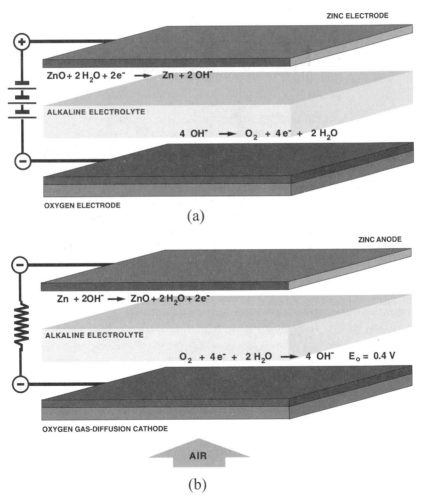

FIGURE 7   Processes in Zn-air battery during (a) charging and (b) discharging (courtesy of Superior MicroPowders).

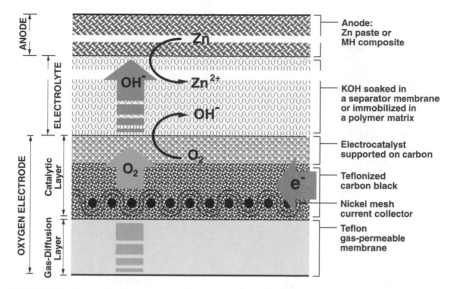

FIGURE 8 Schematic illustration of the construction of a Zn-air battery (courtesy of Superior MicroPowders).

paramount importance for such battery effectiveness. A critical aspect of the performance of the battery is the ability of the air cathode to reduce O_2 to OH^- and transport both the OH^- ions to the anode during discharge and the O_2 to the liquid-solid interface during discharge process.

Figure 9(a) is a schematic representation of the three-phase boundary where the main processes during discharge take place. Oxygen is fed to this zone of three-phase contact preferably by diffusion in the gas phase. Hydroxyl ions are withdrawn from the reaction zone while diffusing through the electrolyte and the useful current uptake is routed through the solid electron-conductive matrix. The electrocatalyst for oxygen reduction (practically its active sites) needs to populate the zone of three-phase contact, being in electrical contact with the solid-state conductor and in diffusional contact with the electrolyte and with the air. Figure 9(b) illustrates schematically the solution to this sophisticated engineering problem employing a porous gas-diffusion layer and a three-phase contact zone. The oxygen gas-diffusion electrode of the hydrophobic type consists of at least two layers and a current collection system:

- **Gas-diffusion layer**—characterized by maximal gas permeability combined with absolute impermeability to aqueous solutions. This represents the layer of hydrophobic pores. This layer is formed when

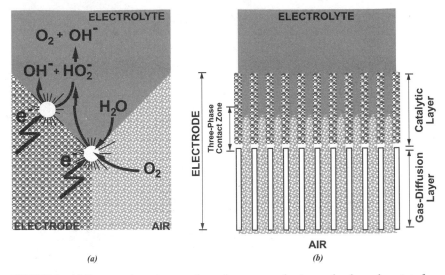

FIGURE 9   (a) Processes in an oxygen electrode: oxygen reduction at the three-phase interface; (b) a catalytic hydrophobic-type gas-diffusion oxygen electrode (courtesy of Superior Micro-Powders).

using a continuous Teflon membrane or by hot-pressing of highly hydrophobized dispersed carbon material (Teflonized carbon black).

- **Catalytic layer**—consisting of a porous body of a conductive matrix (Teflonized carbon black) and highly dispersed electrocatalysts (usually supported on a dispersed carbon material) to yield a meso-heterogeneous system with distribution of hydrophobic pores (or zones) for O_2 supply and hydrophilic pores (wet pores) for electrolyte exposure.
- **Current collector**—in the current technologies is usually a mesh made of inert metal (Ni or alloy) placed in intimate mechanical contact with the pressed matrix of highly dispersed carbon. When a Teflon film is used for a gas-diffusion layer the current collector is incorporated into the catalytic layer. Otherwise, when the gas-diffusion layer is pressed from Teflonized carbon black, the current collector mesh is incorporated in the hydrophobic layer generally closer to its gas-open side.

Figure 9(b) illustrates the importance of the spreading of the three-phase contact zone. By increasing the surface area exposed to both air and electrolyte, the population of active catalytic sites increases. This directly affects the overall catalytic activity of the electrode. In addition, increasing the surface area increases the surface-to-volume ratio in the heterogeneous electrochemical reaction, thus facilitating a reduction of the diffusional barriers. A schematic representation of the steps used in the construction of an existing air electrode is given in Fig. 10. The hydrophilic/hydrophobic profile in the catalytic layer is usually achieved by incorporating two separate sublayers: the electrocatalyst exposed to the electrolyte and the "secondary" Teflonized carbon. The three-phase contact zone is formed somewhere at the boundary between these layers. It has been observed in practice that preparation of these two sublayers from dry powders with a single pressing stage is advantageous over consecutive layer formation-pressing of the individual sublayers. During the dry-powder application, the intercalating zone of partial mixing of the two sublayers is effectively thicker, providing for a larger zone of three-phase contact. With the decrease in the overall thickness of the oxygen electrode, by direct-write deposition, the importance of forming an efficient three-phase contact zone increases. Printing techniques allow for controlled compositional formulation of the individual layers with creation of a tailored gradient of hydrophilic/hydrophobic composition. This is in distinct contrast to the existing

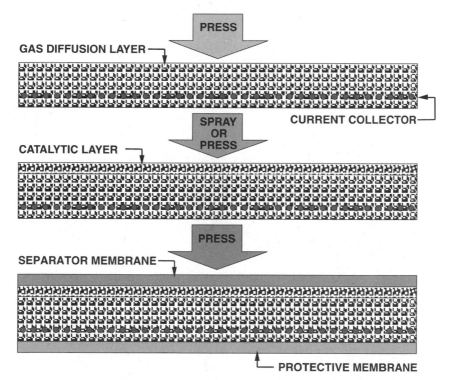

FIGURE 10   Schematic representation of the steps employed in the construction of existing air electrodes (in contrast to a direct-write approach) (courtesy of Superior MicroPowders).

approaches to conventional electrode designs that rely on poorly controlled stepwise changes in composition and properties.

4. MATERIALS FOR METAL-AIR BATTERIES AND PEM FUEL CELLS

The technologies of fuel cells and metal-air batteries are similar in terms of implementation. As a result, the materials used for both metal-air batteries and fuel cells are similar, and fall into three general classes: (1) materials, such as silver, to produce current collectors that can be processed at low temperatures; (2) electrocatalyst powders, comprising metal- or metal oxide-carbon composites; and (3) ion transport and hydrophobic control materials, such as polymer-carbon composites. Some approaches used to produce materials used for fuel cells and metal-air batteries designed for direct-write deposition will be summarized in this section.

4.1. CONDUCTIVE MATERIALS FOR CURRENT COLLECTORS

The materials systems used to direct-write conductive layers as current collectors must be processable at low temperatures to be compatible with the other materials in the battery or fuel cell. In the case of a metal-air battery, this is primarily because the current collector is in contact with a porous layer, most likely a porous Teflon layer, and so the processing of the conductive ink must occur at temperatures below which the pore structure of that material is affected. For the case of PTFE, this temperature is 250 °C for short times (10 min.). In the case of fuel cells, the polymeric materials are generally proton-transporting polymers such as Nafion and hydrophobic layers such as Teflon. Again the major concern is processing of the conductive layer without disrupting the performance of the polymer or pore structure of the layer and this must typically be achieved below 200 °C, and in some cases < 100 °C.

The nature of the conductive materials has been described elsewhere so they will not be described here. However, their direct-write layers specific to this application will be described.

4.2. ELECTROCATALYST POWDERS

Spray-based manufacturing is one example of a process (10) that can produce electrocatalyst powders with a variety of compositions and morphology with

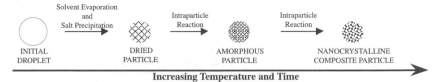

FIGURE 11   Schematic diagram describing steps in the spray-based production of electrocatalyst powders (courtesy of Superior MicroPowders).

controlled hierarchical microstructure in a highly reproducible fashion. This approach allows the use of carbon and the creation of powder batches with microstructures that provide control not only over particle size and size distribution but also over pore structure, morphology, and formation of highly dispersed catalytically active phases on the surface of the carbon (11). This modified process is shown schematically in Fig. 11 and examples of the powder morphology are shown in Fig. 12. This approach offers higher control over the manufacturing and the dispersion of the active species, in contrast to the wet methods for producing these materials (12).

The microstructure of the carbon-supported catalysts comprises highly dispersed catalysts ($\sim$1–3 nm) on primary carbon particles ($\sim$30–50 nm) that are agglomerated into spherical micron-sized porous secondary particles. The material shown in the series of SEM micrographs in Fig. 13 that illustrates this hierarchical structure is PtRu/C. This already demonstrates the ability to produce complex composition with excellent control over microstructure in a

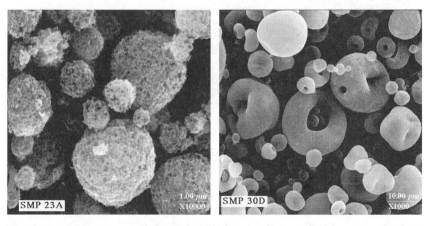

FIGURE 12   Variation in morphology of SMP electrocatalyst powders (courtesy of Superior MicroPowders).

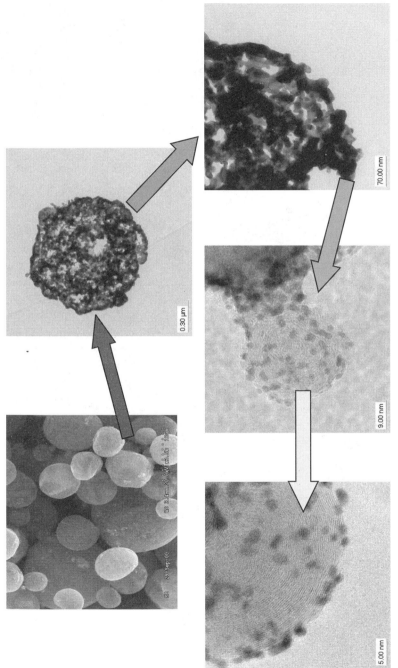

FIGURE 13 SEM micrographs illustrating the excellent control over hierarchical microstructure in the one-step manufacture of PtRu/C (courtesy of Superior MicroPowders).

one-step manufacturing process. The pore structure of these materials and control over particle size and size distribution is such that they have significant performance advantages when deposited into an electrode compared to the performance of existing catalysts.

Spray-based processing has the advantage that a higher degree of dispersion (smaller catalyst particles) can be reproducibly achieved when compared to the alternative existing manufacturing methods. In addition to these materials, the capability to produce a wide range of carbon support catalyst materials as well as self-supported (13) and metal-oxide support materials (11) has been demonstrated which exhibit excellent performance in a number of applications including PEMFCs for residential applications, PEMFCs for space exploration applications, DMFCs, and AFCs (14). Some of the materials are listed in Table 2.

Electrocatalysts for the oxygen reduction reaction in alkaline solution are critical to performance improvement of metal-air batteries. This is of particular importance for introduction of metal-air systems for portable electronics applications. We have developed and introduced several series of economically viable MnO_x/Carbon composite electrocatalysts for oxygen reduction in metal-air (Zn-air) battery systems.

Catalyst morphology and surface area are characteristics that typically have critical impact on the catalyst's performance. The morphology influences the packing density and eventually enables new printing methods, and the surface area influences the type and number of surface adsorption centers, where the active species are formed during the catalyst synthesis. As with the case of the supported metal catalysts, the dispersion of the active metal oxide phase can be controlled as indicated in Fig. 14.

TABLE 2 Variety of Different Electrocatalyst Materials that Can Be Produced by Spray-Based Methods[a]

Support	Catalyst phase	Application
Carbon	Metal oxides	Alkaline fuel cells, metal-air batteries
Carbon	Metal and metal oxide	Alkaline fuel cells, metal-air batteries
Carbon	Metal and metal oxide (e.g., Pt and RuOx)	CO tolerant anodes for PEMFCs
Carbon	Platinum and platinum/element mixtures (e.g., Pt, PtRu)	PEMFCs
Carbon	Platinum at high concentrations on carbon (e.g., 60 wt.%)	DMFC applications
Carbon	Multicomponent alloys with proprietary compositions	PEMFC cathodes and DMFC

[a]Courtesy of Superior MicroPowders.

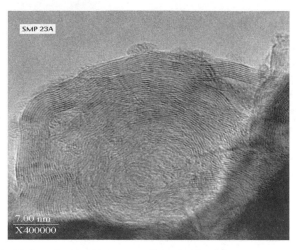

FIGURE 14 High-resolution TEM image of MnO_x/C composite: structure of an individual primary particle (courtesy of Superior MicroPowders).

The understanding of these structure-property relationships has led to improvement of catalytic activity. For example, XPS studies showed that there is a linear correlation between the electrode potential of the catalysts and the average MnO_x crystallite size (see Fig. 15). The combined information on the Mn oxidation state and MnO_x species dispersion derived from the XPS analysis proves to be a valuable source for clarifying the MnO_x/C electrocatalyst structure and for predicting the electrocatalyst performance. Achieving a Mn-oxidation state that is optimal for the electrocatalytic performance is probably the most critical requirement. However, achieving such active species in a highly dispersed form is of no less importance. The higher the dispersion the higher the number of active centers exposed to the electrochemical reagents and so the higher the turnover numbers. The ability to deposit layers using direct-write approaches, which preserve these nanostructures, is critical.

5. DIRECT-WRITE LAYERS FOR BATTERY AND FUEL CELL APPLICATIONS

The goal of fabricating a battery or fuel cell using direct-write methods is to improve the performance while simultaneously reducing the volume or mass of certain components within certain cost constraints. Here, some examples are given of layers and structures that have been deposited by direct-write processes for these applications.

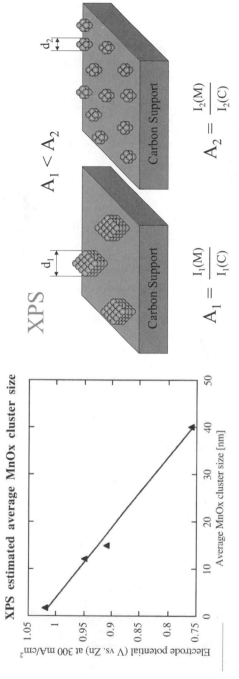

FIGURE 15 Dependence of electrode potential vs. average MnOₓ cluster size as estimated from XPS data.

5.1. CURRENT COLLECTOR

Current collectors in batteries are typically composed of a nickel mesh (6). Nickel mesh has some advantages in terms of its structural rigidity, but has disadvantages in terms of its relatively large size (thickness) and difficulty of integration into a high-volume manufacturing process (e.g., for a fully printed battery integrated into an active structure and miniaturization for ultra-miniature applications). An example of the comparison between a typical nickel mesh and a silver current collector printed onto porous Teflon is shown in Fig. 16.

A direct-write current collector is a viable alternative. The feasibility of direct-write current collectors has been demonstrated; a number of silver-based grid structures have been deposited by a variety of different direct-write deposition methods (15). As described in other chapters, direct-write tools each have their own specific attributes in terms of processing characteristics, feature size, and printing speed (16–19). In addition, the materials requirements are often tool specific. In the remainder of this section, examples of current collectors that have been printed by different direct-write tools for use in direct-write metal-air batteries and fuel cells are presented.

Figure 17 shows a direct-written silver current collector produced using a micro-pen-based tool onto a porous Teflon substrate (17). The deposited layer was processed under conditions that did not affect the porosity of the Teflon and had a resistivity of five times bulk Ag. This silver current collector grid has line widths of approximately 300 microns with spacings of approximately 1 mm. The cross-section of this current collector shows a hemispherical topography with a maximum thickness of about 65 microns. The deposition was conducted under conditions where the intersection of the lines did not lead to doubling of the line thickness.

FIGURE 16   A typical nickel mesh (left) and a SMP silver current collector (right) printed onto porous rolled Teflon (courtesy of Superior MicroPowders).

FIGURE 17 Silver current collector deposited using the CMS Technetronics micropen-based direct-write deposition system (courtesy of Superior MicroPowders and CMS Technetronics).

Figure 18 shows a current collector deposited onto porous Teflon using the plasma-spray-based direct-write process being developed at State University of New York (SUNY) at Stony Brook. This silver current collector showed good conductivity (4–5x silver bulk resistivity) with line widths of ~700 microns

FIGURE 18 Silver current collector deposited using plasma spray-based direct-write deposition system (courtesy of Superior MicroPowders and SUNY, Stony Brook).

FIGURE 19   Silver current collector deposited using the Potomac Photonics laser-transfer-based direct-write deposition system (courtesy of Superior MicroPowders and Potomac Photonics).

and spaces of about 1.5 mm. In this case the layer thickness was ∼5 microns as can be seen in the cross-section.

Figures 19 and 20 show SEM micrographs of the silver current collector deposited onto PTFE using the laser-transfer process being developed at Potomac Photonics (19). In this case, the grid width is about 100 microns with spacing of about 200 microns. The grid has good electrical conductivity (4–8x silver bulk resistivity) and the layer thickness is 3 microns.

Figure 21 illustrates a silver current collector deposited by the aerosol jetting, resulting in silver with a resistivity of 2x bulk silver. The line width is ∼100 microns with layer thickness of 0.8 microns.

As can be observed from these figures, all these direct-write tools can deposit silver current collectors with good conductivity characteristics under processing conditions that are not detrimental to the porosity of the PTFE

FIGURE 20   Higher magnification SEMs of silver current collector deposited using the Potomac Photonics laser (courtesy of Superior MicroPowders and Potomac Photonics).

FIGURE 21 Silver current collector deposited using the Optomec aerosol-based direct-write deposition system (courtesy of Superior MicroPowders and Optomec Design).

substrate. However, each structure has a different morphology, feature size, and thickness, and although these feature sizes do not reflect the range of sizes possible with each of these direct-write tools, it is clear that there will be specific advantages to each tool depending on the device needs.

5.2. ACTIVE LAYERS

In the case of a direct-write gas-diffusion electrode for a metal-air battery or fuel cell, the layers that are deposited onto the surface of the current collector are the active layers responsible for the catalytic reaction of the gases. The advantages of direct writing these layers compared to convention deposition technologies is to achieve specific placement of material, thinner layers, and/or better control over the composition of the layers within a very small distance to achieve control over a gradient in composition.

In order to investigate the ability to tailor the layer characteristics such as specific material placement, gradients in composition, and layer thickness, a syringe dispenser can be employed. This simple tool allows for digital programming to achieve patterned deposition. As with other direct-write tools, the syringe dispenser has a number of attributes that can be expected to lead to a good level of control over these parameters. The thickness of a layer can be controlled by the solids loading of the active material in the ink and the writing speed during deposition. The gradient-in-layer composition can be controlled by the composition of the ink, the writing speed, and the number of sublayers deposited. Some examples of the achievements on layer thickness are illustrated here.

The formulation of an ink containing electrocatalyst powders as well as other powders to control transport processes was produced and deposited with the goal of producing a thin layer of this material onto the surface of a printed silver current collector. The results are shown in Fig. 22. As can be seen from this figure the underlying current collector can be observed around the edges of the deposited active layer. The edge view reveals the characteristic morphology of the spherical particles and the thickness of the layer can be seen as approximately one-particle-layer thick.

At the other extreme designed to achieve much greater layer thickness, the materials and operation of the syringe dispenser can be optimized to deposit much thicker layers. An example of thicker active-layer deposition is shown in Fig. 23. Furthermore, the layer thickness and layer mass can be correlated as indicated in this figure. This is an extremely valuable asset of this tool: to enable the reproducible deposition of layers with controlled thickness over a relatively large thickness range and useful area in a relatively short period of time. The rapid prototyping capability that this tool affords enables rapid development of optimum layer structure and thickness that would be extremely difficult and time consuming and in some cases impossible with a conventional non-direct-write tool.

(a)   (b)

FIGURE 22 (a) Optical micrograph of an active layer syringe deposited onto a printed current collector; (b) SEM showing edge view of the structure (courtesy of Superior MicroPowders).

87

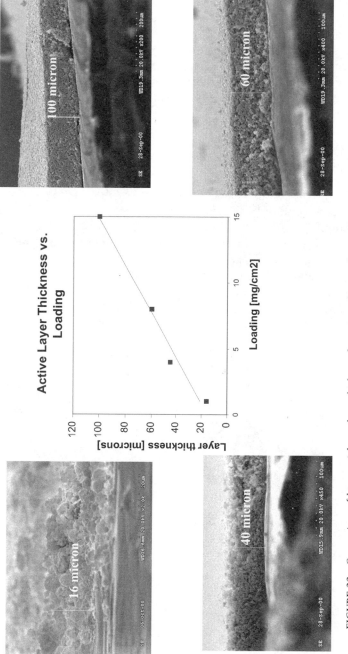

FIGURE 23   Comparison of layer weight vs. layer thickness for syringe dispensed active layers (courtesy of Superior MicroPowders).

As a result of this ability to rapidly determine the optimum structure-performance relationship a complete gas electrode can be constructed using these approaches. An example of a printed gas-diffusion electrode in cross-section is shown in Fig. 24.

This gas-diffusion electrode is composed of a porous gas-diffusion layer on which has been printed a silver current collector. The current collector is shown by the lighter areas in Fig. 24 and has dimensions of 40 micron lines, 15 microns in height with 300 micron spacings. The active layer deposited onto the current collector/gas-diffusion layer comprises materials that catalyze the chemical conversion of the gas, and materials to control the hydrophobicity of this layer. This layer is about 30 microns thick.

The electrochemical performance attributes of direct-write layers can be quantified in terms of the polarization curves. For the case of an air electrode under alkaline conditions, the polarization curves for a variety of different layers using the same materials are shown in Fig. 25. The solid lines are the polarization data measured in oxygen, while the dotted lines are the polarization data measured in air. The structure represented by the green lines with an active layer loading of $20\,mg/cm^2$ was deposited by conventional methods using a technique that could not deposit a lower mass-loading (thinner) layer. The other active layers were deposited using direct-write syringe dispensing with a series of different mass loadings. As it can be seen, the different direct-write mass loadings of 5 and $12\,mg/cm^2$ in air have performance that is similar

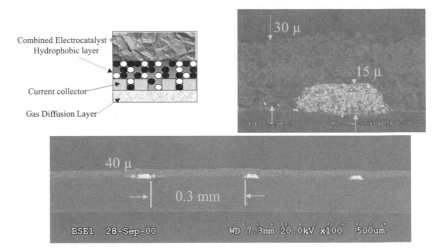

FIGURE 24 Cross-sections and schematic representation of a printed gas-diffusion electrode composed of a gas-diffusion layer, a printed current collector, and printed active layer (courtesy of Superior MicroPowders).

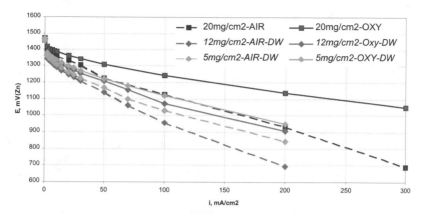

FIGURE 25 Polarization curves for a number of alkaline air electrodes comparing active layers deposited by conventional and direct-write methods as indicated in the caption (courtesy of Superior MicroPowders).

to each other but a lower performance with the 20-mg/cm^2 layer deposited conventionally. This might be expected based on the lower mass of active material present. However, in air, the 5-mg/cm^2 (yellow line) direct-write layer has only slightly lower performance in air compared to a conventionally deposited layer with 4 times more material because the layer structure has improved transport and catalyst utilization characteristics. Thus, in air—the practically useful gas—the performance of these two layers is comparable. The difference between the electrochemical performance in oxygen vs. air (effectively an oxygen-concentration-dependence measurement) reveals information on the diffusion characteristics of the layer. The difference plots in Fig. 26

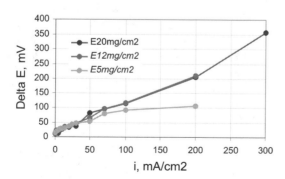

FIGURE 26 Plots of the difference in potential as a function of current density based on the polarization curves illustrated in Fig. 25 (courtesy of Superior MicroPowders).

reveal this difference and show the improved layer characteristics of the direct-write layers.

Similar performance improvements taking advantage of the improved material performance and layer thickness and structure optimization afforded by direct-write techniques can be demonstrated under acid conditions as in a PEMFC. Figure 27 shows the polarization curve for 3 different cathode electrocatalyst materials formulated into an MEA at the same cathode Pt mass loading of $0.2 \, mgPt/cm^2$.

The improved kinetic performance of our best catalyst can be identified from a Tafel plot over the kinetic region of the polarization curve and is illustrated in Fig. 27. The common slope of these lines indicates a similar catalytic reaction mechanism, and the higher position of the line derived from this electrocatalyst indicates a higher catalytic activity. However, it can be seen in Fig. 27 that the performance improvement of the best material proportionately improves at higher current densities. If the performance improvement resulted purely from the catalytic activity of our material, then the performance would not proportionately be higher at a higher current density. The improvements at higher current densities result also from improvements in ohmic and transport properties (see Fig. 5) based on the layer structure afforded by the direct-write deposition process employed.

The MAPLE-DW process, described in detail elsewhere in this book, has also been applied to the deposition of power devices (18,19). MAPLE-DW offers many opportunities for microbattery fabrication, because it can be used to nondestructively deposit hydrated and defective powders, as well as to develop a broad platform of battery materials systems (e.g., alkaline, lithium, polymer) that cannot be fabricated using high-temperature, lithographic, or vacuum-based processes. The CAD/CAM feature of MAPLE-DW can be utilized to rapidly vary and optimize the design of the batteries, plus build series or parallel battery arrays to meet a desired power demand. These DW batteries can also be fabricated directly onto an electronic substrate or solar panels, eliminating some of the packaging weight necessary in a surface-mount battery. MAPLE-DW microbatteries may be applied conformally to nonuniform surfaces and/or integrated with other power sources and electronic components to serve systems with size and/or design restrictions.

6. CONCLUSIONS

This chapter was written to provide the reader with the rationale behind the work in direct-write construction of electrochemical power devices. The goal is to show that there is a strong need for improvement in the performance of batteries, fuel cells, and supercapacitors and that in a number of cases, direct-

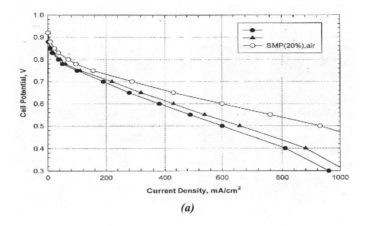

(a)

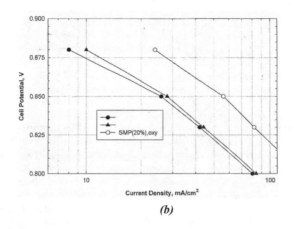

(b)

FIGURE 27   (a) Polarization curve for three different electrocatalyst samples deposited into an MEA structure at a cathode loading of $0.2\,mgPt/cm^2$ measured under identical conditions of atmospheric pressure gases and $50\,^\circ C$; (b) Tafel plot for the data presented in (a) (courtesy of Superior MicroPowders).

write deposition technologies together with specifically designed materials can make a significant impact. Direct-write deposition methods may not be suitable for all device manufacture. In this chapter we have shown that for certain electrochemical power devices, especially those that operate on a principle involving the conversion of gases to electrical energy, a significant benefit can be derived from direct-write processes.

ACKNOWLEDGMENTS

SMP wishes to thank many collaborators who have been involved in the work described in this chapter including CMS Technetronics, Potomac Photonics, Naval Research Laboratory (NRL), Optomec Design Company, and SUNY at Stony Brook who are developing direct-write tools using SMP materials, and Drs. A. John Appleby, Sergey Gamburzev, and Konstantin Petrov at Texas A & M University for electrochemical measurements for batteries and fuel cells. Much of this work was funded through contract funding from the DARPA MICE and NIST ATP programs.

REFERENCES

1. D. Saxman, Business Communications Company, Inc., "The US Battery Industry: Developing Technologies and Markets."
2. R. Kotz, M. Carlen, Principles and Applications of Electrochemical Capacitors, *Electrochim. Acta*, 45 (2000) 2484–2498.
3. J. Larminie, A. Dicks, *Fuel Cell Systems Explained*, John Wiley & Sons, New York, 2000.
4. *Fuel Cell Handbook* (Fifth Edition), E G & G Services Parsons, Inc., US DoE, Oct., 2000.
5. A. Kozawa, *Fuel Cells*, Wiley-VCH, New York, 1996.
6. *Handbook of Batteries*, D. Linden (Ed.), Second Edition, McGraw-Hill, Inc., 1995.
7. B. E. Conway, *Electrochemical Supercapacitors*, Kluwer Academic/Plenum Publishers, New York, 1999.
8. K. Kinoshita, *Electrochemical Oxygen Technology*, John Wiley, NY, 1992.
9. A. J. Appleby, Electrocatalysis of Aqueous Dioxygen Reduction, *J. Electroanal. Chem.*, 347 (1993) 117–179.
10. T. T. Kodas, M. Hampden-Smith, J. Caruso, D. J. Skamser, Q. H. Powell, Metal-Carbon Composite Powders, Methods for Producing Powders and Devices Fabricated from Same, Patent Number US6103393.
11. S. Gamburzev, K. Kunze, P. Atanassova, P. Atanasov, M. Hampden-Smith, T. Kodas, Performance of Non-Precious-Metal Catalyst for Alkaline Fuel Cell, *The Proceedings of 199th ECS meeting*, March 25–29, 2001, Washington, Abs. 25.
12. *Electrocatalysis*, J. Lipkowski and P. N. Ross (Eds.), Wiley-VCH, New York, 1998.
13. S. Gamburzev, K. Petrov, A. J. Appleby, P. Atanasov, R. Bhatia, P. Napolitano, M. Hampden-Smith, T. Kodas, Bifunctional Air Gas Diffusion Electrodes with Self-supported Catalysts of Nickel-cobalt Oxide, *The Proceedings of 198th ECS meeting*, October 22–27, 2000, Phoenix, Abs. 231.
14. S. Gamburzev, A. J. Appleby, K. Kunze, P. Atanassova, P. Atanasov, M. Hampden-Smith, T. Kodas, Performance of PEMFC with Improved Pt Supported on Carbon Catalyst, *The Proceedings of 197th ECS meeting*, May 14–18, 2000, Toronto, Abs. 88.
15. *Materials Development for Direct Write Technologies*, D. B. Chrisey, D. R. Gamota, H. Helvajian, D. P. Taylor (Eds.), Materials Research Society, v 624, 2001.
16. M. Renn, Laser Guided Direct Writing, *Mat. Res., Soc. Symp. Proc.*, Vol. 624, p. 107.
17. K. Church, C. Fore, T. Feeley, Commercial Applications and Review for Direct Write Technologies, *Mat. Res., Soc. Symp. Proc.*, Vol. 624, p. 3.
18. J. M. Fitz-Gerald, D. B. Chrisey, A. Piqu, R. C. Y. Auyeung, R. Mohdi, H. D. Yong, H. D. Wu, S. Lakeou, R. Chung, Matrix Assisted Pulsed Laser Evaporation Direct Write (MAPLE DW): A New Method to Rapidly Prototype Active and Passive Electronic Circuit Elements, *Mat. Res., Soc. Symp. Proc.*, Vol. 624, p. 143.
19. K. M. A. Rahman, D. N. Wells, M. T. Duignan, Laser Direct Write of Materials for Microelectronics Applications, *Mat. Res., Soc. Symp. Proc.*, Vol. 624, p. 99.

CHAPTER 5

The Role of Direct Writing for Chemical and Biological Materials: Commercial and Military Sensing Applications

R. ANDREW MCGILL, AND BRADLEY RINGEISEN
Naval Research Laboratory, Washington, D.C.

PETER K. WU
Southern Oregon University, Ashland, Oregon

1. Introduction
2. Chemical Microsensors
3. Biosensors and Microwell Technology
4. Coating Techniques for Sensing Applications
5. Case Studies
6. Summary

1. INTRODUCTION

The direction of processing materials for chemical and biological applications is moving inexorably toward micron scale devices and dense array platforms in which micromachined arrays are fabricated as single substrates, and not as separate elements which are later assembled into an array. The driving forces for miniaturization are many and include smaller system size, increased analytical throughput, conservation of analyte and support materials, increased ruggedization and quality control, and the option for disposable arrays. This direction makes it increasingly difficult to utilize traditional parallel processing techniques for materials because of the cumbersome need to mask off for every

Direct-Write Technologies for Rapid Prototyping Applications

93

different chemical or biological material needed to assemble arrays of interest, and the inherent material waste involved in each masking step. The arrays involved could utilize 100 or more different coated elements. The processing of so many different materials in dense array formats demands direct-write techniques that can rapidly coat individual array elements in a serial fashion, and with a noncontact-deposition technique that avoids subsequent substrate contamination issues. The use of multiple dispensing heads can increase the speed of the serial processing.

An individual chemical and biological sensor, or microwell, consists of a transducer or substrate coated with one of a range of sorbent coating materials that act to collect an analyte of interest either in a reversible or irreversible binding process. Physicochemical changes in the sorbent coating as a result of analyte binding events are monitored and converted to an electrical signal for display or recording. The range of sorbent materials covers a plethora of types and properties from simple polymers to higher-ordered biological structures including whole live cells with relatively fragile domains that require carefully controlled processing conditions to maintain viability (1–3).

The role of the processing tool used to fabricate devices coated with sorbent material is to grow thin films on the active area of the transducer or substrate with the desired physicochemical properties for the device in question, and with a suitable throughput rate. The desired film properties are wide ranging and not all deposition techniques will meet the required challenges. It is therefore important to clearly define the goals for thin film coating in terms of the desirable growth parameters and the tolerance levels that do not significantly compromise the performance of the finished device. The next step is to examine the specifications of deposition tools that are available and select one that can perform to the level required, and within a desired budget. A suggested list of film parameters to consider is shown in Table 1, and a corresponding list of deposition tool parameters is shown in Table 2 with a comparison of available techniques.

At the prototyping stage it is more important for the deposition tool to offer a wide flexibility to vary film parameters and allow for performance optimization of the sensor or device being developed. For example, during prototyping the device size may not be fixed, and if its size is reduced, especially in an array format, it may become too small for the acquired deposition tool to optimally coat without coating additional elements reserved for other materials. Without resorting to masking procedures, significant problems will result. The tool used in a high-volume manufacturing process must afford the required film properties but does not necessarily have to offer the flexibility a prototyping tool does because the film parameter space is narrower. The capital costs involved for a manufacturing tool must be examined for capital investment and related depreciation cost, maintenance requirements, and operator training level or

TABLE 1   Material Film Parameters to Consider

Thickness range
Thickness accuracy
Thickness precision
Uniformity
Coverage area
   Discrete pattern requirements
   Spreading or shrinkage effects
Chemical integrity
   Chemical structure
   Molecular weight
   Biological higher-ordered structures
Temperature limit above which decomposition occurs
Multilayer bleeding
Conformal coverage
Outgasing
Solubility and solvent selection
Surface adhesion or wetting

---

degree of automation versus the manufacturing throughput rate, and device quality control and yield. Automation is highly desirable for manufacturing cost and reproducibility, but is not a prerequisite for the prototyping stage, although for coating complex and large array structures some level of automation is advantageous for the iterative design and test process.

1.1. MICRON SCALE TRENDS

A trend that is important to consider is the shift in need for coating individual devices that are measured in millimeters to devices that have dimensions of a few tens or hundreds of microns. (The thickness of a hair is about seventy microns for reference.) Deposition tool requirements are changing as a result of this trend and in the future it will not be as easy to utilize relatively simple coating techniques with prototype device applications. The shift in size requirements for some technologies is already current but is being consolidated by the development of more and more micromachined devices and dense arrayed structures, including microelectromechanical systems (MEMS) intended for sensor applications (4), and microwell technology for high throughput pharmaceutical screening and diagnostics, or combinatorial synthesis applications (5–7). In both the commercial and military domains the current emphasis in sensor applications is to provide systems that are handheld size or smaller so that the end user is not burdened by extra weight, and so that analytical measurements can be made in the field without bringing samples back to the laboratory, which costs time and money. In military

TABLE 2 Comparison of Different Techniques Used for Growing Thin Films of Polymers and Biological Materials

Coating technique	Doctor blade	Spin coating	Vacuum sublimation	Dip coat	Molecular aggregation	PLD	MAPLE	Aerosol	In situ polymerization	Ink jet	MAPLE DW	Laser guidance	AFM dip pen	Soft lithography
Direct write	○	○	○	○	○	○	○	◐	◐	●	●	●	●	●
Compatible with broad range of polymeric and biological materials	●	●	○	●	●	○	●	●	○	●	●	●	●	●
Compatible with composites	●	●	○	●	○	●	●	●	○	●	●	●	●	●
Room temperature process	●	●	○	●	●	●	●	●	●	●	●	●	●	●
Atmospheric pressure process	●	●	○	●	●	○	○	●	●	●	●	●	●	●
Works with small amounts of material	○	○	○	○	●	○	●	○	○	●	●	●	●	●
Monolayer thickness control	○	○	●	○	○	●	●	○	●	○	○	●	○	●
Film distribution controlled by wetting of solvent to substrate	●	●	○	●	●	●	●	●	○	●	◐	◐	●	○
Enhanced film-substrate adhesion	○	○	○	○	●	●	●	○	●	○	○	●	○	○
Compatible with noncontact masking techniques	○	○	●	○	○	●	●	○	●	○	●	●	○	●
Multilayer capability without interlayer bleeding	○	○	●	○	○	●	◐	○	●	○	○	●	○	○
Does not require dissolution	○	○	●	○	○	●	●	○	◐	○	◐	○	○	○
Unaffected by material viscosity and temperature effects	○	○	●	○	●	●	●	○	●	○	●	●	○	●
Maintenance considerations (H/M/L)	L	L	M	L	L	H	H	M	L	M	H	H	H	ND
Technology commercially available (●)	●	●	●	●	●	●	○	●	●	●	○	○	○	○
Capital costs (H/M/L)	L	L	M	L	M	H	H	M	M	M	H	H	H	L

● = yes; ○ = no; ◐ = yes/no.

scenarios this is especially important to consider. Locally and at a command post level, decision making in response to a chemical or biological agent release has to be immediate to prevent the loss of life or degradation of performance (8). In a commercial application, the opportunity to provide *in situ* and real-time analytical measurements provides needed information to allow action to be taken before critical measures are needed and also allows the monitoring of chemical and biological releases that can be completely missed if samples for laboratory analysis are taken infrequently. Longer-term trends are demanding that system sizes shrink to the *Dick Tracy* watch size (9) so that expendable and distributed sensor networks can be implemented in the field and buildings to monitor for toxic analytes of interest.

2. CHEMICAL MICROSENSORS

In order to meet the demands of chemical detector field operation, instruments are being reduced to hand-held size and smaller. A wide range of different transducers is being used to develop hand-held chemical detectors, which fall into two main classes. The first is based on technologies that ionize analyte molecules prior to detection such as ion mobility spectrometers (IMS) and photoionization detectors (PID). The second technology class incorporate transducers that are coated with a chemically sorbent coating, such as a chemoselective polymer, as shown in Fig. 1. The latter class of chemical sensors is discussed here together with the relevance of direct-write techniques. Unlike the extreme specificity that accompanies biochemical interactions and binding analyte, chemoselective coatings typically offer some degree of solubility-class selection governed by the intermolecular forces of dipolarity, hydrogen bonding, and van der Waals that are present between analyte and sorbent coating. In order to allow more specificity for analytical applications, a common configuration (shown in Fig. 1b) is employed, which includes an array of cross-reactive sensors, with each transducer element coated with a different chemoselective material (10). Example coatings include polymers, dyes, metal oxides, metals, or functionalized self-assembled monolayers and combinations of these materials. Arrayed responses to analyte produce signal patterns that can be interpreted by pattern-recognition techniques to identify the analyte from a library of training data (11).

When analyte molecules enter the sensor and are exposed to the chemoselective coating, some are sorbed to the coating and a physical change in this material is monitored. There are three main classes of miniature chemical transducers whose measurands include changing mass, optical, and electrical properties of the chemoselective coating after analyte is sorbed. Table 3 categorizes a number of these transducers according to their measurand categories. Representative examples of these devices are discussed in the sections that follow.

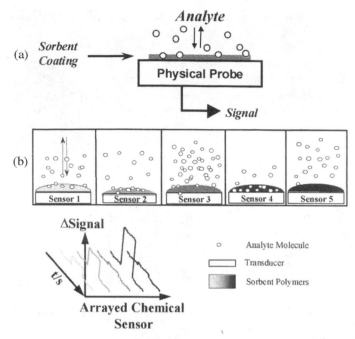

FIGURE 1 Transducer coated with sorbent coating showing reversible analyte uptake (a); Arrayed chemical or biochemical sensors and patterned signal responses (b).

2.1. MASS SENSORS

Mass sensitive devices include acoustic wave or resonator structures such as surface acoustic wave (SAW), quartz crystal microbalance (QCM), acoustic plate mode (APM), flexural plate wave (FPW), thin film resonator (TFR), and cantilever (CL), and structures that respond to bending such as the cantilever

TABLE 3   Chemical and Biological Transducers

Measurand	Mass	Optical	Electrical	Thermal	Magnetic
T R A N S D U C E R	SAW	FOCS	CR	CAL	BARC
	QCM	TIRF	EC		
	APM	SPR	MOS		
	FPW	SERS	CAP		
	TFR	AFP	CELL		
	CL				
	FABS				

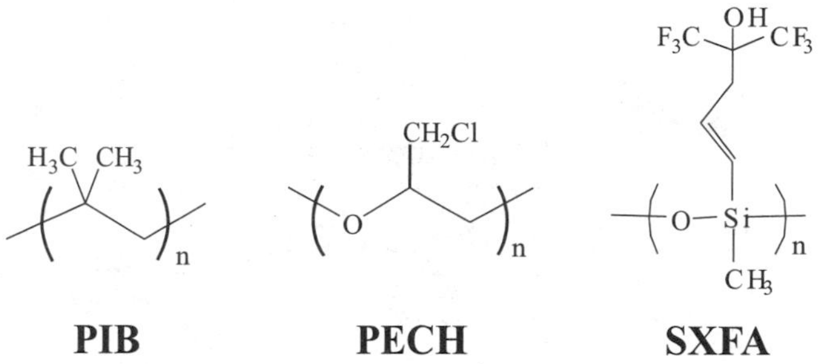

PIB   **PECH**   **SXFA**

FIGURE 2   Sorbent chemoselective polymers.

structures operated in a static mode (12–17). The SAW, QCM, and APM devices are typically made from piezoelectric quartz cut to the appropriate plane. The FPW, TFR, and CL devices are fabricated with micromachining techniques and offer the potential for large dense arrays that can be integrated with electronics on a single substrate. Piezoelectric materials for these devices are grown as thin films, for example, of zinc oxide or aluminium nitride. Polymer coatings for these devices have been developed for a range of vapor and gas applications including chemical warfare gases, explosives, toxic industrial chemicals, and household-related chemicals (1,18,19). Figure 2 shows some example polymers that all exhibit rubberlike properties at room temperature. This is important for rapid uptake of analyte which can result in faster signal kinetics for polymer-coated sensors. Of the examples selected, polyisobutylene (PIB) selectively sorbs hydrocarbons, polysiloxanefluoroalcohol (SXFA) is selective for hydrogen-bond basic analytes, and polyepichlorohydrin (PECH) sorbs a range of organic analytes. None of these polymers sorbs a single compound, and different polymers will absorb analytes to different degrees. When used to coat-sensor arrays, this approach results in a cross-reactive array system. Figure 3 illustrates the main components of a SAW chemical sensor and the reversible response from a vapor exposure. A 250-MHz SAW device is shown in Fig. 4 as an example, and more recent micromachined cantilever devices fabricated with silicon carbide structures are shown in Fig. 5 (Boston Microsystems, Boston, MA). Surface areas of these types of structures vary from $20 \times 50\,\mu m$ to $200 \times 500\,\mu m$.

2.2. OPTICAL CHEMICAL SENSORS

Optically sensitive devices include the fiberoptic chemical sensor (FOCS), total internal reflectance fluorimeter (TIRF), surface plasmon resonance (SPR), and surface enhanced raman spectroscopy (SERS) (10,20). These devices all utilize

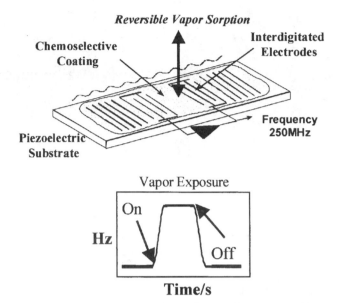

FIGURE 3 Surface acoustic wave (SAW) chemical sensor showing reversible vapor sorption.

FIGURE 4 250 MHz SAW device.

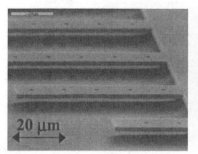

FIGURE 5 Cantilever devices fabricated with micromachined silicon carbide.

a light guide to direct light into a sorbent coating where the analyte is collected. In one example, the FOCS has been demonstrated as a dense array for a chemical sensor application. In this configuration a fluorescent dye can be immobilized within different polymer matrices to represent an array of solubility properties. The distal ends of each fiber in a bundle are individually coated with the different polymer/dye combinations using either photopolymerization or dip-coating techniques. Photopolymerization involves directing UV light through the fiber to initiate the polymerization process. Image guides of 350 μm in diameter have been used that consist of thousands of 3-μm optical fibers bundled together in a coherent fashion, such that each pixel's position is maintained from one end of the bundle to the other. In a novel direct-write application, polymers can be discretely deposited by selectively directing UV light through a pinhole to activate the desired fiber. Other fibers in the array are exposed to the same bath of chemicals but deposition does not occur because they are not exposed to the UV light (21,22).

2.3. ELECTRICAL CHEMICAL SENSORS

Electrically sensitive devices include chemiresistor (CR), electrochemical (EC), metal oxide (MO), and capacitance (CAP)-based sensors. These devices all deliver electrical current to the active sorbent film and monitor the modulation of the electrical properties of the film with analyte sorption. These devices lend themselves to miniaturization and dense array fabrication from micromachined type structures. The CR devices are conceptually simple and consist of either native conductive polymers (23), functionalized nano-gold particles (24), or polymers which are normally nonconductive but have been doped with a sufficient quantity of conducting particles to breach the percolation threshold and allow conduction (25–27). In principle there is no reason why these devices cannot operate on micron-sized substrates in dense arrays. Figure 6

102

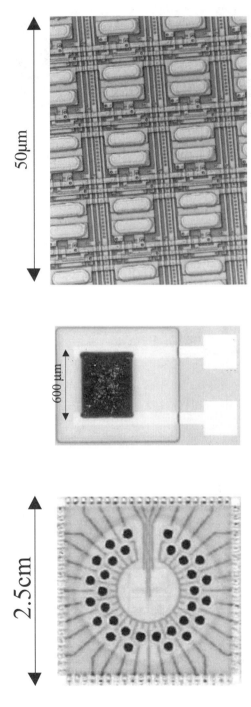

FIGURE 6   Cyrano Sciences sensor arrays fabricated from conducting composite polymer materials.

shows a 32-sensor array coated with different polymer/carbon composite materials. The individual sensor elements are in the range of 1–2 mm in diameter, but future devices with increasingly dense array chips are planned with 500×600-μm sensor elements, and arrays with 10,000 elements each 50×50 μm in surface area (Fig. 6).

3. BIOSENSORS AND MICROWELL TECHNOLOGY

A biosensor is an analytical device that uses an active biological coating interfaced with a transducer to monitor analyte-coating binding events (28–32). The analyte may be biological or nonbiological in character. These devices have wide-ranging opportunities for analytes in clinical (glucose, lactate, cholesterol, viruses), military (TNT, biological warfare agents), environmental (pesticides, heavy metals, waste water), research (DNA, immunoassays, microarrays, microwells), and food safety (*Salmonella*, *E. coli*) applications (29,30,33). There are two basic categories of biosensors: catalytic and affinity-based designs. Catalytic biosensors use active coatings such as enzymes or micro-organisms to selectively sorb analyte and catalyze the reaction of an analyte. The reaction results in a chemical product that can be monitored to determine the concentration of the particular species of interest. Affinity biosensors use receptor molecules to selectively and irreversibly bind an analyte from a sample, but they do not chemically alter these molecules (31). The active elements in affinity biosensors are usually antibodies, nucleic acids, lectins, or cell-membrane receptors. This category of sensor can also be extended to include multisite devices such as cDNA and antibody microarrays and biochips for gene and protein recognition (32). Other novel biosensors that extend beyond these basic definitions include the use of living cells and tissue (34,35) to monitor for environmental toxins or warfare agents. Other biosensor approaches such as the bead array counter (BARC) utilize coated magnetic microbeads, and atomic-force microscope base technology such as the force amplified biological sensor (FABS) (36,37). BARC and FABS are both examples of biosensor technology that involve micromachined devices that can be formed into dense arrays for a wide range of applications.

3.1. TRANSDUCERS FOR BIOSENSOR APPLICATIONS

There are four traditional types of transducers used in biosensors, and each has different requirements to be interfaced with the active biological coating.

Electrochemical sensors incorporate electrodes to measure the current generated from electrode redox events with catalytic reaction products, generated during the catalysis of analyte by the active coating. These devices often require relatively thick films (>1 μm) of active biological material to be deposited onto the electrode and often require a passivation layer to improve sensor performance. Many amperometric devices utilize biocomposite "inks" that are usually screen printed onto the electrode either in single or multiple layers (38,39). Another category of biosensor relies on optically based transducers (40). These devices detect the scatter, absorbance, reflectance, or emission of light as it interacts with the active biological coatings deposited onto an optical waveguide. Calorimetric transducers are used as a way to correlate heats of reaction with analyte concentration (41). These devices detect a change in temperature that results from the binding or catalytic reaction of analyte. The biologically active material can be packed into a flow-through column, and the temperature change of the fluid passed through the device monitored. Another approach utilizes a coated planar device with a temperature sensor integral to the design. Acoustic wave devices are the final category of transducers (42–44). They require that the active biological element be coated onto a piezoelectric substrate such as QCM or APM devices. The biosensor is then monitored for the change in resonant frequency as the analyte interacts with the active biological coating. For all of the types of biosensors described, the interaction between the active element and the transducer is key. The active biological coating must be deposited on the transducer or substrate without denaturing the structure, and it must also adhere well to the transducer if the device is to function properly (45).

3.2. DISPENSING SYSTEMS FOR HIGH-THROUGHPUT SCREENING DIAGNOSTICS

The need for high-throughput pharmaceutical screening and diagnostics, or combinatorial synthesis applications, has helped foster the development of microwell plates (5–7). Recent advances in genomics and combinatorial synthesis have placed increasing demands in this field and spurred rapid evolution of new technologies and modalities. Large arrayed well plates are used for these applications so that in one step, large numbers of pharmaceuticals can be tested for their activity, many biological samples can be analyzed, or combinatorial synthesis performed. Typical well plates contain 1,536 elements, as shown in Fig. 7, and 9,600-element formats are becoming available with assay volumes of down to 200 nl. Each well in this format has dimensions of about 1.0×1.0 mm and a depth of < 10 μm. Every well element in the 1,536- or 9,600 well-plate array is potentially filled with a different solution

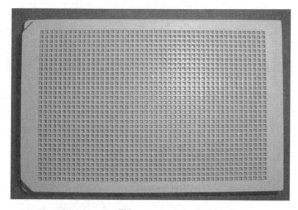

FIGURE 7   Microwell plate (Matriplate™ 1536) from Matrical Inc., showing a 1536 plate array with the dimensions of length (128 mm), height (85 mm), and plate depth of 14.3 mm. Square well dimensions are 1.68 mm on side by 4 mm deep (well volume 7 μl).

requiring a direct-write tool to fill an individual well without cross-contamination from the previous well. For increased efficiencies, the drive toward smaller well sizes will inevitably move commercial products into the micron-size regime. At this point, the substrate surface areas to be coated will limit the effectiveness of currently used deposition tools for microwell plates such as aerosol systems, pin arrays, and displacement pumps. In a recent development at the University of Louisville, researchers have developed micromachined arrays of microreaction wells for genosensor applications. Figure 8 shows part

FIGURE 8   Micromachined microreaction plate array fabricated for genosensor applications at the University of Louisville. Individual wells are 150 μm wide, 120 μm tall, and have 50-μm wall thickness feature sizes. For a 10-cm square microwell plate this would correspond to an array of 250,000 wells.

of an array structure where individual wells are 150 μm wide, 120 μm tall, and have 50-μm wall thickness feature sizes (46). For a 10-cm square microwell plate this would correspond to an array of 250,000 wells.

4. COATING TECHNIQUES FOR SENSING APPLICATIONS

4.1. SOLUTION-BASED COATING TECHNIQUES

Polymeric- and biological-related materials are commonly coated from solution because the dry material forms are much more difficult to process and can result in significant material decomposition. For the large spectrum of devices to be coated with polymers and biological materials, a generically applicable coating technique is desirable. In this chapter it is not the intent to cover in depth all the available techniques or those in development, so only a few sample technologies—including ink jetting and laser-based technologies—will be discussed in detail. In Table 2, a list of relevant technologies is compared with estimated performance capability based on the current developments in selected categories. The user may find this useful as an initial tool to narrow the selection process of the coating deposition tool.

Solution-based coating can employ a wide variety of techniques such as dip coating, syringe displacement, aerosol spray, spin coating, ink jet, AFM dip-pen techniques (47,48), soft lithography (49,50), and laser-based techniques: matrix assisted pulsed laser evaporation (MAPLE) (51–55), MAPLE direct write (MAPLE DW) (56–59), and laser guidance (60–62).

4.1.1. Solvent Evaporation and Wetting Effects on Surface Morphology

For deposition tools in which the solution of a solute material to be deposited makes contact with the substrate, the solute (e.g., a polymer) movement at the substrate surface is controlled by where the solvent moves, and this depends upon a number of factors including solvent evaporation rate, surface wetting of the solvent and substrate surface and the resulting spreading effects, and surface roughness. The wetting effects are difficult to control and require careful surface-cleaning techniques to be applied to ensure reproducibility (63). As a result of various solvent-related effects, thin films are found to contain nonuniformities. For example, the appearance of individual ink-jet spots as round dots with edges that are different in thickness than in the middle is due to solvent evaporation rates being different at the edge of the drop than in the middle and possibly from splash down effects which do not have sufficient time to recover before the solvent evaporates. An example ink-jet spot for a glucose oxidase enzyme coated electrode is shown in Fig. 9.

4.2. Non-Direct-Write Techniques

Dip coating, spin coating, and MAPLE all require a mask technique for discrete coating capability to protect areas where coating is undesirable. For highly dense arrays of devices that are coated as *integrated* array structures (i.e., not individual elements that are later assembled into an array) these techniques become increasingly undesirable as the array number increases.

Langmuir-Blodgett dip coating is limited to specialized materials that contain major polar and nonpolar components to their structure and this allows the application of the Langmuir-Blodgett techniques to form single monolayer structures on a substrate (64). The dip process can be repeated to build up a layered structure. Spin coating is useful for film deposition of relatively thick films of about a micron or thicker at a wafer-level basis but is wasteful in material consumption and difficult to control uniformity over large substrate surfaces.

Of the laser-based techniques, matrix assisted pulsed laser evaporation, or MAPLE, is a relatively new technique which comprises a frozen target consisting of a dilute solution of a polymer or biomaterial to be deposited in a suitable solvent (51–55). A pulsed laser is directed at the target and evaporates the solvent, which in turn, through collective collision, evaporates the large polymer or biomolecule solutes into the gas phase. Once in the gas phase the macromolecules are forward directed and positioning a substrate in their path allows film growth to occur. Solvent is removed through the pumped chamber system. High-quality films can be grown with this technique and can provide precise and accurate thickness control. The technique is pseudodry, in that the coating material arrives at the substrate surface significantly free of the solvent matrix. Shadow mask patterning with MAPLE has been demonstrated down to 20 microns (53).

4.3. Direct-Write Techniques

Direct-write techniques, such as ink jet and aerosol spray coating, are already being utilized commercially to deposit the active elements for some types of chemical and biochemical sensors. More recently developed techniques such as soft lithography (49), MAPLE DW (56–59), laser guidance approaches (60–62), and AFM dip-pen techniques (47,48), have demonstrated the ability to form patterns and three-dimensional structures of different biomaterials (living cells, active proteins, and antibodies) and each may find niche or more general application that displace currently used deposition tools.

4.3.1. Ink Jetting

Ink-jet printers have become a standard desktop item for home and work computers. They are very inexpensive and offer good quality for text and image printing applications. Unfortunately, the incorporation of this technology into laboratory applications, in which it is used to deposit polymers and biological materials, is not straightforward. Ink-jet manufacturers who cater to the paper printing markets have developed specialized dyes and solvent combinations that have optimal physical properties that include viscosity, surface tension, wetability, solubility, and drying rates. Any variability in the physical properties of the ink-jet solution such as viscosity due to temperature, the use of a different solvent or solute, or air bubbles present in the solution can greatly affect the performance of an ink-jet system. Nevertheless, the attractiveness and potential of ink-jet systems for laboratory applications is high. This comes from its inherent noncontact nature, which reduces cross-contamination concerns, its potential for economy, and its speed in being able to move around a substrate surface in a rapid fashion depositing discrete and small spots of material where desired. The resolution limits for ink jet are controlled by the minimum drop volume that can be deposited. Volumes of a few picoliters have been dispensed. A 3-pl drop translates to a circular surface coverage with a diameter of about 20 μm. The circular nature of the as-deposited films is a potential drawback if the substrate to be coated requires a contiguous film. In order to remedy this, multiple overlapping drops need to be deposited and this has impact in the area of film uniformity and ultimate film thickness at the thin end. Several polymeric materials and a range of biomaterials have been successfully deposited by ink-jet technology (65), with the emphasis to date being geared toward microwell technology.

4.3.1.1. Ink-Jet Approaches

There are several types of ink-jet technology but most common is "drop on demand." Most ink jets use thermal technology (Canon and Hewlett-Packard), whereby heat is used to fire ink onto the paper. An ink bubble is created, which bursts out through a nozzle plate, lands on a substrate, and solidifies after solvent evaporation or absorption into the substrate. The drop is initiated by heating the ink to create a bubble until the pressure forces it to burst and hit the paper. For laboratory applications, the advantage of the thermal ink-jet technique is in the formation of small ejection volumes which can be as small as a few picoliters. A 10-pl droplet generates a round spot with a diameter of around 60 μm. Unfortunately the heat involved with temperatures of about 200 °C and the shearing stresses that result from high linear velocities ($10\,\mathrm{m\,s^{-1}}$) could result in solvent reaction with polymer material and can denature biological materials. However, given these drawbacks, certain applications with biomaterials can be found that take advantage of this technique and recently bubble jet technology has been successfully used for microarray fabrication with covalent attachment of DNA (6). Another recent development in ink-jet technology has partnered a positive-displacement pump connected to a syringe and a solenoid time-open valve, the former to provide favorable quantitative pumping characteristics, and the latter to provide the noncontact ejection properties of the dispense element (5). For a 50-μl syringe this arrangement can yield a minimum drop volume of about 10 nl. This technique has been developed for microwell applications but lacks the resolution for MEMS-based technology with sub-100-micron-sized dimensions to be coated. The other common ink-jet technology from Epson utilizes a piezoelectric crystal positioned at the rear of the ink-jet reservoir. When actuated, the piezoelectric crystal flexes and forces a drop of ink out of a nozzle. This technique allows for smaller drop sizes and higher coating resolution. Additional details regarding ink-jet direct-write techniques can be found in Chapter 7 of this book (65).

4.3.2. MAPLE Direct Write

Matrix assisted pulsed laser evaporation direct write, or MAPLE DW, is a laser-based processing technique that was originally designed to fabricate and rapidly prototype mesoscopic electronic devices from composite materials (56,57). This approach, however, is gentle enough to successfully form patterns and three-dimensional structures of a wide variety of organics including chemically sensitive polymers, active proteins and antibodies, as well as living prokaryotic and eukaryotic cells (27,58).

MAPLE DW involves the forward transfer of materials from a UV transparent support to a receiving substrate. The transfers are performed by mixing the active, or sensitive, material in a UV-absorbent matrix, which is then coated onto the UV-transparent support. A focused UV laser pulse is then directed through the backside of the support so that the laser energy first interacts with the matrix at the support interface. A UV microscope objective focuses the laser at the interface and also serves as an optical guide to determine the portion of the matrix to be transferred. Layers of matrix near the support interface evaporate due to localized heating from the laser-material interaction. This vaporization releases the remaining material by gently and uniformly propelling it away from the support. In this manner, MAPLE DW is capable of producing passive electronic devices (i.e., interconnects, resistors, capacitors) and organic structures with line widths under 10 microns. By removing the quartz support and allowing the laser pulse to interact with the device, this approach is also able to micromachine channels and through vias into polymer, semiconductor, and metal surfaces as well as trim deposited structures to meet design specifications. All micromachining and material transfer can be controlled by computer (CAD/CAM), which enables this tool to rapidly fabricate complex structures without the aid of masks or molds.

MAPLE DW also allows the flexibility with the same tool to micromachine various substrates and deposit mesoscopic passive electronic components, and interconnects. This technique, therefore, has the potential to fabricate by a single technique, complete prototype systems on a single substrate including the sensor array, support electronics, and battery.

4.4. PROCESSING MATERIALS DIFFICULT TO DISSOLVE

It is worth noting that not all materials can be dissolved and processed in common solvents. For these materials, solvent-based direct-write techniques discussed here are not applicable, and alternative techniques are sought. In certain instances, the actuation of transducers can allow for selective film growth in the desired area from monomer material. For example conducting polymers such as polyaniline can be electrochemically grown *in situ* to a conducting electrode, which after film growth results in a chemiresistor sensor (23). In another application, metal-oxide films have been selectively grown on heated micro hotplates within a large array of micromachined device (66). The preceding examples are remarkable techniques within their domain but their widespread application is limited. Elsewhere in this chapter and this book, laser processing of materials in solvent matrices is discussed, however pulsed laser evaporation (PLE) of solid polymers is possible for some materials and has been demonstrated with reasonable success for Teflon™, (67) and other polymers (68,69). For increasingly fragile materials such as biological materials, PLE will probably never develop to the point where biological materials can be processed without denaturation. In a recent development though, tunable infrared laser processing of polymers has been successfully demonstrated, so that a laser wavelength is selected that corresponds to a significant absorption band in the infrared spectra of the polymer (70). This technique has been applied with very promising results to processing polymer materials that previously resulted in significant polymer decomposition under other laser wavelength processing conditions.

5. CASE STUDIES

In order to illustrate the complex requirements for coating sensing technologies, and how they greatly differ, four case studies are presented that relate the influence of substrate size and array configuration to the need for direct-write application. The examples chosen cover recent work in chemical sensing, biochemical sensing, and protein microarrays at the U.S. Naval Research Laboratory.

5.1. SORBENT POLYMER-COATED RESONANT TRANSDUCERS

A number of resonant chemical sensors have been developed for the detection of toxic gases and these include the surface acoustic wave (SAW), flexural plate wave (FPW), and cantilever (CL) devices, among others. Resonant devices normally consist of a piezoelectric slab, such as quartz or zinc oxide, which is lithographically patterned to provide two sets of electrodes. When an alternating voltage is applied to the electrodes, the piezo-material can be made to expand and contract. A wave is propagated along the surface by this action and the resulting oscillation frequency can be readily monitored.

The application of SAW devices to gas sensing applications takes advantage of a range of commercial devices utilized for the cellular phone and other telecommunications markets. These devices oscillate at very precise frequencies and are very sensitive to any additional mass deposited on their surface. Coating a SAW device surface with a thin layer of $\sim 50\,nm$ of chemoselective polymer over the entire crystal surface allows this device to operate as a chemical sensor. Vapor or gas is exposed to the sensor surface either by passive air sampling or through actively pumped air. Analyte vapor molecules are selectively and reversibly absorbed into the polymer and perturb the velocity of the wave resulting in a detectable shift in device frequency. The sensitivity of the SAW sensor is proportional to the amount of polymer coating deposited on the device. As a result, it is desirable to deposit reproducible amounts of polymer on each device in different systems if the same performance is required. This places the burden on the deposition technique to provide high-quality film coatings with accurate and precise control over thickness.

Normally SAW sensors are operated as arrays of up to ten different polymer-coated devices and these devices are mounted individually into a package connected to the necessary pneumatics for air flow. This arrangement allows the individual SAW sensors to be coated one-by-one with their respective polymer films, and for high-volume processes, a chuck can be loaded with a few hundred devices to allow batch coatings. A parallel processing technique requires that the deposition tool cover the entire area to be coated uniformly. For a direct-write processing technique, the deposition process can be more easily adjusted to take into account any variations in film thickness as the substrate is traversed. The issues of thickness control become even more critical when you consider that the arrayed responses from different polymer-coated devices generate a "nose" print that can identify a gas but has to be analyzed by a pattern recognition technique to do so. If one analytical system or detector was used to train an entire batch of many detectors, then the coating thicknesses of all the devices in all detectors need to be as close as possible. Otherwise, the pattern recognition algorithm will not perform optimally and may have to be adjusted for each analytical instrument which is not a practical thing to do. Additionally, the signal response kinetics to a gas exposure and the quality factor of the SAW sensor are partially related to the uniformity of the polymer film and can be improved by increasing uniformity. For increasingly mature systems integrated arrays are sometimes used and this design makes parallel processing techniques less attractive; if one sensor in the integrated array is damaged or does not perform optimally for some reason, the entire array will have to be thrown away. This in turn places increased demands on deposition tool control in the deposition of polymers and other materials. These effects can significantly affect the performance yield of a manufacturing process.

More recent resonant devices such as the FPW and CL have been manufactured by micromachining techniques. In principle they have similar operational requirements to SAW sensors, especially in regard to the sorbent film properties such as film thickness and uniformity control. Because the FPW and CL devices are fabricated from micromachined substrates such as silicon, the opportunity arises to integrate the devices with electronics fabricated by similar techniques. In addition, it is possible to fabricate devices that are significantly smaller than SAW devices. The possibilities of dense array formats are desirable with micromachined substrates. In this format, direct-write techniques become increasingly attractive. In the FPW device, the micromachined pit of the device has been coated by an ink-jet technique (71), and in another example arrays of micromachined cantilevers have been coated by the MAPLE technique (72). The ink-jet technique does not offer the finesse provided by the MAPLE for film thickness control, however it allows for more flexibility in dense array applications that require coating with many different materials.

5.2. Enzyme-Based Sensors

5.2.1. Glucose Sensor

Roughly 1 to 2% of the world's population is diabetic and experts predict that diabetes-related cases of blindness, kidney and heart failure, and gangrene could be reduced by up to 60% through stringent personal control of blood glucose levels. Therefore, there is a large commercial market for an inexpensive, disposable glucose sensor. Yellow Springs International, Inc. (YSI) is developing a microfluidic glucose sensor designed on a flexible plastic substrate that would satisfy the need for millions of diabetics to personally monitor their blood glucose level. This device is based on the electrochemical detection of the oxidation of glucose by glucose oxidase (GOD), and uses a miniature electrode to measure the current produced by this enzymatic reaction. Figure 9 shows the deposition of cross-linked GOD onto the miniature electrode by an ink-jet technique. The deposited enzyme retains glucose sensitivity, but overall there are several problems with using this deposition method. Ink jet is a solvent-based approach and therefore can only deposit materials in a circular (i.e., droplets) pattern. In this case, the inability of the ink-jet apparatus to deposit different patterns limits the electrode coverage, and therefore the output signal, significantly. The ink-jet-deposited GOD film is also not adequately adherent to the platinum electrode, resulting in discontinuities during stress tests of the flexible substrate. Alternative direct-write techniques such as MAPLE DW applied to the glucose sensor could optimize the sensor performance by providing improved adhesion control and shape of the deposited GOD films by allowing them to conform to the shape of the electrode.

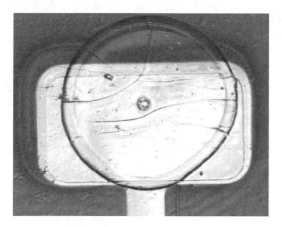

FIGURE 9   Glucose sensor; glucose oxidase (GOD) was deposited onto a miniature platinum electrode using ink jet.

5.2.2. Tissue-Based Dopamine Sensor

MAPLE DW has been used to fabricate an amperometric dopamine sensor using a plant-tissue-sorbent coating. Dopamine, also known as catecholamine, is a catechol derivative and is a neurotransmitter in the brain (73). Dopamine concentration, therefore, can be an indicator for brain activity or stimulation.

The determination of dopamine concentration using electrochemical methods and a carbon paste electrode was first achieved by Adams (74) and Ponchon et al. (75). The use of a banana-tissue-based dopamine sensor was first developed by Sidwell and Rechnitz (76) and later simplified by Wang and Lin (77). Polyphenol oxidase (PPO), an enzyme found in banana tissue, is used for promoting oxidation of dopamine to dopamine quinone. The transduction mechanism is the electrochemical reduction of the dopamine quinone back to dopamine, which requires two electrons/molecule from the electrode. In this sensor, the PPO is the BE, and the signal is the electrical current at the electrode.

Figure 10 shows a miniature dopamine amperometric sensor fabricated by MAPLE DW, in which a PPO/mineral oil/graphite composite was deposited onto a Pt microelectrode, on a flexible Kapton substrate. A calibration curve for this sensor was obtained using catechol, a dopamine stimulant, and provided comparable data to a macroscopic version of this sensor (78). In this application, MAPLE DW has been used to good effect in the transfer of a tissue-based composite with good resolution onto flexible substrates (79).

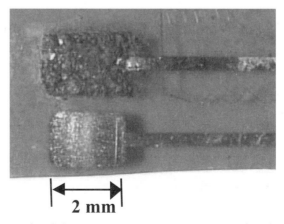

2 mm

FIGURE 10   Tissue-based dopamine sensor; MAPLE DW was used to deposit a polyphenol oxidase (PPO) composite onto a miniature platinum electrode. This laser-transfer approach enables complete electrode coverage and produces a functioning dopamine sensor.

5.3. Protein Microarray

Microarray technology is the foundation for high-throughput screening experiments for genomics, proteomics, and pharmaceutical applications. The basic principle of a microarray is to deposit thousands of immobilized biological molecules, each selective to a single analyte, in high densities on a single substrate. Current microarrays are fabricated using a manual pin to deposit roughly 200-μm diameter spots of biomolecules onto substrates such as functionalized glass slides and gel-coated slides. These approaches require several microliters of starting material and are not readily suitable for the formation of spots with high concentrations of molecules needed for higher sensitivity.

Improved or new deposition tools are needed. MAPLE DW provides a potential candidate for fabricating high-density, high-concentration cDNA and protein microarrays. This is because MAPLE DW offers the dispensing volumes of between 1 and 10 pl aliquots of active biomolecular solutions. The resulting spot sizes can be as small as 5 or 10 μm in diameter. MAPLE DW also uses less than 100 nl of starting material and allows this relatively large volume to be concentrated into smaller surface coverage areas than other techniques. Figure 11 is a single-coating microarray of fluorescently labeled (Cy5 dye) biotinylated BSA that has been fabricated by MAPLE DW. The consistent spot size of

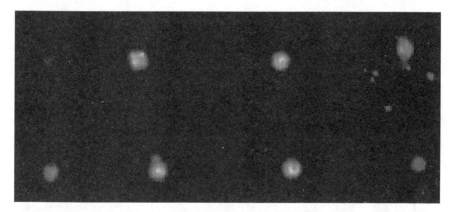

FIGURE 11 Protein microarray; MAPLE DW was used to deposit biotinylated bovine serum albumin onto a nitrocellulose-coated glass slide. Cy5-labeled streptavidin was then used as a fluorescent tag to image the microarray under UV exposure. Laser-transferred spots resulted in 50-μm spot sizes, which are four times smaller than spots achieved by commercially available techniques.

50 μm in diameter attests to the reproducibility of the MAPLE DW process. This particular array is fabricated on a nitrocellulose-coated glass slide and was interpreted with a GenePix 4000B (Axon Instruments, Inc.) fluorescent reader (80). The result demonstrates that laser transfer can be used to form high-density microarrays. MAPLE DW offers a generic capability to deposit biological materials, and future experiments will take advantage of this to form improved gene and protein recognition microarrays.

6. SUMMARY

The processing of materials for chemical- and biological-sensing applications by direct-write techniques is complementary to the micromachining developments that are impacting the form factor of a broad range of applications including chemical and biological sensors, and microwell-related technologies. Chemical-sensor applications utilize cross-reactive arrays in which different sorbent materials are coated in order to offer array selectivity to a range of different analytes. For inverse reasons, arrays of biosensors are needed to monitor different analytes, because individual sensor elements are highly selective. There are many driving forces pushing for sensing device miniaturization and they include smaller system size, increased analytical throughput, conservation of analyte and support materials, increased ruggedization and quality control, and the option for disposable arrays. The desire to create larger arrays of sensor elements is made possible by the micromachined miniaturization of sensor arrays. The dense array formats favor direct-write techniques because it is increasingly difficult to utilize traditional parallel processing techniques because of the cumbersome need to mask off for every different chemical or biological material needed to assemble arrays of interest. The inherent material waste involved in each masking step is undesirable and this issue becomes amplified as the cost of the material to be deposited increases. Typical sensor arrays involve around 10 sensor elements, but future array sizes could utilize thousands of different coated elements. Microwell plate technologies developed for high-throughput pharmaceutical screening and diagnostics, or combinatorial synthesis applications, are commercially produced with arrays of 9,600 plates. In the future, microwell arrays of 250,000 will be available. The processing of so many different materials in these dense array formats demands direct-write techniques that can rapidly coat individual array elements in a serial fashion and with a noncontact deposition technique that avoids subsequent substrate contamination issues. Of course, the use of multiple dispensing heads can speed the serial processing.

ACKNOWLEDGMENTS

The authors are grateful to Rick Mlcak, Beth Muñoz, Stephen Sunshine, Kevin Walsh, and Don Mole for providing array or direct-write-deposited structures or photographs, as well as for related discussions. Thanks also to Eric Houser, Daniel Bubb, Jim Horwitz, Alberto Piqué, Douglas Chrisey, Dave Krizman, Barry Spargo, Matt Brooks, Joanne Jones-Meehan, and Todd Mlsna for reading various parts of the manuscript or related discussions. This work was supported in part through the Office of Naval Research, and by DARPA with funds administered by Bill Warren.

REFERENCES

1. *Polymer Films in Sensor Applications: Technology, Materials, Devices and Their Characteristics,* Ed. Gabor Harsanyi, Technomic Pub Co., Lancaster, Pennsylvania, 1995.
2. *Sensors, A Comprehensive Survey: Volume 2 Chemical and Biochemical Sensors Part I, and Part II,* Eds. W. Gopel, J. Hesse, and J. N. Zemel. VCH, New York, 1991.
3. *Biochemistry,* 4th ed. L. Stryer, W. H. Freeman & Company, New York, 1999.
4. *An Introduction to Microelectromechanical Systems Engineering,* N. Maluf, Artech House, Boston, MA, 2000.
5. T. C. Tisone, "Dispensing systems for miniaturized diagnostics," IVD Technology, May 1998, p. 40 (http://www.devicelink.com/ivdt/archive/98/05/011).
6. T. Okamoto, T. Suzuki, and N. Yamamoto, Microarray fabrication with covalent attachment of DNA using Bubble Jet technology, Nat. Biotechnol., 18 (2000) 438–441.
7. L. A. Fisher, Geysen H., et al., Characterization of an inkjet chemical microdispenser for combinatorial library synthesis, Anal. Chem., 69 (1997) 543–551.
8. Proceedings of the First Joint Conference on Point Detection for Chemical and Biological Defense, 23–27th October 2000, Williamsburg, Virginia.
9. A. Johnson-Winegar, The DOD Chemical and Biological Defense Program, Proceedings of the First Joint Conference on Point Detection for Chemical and Biological Defense, pp. 28–40, 23–27th October 2000, Williamsburg, Virginia.
10. K. J. Albert, et al., Cross-reactive chemical sensor arrays, Chem. Rev., 100(7) (2000) 2595–2626.
11. R. E. Shaffer, S. L. Rose-Pehrsson, and R. A. McGill, A comparison study of chemical sensor array pattern recognition algorithms, Anal. Chim. Acta, 384 (1999) 305–317.
12. J. W. Grate, Acoustic wave microsensor arrays for vapor sensing, Chem. Rev., 100(7) (2000) 2627–2648.
13. E. A. Wachter and T. Thundat, Micromechanical sensors for chemical and physical measurements, Rev. Sci. Instrum., 66(6) (1995) 3662–3667.
14. M. K. Baller, H. P. Lang, et al., A cantilever array-based artificial nose, Ultramicroscopy, 82(1) (2000) 1–10.
15. P. Kobrin, C. Seabury, C. Linnen, A. Harker, R. Chung, R. A. McGill, and P. Matthews, Thin Film Resonators for TNT Vapor Detection, Proc. SPIE Aerospace/Defense Sensing and Controls, 13–17 April 1998, Orlando, Florida, Vol. 3392, 418–423.
16. B. Cunningham, et al., Design, Fabrication, and Vapor Characterization of a Microfabricated Flexural Resonator Sensor and Application to Integrated Sensor Arrays, Sensors and Actuators, B 73(2), 112 (2001).

17. D. S. Ballentine, R. M. White, S. J. Martin, A. J. Ricco, E. T. Zellers, G. C. Fryer, and H. Wohltjen, *Acoustic Wave Sensors: Theory, Design and Physicochemical Applications*, Academic Press, Boston, 1996.
18. R. A. McGill, M. H. Abraham, and J. W. Grate, Choosing polymer coatings for chemical sensors, CHEMTECH, 24(9) (1994) 27.
19. E. J. Houser, et al., Rational Materials Design For Sorbent Coatings for Explosives: Applications with Chemical Sensors, Talanta, 54, 469 (2001).
20. *Handbook of Biosensors & Electronic Noses, Medicine, Food, and the Environment*, Ed. E. Kress-Rogers, CRC Press, Boca Raton, Florida, 1996.
21. K. J. Albert and D. R. Walt, High-speed fluorescence detection of explosives-like vapors, Anal. Chem., 72(9) (2000) 1947–1955.
22. J. A. Ferguson, F. J. Steemers, and D. R. Walt, High-density fiber-optic DNA random microsphere array, Anal. Chem., 72(22) (2000) 5618–5624.
23. K. C. Persaud, et al., A smart gas sensor for monitoring environmental changes in closed systems: results from the MIR space station, Sens. Actuators, B, 55(2) (1999) 118–126.
24. H. Wohltjen and A. Snow, Colloidal metal-insulator-metal ensemble chemiresistor sensor, Anal. Chem., 70(14) (1998) 2856–2559.
25. E. J. Severin, B. J. Doleman, and N. S. Lewis, An investigation of the concentration dependence and response to analute mixtures of carbon black/insulating organic polymer composite vapor detectors, Anal. Chem., 72(4) (2000) 658–668 (http://cyranosciences.com/).
26. E. Severin, Cyrano Sciences Sensor Technology: The Heart of the Cyranose 320 Electronic Nose. *http://cyranosciences.com/technology/sensor.html*.
27. R. A. McGill, A. Pique, D. Chrisey, J. Fitzgerald, V. Nguyen, and R. Chung, Laser Processing of Polymers and Conductive Materials for the Fabrication of Conductive Composite Coatings: Applications with Chemical Sensors, Proc. 6th International Conference on Composites Eng., 27 June–3 July 1999, Orlando, Florida, 563–564.
28. B. R. Eggins, *Biosensors: An Introduction*, John Wiley and Sons, Inc., New York, 1996, 1–117.
29. M. Mehrvar, C. Bis, J. M. Scharer, M. Moo-Young, and H. Luong, Fiber-optic biosensors—Trends and advances. Anal. Sci., 16, 677 (2000).
30. J. F. Liang, Y. T. Li, and V. C. Yang, Biomedical application of immobilized enzymes. J. Pharm. Sci., 89, 979 (2000).
31. K. R. Rogers, Principles of affinity-based biosensors. Mol. Biotechnol., 14, 109 (2000).
32. T. Vo-Dinh and B. Cullum, Biosensors and biochips: Advances in biological and medical diagnostics. Fresenius, J. Anal. Chem., 366, 540 (2000).
33. C. A. Rowe, et al., Array biosensor for simultaneous identification of bacterial, viral, and protein analytes, Anal. Chem., 71(17) (1999) 3846–3852.
34. A. T. Capitano, M. J. Powers, A. Sivaraman, P. T. So, and L. G. Griffith, Biophys. J., 80, 638, Part 2 (2001).
35. S. M. O'Connor, J. D. Andreadis, K. M. Shaffer, W. Ma, J. J. Pancrazio, and D. A. Stenger, Immobilization of neural cells in three-dimensional matrices for biosensor applications. Biosens. Bioelectron., 14, 871 (2000).
36. D. R. Baselt, G. U. Lee, M. Natesan, S. W. Metzger, P. E. Sheehan, and R. J. Colton, Biosens. Bioelectron., 13, 731 (1998).
37. D. R. Baselt, G. U. Lee, and R. J. Colton, Biosensor based on force microscope technology. J. Vac. Sci. Technol., B 14, 789 (1996).
38. J. Wang, P. V. A. Pamidi and K. R. Rogers, Sol-gel-derived thick-film amperometric immuno-sensors. Anal. Chem., 70, 1171 (1998).

39. M. Albareda-Sirvent, A. Merkoçi, and S. Alegret, Configurations used in the design of screen-printed enxymatic biosensors. A review. Sens. Actuators, B, 69, 153 (2000).
40. D. G. Myszka, Survey of the 1999 surface plasmon resonance biosensor literature. J. Molec. Recog., 12, 390 (1999).
41. B. Xie, K. Ramanathan, and B. Danielsson, Trends Anal. Chem., 19, 340 (2000).
42. X. Su, H. T. Ng, C.-C. Dai, S. J. O'Shea, and S. F. Y. Li, Analyst, 125, 2268 (2000).
43. Z. X. Gao and F. H. Chao, Chinese J. Anal. Chem., 28, 1421 (2000).
44. S. Tombelli, M. Mascini, C. Sacco, and A. P. F. Turner, Anal. Chem. Acta, 418, 1 (2000).
45. S. Alegret, E. Fabregas, F. Cespedes, A. Merkoci, S. Sole, M. Albareda, and M. I. Pividori, Quim. Anal., 18 [Supp. 1], 23 (1999).
46. H. C. Shinde, Development of Microfabricated Micro-well Structures using SU-8 Negative Photoresist, Master of Engineering Thesis (K. M. Walsh, director), University of Louisville, May, 2001.
47. R. D. Piner, J. Zhu, F. Xu, S. H. Hong, and C. A. Mirkin, Science, 283, 661 (1999).
48. S. H. Hong and C. A. Mirkin, Science, 288, 1808 (2000).
49. Y. N. Xia, and G. M. Whitesides, Ann. Rev. Mat. Sci., 28, 153 (1998).
50. J. Lahiri, E. Ostuni, and G. M. Whitesides, Langmuir, 15, 2055 (1999).
51. R. A. McGill, R. Chung, D. B. Chrisey, P. C. Dorsey, P. Matthews, A. Piqué, T. E. Mlsna, and J. L. Stepnowski, Performance optimization of surface acoustic wave chemical sensors, IEEE Trans. Ultrasonics, Ferroelectrics and Freq. Control., 45(5), 1370 (1998).
52. A. Piqué, R. A. McGill, D. B. Chrisey, D. Leonhardt, T. E. Mlsna, B. J. Spargo, J. H. Callahan, R. W. Vachet, R. Chung, and M. A. Bucaro; Growth of organic thin films by the Matrix Assisted Pulsed Laser Evaporation (MAPLE) technique, Thin Solid Films, 354 (1999) 1–6.
53. A. Piqué, R. A. McGill, and D. B. Chrisey, A new way to deposit organic thin films, The Industrial Physicist, 6(5) (2000) 20–23.
54. B. R. Ringeisen, et al., Novel Laser-Based Deposition of Active Protein Thin Films, Langmuir, 17(11), 3472 (2001).
55. D. M. Bubb, et al., The Effect of the Matrix on Film Properties in Matrix-Assisted Pulsed Laser Evaporation of Polyethylene Glycol. Applied Physics (in press).
56. A. Piqué, et al., Appl. Phys. A-Mat. Sci. Proc., 69, S279 (1999).
57. D. B. Chrisey, et al., Appl. Surf. Sci., 154, 593 (2000).
58. B. R. Ringeisen, D. B. Chrisey, A. Piqué, and R. A. McGill, "Generation of Living Cell and Active Biomaterial Patterns by Laser Transfer", Patent Pending (2000).
59. J. Fitz-Gerald, P. D. Rack, B. R. Ringeisen, see Chapter 17 in this book.
60. D. Odde and M. Renn, Biotech. Bioeng., 67, 312 (2000).
61. M. Renn and D. Odde, TIBTech, 17, 383 (1999).
62. D. Odde and M. Renn, "Laser Guidance for Direct Write of Passive Electronics and Living Cells", in Direct-Write Technologies for Rapid Prototyping Applications, Eds. A. Piqué and D. Chrisey. Adacemic Press, Boston, 2002.
63. J. W. Grate and R. A. McGill, Dewetting effects on polymer-coated surface acoustic wave vapor sensors, Anal. Chem., 67 (1995) 4015–4019.
64. M. C. Petty, Langmuir-Blodgett Films: An Introduction, Cambridge University Press, New York, 1996.
65. D. Wallace, "Direct Writing Using Ink-Jet Techniques", in Direct-Write Technologies for Rapid Prototyping Applications, Eds. A. Piqué and D. Chrisey. Academic Press, Boston, 2002.
66. T. A. Kunt, T. J. McAvoy, R. E. Cavicchi, and S. Semancik, Optimization of temperature programmed sensing for gas identification using micro-hotplate sensors, Sens. Actuators, B, 53(1) (1998) 24–43.
67. G. B. Blanchet, C. R. Fincher, C. L. Jackson, S. I. Shah, and K. H. Gardner, Laser ablation and the production of polymer films, Science, 262 (1993) 719–721.

68. D. M. Bubb, et al., Vapor deposition of intact polyethylene glycol thin films by pulsed laser deposition, Appl. Phys. A, Materials Sciences Processing 73(1), 121 (2000).
69. J. Robertson, Adv. Phys., 35 (1986) 317–374.
70. D. M. Bubb, et al., Resonant IR-pulsed laser deposition of polymer films using a free-electron laser, J. Vac. Sci. Technol., A, in Press.
71. Iris Bloom, personal communication.
72. D. M. Bubb, et al., in Preparation.
73. J. A. Stamford and J. B. Justice Jr., Anal. Chem., 68 (1996) 359.
74. R. N. Adams, Anal. Chem., 48 (1976) 359.
75. J. L. Ponchon, R. Cespuglio, G. Gonon, M. Jouvet, and J. F. Pujol, Anal. Chem., 51 (1979) 1483.
76. J. S. Sidwell and G. A. Rechnitz, Biotech. Lett., 7 (1985) 419.
77. Wang and M. S. Lin, Anal. Chem., 60 (1988) 1545.
78. B. R. Eggins, *Biosensors: An Introduction*, John Wiley & Sons and B. B. Teubner, New York, 1996.
79. P. K. Wu, B. R. Ringeisen, J. Callahan, M. Brooks, D. M. Bubb, H. D. Wu, A. Piqué, B. Spargo, R. A. McGill, and D. B. Chrisey. The deposition, structure, pattern deposition, and activity of biomaterial thin-films by matrix-assisted pulsed-laser evaporation (MAPLE) and MAPLE direct write. Thin Solid Film, in Press.

PART II

Materials

The biggest challenge to the development of direct-write tools for the rapid prototyping of electronics, sensors, and other functioning devices is the ability to fabricate structures composed of materials that are totally dissimilar to each other, but whose intrinsic properties and functionality must be maintained in order for the deposited structure to perform the desired function. Although the material transfer approach of a particular direct-write tool is certainly important, it is the starting materials that are key to the whole process. There are many different approaches to the use of direct-write materials, but all techniques are nevertheless dependent on high-quality starting materials. Many of the starting material requirements are similar to those applied to the direct-write approach, as in resolution, flexibility, discrete multilayers, and reproducibility. Typically, the starting materials have special chemistries that permit particularly low processing temperatures for flexible polymer or other temperature-sensitive substrates. Electronic starting materials for laser transfer or for dispensing direct-write techniques are usually combinations of electronic powders (nanopow-ders, flakes, powders with modified surface coatings) and metallo-organic precursors, as well as of various binders, vehicles, solvents, dispersants, and surfactants. Ultimately, the goal is to develop new materials approaches based on the needs of the direct-write technique (transfer method) and the required electronic, sensor, power generation, or other device performance. Part II covers electronic, battery, and display materials and the specific requirements placed on them based on several different additive direct-write techniques.

CHAPTER 6

Advanced Materials Systems for Ultra-Low-Temperature, Digital, Direct-Write Technologies

KAREL VANHEUSDEN, PAOLINA ATANASSOVA, JAMES CARUSO,
HUGH DENHAM, MARK HAMPDEN-SMITH, KLAUS KUNZE, TOIVO KODAS,
ALLEN SCHULT, AND AARON STUMP

Superior MicroPowders, Albuquerque, New Mexico

1. INTRODUCTION

The motivation for the development of the materials described in this chapter, in particular, the need for materials that can be directly written onto organic substrates at low temperatures to give materials with state-of-the-art performance, is reviewed. The existing options for these materials, such as polymer thick film, are discussed and the constraints on materials imposed by emerging digital direct write deposition systems are reviewed. We then outline a family of new materials chemistry approaches that allow ultra-low-temperature deposition of high-performance, high-reliability materials onto organic substrates. This is followed by specific discussions of results for conductors, resistors, dielectrics, and phosphor materials.

1.1. DRIVERS FOR MATERIALS CHEMISTRY AND TOOL INNOVATION

The recent surge in the demand for wireless components driven by the growing demand for mobile phones and palmtop devices with Internet capability is driving a revolution in the ways electronic devices are designed and packaged. Integration of multiple passive components and active devices into highly integrated hybrid systems is required to offer both cost and size reduction. At the same time, these devices must often perform in the demanding GHz frequency range. Yet another constraint is the requirement for many devices to straddle the size range between conventional microelectronics (submicron range) and traditional surface mount components (10-mm range).

The driving force behind these requirements is integrated circuit (IC) performance which doubles and quadruples in shorter and shorter time spans, as Moore's law continues to hold. New speed and performance records are set every year; microelectronics is fast becoming nano-electronics, and entirely new types of devices are emerging. However, interconnection technology, passive components, and even active components such as displays and power sources (batteries) have evolved very slowly in comparison. The printed circuit board industry has really never caught up to the demands of electronic devices, and the mismatch in dimensions currently stands at close to a factor of 100. We are now at a stage where packaging technology is often the real bottleneck, not the integrated circuit.

Various packaging approaches have been developed to address this problem. Flip-chip, for example, is an interface technology that offers a platform for next generation high-density interconnects, as an alternative for the predominant standard of wire-bonded packages. Flip-chip technology provides the substrate direct electrical access to the die. The nearly countless combinations of flip-chip technologies can be reduced to the four principles shown in Fig. 1. Flip-chip and related technologies require fine line connections over various organic and inorganic materials and complex topographies, a need that can be met by the combination of tools and materials discussed in this chapter.

On top of the demand for higher density and higher speed at lower cost, rigid substrates like alumina and FR4 are often no longer satisfactory due to the fast growing demand for low-cost flexible high-performance substrates. Flex circuits enable applications such as "flip-out" parts in cell phones, palm

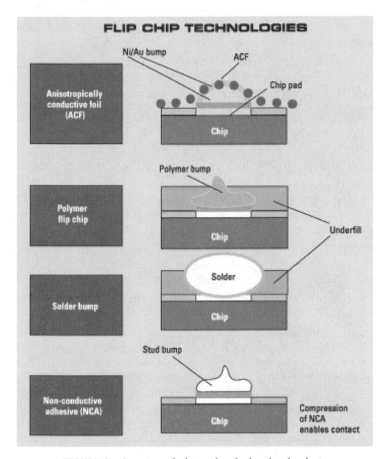

FIGURE 1   Overview of advanced wafer bond technologies.

tops, and other portable electronic devices, medical applications with miniaturized probes, flexible smart cards, and radio frequency identification (RFID)
tags. The benefits of flex circuits include lighter weight, lower profile, lower
cost, and the overall conformal nature which often eliminates the need for
bulky cables or interconnects. Flex circuits are currently being fabricated on
polyimide and polyester substrates through etching of copper or aluminum, or
by direct printing of relatively low-performance polymer-based conductive
inks. The myriad materials solutions currently available for high-density
interconnects really underlines the urgent need for an innovative approach
that provides an integral interconnect solution for a wide range of present
and future applications. Materials systems designed for high-resolution, low-
temperature, direct-deposition technologies can facilitate a direct and more

efficient interface between evermore compact integrated solid-state devices and the outside world.

1.2. DATA DRIVEN, DIRECT-WRITE, LOW-TEMPERATURE MATERIALS DEPOSITION APPROACHES

Because of the developments and needs outlined above, research and development of computer aided design/computer aided manufacturing (CAD/CAM) tools to enable the rapid prototyping, miniaturization, and three-dimensional fabrication of customized electronic components (resistors, conductors, inductors, capacitors, transformers, sensors, batteries, solar cells, and antennae) on both planar and nonplanar or conformal substrates is quickly gaining momentum (1). This increasing interest in digital direct write (data-driven materials deposition, DDMD) (2) technologies is propelling the development of myriad maskless computer-driven deposition systems. These include inkjet printing and other droplet delivery approaches (3), thermal spray deposition (4), microsyringe or pen dispensing, various laser-based direct write technologies including laser chemical vapor deposition (LCVD), matrix assisted pulsed laser evaporation (MAPLE) (5), and others. These technologies offer the potential of fast prototyping as well as low-cost small batch, and in some cases large batch, manufacturing and unparalleled levels of integration and density.

The military has also recognized the importance of the approaches just described, as reflected in the DARPA Mesoscopic Integrated Conformal Electronics (MICE) program. The requirements of this program closely reflect the commercial requirements. For example, direct deposition and low-temperature processing of conductors for interconnects is targeted to result in a performance of no less than half the conductivity of the equivalent bulk conductor (silver, copper, gold, platinum). For dielectrics, k values in the 1,000 range are targeted while maintaining reasonably-low high-frequency loss and adequate temperature coefficients of capacitance (TCCs), up into the GHz range. Material deposition and processing are focused on high-performance substrates that cannot sustain processing temperatures above 250 °C (polymers, plastics, etc.).

1.3. APPLICATIONS

Direct write technologies have tremendous potential in the booming markets of mass consumer electronics where they can enable the fabrication of highly compact, flexible, conformal, ultra-lightweight assemblies and, especially, allow for rapid product introductions and enhancements, which are becoming increasingly critical to stay competitive in the global marketplace. The additive nature of these new approaches makes them desirable over the traditional subtractive approaches (etching to define a pattern) from both a cost and an environmental perspective. It has even been suggested that direct write technologies will ultimately allow manufacturers to eliminate dedicated CVD and sputtering chambers and many photolithographic techniques and materials (1). This would enable electronic device and component fabrication in a clean room environment under ambient conditions, thereby making present high-cost, high-maintenance, volume-throughput-limiting, vacuum environments obsolete.

A typical example of what these novel technologies will enable is System on Card (SoC), also known as Smart Card products. These are essentially a small version of a personal computer in the configuration of a credit card. Figure 2 shows a prototype sample of an SoD with LCD, biometric sensor, and ISO module from Infineon Technologies (Munich, Germany). These features would offer numerous advantages and convenience for the customer, so the market volume could rise enormously.

Although significant progress has been made in terms of improved component tolerances and unique patterning capabilities for many of the emerging direct write technologies, further progress is required to address reproducibility, quality, accuracy, yield, robustness, and product reliability in rugged

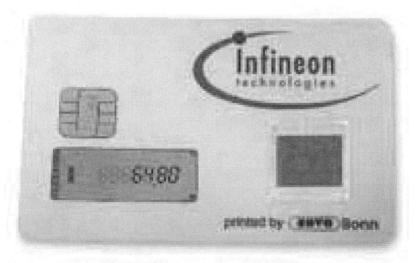

FIGURE 2   System on Card providing limited computing along with communications in a card.

environments necessary for mass production of complex products containing hundreds of passive and active components and very high densities of interconnects.

1.4. EXISTING OPTIONS FOR MATERIALS SYSTEMS FOR DIRECT-WRITE TECHNOLOGIES

Much of the success of any low-temperature, direct-write technology hinges on the ultimate material performance that can be achieved after deposition and postdeposition processing are completed. The established and commercially available thin-film and thick-film source materials discussed next and their recommended postdeposition processing such as sintering are often incompatible with the most promising and innovative direct write technologies and the associated materials processing conditions. Such incompatibility requires the development of novel and tool-specific materials that can be deposited and fully processed under very specific thermal processing conditions. Furthermore, these materials must permit direct multilayered deposition of components with feature sizes down to 10 µm utilizing the aforementioned CAD/CAM tools, while maintaining stringent materials standards in terms of performance, adhesion, compositional and mechanical robustness, and stability in extreme environments. In order to meet these challenging demands on the precursor materials without compromising the ultimate material performance, materials systems are being engineered towards specific deposition tools and applications. This requires innovative approaches, combining experience from chemistry and materials science, guided by a fundamental understanding of how the microscopic features impact the ultimate material and device performance. These issues are discussed next.

The existing options for deposition of materials onto surfaces include thin film, high-temperature inorganic thick film, and polymer thick film, and are summarized in Fig. 3. Several aspects of these technologies are important to this discussion. From the perspective of deposition technologies, polymer thick film and inorganic thick-film technologies have been developed for screen printing while the thin-film approaches utilize subtractive approaches. For processing conditions, these technologies are inadequate. Conventional inorganic thick-film processing is based on sintering of metal, metal oxide, glass, and other powders, which requires temperatures greater than roughly 600 °C, making them unusable for organic substrates. In addition, the feature sizes obtainable by this approach are greater than about 100 microns as defined by screen printing (except when photodefinable approaches are used). Polymer thick film has similar limitations on feature size, but allows processing at low temperatures compatible with organic substrates. The drawback here, however,

Processing
Temperature

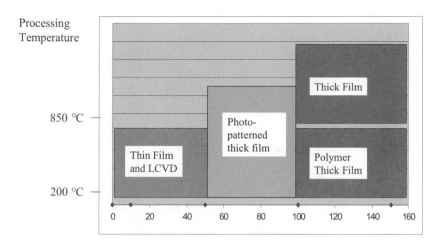

Feature Size (μm)

FIGURE 3   Overview of today's mainstream thin-film and thick-film technologies. Courtesy of Superior MicroPowders.

is the relatively poor performance and reliability afforded by loaded polymers. Thin-film approaches provide fine feature size and low processing temperature, but at the expense of high cost and complexity.

The issue of cost is more clearly seen in Fig. 4, which shows how the costs, performance/reliability, and processing temperatures of PTF, inorganic thick-film, thin-film, and super-low-fire (SLF) materials compare.

Polymer thick film has low cost and allows processing on organic substrates, but provides relatively poor performance and reliability due to the nature of these loaded polymer compositions. Cermet has good performance and reliability, but has higher cost than PTF and requires processing on ceramic substrates that can withstand the high processing temperatures. Thin film has low temperature processing and excellent reliability and performance, but high process costs. The SLF materials discussed here have the potential to provide the combination of relatively low cost, high performance and reliability, and ability to deposit materials onto organic substrates.

These approaches will now be discussed in more detail to provide adequate perspective for the discussion of SLF materials.

1.4.1. Thin Film

There is a strong trend toward thin-film interconnections in high-density packaging. Thin film is often the preferred technology for high-density interconnect and RF/microwave applications, because it allows the production

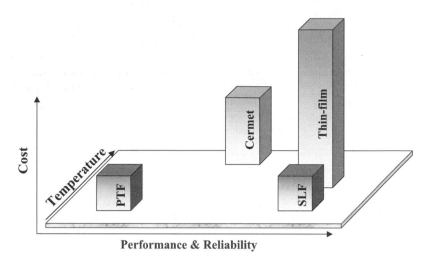

FIGURE 4 Costs, performance/reliability and processing temperatures chart comparing various hybrid micro-electronics technologies.

of interconnect feature sizes less than 100 microns and densities that approach those typical in integrated circuits. The high purity and uniformity of thin-film layers ensures that the performance of thin-film materials approaches that of the bulk source conductor material, resulting in significant reductions in interchip signal delays, reproducibility, and performance reliability. Very small feature sizes are achieved through photolithographic procedures requiring separate masking steps and etching techniques similar to those used in IC manufacturing.

This performance comes at a significant cost as thin-film technology uses a deposit-and-etch approach to produce the desired patterns. This subtractive approach is lengthy, requiring many separate processing steps, and is performed on discrete batches of substrates. This lithographic approach generally prevents the introduction of cost effective (nonvacuum) continuous-flow processing, similar to the techniques utilized in thick-film processing. Another obvious drawback is the inefficient use of source material, which is inherent to the subtractive nature of thin-film technology. Finally the production of large streams of heavy metal waste and other undesirable compounds is a growing concern from an environmental standpoint.

1.4.2. Inorganic Thick Film

Thick-film processes have a long history and are known for their simplicity and low cost compared to their thin-film counterparts. Technologies such as low-temperature co-fire ceramic and screen printing allow high-volume through-puts and minimize material costs due to their additive nature. These technologies further allow the production of multilayer hybrid circuits, which can incorporate functional ceramics such as capacitors, inductors, varistors, thermistors, and sensors. Thick-film technology relies heavily on paste formulations, which consist of a functional phase, a permanent binder phase (typically a glass), and a vehicle, which acts as a carrier agent and provides the appropriate rheology. Over the past several decades significant improvements have been accomplished in these materials in terms of the electrical performance and physicochemical compatibility with other components (6).

The major limitation of traditional thick-film technology is the requirement to perform a high-temperature sintering treatment to obtain the desired mechanical strength and electronic performance. This involves firing the materials at high temperatures, typically above 800 °C, to obtain the desired performance. Sintering is essentially a microwelding process, bonding individual particles together into a coherent, predominantly solid structure. This is accomplished through necking, induced via mass transport events that occur predominantly on the atomic level. Figure 5 shows the necking process for Ag particles as a function of sintering time. The need for sintering poses a major limitation, as it requires the use of expensive high-temperature ceramic substrates. Another serious limitation is the rather large minimum feature size of about 100 μm that can be obtained with traditional additive printing approaches.

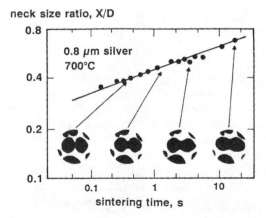

FIGURE 5 Necking vs. sintering time for 0.8-μm Ag particles at 700 °C. The shape of the particles is shown at 4 points during the sintering cycle. From *Sintering Theory and Practice*, Randall M. German, Copyright © 1996. Reprinted by permission of John Wiley & Sons, Inc.

1.4.3. Polymer Thick Films

Polymer thick film (PTF) technology has been traditionally used in low-performance, low-cost hybrid circuits on organic substrates. This technology is based on manufacturing insulators, conductive tracks, and resistors by screen-printing inks/pastes consisting of thermoplastic polymers with conductor or insulator materials and solvents on the substrate (7). Polymer thick films offer great versatility in that simply changing their composition makes it possible to obtain materials with very different electrical characteristics and excellent adhesion to organic substrates. As a result they can be tailored to many applications, such as low-temperature processing and even flexible substrates. Other advantages of this technology are obvious. As an additive technology, hybrid circuits, incorporating embedded resistors, capacitors, and inductors, can be readily produced, saving materials and board space, thus saving cost. The ability to screen print and process subsequent films allows the implementation of multilayer designs using single-layer boards.

By using an organic resin with silver, it is currently possible to obtain conductive inks, which show resistivities down to 7–10 times bulk Ag performance (7). For this reason the use of polymer thick films is precluded in power and high-frequency applications when low resistivity is required. Other limitations of this technology are the inability to solder directly to most polymer conductor lines, the overall lower reliability compared to pure functional materials, and poor performance at high frequencies.

Typical dielectric constants for polymers range from 2.5 to 6, and have high loss compared to inorganic materials. Polyimide, for example has a dielectric constant of 3.5 and a loss tangent of 0.01 (8). Higher-k polymer thick films can be obtained by adding a high-k ceramic filler to the polymer. However, mixing rules for multicomponent dielectric layers, discussed in detail later in this chapter, dictate the intrinsic limitations of this approach. Hence there is very limited use of PTF capacitors in obtaining stable, high-performing components.

1.4.4. Photopatternable Thick Film

As with all thick-film technologies, the resolution one can achieve through screen printing is an obvious limitation. The need for increased density continues to drive innovations in this market segment, hence the success of photopatternable materials which are a hybrid between thick film and classic printed wiring board photoresist technology. This material mixes functional materials such as metal powders or glass powders with organic components used in photoresist films to provide enhanced resolution down to 40 μm or less. The processing flow is shown in Fig. 6. The films are processed as thick films. Subsequently, photolithographic techniques are applied, including UV light exposure and development in aqueous solutions. Firing is finally applied to remove the organic components used in the imaging process. Large areas are

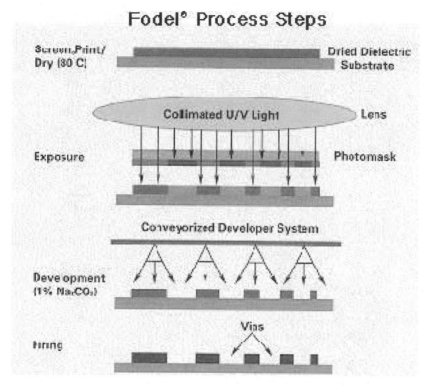

FIGURE 6   Overview of FODEL processing steps.

usually deposited using screen printing and then "photo trimmed." This ensures high resolution and thickness uniformity which conventional screen printing of different feature sizes cannot. One obvious drawback of this technology is that firing is required to get rid of the organic components. This puts serious limitations on the material performance that can be achieved at low processing temperatures, and thus on substrate versatility. In addition, it is a subtractive technology, which unavoidably comes with a significant increase in materials costs.

2. DEPOSITION METHODS AND ASSOCIATED MATERIALS REQUIREMENTS

The characteristics of the various available and emerging printing approaches impose additional requirements on top of the materials-processing-driven needs discussed. The requirements for screen-printable materials have been well established and are discussed elsewhere (6). What follows is a discussion of the materials requirements for emerging direct write technologies: ink jetting, MAPLE, aerosol jetting, and others.

2.1. INK JETTING

The term inkjet printing covers a wide range of nonimpact printing techniques, all of which project drops or columns of ink onto a printing surface. Application of inkjet technology in thick-film hybrids is a fairly new area of research and development, although achievements in this area have been reported as early as 1987 by Teng and Vest (9). The advantages of drop-on-demand printing are obvious. The discrete aspect allows for a high level of automation and makes it inherently compatible with digital technology (pixel maps). Another huge advantage of drop-on-demand printing is its speed and scalability, allowing the use of high-frequency multiple nozzles grouped into staggered array inkjet systems. Different types of functional inks can be printed in one printing process. This concept is similar to the well-known color inkjet printer.

Different drop-on-demand technologies have been developed in the past for traditional inkjet printing. Hewlett-Packard developed the thermal inkjet system where the ink in the orifice is locally heated, causing the volatile liquid to evaporate locally, creating a small gas bubble. The sudden bubble expansion will force a droplet to be ejected, hence the term bubble jet (10). Thermal jet technology may be incompatible with the printing of some functional materials because the high temperatures involved in the bubble formation may trigger decomposition reactions of the functional precursors present in many formulations. This will eventually result in nozzle clogging.

Energizing piezoelectric materials electrically appears to be the most attractive technique for jet printing of hybrid circuits. This creates an acoustic pressure wave in the capillary tube, ejecting a single droplet past the nozzle plate. The success of direct jet printing depends crucially on the preparation of suitable inks, which essentially are well-dispersed dilute suspensions of solid particles in a liquid. Increasing the volume fraction of the solid phase is often desirable from a materials standpoint. However, for proper operation the volume fraction of the solid phase should be kept to a minimum because the inks must be able to flow at high speed through a nozzle that can be as small as 30 μm in order to form fine droplets (11). The ink viscosity requirements for most commercially available inkjet print heads range between $\eta = 10$ and 20 cP. The need for such low-viscosity formulations makes it highly desirable to use a functional liquid matrix—a metallo-organic molecular solution—that converts into a functional phase upon heating (9). For applica-

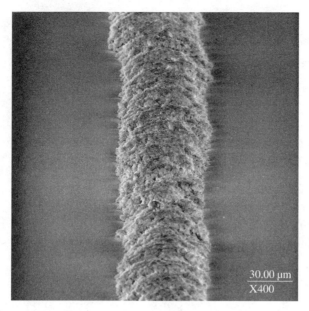

FIGURE 7   SEM image of an inkjet printed phosphor material. Courtesy Superior MicroPowders.

tions such as ceramic jet printing where a solid phase is a necessary additive to the ink, nonagglomerated spherical particles with a narrow particle size distribution in the submicron region are desirable. The liquid matrix should be volatile enough to ensure fast drying. In addition, the ink should be of low-viscosity, high-functional phase content, and have appropriate surface tension to avoid spreading after printing. The surface properties of the substrates onto which the ink is jetted are also critical parameters that will dictate the rheology of the ink to be used. Finally, thermal stability and shelf life of the inks have to be considered at the same time as lowering the chemical conversion tempera-ture, as some substrates currently in use require processing temperatures not to exceed temperatures as low as 150 °C. Figure 7 shows an example of an inkjet-printed functional metal oxide material.

2.2. THERMAL/PLASMA SPRAY DEPOSITION

Thermal spray deposition is an established method for depositing dense thin layers (e.g., > 10 µm) of metals, metal alloys, ceramics, dielectrics, polymers, and composites of these materials (12). Common applications of thermal spray include surface wear and corrosion protection, where the deposition equipment can often operate under atmospheric conditions and be portable for use in the field. There is a wide range of other applications such as gas-turbine components, automotive components, pulp and paper processing, printing, the steel industry, and biomedical parts. Recently there has been growing interest in the use of thermal spray technology for new applications such as fabrication of meso-electronic multilayers, sensor systems, and embedded sensors (4).

Thermal spray is, in one manifestation, a continuous melt-spray procedure where the materials are fed into the plasma or flame as fine powders, sintered rods, or wires. Most materials can be deposited by using a feedstock of fine particles (1–50 μm diameter) that is partially or fully melted through the use of a combustion flame or a thermal-plasma arc. The droplets are accelerated to high velocities and intercepted by a substrate where they rapidly solidify after spreading out in a thin layer. The particle size for the powder feed ranges from 1–20 μm with a requirement for substantially nonagglomerated particles, preferably spherical with good control over particle size, particle-size distribution, and microstructure. The chemistry of the deposited films is essentially determined by the composition of powder feedstock, requiring excellent control over the composition to obtain good deposits. Electronic-grade feedstock materials have to be developed to enable thermal spray deposition of high-performance electronic components. More than one powder material can be fed into the process, allowing for deposition of composite materials. The physical characteristics and the properties of the coatings depend primarily on the particle impact conditions such as velocity and particle temperature, but substrate temperature and reactivity have also been shown to be key variables in controlling the microstructure and film properties. The particle velocity in the flame can be as high as several hundred meters per second and the flame temperature can range from about 3,000 °C for a combustion spray process to as high as 25,000 °C for plasma spray.

Electro-ceramic thin films deposited using thermal spray processes often exhibit defects such as pores and microcracks. Nonetheless, preliminary data indicate dielectric constants above 100 (4). Silver features have been deposited successfully with resistivity values as low as four times the bulk resistivity of Ag (4).

2.3. Microsyringe Dispensing

The syringe dispensing tool is perhaps the most straightforward approach to direct deposition of functional materials. A continuous flow of ink or paste is forced through a small tube or syringe and directly deposited onto the substrate while the pen moves relative to the substrate forming the desired written pattern. The deposited material is subsequently heat treated to evaporate volatiles and transform into a functional phase. The ability to produce designs with this approach is virtually unlimited. Given the limitations of manufacturing syringes with ultrafine inner and outer diameters, the ultimate limits of micropen dispensing will be those imposed by the source materials. In the case of thick-film paste materials, 150-μm-wide lines and 25-μm spaces are currently state-of-the-art for commercial tools. There is a difference between achieving these tolerances over a short line segment versus a long continuous line or space due to start-stop issues. The width limits are those imposed by the size of the particles in the pastes and the flow and wetting characteristics. Smaller feature sizes may be achieved when using inks that contain no particles so only the flow and wetting characteristics are controlling the minimum feature sizes in this case.

The use of advanced inks and pastes that incorporate state-of-the-art chemistry and particle technology will enable the next generation of syringe-based direct write tools. When incorporating a solid phase into a paste for syringe dispensing, spherical particles are highly desirable because of their ball-bearing-like behavior, allowing much higher solid loading while maintaining unmatched control over viscosity. Figure 8 shows an SEM image of a microsyringe-dispensed Ag pattern using a paste based on spherical Ag

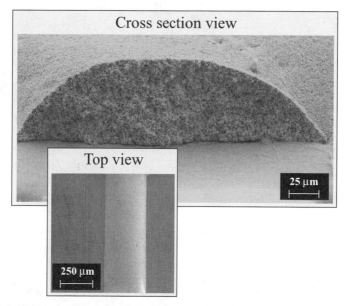

FIGURE 8   SEM image of a syringe dispensed Ag-conductor line on polyimide. Courtesy Superior MicroPowders and CMS Technetronics.

particles in conjunction with a low-temperature Ag precursor. The paste was deposited on a polyimide substrate and heated to 250 °C for 10 minutes (13). When using ultrafine tips, effects such as filter pressing need to be considered. Filter pressing occurs when the liquid phase in the paste flows faster than the solid phase (particles), ultimately resulting in buildup of solid and clogging of the tip. To prevent filter pressing, additives are used that allow the material to flow better and prevent the particles from collecting in closely packed groups. The additives include a surfactant that allows the particles to be better wetted and assist in the dispersion of the particles in the paste. Thickeners are added to pastes to prevent settling of particles during flow and also help the paste to maintain its shape after dispensing. The emerging ability to produce nonagglomerated highly spherical particles with control over size (0.3–1.5 μm), purity, and crystal structure, will allow lowering of the line resolution down to 50 μm. New ink formulations incorporating low-temperature chemistry and engineered rheology are also being developed. It is envisioned that these systems will allow direct writing of minimum feature sizes down to the 10-μm range (13).

2.4. MATRIX ASSISTED PULSED LASER EVAPORATION-DIRECT WRITE (MAPLE-DW)

The direct write technique called matrix assisted pulsed laser evaporation (MAPLE) uses a focused ultraviolet laser pulse to transfer material from a coating on a transparent carrier onto a substrate. The laser impacts the material to be transferred from the back at the carrier/material interface through the transparent carrier. The material is designed to absorb the laser energy causing local evaporation at the interface and propulsion of a discrete packet of material toward the substrate. By using a sequence of laser pulses while moving one or both of the carriers (often called the ribbon) and the substrate, a desired pattern can be directly written. Feature sizes of 20 μm have been demonstrated (5). When laser trimming is utilized in conjunction with laser deposition, feature sizes below 10 μm can be obtained. Minimum feature size is ultimately limited by the diffraction limit of the laser wavelength employed, the quality of the optical components, and, last but not least, *the detailed physical and compositional properties of the material being transferred.*

The material performance and resolution obtained after transfer both depend critically on the rheology of the ribbon coating material at the moment of the transfer. The ribbon coating material is typically a paste or ink. These formulations face similar challenges as those for other direct write approaches such as ink jetting or paste dispensing. The ability to produce

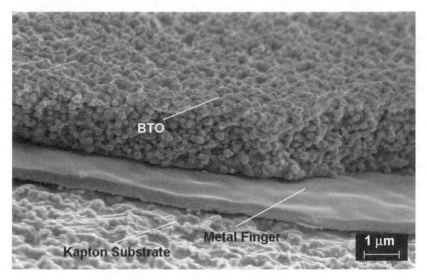

FIGURE 9   SEM micrograph of a laser transferred dielectric layer. Courtesy of Superior Micro-Powders, Naval Research Laboratory, and Potomac Photonics, Inc.

nonagglomerated highly spherical particles with control over size and size distribution (0.3–1.5 μm), purity, and crystal structure, will enable the transfer of paste resulting in better performance and smaller minimum feature sizes. The addition of low-temperature chemistry to the liquid phase lowers the postdeposition processing temperature. In addition, the paste rheology and paste formulation must meet a number of requirements that are specific to the laser transfer process. The coating has to be optimized to avoid the spreading of material packets during transfer resulting in loss of cohesiveness. This will not only reduce resolution, but will also have a negative impact on the materials performance that can be achieved by postdeposition processing. An additional complication, which is inherent to the ribbon approach, is the need for mechanical strength and the requirement for extended shelf life of the ribbon. This is due to the large surface area of the coating material and the possible associated evaporation of volatile components in the liquid matrix material. Figure 9 shows a dielectric layer that was deposited using laser transfer.

2.5. LASER GUIDANCE AND FLOW GUIDANCE

In another approach laser guidance and flow guidance methods are used to deliver materials onto a substrate (3). The precursor materials are generated in an aerosol generator. The aerosol droplets are either laser guided through a thin optical fiber or flow guided through an orifice and finally delivered to the substrate. In contrast to other delivery systems discussed in this chapter, the requirements for precursor formulations are somewhat different. Aerosol generation of droplets (14) containing the desired materials for delivery starts with aqueous or organic solutions of molecular precursors or suspensions of powders that are comprised of high-quality particles in the micron-to-submicron range. Other droplet generation methods are also possible (14). For processing simplicity, aqueous solutions or suspensions are preferred. With regard to precursor formulations, the most important requirements are high solubility and low conversion temperature. The addition of other additives such as binders and vehicles, as required in most cases for inks and pastes, is normally not required for the aerosol delivery system. Furthermore, the process uniquely allows for precursor modification during delivery to the substrate by either changing time and temperature history or by laser treatment. Hence, precursor material can be delivered either in wet or dry form, unconverted or converted, or any state within. This makes tailoring of the materials that are ultimately deposited on the substrate feasible and gives the process a great versatility.

2.6. LASER CHEMICAL VAPOR DEPOSITION (LCVD)

Laser direct write techniques have been developed and applied to the deposition of μm-sized conductive lines, mainly for microelectronic applications such as mask repair, circuit customization, or localized doping (15). Laser chemical vapor deposition (LCVD) is a thermal technique for film growth that uses the interaction of a CW laser with an absorbing substrate in the presence of reactive molecules. The primary role of the laser is to heat the substrate and induce pyrolysis of the gaseous metal-containing precursor and this can be carried out at any laser wavelength absorbed by both the substrate and the growing film. Laser chemical processing can avoid high temperatures by using a short pulse or by rapidly scanning the laser beam to heat only the small spatial region of the substrate that is being processed. The exact dimensions of the heated zone are a function of the substrate thermal diffusivity, the optical absorption length in the substrate, and the laser pulse duration.

The substrate is placed at the focal point of the laser beam where the laser beam is perpendicular to the substrate surface. The other placement functions can be either optically solved by acousto-optic deflectors or mechanically, either by translation stages or by galvanometric mirrors. Prior to film deposi-

tion, a substrate is placed in a vacuum chamber and after evacuation down to
< 1 m Torr the chamber is filled with a static charge of reactant gas with
pressures ranging from 0.1 to 100 Torr. Using the translation stages, the
substrate is moved in the focal plane of the lens at spot-scanning speed of up to
1 mm/s. After processing steps that consist of depositing lines onto the
substrate, the chamber is evacuated and the sample is removed via a quick
access door or a load-lock.

LCVD of metal films from the gas phase requires meeting certain prerequi-
sites for vapor pressure, stability, and activation energy for decomposition of
the gaseous precursors (16). The use of bidentate β-diketonate compounds
that form volatile compounds and can undergo both thermal and photochemical reduction to metals has been extensively used in CVD for thin-film
growth. Examples of metals that have been deposited from β-diketonate
compounds with LCVD include Au, Pt, Ag, W, and Cu. As an example, gold
precursors with the general formula Me_2Au(β-diketonate) have vapor pressures from 0.0085 Torr (at 35 °C) for Me_2Au(acac) to 0.7 Torr (at 25 °C) for
Me_2Au(hfac) and have been shown to deposit very pure metal films with low
impurity levels at temperatures as low as 200–300 °C (16).

3. SUPER-LOW-FIRE INKS AND PASTES

3.1. SLF MATERIALS REQUIREMENTS

To fill the gaps just outlined, while at the same time providing materials that
are compatible with the deposition techniques discussed, super-low processing
temperature pastes and inks are being developed for conventional as well as
direct write deposition of functional materials such as conductors, resistors,
dielectrics, battery ingredients, and other materials. The main requirements for
conductor materials are the ability to provide printing (in some cases through
direct deposition) of feature sizes down to as low as 10 μm, a maximum
processing temperature of 200–300 °C and often even lower, conductivity
values close to those of bulk material (usually less than 3 times bulk
resistivity), good adhesion to the organic substrate, and high mechanical
strength. Compatibility with a wide variety of substrate materials is also
desired, including rigid and flexible polymers, solid state components, metal
contact landing pads, and solder bumps. For resistor and dielectric materials
systems, good control over layer thickness is critical to achieve narrow
tolerances. As with conductors, compatibility with low-T substrates
(200–300 °C) is required. In addition, changes in performance parameters
with temperature, expressed in terms of parameters such as temperature

142

coefficient of resistance (TCR) and capacitance (TCC), should be below the 100 ppm/°C range.

Finally, it must be emphasized that each deposition technique requires materials with its own set of characteristics including viscosity (when the material is a liquid), surface tension, vapor pressures of constituent materials, etc.

3.2. GENERAL APPROACH

The deficiencies of PTF, inorganic thick-film, and thin-film approaches suggest an alternate family of approaches to low-temperature deposition of materials. It is clear, as can be seen in Fig. 5, that sintering of metals and ceramics occurs at temperatures that are far too high to allow processing on organic substrates. The approaches that have been developed to overcome this problem involve combining materials chemistry and laser processing. The general approach illustrating the interplay between the different phenomena is outlined in Fig. 10. A dispense approach is used to deposit the material onto a surface. After drying, low-temperature materials chemistry and/or laser processing is used to react and densify the material. The materials issues are shown in more detail in Fig. 11.

A source material system based on mixtures of microengineered solid particles, solvents, thickeners, dispersants, surfactants, adhesion promoters, and chemical precursors to a solid phase is printed/dispensed. After deposition the transferred material is typically heated below 200 °C to induce evaporation of volatile components resulting in a dry deposit. At this stage the deposit

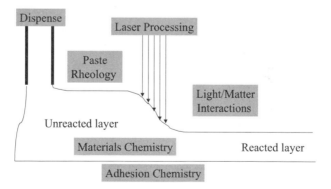

FIGURE 10 Graphical representation of the general approach illustrating the interplay between the different phenomena that occur during the deposition process and during material processing. Courtesy of Superior MicroPowders.

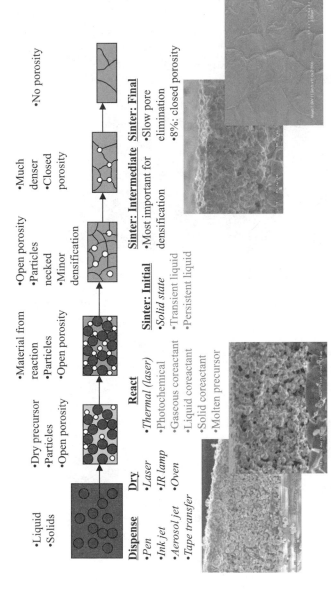

FIGURE 11 Materials issues that play a role in ULF processing. Courtesy of Superior MicroPowders.

FIGURE 12   SEM of thermally processed Ag paste, showing the densification and "chemical welding" of individual particles due to chemical conversion of the precursor phase. Courtesy of Superior MicroPowders.

contains primarily particles and/or molecular precursors. In the case of molecular precursors, further heating, often at higher temperatures, results in conversion of the molecular precursors to products (metals, ceramics, etc.). The reacted precursor forms a solid phase resulting in particle necking, akin to the effect of sintering. This effect is shown in Fig. 12 where a silver precursor was added to a thick film of silver particles and subsequently processed at 200 °C for 5 minutes. In principle, the processing temperature for such systems can be reduced to any given temperature as long as chemical conversion of the precursor to the desired solid phase is accomplished.

At this stage the deposit is likely porous to some extent and may have to be sintered to provide optimum performance. This can be carried out using laser sintering that effectively localizes heating to the materials to be sintered while not degrading polymeric (organic) substrates (3,5,13).

3.3. INKS AND PASTES

The starting point for most depositions is an ink or paste formulation that contains both a solid and a liquid phase. In general we will define ink formulations to be lower viscosity formulations that can be applied by a method similar to inkjet printing. Paste formulations are thicker and are usually applied by techniques such as tape casting, screen printing, or pen dispensing. For deposition of these materials, there are issues associated with the substrate materials that center around wetting and adhesion, as well as issues associated with dispensing the fluid, primarily rheology. The issues of material compatibility, wetting, and adhesion are usually considered before inks or pastes are formulated, and are mentioned here for completeness, not for detailed discussion. The viscosity of inks and pastes determines not just aspects such as spreading, or slumping of deposited features, but also packing density of the particles. These parameters in turn determine deposited characteristics such as line cross-section, line width, but also the final composition and performance of the converted material. For this reason, a brief synopsis of rheology follows.

Rheology (17) is a description of the way a fluid behaves, not just its viscosity but the way that it responds to changes in stress. Essentially there are three categories of rheological behavior. If the shear rate is linear with applied stress, the fluid is said to be Newtonian and the viscosity is independent of the shear stress. This is most common in low viscosity fluids with few particles that could interact to cause changes in the behavior. If there is a change in the shear rate with shear stress, the fluid is non-Newtonian, either plastic or pseudoplastic. In these cases, there are particle interactions that either increase or prohibit flow as the stress is increased. If the viscosity increases with stress the fluid is called dilatent, while if the viscosity drops the fluid is shear thinning. The latter is the most preferred for pastes as it is desired to have them flow readily when forced through a screen or pen tip, but then have them thicken and hold their shape when the stress is removed. The third area of rheology deals with time-dependent behavior and is called thixotropy. In this regime, fluids take time to recover their unstressed state.

Ink formulations face the challenge of maintaining a fairly low solids loading, to keep the viscosity low, while still being able to produce reasonable properties in the final part. Because the properties of many formulations are dominated by the powder portion, this is not a trivial issue. Standard inkjet heads require ink viscosities between $\eta = 10$ and $20\,cP$ (1 and $2 \times 10^{-2}\,N\,s\,m^{-2}$) and a surface tension of about $3 \times 10^{8}\,N\,m^{-1}$ at jetting temperature (10).

The ideal behavior for paste formulations would be shear thinning with a yield point. Viscosities are typically many orders of magnitude higher than for inkjet inks. This pseudoplastic behavior allows for the paste to hold its shape under the influence of gravity and at the same time readily flows through the dispensing system when sheared. The surface tension of a material plays a role in the wetting characteristics. The surface tension of the paste is related directly to the chemical composition of the paste and can be controlled, to a certain extent, via chemistry. The ability to wet a surface will also have an impact on the final shape of the finished product. Flow after deposition, flow during the heat treament cycle, and bleeding-out of precursor/liquid vehicle, all relate to the chemistry and the ability to wet a surface.

During the flow of a paste through a narrow channel such as a mesh or syringe, physical separation of vehicle and solid can occur. When the vehicle flows faster than the solid particle this phenomenon is referred to as filter pressing. The accumulative effect of filter pressing is that the viscosity of the paste in the narrow channel will increase to a point where clogging of the channel will occur. To reduce or avoid filter pressing, the viscosity of the vehicle should be matched to the paste viscosity, or the particle loading. Thickening agents can be added to the vehicle to accomplish this. Furthermore, the rheological behavior of suspensions is strongly influenced by interparticle interactions, which are either electrostatic or steric in nature. The effect of dispersants is a dramatic improvement of the dispersion of the solid particles in the liquid vehicle. This will prevent clogging, improve rheology, and increase green density and homogeneity after drying.

3.4. DRYING

Once a paste or ink has been deposited, the vehicle must be removed before any high-temperature processing can be performed. Traditional thick-film paste processing calls for a drying step prior to the firing treatment. Drying is required to evaporate the liquid vehicle phase from the deposited paste. Most commonly used solvents have boiling points in the range from 180 to 250 °C. It is important to avoid rapid heating at high temperatures, as splattering or even burning of the vehicle may occur. Airflow drying can be used to speed up the drying process. In the case of the materials systems used in conformal writing, the firing step is omitted altogether and the drying process often coincides with the chemical conversion of the precursor phase in the vehicle. A detailed understanding of the complex chemical interactions that occur between the solvents and the precursors can greatly aid in improving formulations in terms of solid yield, film uniformity, and conversion temperature. Even though chemical conversion occurs at a certain temperature, annealing at a higher temperature may still be required to induce a transformation into the proper crystalline phase. This is certainly true for more complex ceramic compositions. When low-temperature substrates are used, rapid thermal annealing utilizing infrared lamps or laser processing can be very useful to accomplish proper conversion.

3.5. Precursor Chemistry

The application of passive electronic components on flexible low-temperature substrates such as polyimide requires new approaches and concepts for the development of suitable precursor chemistries and formulations. The concept for formulating compositions for direct deposition of conductor, resistor, and dielectric circuit components with direct write technology builds on the availability of suitable molecular precursors that can be converted to functional components in a paste or ink. For pastes, a typical mixture would consist of particles, a molecular precursor to the functional phase, vehicle, binder, and additives. The requirements in terms of precursor chemistry can be quite different depending on whether a low-viscosity ink or a thick-film paste is needed, and depending on the specific writing tool used. Both mixtures of molecular precursors with and without particles can be formulated for inks. Other writing methods, such as aerosol-assisted deposition (14), rely on suitable formulations of molecular precursors that can be either aqueous or organic solvent based. In any case, the particular precursor chemistries used can be quite different for these formulations with regard to compatibility with other constituents and processing conditions. Processing procedures, especially when precisely controlled by computer, will need to be associated with specially formulated compositions of the materials to be processed. However, the flexible substrate materials used require very low precursor conversion temperatures. In order to use the molecular design of precursor materials to its full advantage for direct write applications, a good understanding of decomposition mechanisms during thermal and photochemical treatment and for surface reactions occurring on substrate and particle surfaces is needed.

In the past, significant R&D progress has been made in the development of metal organic precursors for printing conductors, dielectrics, and resistors (9,18,19). In more recent work, silver was formed through converting a silver precursor solution in polyamic acid at low temperatures (18). When designing molecular precursors for direct write applications on low-temperature substrates such as polyimide, polyester, polyethylene, polypropylene, paper, and the like, the following aspects have to be addressed. The chemical precursor to the functional phase should convert to the final functional material at low temperature. The formulations should be easy to synthesize, environmentally benign, provide clean elimination of inorganic or organic ligands, and be compatible with other paste/ink/solution constituents. Other factors are solubility in various solvents, stability during the delivery process, homogeneous film formation, good adhesion to the substrate, high functional phase content, and shelf life. Depending on the type of laser used for precursor conversion, the precursor material needs to be highly absorptive at the laser wavelength being used to promote efficient laser energy coupling allowing for decomposition at low laser power. This will prevent substrate damage during the laser writing process.

The metal-ligand bond is a key factor in designing the metal organic precursors. For conductors and conductive phases in low-Ohm resistors, this bond should be reactive enough to permit complete elimination of the ligand during formation of metallic features for conductors like silver, gold, nickel, copper, palladium, platinum, or alloys of these elements. Typical precursor families include metal carboxylates, alkoxides, and β-diketonates comprising at least one metal oxygen bond. Other precursors like thiolates and amines can be specifically tailored to the required characteristics depending on the metal.

Direct deposition of electro-ceramic materials for dielectric, ferrite, and resistor applications requires precursors that are able to undergo clean and low-temperature transformation to single oxides or mixed oxides. This will be required to mimic the compositions currently being used in the electronic industry. The general chemistries needed for these reactions can be borrowed from the large amount of ongoing research that has focused, particularly in the last 15 years, on the development of dielectric and resistor and inductor materials using routes other than solid state synthesis. Typical chemistries involved for these metal-oxide-based formulations are condensation, polymerization, or elimination reactions of alkoxides typically used in sol gel processes. Other typical routes involve ether, carboxylic anhydride, or ester elimination (19).

The crucial challenge in all these approaches is to find the specific combination of precursors, additives, and solvents for the successful conversion to the final material at low temperatures. Although this has been achieved with some success for metallic functional materials, a lot of development work remains for oxide-based components. Even if a conversion at low temperatures with complete elimination of byproducts can be achieved, the metal oxide materials may still need some higher temperature treatment for proper crystallization and consolidation. In contrast, important metals like silver, gold, palladium, and copper have been deposited using carefully designed metal precursors at temperatures well below 200 °C, in some cases even below 150 °C with good adhesion to polyimide substrates (20). The lower deposition temperatures required for complex mixed metal oxides would result in structures with materials that have controlled stoichiometries and in some cases would afford kinetic routes to new meta-stable crystal structures.

3.6. ROLE OF PARTICULATES

From a particulate materials development perspective, a key to enabling a significant reduction in minimum feature sizes is improved control over ink and paste performance. This is, in turn, at least partially dictated by particle size, particle shape, and the shape and spread of the particle size distribution, along with the extent of particle agglomeration, solid dispersion, and loading.

Another consideration when using ultra-low-fire formulations containing solid particles is that the particles possess the final desired physical properties. Optimization of the intrinsic properties of the particles is crucial because recrystallization and annealing of crystal defects during thermal processing is no longer an option at the typical processing temperatures ($< 300\,°C$) that these materials are designed for. This is illustrated in Fig. 13, which shows measured dielectric constants for a variety of dielectric materials. All of the materials in Fig. 13 are high-k dielectric powders, but the dielectric constant varies dramatically depending on composition and prior powder processing conditions. The losses (not shown) also vary dramatically for the different starting materials and must be optimized in the particles.

These material considerations have pushed the demand for advanced particle production technologies that can provide high-quality powder. An example of one such approach is spray pyrolysis. This emerging technology produces powders with superior properties that conventional methods, such as solid state or liquid precipitation, cannot duplicate. A schematic process flow is shown in Fig. 14. The use of spray pyrolysis allows for the formation of nonagglomerated spherical particles, as well as extreme control over particle size and particle size distribution (14). The spherical particle size can range

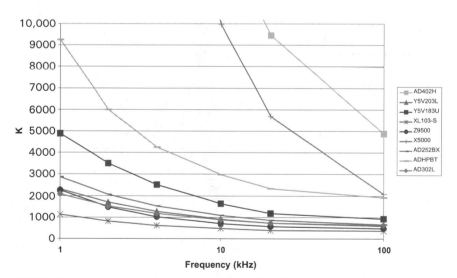

FIGURE 13 Screening of commercially available dielectric powders performed by Superior MicroPowders. Courtesy of Superior MicroPowders.

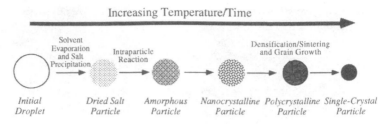

FIGURE 14 Schematic of the spray pyrolysis process. From *Aerosol Processing of Materials*, T. T. Kodas and M. Hampden-Smith. Copyright © 1999. Reprinted by permission of Wiley-VCH.

from 0.3 to 20 μm, with very little batch-to-batch variation. This method also provides better control over particle microstructure, including crystallinity, impurity levels, density, and surface properties, all of which are crucial in performance-driven applications.

This approach can provide enhanced packing of particles by providing tailored particle-size distributions as shown in Fig. 15 and Fig. 16. By providing a bimodal particle-size distribution, the green density of the structure before firing is increased, leading to a higher probability of obtaining a dense final structure.

Spray pyrolysis has also been shown to have the ability to produce powders unattainable by more traditional methods that include complex multiphase composites. Figure 17 shows an SEM micrograph of a three-dimensional

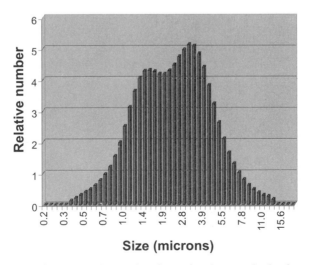

FIGURE 15 Particle size distribution data for a phosphor powder batch optimized for best packing characteristics. $d_{10} = 0.94$ μm, $d_{50} = 2.23$ μm, $d_{90} = 4.88$ μm. Courtesy of Superior MicroPowders.

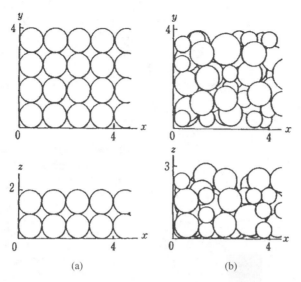

FIGURE 16 A comparison between the packing of mono-sized (a) and a trimodal size (b) distribution particle batch at different layer thicknesses. From K. Urabe, *Jpn. J. App. Phys.* **19**, 885 (1980).

FIGURE 17 SEM micrograph of a three-dimensional interconnected precious-metal ceramic composite produced by spray pyrolysis. Courtesy of Superior MicroPowders.

interconnected precious-metal ceramic composite produced by spray pyrolysis. This type of particle is an excellent example of the unique particle micro-structures that can be achieved through the use of spray pyrolysis. Using more traditional powder technologies would leave virtually no opportunity to develop such unique microstructures. As an example of the flexibility of the spray pyrolysis process, complex multicomponent dielectric compositions are being produced in a single step and each particle is compositionally identical. These particles are spherical, dense, and have a controlled particle size and size distribution. The ultimate control over phase and composition allows the incorporation of dopants down to the ppm level, resulting in a high level of control over material performance on the particle level. Spray pyrolysis can generate particles that are ready for low-temperature deposition and processing into a functional component.

3.7. LOW-MELTING GLASS ADDITIVES

Historically, glasses have been widely used in the inorganic thick-film industry to lower ceramic densification temperatures (6). Mixing glasses and ceramics has many technological advantages. In low-k dielectric applications ($k \sim 10$), typically used as insulator dielectrics such as ceramic filled glass, a significant amount of glass is mixed with one or more refractory oxides. During firing of these materials the glass melts, resulting in a hermetic structure. The ceramic oxide components provide dielectric performance and structural strength. The melting temperature of these glass additives is typically greater than 600 °C. In some cases low-temperature sealing glasses or overglaze glasses are used with melting temperatures in the range of 300–400 °C. When such low-melting-point glasses are used, entrapment of organic vehicles inside the thick film becomes a real concern. In resistor formulations such as cermet resistors, similar glasses are used as the nonconductive matrix structure, surrounding the conductive phase. Finally, glass additives are also used in small amounts in high-k capacitor dielectric formulations. During firing the glass component melts and the liquid glass phase provides capillary pressure and a diffusion path for the ceramic phase. The amount of glass that is used in these for-mulations is a delicate balance between maximizing sintering aid and mini-mizing low-k glass contamination in the processed film to achieve a high dielectric constant.

The chemistry of glasses used in thick-film dielectrics covers a broad range of formulations, depending on the specific application. Relevant parameters include softening point, viscosity, melting point, crystallization temperature, coefficient of thermal expansion, the solubility of various components in the liquid phase, electrical conductivity, and dielectric performance. Lead is very commonly used in these formulations to lower the melting temperature. The ceramic components and dopant additives in thick-film formulations usually partially dissolve into the glass during firing, changing glass properties such as melting point and viscosity.

Low-melting glasses continue to prove their usefulness and versatility in the low-fire compositions for the super-low-fire formulations discussed in this chapter (13). Glasses with softening points as low as 280 °C are currently available and have been added to functional dielectric and resistor formulations, resulting in lower processing temperatures while maintaining performance that can be competitive with existing formulations that are fired at much higher temperatures (13,20). The use of a process similar to selective laser sintering can be exploited to selectively melt or soften the glass from the top, without heating the substrate to the melting point of the glass. In an ideal scenario, these glasses will soften after the organic constituents have been volatilized and the organic precursor chemistry is complete. This will allow the glass to infiltrate the pores that remain in the thick film, replacing air with glass. The true challenge lies in fine-tuning the glass composition to have properties that are compatible with the desired properties of the material system that they are added to. For example, a high-k glass is preferred for high-k dielectric formulations because it will replace air with glass during melting. This will boost the dielectric performance of the final product, as will be discussed in more detail. Microwave dielectric formulations on the other hand call for a low-loss glass.

3.8. LASER SINTERING

Lasers have many inherent properties that make them attractive for use in direct write applications. The spatial coherence of laser light is compatible with ultrasmall feature sizes. Phase and time coherence offer high temporal resolution and make it possible to overcome competing heat-dissipating mechanisms. The monochromaticity makes it possible to control the depth resolution during heat treatments, to perform composition selective heating, and enables selective nonthermal excitations that can trigger specific chemical reactions. The big thrust for the use of laser processing in direct write applications is the promise of local processing (21). This has the potential of inducing chemical reactions or heating a deposited material well above the temperature that would be tolerated with conventional heating. Conventional heating is limited by the heat tolerance of the substrate, which can be as low as 150 °C for many novel substrate materials in use today. However, lasers do not induce miraculous materials transformations. Laser processing can be roughly classified into nonreactive and chemical processing.

Under nonreactive processing one can classify traditional processes such as trimming and cutting which require high power lasers such as CO_2 and Nd:YAG lasers. Innovative processes in this class are laser induced structural transformations including surface hardening, amorphization, defect annealing, recrystallization, local melting, and alloying. These processes require high control over laser intensity and dwell time. A typical illustration of the commercial success of this type of controlled laser transformation is the post-treatment of ion-implanted silicon, where rapid thermal cycling with a laser allows recrystallization of lattice defects without dopant redistribution (22). Longer beam dwell times are required if a certain amount of diffusion induced mass transport is required to obtain the proper transformation. Also, the delivered energy declines with an inverse-fifth-power dependence on the powder depth, meaning that it is only a near surface effect (21). Consequently, the surface heating is high but slow heat transport through the powder bed limits the depth of sintering. Adding to the challenge of laser sintering on low-T substrates is the fact that beam dwell times need to be limited if one wants to contain heat transport and avoid excessive heating of the underlying substrate. Such challenges limit the transformations that laser processing can realistically achieve. Laser processing of materials deposited on low-temperature substrates can induce defect annealing and recrystallization, but it is a major challenge to reach the levels of mass transport typically achieved with sintering-induced densification of particle beds on ceramic substrates.

The object of laser-induced chemical processing in direct write technologies is the modification of deposited materials at temperatures above those the substrate could tolerate. The laser-induced activation or enhancement of the reaction can be pyrolytic (photothermal) or photolytic (photochemical) in nature (21). In the past, laser chemistry has been successfully applied to surfaces and thin-film materials. Laser-induced chemical reactions in thick-film materials pose additional challenges to control reaction rates, diffusion, and homogeneity over the entire cross-section of the deposited layer, hence the importance of controlling frequency, power, and dwell time. To facilitate this approach, paste formulations can be tailored and additives can be used that absorb specific photon energies that do not interact with the substrate or other constituents of the ink or paste. Another way to improve reaction control and avoid overheating the substrate is the use of photolytical chemistry, which can be triggered by a specific laser frequency.

Laser sintering has been examined before for increasing the density of deposits (23). A large body of literature provides insight into the issues associated with sintering (24). The basic concept of laser sintering is to induce densification by shining laser light onto the material. If the light is at least partially absorbed, it will be converted to localized heat, which will then induce local densification in the material. The primary issues are associated with the intensity/energy density of the laser beam, the time scale of heating, and varying reflectivities and thermal diffusivities of materials.

Both pulsed and continuous wave lasers can accomplish laser heating. Common continuous wave lasers are argon ion and CO_2, while common pulsed lasers are YAG and excimers lasers. The time over which the material is heated roughly corresponds to the dwell time of the CW laser beam (spot size, scan speed) or the pulse length of the pulsed laser. Simple calculations indicate that pulse lengths of milliseconds to nano-seconds or less are required to establish a significant duration and depth of heating.

At the most common laser wavelengths corresponding to a YAG laser (~ 1.0, 0.5, and 0.3 microns), the reflectivity varies considerably not just with the identity of the metal, but also with the details of the surface of the specific metal. Figure 18 shows the reflectivities of several metals as a function of wavelength. This leads to significant issues associated with controlling deposit temperature because temperature is directly related to reflectivity.

Another consideration is that once laser energy is absorbed at the surface of a material, the penetration depth of the heat varies dramatically from material to material because of their differing thermal transport properties. Figure 19

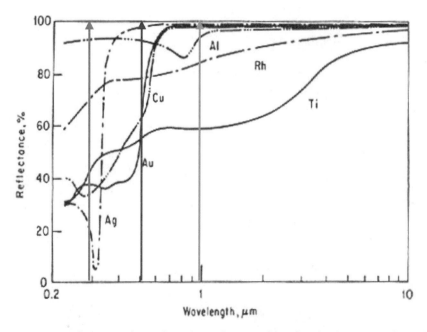

FIGURE 18 Reflectivities of several metals as a function of wavelength. The arrows indicate the various laser frequencies that are commonly used for laser postprocessing.

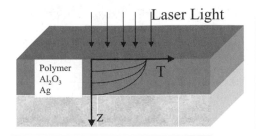

Material	δ normalized
Ag	43 μm
Au	35 μm
Al_2O_3	8 μm
Glasses	2 μm
SiO_2	2 μm
Polymers	1 μm

Analytical solution for short thermal penetration depths relative to material

$$\delta = 4\ (\alpha t)^{1/2}$$
$$= 4\ (k/\rho C_p t)^{1/2}$$

• Varying thermal penetration depths for constant laser pulse length!
• Top surface heated longer than bottom!

FIGURE 19   Illustration of how penetration depth of the heat can vary dramatically from material to material. Courtesy of Superior MicroPowders.

shows the normalized thermal penetration depth for various materials. Heat in general penetrates deeper in materials with higher thermal diffusivities, which causes problems when a laser strikes surfaces with differing thermal properties.

Selective laser sintering was developed as a method for solid free-form fabrication of three-dimensional parts. One process involves spreading a layer of powder evenly over an area. A laser is then used to selectively melt the powder in a pattern that is representative of one layer of the desired part. The melted region becomes a solid layer while the untreated powder provides support for subsequent layers. A second layer of powder is then spread over the entire area and the laser is used to melt the second layer. The process continues, building the part layer by layer until the final shape is complete. While the process really involves selective laser melting, it has been dubbed selective laser sintering because ceramic parts can be built by this method.

The selective laser sintering process is traditionally used with only one material, but the combination of a ceramic powder and a low melting glass allows for new applications for laser melting. Once a direct write tool has deposited a mixture of ceramic oxide powder and glass, a laser may be employed to densify the structure by melting the glass phase. The proper balance of oxide powder and glass must be achieved as should the proper size distribution of both particulate phases. For high-*k* dielectric applications the glass would ideally be minimized so that the high-*k* performance of the dielectric powder is maximized. For high-ohm resistors the glass phase may be the majority of the composition so that the conduction between the conductive oxide particles is limited by the insulating glass phase. As the glass phase is melted it needs to wet the oxide powder and assist in densification.

4. CONDUCTORS

Conductor materials are indispensable in hybrid microelectronics—they serve as the "wiring" that connects all the components to form a functional circuit. Conductors further serve various other functions such as electrodes, bond attachment pads, and die bond attachment. That increased packing densities and better compatibility with the micron scale of imbedded chip components demand smaller feature sizes and improved line resolutions has already been discussed. The pervasive nature of conductor materials in the electronic industry both in terms of sales and as a critical component means that it is the enabling material in any of the novel direct write technologies that are discussed in this work.

Conductor materials systems for direct write applications on low-T substrates have compositions that have certain similarities to formulations used today in traditional thick-film hybrid technology. They range from thick pastes with high solid loading to low-viscosity inks with little or no solid particles. Parameters that play a role in determining the best-suited material for a certain application include material cost, attainable minimum feature size, conductivity of the fully processed material, solder leach resistance, resistance to electromigration, long-term stability in hostile environments, and, last but not least, the physicochemical compatibility with the substrate and other passive and active components. Needless to say, all these parameters are correlated and tradeoffs will have to be made.

Polymer thick film provides a potential solution for certain applications but its performance in terms of conductivity (7–10 times bulk values) and reliability are often insufficient. The better solution is an inorganic, ultra-low-temperature (200 °C) fireable paste. The basic constituents of such a conductor material system that is currently being developed for direct write on low-temperature conformal substrates are a metallic solid particle phase and a liquid vehicle with the addition of a metallo-organic precursor. The precursor will decompose during the heating process to form a metallic phase and "weld" the metallic particles together. An example of this approach is Parmod™, developed by Parelec, Inc., USA. Recent improvements in solid yield of the precursors and further lowering of the processing temperature well below 200 °C have made this a particularly attractive solution for conformal write applications and a wide variety of low-temperature substrates such as smart cards. The SEM image in Fig. 12 illustrates the effect of chemical welding in a converted Ag paste. This particular paste was prepared using Ag particles and a metallo-organic precursor to Ag, processed at 250 °C on a polyimide substrate.

The rheology of the conductor paste or ink is also a crucial parameter. The viscosity will determine which method can be used to dispense the material and what feature sizes can be obtained. As a general rule, smaller solid particles allow for better resolution control. Depending on the application, sheer thinning, bleeding of the liquid vehicle after deposition, and other postdeposition deformation and compositional changes need to be considered. Thickeners and surfactants can be added but can have a negative impact on the conductivity of the finished product when low-temperature processing is required. As has been discussed, spherical particles are highly desirable for controlled rheology and optimum solid loading. To obtain good conductivity, the particles need to be phase pure. The best conductors are the group IB metals (Cu, Ag, Au). Pd and Pt are also good conductors and alloys are often formed to enhance overall performance or improve the compositional stability.

In addition, particle-size-distribution and particle shape will have a significant impact on the conductivity of a fully processed film. For example, the packing density is improved by using a bimodal distribution of particles instead of a monomodal distribution. The use of flaked particles gives better conductivity than the use of perfectly spherical particles due to the higher surface/contact area between the particles, but the flowability is decreased relative to the spherical particle. These considerations are especially imperative when processing at low temperatures is required and sintering-induced densification is not an option. As discussed, the conductivity of low temperature processed films can only be improved by the conversion of the metallic precursor. The higher the solid yield of the precursor and the connectivity between the individual particles, the better the final conductivity will be.

5. RESISTORS

Traditional hybrid technologies have serious limitations for today's demanding resistor applications. Screen-printed thick-film resistors, for example, have wide-ranging values, but their short current paths and the inherent limitations of the screen-printing resolution severely compromise their performance characteristics. Laser trimming is thus often required to fine-tune the values within the desired range. Thin-film resistors, while capable of high precision, are expensive to design and manufacture and have their own limitations in terms of obtainable resistance values. The high-precision, high-resolution direct write technologies that are currently being developed are proving to be an enabling tool to produce very well-controlled resistance values, spanning the entire spectrum from low to high ohm.

Most resistors for integrated electronic applications are required to be ohmic, to display small deviations from their predetermined value (tolerance), and to have small temperature coefficients of resistance (TCR). TCR is an expression of change in resistance due to change in temperature and it is expressed in parts per million per degree Celsius (ppm/$^\circ$C). The TCR of given conductive and semiconductive materials can be either positive (increasing resistance with increase in temperature) or negative (decreasing resistance with increasing temperature). The major demand for resistors in electronic applications lies in the resistance range from $10^3 - 10^8$ Ω. This is a serious challenge, because pure materials with suitable and reliable electrical behavior have typical resistivities below 10^{-6} Ωm. Unfortunately there are no pure single phase materials that provide optimum properties for ohmic resistors (6). The key to achieving a resistor with a specific resistivity and low TCR lies in tailoring composition and microstructure of the final product. Two general strategies have proven their success to tailor the resistivity of materials. First, the conductivity can be lowered by diluting a conductive material with an insulative phase. Alternatively, very thin and/or elongated conductive paths can be formed and packaged for stability and reproducibility. Thin-film resistor technology is often based on the latter approach of producing resistors, while thick-film materials are often mixtures of a conductive and an insulative phase (6).

Materials systems have been developed to date for thin- and thick-film technologies. Thick-film technology has seen the development of both polymeric type resistors as well as cermets, while thin films rely mainly on vapor deposition and lithography of metal alloys. Polymeric resistors generally consist of a polymer matrix with a conductive particulate phase of carbon or silver alloy. Cermet-type resistors have gained popularity as they offer better performance and tolerance than the polymeric types while offering a wide range of resistivities. A typical cermet resistor paste or ink consists of a conductive phase, insulative matrix phase, organic carrier, binder, wetting agents, and dispersants. The conductive phase may be a metal or semiconducting oxide. RuO_2 and other ruthenates are very common conductive phases for high-ohm resistors while noble metals or alloys thereof are popular choices for lower-ohm resistors. The insulating phase is usually some type of lead borosilicate glass frit, which consolidates the structure after thermal processing and dilutes the material to obtain a certain resistivity range. The organic carrier is crucial in creating a paste with correct rheology for printing applications such as silk screening. Dispersants and binders also help to create a stable suspension, which will form a solid green compact upon drying. Other "dopants" are commonly added to commercial pastes to impart better electrical properties upon firing. Cermet inks are typically processed at 850 °C and above (6).

To develop resistor formulations for direct write applications it will be useful to tap into the technological advancements made in both fields of thick- and thin-film resistor technology. However, the selection of materials cannot merely be driven by the reproduction of material systems that work well in conventional processing. In conventional processing of thick films, for example, the 850 °C firing process is the crucial step, as it fuses the various compounds together and produces the desired microstructure and composition. This limits the choice of conductive phase for air-fireable resistors systems to metallic oxides based on elements such as Ru and Ir. Both are thermodynamically stable in air and in the various insulating matrices at these high temperatures. Formulations for direct write onto conformal substrates, on the other hand, have to be designed for processing at temperatures below 400 °C. At such low temperatures, compositional stability of the constituents is much less of an issue. The real challenge is the formation of the proper conductive, insulative, or semiconductive phases without the ability to rely on high-temperature diffusion and sintering. The use of low-temperature precursor chemistry combined with laser processing is crucial here to obtain the proper microstructural compositions. This difference in processing temperature and processing conditions is prompting materials developers to use a fresh approach and calls for experiment with new types of resistor compositions.

6. DIELECTRICS AND FERRITES

6.1. DIELECTRIC MATERIALS

Dielectric materials have a wide variety of applications in electronic circuits. They are used to provide electrical insulation as well as to facilitate the temporary storage of electrical charge. The dielectric constant, dielectric loss factor, and dielectric strength determine the suitability of the dielectric material for a specific application. Variations in dielectric properties with frequency, temperature, and a range of environmental conditions such as humidity also play a big role in determining the usefulness of any particular material composition. The novel conformal direct write technologies that are the topic of this book will enable the direct integration of charge storage devices where current chip capacitors solutions are not adequate. One example of this is in cases when the proximity of the device is crucial, such as a decoupling capacitor directly written onto an IC. Insulating inorganic ceramic thick film layers deposited directly onto low-temperature substrates is another class of applications that conformal direct writing will enable.

Dielectric properties result from the short-range motion of charges under applied electric fields, hence the importance of local order parameters such as composition, crystal structure, grain size, particle size, and porosity of the dielectric film. In traditional state-of-the-art ceramic processing a thick-film paste is typically deposited by tape casting or screen printing and subsequently fired. Annealing above 850 °C for extended times up to several hours is needed to induce crystallization and densification of the thick film. During this sintering process, ionic diffusion occurs between the individual particles, resulting in necking and eventually in fully dense polycrystalline film. The organic vehicle is removed during the early stages of the heating cycle while the sintering aid, usually a glass additive, will melt at the peak firing temperature and facilitate the interparticle diffusion.

The ideal dielectric material is a defect-free single crystal of high purity. Real world dielectrics such as those based on pastes and inks are polycrystalline and often even polyphase materials. When mixing several phases, the dielectric behavior of the film becomes a function of the constituent phases, and the detailed morphology of the mixture. K. Lichtenacker proposed a simple and useful rule known as the logarithmic mixing rule (25):

$$\text{Log } \kappa = \Sigma_i v_i \log \kappa_i$$

where κ_i and v_i are the dielectric constant and volume fraction of phase i, respectively. Maxwell has derived a more elaborate relationship for the dielectric constant in the case of a homogeneous dispersion of spherical particles of one dielectric constant in a matrix of another dielectric constant. This equation is particularly useful for dielectric formulations which contain spherical dielectric particles with a high-k value relative to the matrix in which they are dispersed (air). Figure 20 shows that in such cases the Lichtenacker rule is a good approximation. The shape of the curves illustrate the large impact porosity has on the overall dielectric constant of the dielectric film. It can be concluded from the curves shown in Fig. 20 that particle necking is highly desirable. If a high degree of necking can be achieved, the mixture undergoes a transformation from a dispersion of high-k spheres in a low-k matrix to a dispersion of low-k spheres (pores) in a high-k matrix. The solutions for the Maxwell equation in Fig. 20 illustrate that this transformation improves the dielectric constant dramatically.

In addition to the dielectric constant, dielectric loss (tan δ) is of great concern, especially for high-frequency applications. Here lies one of the main advantages of using ceramics as their loss is smaller compared to that of other materials such as polymers. Other than intrinsic loss, which can only be

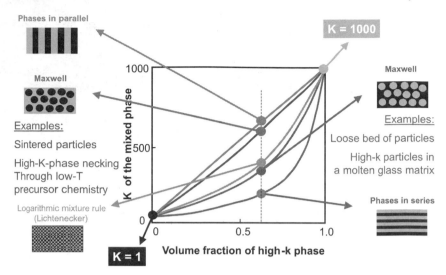

FIGURE 20 Mixture relations for the dielectric constant of two-component dielectric layers. Courtesy of Superior MicroPowders.

controlled by modifying the ceramic phase, loss from defects and impurities needs to be controlled. A significant source of loss in polycrystalline and polyphase aggregates is that associated with surface states and interfacial space charge. This loss occurs at interfaces where two materials with different electrical and chemical characteristics join, resulting in charge and impurity accumulation. Hydroxyl termination at porous ceramic interfaces is a typical example of this type of loss source. Figure 21 illustrates the detrimental effects that hydroxyl groups have on performance. A surface treatment is performed to reduce the density of hydroxyl groups in a porous dielectric layer. The negative impact of porosity on both dielectric constant and loss justifies the big emphasis on sintering in traditional ceramic processes to obtain dense crystalline films and reliable electrical performance.

Typical dielectric compositions such as $BaTiO_3$, $PbZr_{0.5}Ti_{0.5}O_3$ and $PbMg_{1/3}Nb_{2/3}O_3$ all crystallize in the perovskite phase. These dielectrics all have high k values. On the higher-k end of the spectrum, materials such as the relaxor type PMN composition are useful for multilayer chip capacitor components, while other materials such as BST are tailored for DRAM memory applications. Higher k values usually come at the expense of a lower quality factor Q ($= 1/\tan \delta$) values and a higher temperature coefficient of capacitance (TCC), which makes these ferroelectrics undesirable for microwave applications. In addition, many high-k compositions contain Pb, which may pose environmental problems.

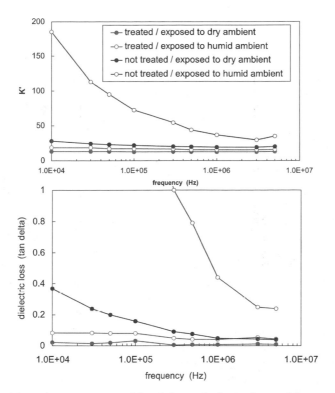

FIGURE 21 Dielectric constant and loss before and after surface modification treatment. The treatment reduces the density of hydroxyl groups inside the porous dielectric layer. Courtesy of Superior MicroPowders.

The use of dielectrics in microwave applications poses a critical need for low dielectric loss or high Q in combination with certain requirements for high k and TCC. Specific compositions have been developed to meet the demanding specifications of microwave applications. Materials such as $Zr_{0.7}Sn_{0.3}TiO_4$, $CaZr_{0.98}Ti_{0.02}O_3$, $SrZr_{0.94}Ti_{0.06}O_3$, and $BaNd_2Ti_5O_{14}$, although excellent in terms of microwave performance, pose quite a challenge when incorporated into a low-temperature direct write process. These compositions have rather complex crystal structures, which only fully form at high temperatures (typically $>1,000\ °C$) and can become unstable when annealed at lower temperatures (26). Depending on the thermal cycle the particles experience during deposition, phase separation may also occur. Alternatively, pyrochlore materials such as $Sr_2Nb_2O_7$ and $Pb_2Ta_2O_7$ are attractive microwave materials, due to their less complex crystal structure, higher stability, and their lower processing temperatures.

6.2. FERRITE MATERIALS

Many of the magnetic characteristics of ceramic materials such as polarization are analogous to their dielectric characteristics. Commercial ferrite applications usually require a high permeability. Short magnetic switching times are also highly desirable. The use of ceramic magnetic materials with ultrafast switching times in memory applications is a crucial component in today's data storage technology. Ceramic magnetic materials are highly desirable in the fast-growing area of high-frequency solid-state devices. The higher resistivity of these ferromagnetic oxides gives them a decisive advantage over magnetic metals. As with dielectric materials, lowering the high-frequency loss is a big challenge and many of the properties are sensitive to the effects of heat treatment and composition. For instance, a surplus or deficiency of Fe ions of a few percent can change the resistivity of a magnetic ceramic by several orders of magnitude. Eddy-current losses can be controlled by improving the resistivity of the ferrite. In a more general sense, phase purity, proper oxidation state, large grain size, and low porosity all contribute strongly to lowering the loss in ferrites.

As with dielectrics, ferrite compositions should be tailored to their specific application. A typical material for high-permeability low-frequency applications is manganese–zinc ferrite. Losses for this composition are relatively high and useful frequencies are limited to about 0.1 MHz. Low-loss high-frequency ferrites are often compositions containing nickel and zinc. Frequently used microwave materials are nickel–zinc ferrites and yttrium iron garnet (YIG)-based formulations.

6.3. DIELECTRIC AND FERRITE FORMULATIONS FOR CONFORMAL DIRECT-WRITE APPLICATIONS

The major challenge in developing ceramic materials systems for low-temperature applications consists of lowering the postdeposition processing temperature without compromising electrical performance. At temperatures below 800 °C one can no longer rely on crystal growth and sintering-induced diffusion to densify a bed of ceramic particles. Alternative strategies need to be developed and refined to modify internal surfaces, reduce porosity, and achieve densification. The appropriate crystal structure of the final film can be obtained by selecting high-quality crystalline particles. Dense films can be achieved through routes such as low-temperature melting additives. Another approach is the development of low-temperature liquid or sol gel precursor chemistry. A high ceramic yield is desirable in this approach to avoid excessive shrinkage and subsequent porosity and cracking during the thermal processing. Another approach involves optimizing solid loading by improving the green density through engineering particle shape and particle-size distribution. Laser postdeposition processing is under active investigation for its ability to apply very localized heating pulses and induce crystallization and densification. Typically, a well-balanced combination of all these approaches can result in the development of a material system that will enable deposition of dielectric layers on substrates with a maximum temperature tolerance as low as 250 °C while maintaining the required electrical performance.

In the development of a formulation for direct write low-temperature conversion to a high-frequency dielectric, a number of factors must be taken into account to achieve excellent performance characteristics. The formation of carbon during the conversion of a metal-organic precursor will lead to a high degree of dielectric loss. Many high-k dielectric compositions contain barium. When processed in air, barium precursors are susceptible to formation of barium carbonate, which, once formed, cannot be converted to an oxide below 1,000 °C. Therefore barium carbonate formation should be avoided. As shown, hydroxyl groups are known to be an important source of loss in dielectric metal oxides. The condensation reactions to convert metal hydroxides to metal oxides are not complete until 800 °C (for isolated surface hydroxyl groups) and so the best strategies would involve precursors that avoid hydrolytic-based chemistry such as sol-gel-based hydrolysis and condensation routes.

As discussed, the incorporation or porosity is seriously detrimental to the performance of a dielectric layer as a result of the contribution of the dielectric properties of the material trapped inside the pore (especially if this is air). Therefore porosity must be avoided. The metal oxide phases that lead to the desired dielectric properties require that the material be highly crystalline. These metal oxides crystallize at high temperatures and so a strategy that relies on an ink or paste that only contains a precursor to the final phase will have both a ceramic yield and a poor crystallinity problem. Conversely, a strategy relying on only particulate material will likely provide high porosity if processed below 300 °C. Recent development work shows a promising dielectric formulation strategy that addresses all these issues and is processable by direct write deposition techniques at low temperatures to form high-performance dielectric features (20). The formulation is designed to contain a large volume and mass fraction of highly crystalline high-performance dielectric powder such as $BaTiO_3$ or $BaNd_2Ti_5O_{14}$ that has the desired k-value, has a low temperature coefficient, and has a low loss. The formulation also contains a smaller mass/volume ratio of precursor to another material for which precursors are available. These precursors should have the following characteristics:

- Avoid the intermediate formation of hydroxyl groups.
- Have ligands that are designed to react preferentially in a way to give a single-phase complex stoichiometry product rather than a mixture of a number of different crystalline phases.

ZST or TiO$_2$

BaTiO$_3$

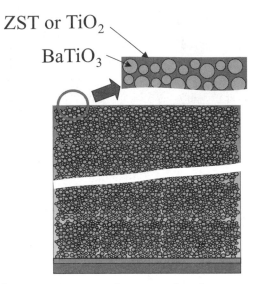

FIGURE 22 Schematic representation of a strategy for a low-temperature high performance dielectric formulation. Courtesy of Superior MicroPowders.

- Can be processed to form a crystalline phase at low temperatures.
- Have high ceramic yield.
- Result in a dielectric with a reasonable k-value of capacitance, low loss, and small temperature coefficient contribution.

An example of such a target phase is TiO$_2$ or Zr$_{0.40}$Sn$_{0.66}$Ti$_{0.94}$O$_2$. This strategy is illustrated in Fig. 22. Effectively the materials that require high-temperature processing have been processed prior to deposition to achieve high crystallinity and good dielectric properties and because they are present in a high mass fraction, they dominate the performance of the layer.

7. PHOSPHOR MATERIALS FOR INFORMATION DISPLAY TECHNOLOGIES

One of the largest barriers to the introduction of new display technologies and market/performance expansion of existing display technologies is associated with the manufacturing cost of the displays. There are major financial issues in the number of steps and their efficiency, the yield of the displays, and the efficiency in the use of materials. As an example, the high manufacturing cost and the low yield of plasma display panel (PDP) production has played a significant role in the delay of the introduction of this product to the consumer market. A lot of these problems are related to the current inability of printing the conductor, dielectric, and luminescent layers to the required fine feature sizes. Revolutionary cost savings can be achieved by improving direct write deposition of high-performance materials. This is particularly true in the case of the color pixels, where smaller and more efficient phosphor powders and their incorporation into liquid vehicles will enable direct write deposition.

Phosphor particles produced by a spray-based route have a number of unique characteristics that make them highly preferable over phosphor particles produced by conventional powder manufacturing methods. These methods are either solid state- or liquid solution-based. In these conventional approaches the final particle size is controlled by a milling step, which is detrimental because it introduces surface defects and impurities that decrease the luminous efficiency of the phosphor. Spray pyrolysis avoids all of these problems simply because the particles are immediately solidified into their final high-purity and spherical morphology. The unique spherical morphology further provides the ability to produce formulations such as stable inks for inkjet printing that cannot be achieved with other powders. In addition, the high control over particle shape, particle size, and their statistical distributions allows the formation of thinner and yet more densely packed beds of particles. As phosphor material development is making great progress, manufacturers will ultimately be limited by the intrinsic performance as dictated by quantum efficiency of each material. Herein lies the great advantage of the ability to microengineer particles. It will be the enabling step in further improving the overall performance of the next generation of low-voltage phosphor screens.

In contrast to the liquid crystal display, the PDP market is an emerging Flat Panel Display (FPD) market where there is a premium on the cost and efficiency of use of materials. Of particular importance are the results of testing of phosphor powders produced by spray pyrolysis in plasma TVs, which show that there is a strong value proposition in the more efficient use of spherical phosphor powders produced by spray pyrolysis compared to the use of phosphor powders made by conventional methods. Figure 23 shows a plot of luminous efficiency as a function of phosphor-powder-layer thickness measured in a plasma TV. The dark-blue data is based on light output from the front of the display using blue phosphor powders with the light-blue data representing an accelerated life test. The data represented by the dark- and light-gray striped columns is obtained from the best available blue phosphor powder manufactured by conventional methods, which was used as reference. The reference powder was printed at 10 microns, its minimum layer thickness. The minimum layer thickness is limited by the particle size and size distribu-

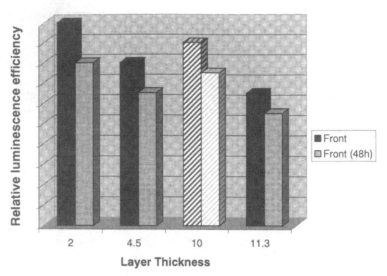

FIGURE 23 Plot of luminous efficiency versus layer thickness for a reference (striped gray) phosphor powder and for a blue plasma TV phosphor powder in a plasma TV produced at Superior MicroPowders (dark). Courtesy of Superior MicroPowders.

tion of this powder. Thinner layers are incomplete and therefore cannot be used. In contrast, due to the smaller particle size and tailored particle-size distribution of the phosphor powder produced by spray pyrolysis, the phosphor powder layer can be printed as thin as 2 microns (20). Figure 23 compares the panel luminous efficiency of phosphor powder layers with thicknesses of 11.3, 4.5, and 2.0 microns. It is clear that as the layer gets thinner the brightness of the display increases to a value above that of the reference material. This clearly indicates that a better performance is achieved with approximately 5 times less phosphor powder.

Phosphor powders have also been printed by a number of different printing methods based on the inks and pastes as described in the formulations section here (20). In particular, recent focus has been on printing methods that enable deposition of features with sizes below that which can be achieved using more conventional approaches such as screen printing. Therefore features with sizes below 100 microns have been the focus of these feasibility demonstrations using direct write or digital methods such as inkjet printing, syringe dispensing, and electrostatic printing. A number of different features comprising phosphor powders that have been inkjet printed (20) are illustrated in Fig. 24 and Fig. 25 and an electrostatic printed layer in Fig. 26.

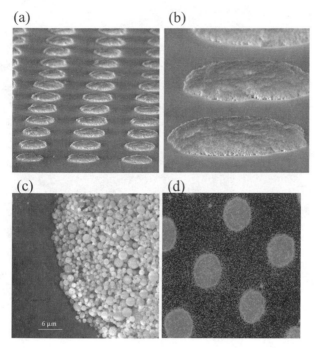

FIGURE 24 SEM micrographs (a), (b), and (c) of inkjet printed 70 micron spherical phosphor powder pixels on glass at different magnifications. (d) Optical micrograph of the same features comprised of Zn_2SiO_4 : Mn powder under photoluminescence excitation. Courtesy of Superior MicroPowders.

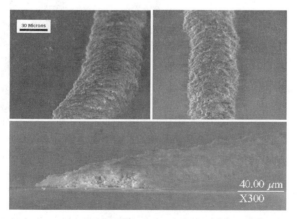

FIGURE 25 SEM micrographs of inkjet printed SMP spherical phosphor powder line ~40 microns wide. Courtesy of Superior MicroPowders.

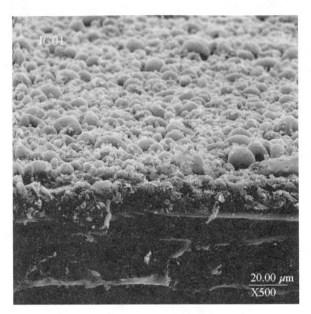

FIGURE 26   SEM micrograph of electrostatically printed spherical powder layer produced by spray pyrolysis. Courtesy of Superior MicroPowders.

8. MATERIALS FOR METAL-AIR BATTERIES AND PROTON EXCHANGE MEMBRANE FUEL CELLS

The market for portable and decentralized power generation is generally viewed as having a tremendous growth potential. The battery industry alone is estimated to grow to $60 billion by 2004. Batteries are largely used to power portable electronic devices while nonportable devices are powered by connection to the electrical grid with electrical power supplied by power stations. With the increasing demand for high-reliability electrical power, fuel cells are also gaining momentum as potential sources for distributed power generation from devices as small as cell phones to as large as ships. The strategic importance of thin-form batteries is in line with the evolution toward higher levels of device integration in the electronics industry. Direct deposition of these conformal power-generating and storage devices is absolutely essential to ensure that electrical power is available to operate the devices produced by the emerging direct write technologies. A detailed discussion of direct write techniques for power devices can be found elsewhere in this book (see Chap. 4).

The technologies of fuel cells and metal air batteries are extremely close in terms of implementation. As a result, the materials used for both metal-air batteries and fuel cells are often similar. The materials required for gas diffusion electrodes fall into three general classes: conductor materials for current collectors that can be processed at low temperatures; electrocatalytic powders, comprising metal or metal-oxide-carbon composites; and polymer-carbon composites. Direct write of fuel cell components and of thin-form batteries offers the potential to improve the performance while simultaneously reducing the volume or mass of certain components within certain cost constraints. In the case of a direct write gas diffusion electrode for example, the active layers responsible for the catalytic reaction of the gases are deposited directly onto the surface of the current collector. The advantages of direct writing these layers include specific placement of material, thinner layers and better control over the composition of the layers within a very small distance to achieve control over a gradient in composition.

8.1. CONDUCTIVE MATERIALS FOR CURRENT COLLECTORS

The materials systems used to direct write conductive layers as current collectors must be processable at low temperatures to be compatible with the other materials in the battery or fuel cell. In the case of a metal-air battery, this can be Teflon, with a processing temperature ceiling of 250 °C for short times (10 min.). In the case of fuel cells, the polymeric materials are generally proton-transporting polymers such as Nafion and hydrophobic layers such as Teflon. Again the major concern is processing of the conductive layer without disrupting the performance of the polymer or pore structure of the layer and this must typically be achieved below 200 °C. These requirements are completely in line with the requirements for direct deposition of conductor materials on low-T substrates such as polyimide for those electronic circuit applications discussed here.

8.2. ELECTROCATALYST POWDERS

The spray-based manufacturing process shown in Fig. 14 is capable of producing electrocatalyst powders with a variety of compositions and morphology with controlled hierarchical microstructure in a highly reproducible fashion (20). Here, spray-based powder-production process is modified to favor the use of carbon and the creation of powder batches with microstructures that provide not only control over particle size and size distribution but also pore structure, morphology, and formation of highly dispersed catalytically active phases on the surface of the carbon. The microstructure of the carbon-supported catalysts comprises highly dispersed catalysts ($\sim 1-3$ nm) on primary carbon particles ($\sim 30-50$ nm) that are agglomerated into spherical micron-sized porous secondary particles. The pore structure of these materials and control over particle size and size distribution is such that they have significant performance advantages when deposited into an electrode compared to existing catalysts as will be described.

Catalyst morphology and surface area are characteristics that typically have critical impact on the catalyst's performance. The morphology determines the packing density (and eventually enables new printing methods) and the surface area determines the type and number of surface adsorption centers, where the active species are formed during the catalyst synthesis.

9. CONCLUSIONS

The development of novel materials systems for direct depositions on unconventional substrates is the enabling technology that will cut across the various direct deposition platforms. Commercially available traditional materials are simply not compatible with the novel tools that are being developed for direct deposition. These materials will have to be tailored to the deposition technique, and will often require completely new material approaches to enable high-speed deposition, improved resolution, and enhanced performance. Postdeposition processing on novel substrates imposes additional requirements such as chemical compatibilities and extremely stringent processing limitations such as firing below 200 °C, or compatibility with laser sintering. It is likely that the newly developed conductor formulations (and resistors to some extent) will find industry acceptance most readily. This is due to the huge existing market for interconnection solutions in the electronics industry and the rapidly expanding gamut of new applications and the low-temperature, low-cost substrate materials they are exploring. Direct deposition of dielectrics and phosphor materials is already finding its way into applications such as flat panel displays, and its use will grow rapidly in the future. Finally, the strategic importance of directly written thin-form batteries fits naturally within the overall strategy of higher levels of device integration in the portable electronics industry. It is obvious that these power-generating and storage devices are absolutely essential to operate the devices produced by the emerging direct write technologies. Their thin-form nature and the capability to directly deposit them on a wide variety of surfaces will further reduce cost and allow integration into numerous portable and even disposable devices.

ACKNOWLEDGMENTS

The authors would like to thank the editors, researchers at the Naval Research laboratory, Potomac Photonics, Inc., Optomec, Inc., CMS Technetronics, Stony Brook Campus of the State University of New York, CFD Research Corporation, Sandia National Laboratory, and Revise Inc., for countless stimulating conversations. A special thanks goes to Dr. William L. Warren and Jeff Bullington at DARPA-DSO, for their invigorating involvement in the direct write effort.

REFERENCES

1. MRS Proceedings Vol. **624** *Materials Development for Direct Write Technologies*, edited by D. B. Chrisey, D. R. Gamota, H. Helvajian, and D. P. Taylor, (Materials Research Society, Warrendale, PA, 2000).
2. "Material systems used by micro dispensing and ink jetting technologies," J. Zhang, I. Shmagin, J. Skinner, J. Szczech, and D. Gamota, in MRS Proceedings Vol. **624** *Materials Development for Direct Write Technologies*, edited by D. B. Chrisey, D. R. Gamota, H. Helvajian, and D. P. Taylor, (Materials Research Society, Warrendale, PA, 2000), pp. 41–46.
3. "Laser guided direct writing of electronic components," M. J. Renn, M. Essien, B. H. King, and W. D. Miller, in MRS Proceedings Vol. **624** *Materials Development for Direct Write Technologies*, edited by D. B. Chrisey, D. R. Gamota, H. Helvajian, and D. P. Taylor, (Materials Research Society, Warrendale, PA, 2000).
4. "Thermal spray techniques for fabrication of meso-electronics and sensors," S. Sampath, H. Herman, A. Patel, R. Gambino, R. Greenlaw, and E. Tormey, in MRS Proceedings Vol. **624** *Materials Development for Direct Write Technologies*, edited by D. B. Chrisey, D. R. Gamota, H. Helvajian, and D. P. Taylor, (Materials Research Society, Warrendale, PA, 2000), pp. 181–188.
5. "Laser direct writing of phosphor screens for high-definition displays," J. M. Fitz-Gerald, A. Piqué, D. B. Chrisey, P. D. Rack, M. Zeleznik, R. C. Y. Auyeung, and S. Lakeou, Appl. Phys. Lett. 76, pp. 1386–1388 (2000).
6. *Hybrid Microelectronics Handbook*, edited by J. E. Sergent and Charles A. Harper, (McGraw-Hill, New York, NY, 1995), Chapter 3.
7. *Polymer Thick Film*, Ken Gilleo, (Van Nostrand Reinhold, New York, NY, 1996).
8. *Micro-Electronics Packaging Handbook*, R. R. Tummala and E. J. Rymaszewski, (Van Nostrand Reinhold, New York, NY, 1989), p. 129.
9. "Liquid ink jet printing with MOD inks for hybrid microcircuits," K. F. Teng and R. W. Vest, *IEEE Transactions on Components, Hybrids, and Manufacturing Technology*, Vol. CHMT 12(4), pp. 545–549, 1987.
10. See, e.g., *Ink Jet Technologies II*, E. Hanson, (Society for Imaging Science and Technology, Springfield, VA, 1999).
11. "Optimization of dispersion and viscosity of a ceramic jet printing ink," W. D. Teng, M. J. Edirisinghe, and J. R. G. Evans, J. Am. Ceram. Soc. 80, pp. 486–494 (1997).
12. "Thermal-spray processing of materials," S. Sampath and R. McCune, in MRS Bulletin 25 (7), pp. 12–42 (2000).
13. CMS Technetronics, Inc., private communications. For a related publication, see, "Direct-write techniques for fabricating unique antennas," R. M. Taylor, K. H. Church, J. Culver, and S. Eason, in MRS Proceedings Vol. **624** *Materials Development for Direct Write Technologies*,

edited by D. B. Chrisey, D. R. Gamota, H. Helvajian, and D. P. Taylor, (Materials Research Society, Warrendale, PA, 2000), pp. 195–198.

14. *Aerosol Processing of Materials*, T. T. Kodas and M. Hampden-Smith, (Wiley-VCH, New York, NY, 1999).
15. *Laser Microfabrication*, D. J. Ehrlich and J. Y. Tsao, (Academic Press, Inc., New York, NY, 1989).
16. *The Chemistry of Metal CVD*, T. Kodas and M. Hampden-Smith, (VCH Verlagsgesellschaft, Weinheim, Germany, 1994).
17. See, e.g., *Foundations of Colloid Science Vol. II*, Robert J. Hunter, (Clarendon Press, Oxford, UK, 1989), pp. 992–1036.
18. "Inverse CVD: A novel synthetic approach to metallized polymeric films," R. E. Southward and D. W. Thompson, Adv. Mater. 11, pp. 1043–1047 (1999).
19. "Chemical aspects of solution routes to perovskite-phase mixed-metal oxides from metal-organic precursors," C. D. Chandler, C. Roger, and M. J. Hampden-Smith, Chem. Rev. 93, pp. 1205–1241 (1993).
20. Superior MicroPowders Internal R&D data.
21. See, e.g., *Chemical Processing with Lasers*, Dieter Bauerle, (Springer-Verlag, New York, NY, 1986).
22. *Laser Annealing of Semiconductors*, J. M. Poate and J. W. Mayer, (Academic Press, New York, NY, 1982).
23. "Laser rapid sintering of metallo-organic derived oxide films," R. F. Louh and R. C. Buchanan, MRS Proceedings Vol. 201, (Materials Research Society, Pittsburgh, PA, 1991), pp. 525–534.
24. *Sintering Theory and Practice*, R. M. German, (John Wiley & Sons, New York, NY, 1996).
25. *Electroceramics*, A. J. Moulson and J. M. Herbert (Chapman & Hall, New York, NY, 1990), pp. 79–82.
26. "Dielectric characteristics of zirconium tin titanate ceramics at microwave frequencies," K. H. Yoon and E. S. Kim, Mat. Res. Bull. 30, pp. 813–820 (1995).

PART III

Direct-Write Techniques

There are many ways to direct write or transfer material patterns. The range of additive techniques includes plasma spray, laser transfer and particle guidance, and both common and novel material dispensing techniques including ink jets, micropens, and AFM dip-pens. Subtractive techniques include the use of electron and focused-ion beams as well as laser micromachining. Of these, all but the last are additive techniques that depend on high-quality starting materials. Successful direct-written electronic materials should have electronic properties comparable to conventional thick-film processed material techniques like screen printing. But CAD/CAM direct-write techniques offer increased flexibility and speed for rapid prototyping, product and material development, and small-lot manufacturing. Each direct-write technique has its own merits and shortcomings and achievable length scales range from nanometers to millimeters. Some requirements apply generally such as the deposition of fine features, low prototyping and production costs, high manufacturing yields, fast prototyping and production times, manufacturing flexibility, and low capital investment.

CHAPTER 7

Direct Write Using Ink-Jet Techniques

DAVID B. WALLACE, W. ROYALL COX, AND DONALD J. HAYES

MicroFab Technologies, Inc., Plano, Texas

1. INTRODUCTION

Ink-jet printing technology can reproducibly dispense spheres of fluid with diameters of 15–200 μm (2 pl to 5 nl) at rates of 0–25,000 per second for single droplets on demand, and up to 1 MHz for continuous droplets. Piezoelectric dispensing technology is adaptable to a wide range of material dispensing applications, such as biomedical reagents, liquid metals, and optical polymers. Ink-jet-based deposition requires no tooling, is noncontact, and is data driven: no masks or screens are required and the printing information is created directly from CAD information and stored digitally. Being data driven, it is thus flexible. As an additive process with no chemical waste, it is environmentally friendly.

Direct-Write Technologies for Rapid Prototyping Applications
Copyright © 2002 by Academic Press. All rights of reproduction in any form reserved.

2. HISTORY

Ink-jet printing technologies have been developed over the past forty years for low cost and high-quality office printers, for industrial marking applications, and for specialty applications. Color ink-jet printers dominate the printer market in the home and small office segments; most packaged consumables such as drugs, soft drinks, and snack foods have date and lot codes printed on them using industrial ink-jet printers; cardboard boxes used for shipping products frequently are bar coded by ink-jet printers, enabling automated shipping, warehousing, etc.; and large format ink-jet printers, the fastest growing segment in the ink-jet printer industry, have enabled low cost, rapid response signage.

As early as 1968 (1), the potential of ink-jet printing technology as a digital, high-resolution, noncontact, direct-write fluid deposition method was recognized. Electronics manufacturing, medical diagnostics, and solid free-form-fabrication applications drove initial developments of ink-jet printing technology for novel applications during the 1980s. More recently, photonics, MEMS (Micro Electromechanical Systems), wireless communication, and portable electronics are driving forces in the adaptation of ink-jet printing technology.

This chapter is intended to provide an overview of ink-jet printing technology as a direct-write, rapid prototyping technology. We will start with a general description of ink-jet printing technology, the physics behind droplet generation methods, and important characteristics of these methods. Next, practical considerations for fluid property requirements, image formation (i.e., fluid/substrate interaction), and throughput will be discussed. Specific examples of ink-jet deposition in electronics, photonics, and biomedical applications will then be discussed in detail. Finally, a discussion of current commercially available direct-write ink-jet systems and future trends will conclude the chapter.

3. BACKGROUND ON INK-JET TECHNOLOGY

3.1. CONTINUOUS MODE INK-JET TECHNOLOGY

The phenomena of uniform drop formation from a stream of liquid issuing from an orifice were noted as early as 1833 by Savart (2) and described mathematically by Lord Rayleigh (3,4) and Weber (5). In the type of system that is based on their observations, fluid under pressure issues from an orifice, typically 50–80 µm in diameter, and breaks up into uniform drops by the amplification of capillary waves induced onto the jet, usually by an electro-mechanical device that causes pressure oscillations to propagate through the fluid. The drops break off from the jet in the presence of an electrostatic field, referred to as the charging field, and thus acquire an electrostatic charge. The charged drops are directed to their desired location—either the catcher or one of several locations on the substrate—by another electrostatic field, the deflection field. This type of system is generally referred to as "continuous" because drops are continuously produced and their trajectories are varied by the amount of charge applied. Theoretical and experimental analysis of continuous ink-jet devices, particularly the process of disturbance growth on the jet that leads to drop formation, has been fairly extensive (6,7). Continuous mode ink-jet printing systems produce droplets that are approximately twice the orifice diameter of the droplet generator. Droplet generation rates for commercially available continuous mode ink-jet systems are usually in the 80–100-kHz range, but systems with operating frequencies up to 1 MHz are in use. Droplet sizes can be as small as 20 μm in a continuous system, but 150 μm is typical. Droplets as large as 1 mm (~0.5 μl) have been observed.

Figure 1 shows a schematic of this type of ink-jet printing system, and Fig. 2 shows a photomicrograph of a 50-μm-diameter jet of water issuing from a droplet generator device and breaking up due to Rayleigh instability (i.e., continuous mode ink-jet) into 100-μm-diameter droplets at 20,000 per second.

Continuous mode ink-jet systems are currently in widespread use in the industrial market, principally for product labeling of food and medicines. They

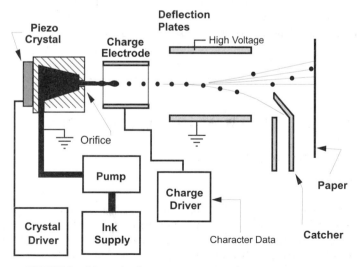

FIGURE 1   Schematic of a continuous type ink-jet printing system.

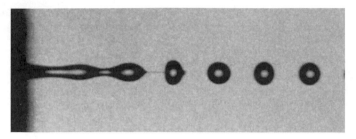

FIGURE 2   A 50-μm jet of water breaking up due to Rayleigh instability into 100-μm droplets at 20 kHz.

have high throughput capabilities, especially array continuous mode systems, and are best suited for very high duty cycle applications. Few continuous mode ink-jet systems are multicolor (multifluid), but two-color systems are in use. Because they require unused drops to be recirculated or wasted, the potential is limited for using continuous mode ink-jet technology for novel applications, such as rapid prototyping processes. Notable exceptions include MIT's 3D printing rapid prototyping technology (8), metal jetting technology, and medical diagnostic test strip production. The latter two are discussed in Section 7. *Direct-Write Applications.*

3.2. DEMAND MODE INK-JET TECHNOLOGY

In the 1950s, the production of drops by electromechanically induced pressure waves was observed by Hansell (9). In this type of system, a volumetric change in the fluid is induced by the application of a voltage pulse to a piezoelectric material that is coupled, directly or indirectly, to the fluid. This volumetric change causes pressure/velocity transients to occur in the fluid and these are directed so as to produce a drop that issues from an orifice (10–12). Because the voltage is applied only when a drop is desired, these types of systems are referred to as "drop-on-demand," or "demand mode."

Figure 3 shows a schematic of a drop-on-demand-type ink-jet system, and Fig. 4 shows a MicroFab drop-on-demand ink-jet device generating 60-μm diameter drops of butyl carbitol from a device with a 50-μm orifice at 2,000 per second. Demand-mode ink-jet printing systems produce droplets that are approximately equal to the orifice diameter of the droplet generator (13). As Fig. 3 indicates, demand-mode systems are conceptually far less complex than continuous mode systems. On the other hand, demand-mode droplet genera-tion requires the transducer to deliver three or more orders of magnitude

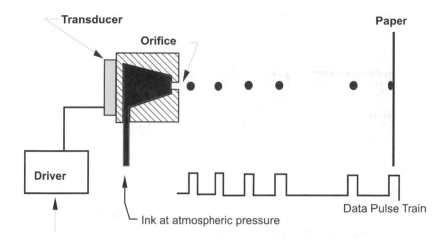

FIGURE 3   Schematic of a drop-on-demand ink-jet printing system.

greater energy to produce a droplet, compared to continuous mode, and there are many "elegant" (i.e., complex) array demand-mode printhead designs (14–17).

A recent demand-mode droplet generation technology uses focused acoustic energy to cause a droplet to be emitted from a free surface. This novel "orificeless" technology, developed for high throughput printers/copiers (18), has been employed in industrial processes for adhesive coating, and in NASA's liquid metal droplet free-form fabrication efforts (19). Figure 5 shows a schematic of a free surface droplet generator.

In many commercially available demand-mode ink-jet printing systems today, a thin-film resistor is substituted for the piezoelectric drive transducer. When a high current is passed through this resistor, the ink in contact with it is

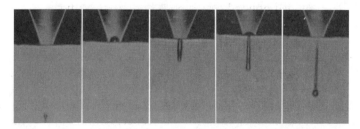

FIGURE 4   Drop-on-demand-type ink-jet device generating 60-μm-diameter drops at 4 kHz. Sequence from left to right spans 130 μs.

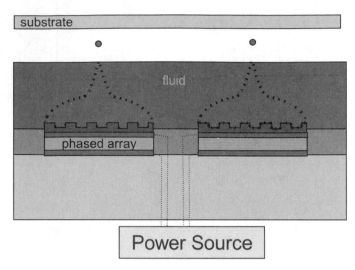

FIGURE 5   Schematic of free surface droplet generator.

vaporized, forming a vapor bubble over the resistor (20). This vapor bubble serves the same functional purpose as the piezoelectric transducer. This type of printer is usually referred to as a thermal ink-jet printer.

One of the characteristics of ink-jet printing technology that makes it attractive as a precision fluid microdispensing technology is the repeatability of process. The images of droplets shown in Fig. 2 and Fig. 4 were made by illuminating the droplets with an LED that was pulsed at the droplet generation frequency. The exposure time of the camera was ~1 second, so that the images represent thousands of events superimposed on each other. The repeatability of the process results in an extremely clear image of the droplets, making it appear to be a high speed photograph. To further illustrate this point, Fig. 6 shows two 60-µm-diameter jets of water breaking up into 120-µm-

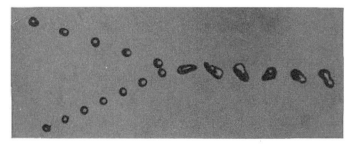

FIGURE 6   Two streams of 120-µm water droplets merging into a single droplet stream at 20 kHz.

diameter droplet streams at 20,000 per second, and being caused to merge into a single droplet stream. Again, this image was created using a strobed LED and an ~1-second exposure time, resulting in 20,000 events being super-imposed in the image. Not only is the droplet formation process so repeatable that the image of the droplets is sharp, but also, when the droplets are caused to merge, the formation of the highly contorted merged droplets is seen to be just as repeatable.

Drop-on-demand ink-jet systems have been used primarily in the office printer market and have come to dominate the low-end printer market (HP's DeskJets, Cannon's Bubble Jets, and Epson's Stylus). Demand-mode ink-jet systems have no fluid recirculation requirement, and this makes their use as a general fluid microdispensing technology more straightforward than the use of continuous mode technology. Thermal demand-mode ink-jet technology systems can achieve extremely high fluid-dispensing performance at a very low cost. However, this performance/cost has been achieved by highly tailoring the ink: thermal ink-jet systems are restricted to fluids that can be vaporized by the heater element (without igniting the fluid) and their performance/life can be degraded drastically if other fluids are used. In practice, thermal ink-jet systems are limited to use with aqueous fluids.

Because piezoelectric demand-mode ink-jet technology does not require recirculation of the working fluid, does not create thermal stress on the fluid, and does not depend on a thermal process to impart acoustic energy to the working fluid, it is the most adaptable of the ink-jet printing technologies to fluid microdispensing in general, and in particular to rapid prototyping applications. As a noncontact printing process, the volumetric accuracy of ink-jet dispensing is not affected by how the fluid wets a substrate, as is the case when positive displacement or pin transfer systems "touch off" the fluid onto the substrate during the dispensing event. In addition, the fluid source cannot be contaminated by the substrate, or contamination on the substrate, in a noncontact dispensing process. Finally, the ability to free-fly the droplets of fluid over a millimeter or more allows fluids to be dispensed into wells or other substrate features (e.g., features that are created to control wetting and spreading).

4. JETTING MATERIALS

Generally, the fluid property requirements for demand-mode ink-jet dispensing are as follows: the viscosity should be Newtonian and less than 40 cp, and the surface tension should be greater than 20 dynes/cm. Very low viscosities can lead to difficulties with satellite formation and lack of acoustic damping, but organic solvents such as methanol with viscosities less than that of water (1 cp) can be jetted. Very high surface tensions present unique difficulties, but the solders discussed in Section 7. *Direct-Write Applications* have surface tensions greater than 400 dyne/cm, or roughly 6 times that of distilled water.

If the fluid is heated or cooled, the above properties are required *at the orifice.* Higher viscosities can be tolerated in the fluid delivery system *if* this does not create a pressure drop that limits the desired maximum frequency. For high-density fluids such as molten metals, the fluid properties should be converted to kinematic values to determine if the fluid properties are acceptable.

Newtonian behavior is not strictly required, but the fluid properties at the orifice flow conditions must be less than 40 cp. Thus a shear thinning fluid could have a low shear rate viscosity much higher than the 40 cp. Viscoelastic behavior causes significant performance problems by increasing the amount of deformation the fluid can withstand without breaking off from the orifice. This is illustrated by the lubricating wax shown in Fig. 7 which forms a tail that atomizes instead of being pulled up into the main drop (see Fig. 4 for comparison).

Particle suspensions, such as pigmented inks, are acceptable as long as the particle/agglomerate size and density do not cause the suspension to depart from the aforementioned fluid properties range. Particles that are >5% of the orifice diameter (e.g., human liver cells, see Fig. 8) will cause at least some instability in drop generation behavior, but still may be acceptable in low concentrations.

The "window" of fluids and suspensions that can be dispensed using ink-jet technology has been stretched by heating, cooling, stirring, wiping, purging, pre-oscillating, diluting, and other methods. However, this window is unavoidably narrowed as orifice diameter decreases, frequency increases, and number of jets in an array increases.

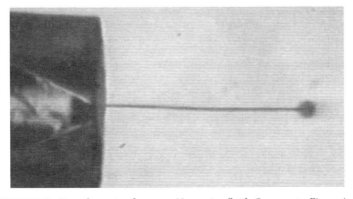

FIGURE 7   Drop formation for a non-Newtonian fluid. Compare to Figure 4.

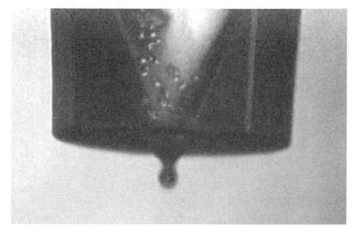

FIGURE 8   Human liver cells >10 μm being dispensed in an aqueous buffer.

The diversity of fluids that have been dispensed using ink-jet technology is impressive, given the fluid property restrictions just described. Inks by themselves represent a broad class of materials. Dye-based aqueous suspensions are the most commonly used, but formulations of aqueous/pigment, volatile solvent (e.g., methyl ethyl ketone)/dye, volatile solvent/pigment, and low-volatility solvent/dye are all in common use. Table 1 lists specific materials that have been dispensed using ink-jet technology. Some of those listed have a wide latitude in their formulation, and thus represent a broad class of materials. Particle suspensions, including inks, inherently fall into this category. Polymers do also, but the knowledge, intellectual property, equipment, or budget required to tailor polymeric formulations to ink-jet dispensing can make their use problematic. Metals, oligonucleotides, and precursor formulations used for chemical synthesis (e.g., peptides, DNA) represent fluids for which there is less latitude in modifying the fluid formulation to adapt to an ink-jet dispensing method.

5. PATTERN/IMAGE FORMATION: FLUID/SUBSTRATE INTERACTION

Except for the cases where ink-jet technology is used to meter fluid, as in filing a microwell plate for a high-throughput drug screening application, ink-jet deposition processes are used to produce a desired pattern of material onto a substrate. The interaction between the fluid formulation, jetting parameters

TABLE 1 Materials that Have Dispensed Using Ink-Jet Printing Technology

Electronic/Optical Materials
   solders
   fluxes (low solids and tacky)
   photoresists
   epoxies (UV and thermal cure; oligomeric and filled systems)
   polyimides
   electroactive polymers (light-emitting and conductive)
   cyanoacrylates
   organometallics
Biological Fluids
   DNA
   nucleic acids
   amino acids
   proteins (antibodies and antigens)
   lipids (cholesterol, steroids)
   biodegradable polymers (PEG, PLA; bioactive molecules embedded)
Organic Solvents
   alcohols
   ketones
   aliphatics
   aromatics (xylene, toluene)
   dipolar solvents (NMP, DMF)
Other
   sol-gels
   thermoplastics
   thermosets
   acrylics
   >1M salt solutions
   photographic developer
   fuels
   aqueous adhesives
   odorants (water and alcohol based)
Particle Suspensions
   pigments
   cells
   latex spheres
   metal particles (Ag, W, Cu)
   Teflon
   phosphors
   ferrites
   zeolyts for catalysts

(drop size, velocity, frequency), substrate characteristics, printing grid (dots per inch), and printing sequence (interleave, overprinting, color sequence, etc.) is the "battlefield" in the development of all ink-jet printing systems, whether they use an aqueous ink or solder for electronic assembly. For the more familiar case of liquid ink on paper, the porosity of the paper and the low viscosity of the ink represented a major challenge in the initial development of ink-jet printers. Rapid spreading of liquid ink through the fibers can cause the spot size to become much larger than the drop size, decreasing the optical density of the spot and resulting in irregular spots that degrade the quality of characters, lines, etc. Ink formulations that produce good print quality on a wide range of papers has been a cornerstone in the wide acceptance of ink-jet printers in the marketplace.

Use of ink-jet printing technology for rapid prototyping applications produces most of the same fluid/substrate-interaction issues, in addition to unique ones. Most rapid prototyping applications of ink-jet technology deposit liquid onto a nonporous substrate (similar to printing an overhead transparency). Control of the spreading is essential if the desired resolution is to be obtained. Phase change inks (Dataproducts, Tektronix printers) were developed for conventional ink-jet printers for precisely this reason, because they solidify quickly after impact. For rapid prototyping applications, solders for electronic manufacturing and thermoplastics for free-form fabrication are notable examples of phase change materials. Figure 9 illustrates a solder drop that has been deposited in molten form and solidified after impact (note the ridges produced by freezing the oscillations that occur due to impact (21)). The control of spreading due to solidification is a positive aspect of phase change materials if one is trying to limit spreading and obtain the smallest spot for a given drop size. However, if one is trying for deposition of a uniform layer, solidification into a bump is a problem, not a feature. For

FIGURE 9 60-μm solder bump deposited as a liquid onto copper using ink-jet printing technology. Left is model.

instance, phase change inks make very dense/saturated prints and very sharp characters, but the rounded shape of the individual bumps causes diffraction when light is project through overhead transparencies, resulting in total distortion of the image. To correct this problem, Tektronix flattens the bumps into a smooth surface on their phase change printers.

At the other end of the spectrum, many organic liquids, such as isopropanol, acetone, and acetonitrile, have very low viscosity, low contact angle/ surface tension, and are volatile. The ability to wet most surfaces and their low viscosity allows these fluids to spread rapidly. As with a phase change material's lack of spreading, this too is either a feature or a problem, depending on the application. If one is trying to write a conductor using an organometallic ink, or create a pixel in a light-emitting polymer display, it is definitely a problem. In many cases, surface features, such as the wells commonly used in phosphor-based flat panel displays, provide a barrier to spreading and help physically define the feature size. In other cases, surface treatments, such as plasma cleaning or application of a nonwetting coating, are used to control spreading.

Volatility of a solution with dissolved or suspended solids can cause operational issue, and ink drying is one of the most common failure modes with office ink-jet printers. In addition, for most direct-write applications of ink-jet technology, volatility can cause nonuniform distribution of the solid on

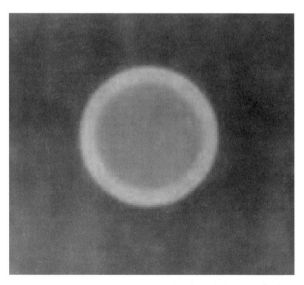

FIGURE 10   75-μm light-emitting polymer spot with non-uniform distribution due to solvent volatility.

the substrate (usually nonporous) after drying (22), similar to the ring that coffee stains form on clothes. Solutions to this problem have been many and diverse: reactive substrate, covalent binding of the solid to the substrate, cosolvents that are lower volatility, UV or thermal cross-linking, etc. Figure 10 shows a 75-μm light-emitting polymer spot with nonuniform distribution due to solvent volatility.

Pattern or image formation in its simplest implementation can be just the selection of pixel (picture element) size and spacing, then using the ink-jet dispenser to fill up the desired pixels. Or, in the case of a vector graphics analogy, simply selecting the spacing between drop dispenses along a line, as illustrated by the flux printed in Fig. 11. However, even the lowest cost ink-jet printers have complicated print modes that are used to increase print quality. Rows of spots are interleaved to high coherent errors from a single jet; colors are printed so as not to bleed together in wet state; multiple passes are made over an area to increase color saturation of the printed area; and operating frequency is decreased for high-quality printing. All of these methods, and more, have applicability to nonink direct-write applications.

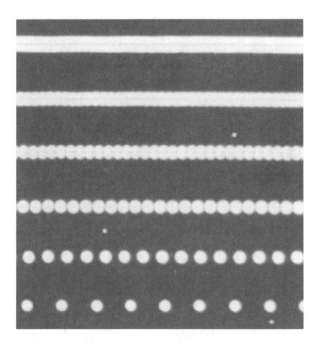

FIGURE 11   Low solids flux spots, ink-jet printed as dots and lines.

6. THROUGHPUT CONSIDERATIONS

As discussed, droplet generation rates for continuous ink-jet technology are on the order of 80 kHz, and rates for demand-mode technology are on the order of 8 kHz. For office ink-jet printers, the maximum droplet rate, the number of channels in the printhead, and the printing resolution combine to determine the nominal throughput. For many direct-write applications, this is not the case. If multiple drops are required per spot, as is the case with some of the micro-optics applications discussed here, then the printhead will usually move to a new location and stop before dispensing. The acceleration associated with this procedure limits the effective maximum throughput to approximately five dispenses per second for most applications.

Even when printing while the printhead is in motion, which is how office ink-jet printers work, the effective dispensing rate can be far below the maximum. If the distance between dispensing sites is fairly large (e.g., 200 µm or more), translational speed limits can be encountered before drop generation limits are encountered.

7. DIRECT-WRITE APPLICATIONS

7.1. SOLDER JETTING

Deposition of small quantities of solder onto the interconnect pads of integrated circuits or chip-scale-packages is a large rapidly growing application in electronic assembly, driven by flip-chip and other space/weight-saving electronics packaging developments (23,24). Photonics assembly processes that use surface-tension-driven self-centering to enable alignment of optical components to <1 µm are also beginning to be used (25). Ink-jet technology is one method used to deposit solder bumps for these applications, and its use has been explored by several organizations (26,27). Continuous-mode metal jetting technology has been developed by industry (28), academia (29,30), and national laboratories (31) as a very high throughput method of depositing solder bumps or producing metal spheres and balloons (see Fig. 12). Demand-mode solder jetting systems using both electrodynamic (32,33) and piezo-electric (34,35) actuators have been developed that avoid the drawbacks to continuous-mode systems (long flight paths and large material usage). Piezo-electric-driven solder jetting developments, exemplary of this application, will be discussed.

Operation of piezoelectric demand-mode ink-jet devices at temperatures above 200 °C is one of the principal challenges in developing solder jetting

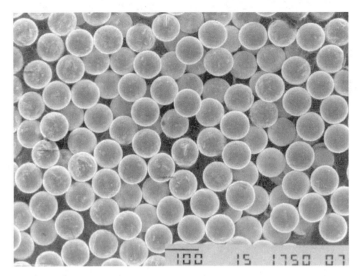

FIGURE 12   90-μm spheres of Sn63/Pb37 solder formed at 8 kHz and 220 °C using continuous-mode ink-jet technology.

technology. In addition to selecting materials, designs, and assembly processes that are compatible with these operating temperatures, unique drive waveforms have been used for piezoelectric devices at elevated temperatures (36).

Operating characteristics for solders dispensed using piezoelectric demand-mode systems include formation of spheres with diameters of 25–125 μm; drop formation rates (on demand) up to 1,000 per second; deposition onto pads at up to 600 per second; and operating temperatures to 320 °C. The solder dispensed has been primarily eutectic tin–lead (63Sn/37Pb), but a number of other solders have been demonstrated, including high lead (95Pb/5Sn), no leads (96.5Sn/3.5Ag; indium; 52In/48Sn), and low-temperature bismuth solders.

Figure 13 shows results from printing solder onto an 18 × 18 test coupon with 100-μm-diameter pads on 250-μm centers. The deposited solder volume is equivalent to a drop diameter of 100 μm. Note that the bump shape shown in Fig. 13 is a consequence of rapid (<100 μs) solidification (21). The instantaneous droplet rate for these tests was 400 per second and the pattern was printed by rastering the substrate in the horizontal direction of the figures. An average placement error of 10 μm was achieved in these tests, which is close to the accuracy limitations imposed by the positioning and alignment systems of the platform employed.

In addition to the digital control of the number of drops deposited by a solder jetting system, the drive waveform can be employed to modulate the

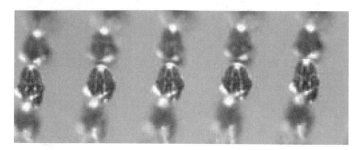

FIGURE 13   100-μm-diameter solder bumps placed onto 100-μm pads on 250-μm centers at 400 per second on the MPM solder jet feasibility platform.

volume of a drop over an approximately 2 : 1-diameter (8 : 1-volume) range. Figure 14 shows a solder jet device producing 62-μm-diameter droplets of solder at a rate of 120 Hz. The image on the left in this figure shows the droplet being formed while it is still attached to the orifice of the dispensing device, and the image on the right shows the drop approximately 1 ms later, after it has broken free from the dispenser.

Figure 15 shows the same device operating moments later, again at 120 Hz. In this figure, a drive waveform that extends the drop formation process over a

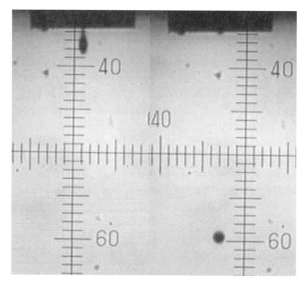

FIGURE 14   Drop formation process for a solder at two times during the process. Drop rate is 120 Hz and drop size is 62 μm.

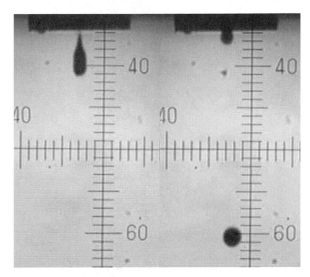

FIGURE 15   Drop formation process for same device shown in the previous figure, but using a different drive waveform. Drop rate is 120 Hz and drop size is 106 μm.

significantly longer time period is being used. By doing this, a considerably larger droplet is produced. In this case, the diameter is increased to 106 μm. The volume modulation using this method is continuous over the entire range of achievable volumes. This capability could be used to allow bump size to be changed under software control, either for product change over, or for application of variable-sized bumps onto a single substrate.

Electronic and photonics assembly applications of solder jetting technology would optimally use a single drop per site in order to maximize throughput. However, multiple drops may be employed to increase the volume delivered to single site. Figure 16 illustrates a case in which 8 nominally 50-μm-diameter droplets have been printed onto an integrated circuit pad 125 μm in diameter. In this case, the individual solder droplets have frozen before the impact of the next drop, producing a tower like object. Figure 17 and Fig. 18 illustrate how this process can be extended to produce simple geometric structures. It is easy to envision extension of these simple geometries into more complex shapes, especially because solid free-form-fabrication systems employing ink-jet methods have been commercially available for some time (37,38). However, the wetting, spreading, and fusion processes that must be controlled in free-form fabrication are significantly more complex for solders than for the thermoplastics employed in commercial free-form-fabrication products.

194

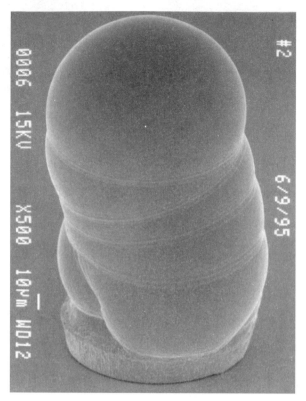

FIGURE 16 Tower of eight 50-μm-diameter solder drops placed onto a 125-μm-diameter pad on an integrated circuit.

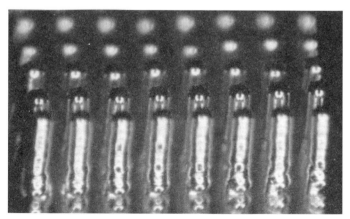

FIGURE 17 Towers created by dispensing fourteen 50-μm-diameter droplets on top of each other at 240 Hz.

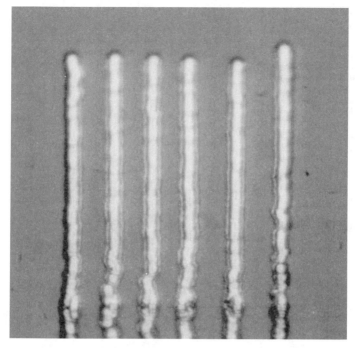

FIGURE 18   25-μm-diameter towers on 50-μm centers of 63/37 created using MicroJet Technology.

7.2. PHOTONIC ELEMENTS AND SENSORS

Information transmission and processing systems (Datacom and Telecom) will increasingly rely on lower-cost, higher-performance optical interconnect technologies to achieve greater speed and parallelism. To meet this need, organic optical materials are becoming accepted and used in the fabrication of micro-optical elements. Direct write of refractive microlenses and waveguides by ink-jet-printed (39,40) optical epoxies have higher thermal durability than the PMMA photoresist used in photolithographic methods (41), and can be fabricated directly onto optical components of arbitrary geometry (42–44).

Optical materials ranging from epoxies (45) to thermoplastics have been utilized in ink-jet-based micro-optics printing. The most stringent requirement for these materials is the fluid viscosity threshold of less than 40 cps, which is typically achieved by heating the printhead to 130–175 °C. The volume of a printed lenslet is a digital function of the smallest droplet size which can be generated efficiently, and its aspect ratio (diameter/sag) is determined by the degree of spreading of the deposited fluid on the target substrate prior to solidification. For UV-cured optical formulations, control of microlens aspect ratio is achieved by one or more of three methods: by applying a low-wetting optical coating of the requisite free-energy level to the substrate prior to printing; by heating the substrate during printing; or by creating physical features on the substrate that control spreading. Current hemispherical microlens-printing accuracies and reproducibilities are on the order of 1 and 2% of nominal values for diameter and focal length, respectively, with relative and absolute placement accuracies of about 2 and 5 μm, respectively.

7.2.1. Refractive Microlenses

Refractive microlens configurations which may be printed using ink-jet processes range from convex/plano hemispherical, hemi-elliptical, and square (46), to convex–convex. The latter configuration, illustrated in Fig. 19, was fabricated by printing two plano/convex lenslets coaxially on opposite sides of a 125-μm-thick glass substrate. This microlens geometry, which would be more challenging to fabricate by conventional photolithographic methods, could potentially be utilized to reduce focal spot size in, for example, optical recording applications. The direct-write method may also be employed to print microlenses of hemi-elliptical and square footprint, as illustrated in Fig. 20 and Fig. 21 where adjacent droplets are printed along one and two axes, respectively, and allowed to flow together prior to solidification and curing. The elliptical and square lens configuration could be useful in edge-emitting diode

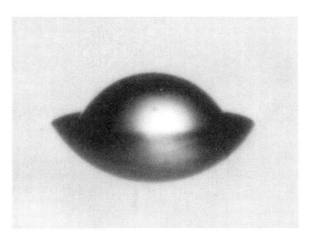

FIGURE 19   Two microlenses printed on opposite sides of the same substrate (top lenslet is 625 μm diameter).

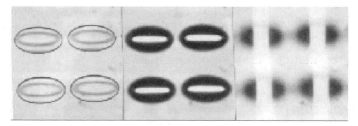

FIGURE 20   Printed hemi-elliptical microlenses 284 μm long, shown in substrate, fast-focal, and slow-focal planes.

laser collimation and light-collection for CCD (charged-coupled device), respectfully.

7.2.2. Refractive Microlens Arrays

Arrays of thousands of microlenses are being ink-jet printed for use as free-space optical interconnects in massively parallel, VCSEL-based, photonic switches under development in conjunction with the DARPA VCSEL-based Interconnects in VLSI Architectures for Computational Enhancement (VIVACE) program (47). These arrays are printed onto 3-inch-diameter thin quartz wafers in 12-group patterns, where each group consists of 2 each 34 × 34 identical arrays of 300-μm diameter, 60-μm sag microlenses, giving a total of 13,872 lenslets per wafer. The microlenses in each array are printed on 500-μm centers, and the 2 arrays are offset from each other by 250 μm along a diagonal, as shown in Fig. 22. The microlens array is aligned to a similar VCSEL/PD array with identically interlaced patterns of vertical cavity surface emitting lasers and photodetectors, where the two lenslets within each pixel

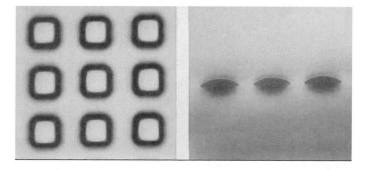

FIGURE 21   Printed 300-μm square microlenses shown in focal plane and in profile.

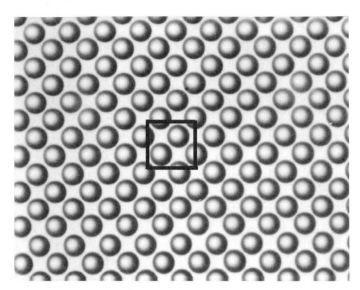

FIGURE 22   Portion of printed interlaced arrays of 300-μm-diameter lenslets for use in "smart-pixel"-based datacom switch.

area serve to collimate the beam from a VCSEL emitter and focus a returning beam into the adjacent photodetector. Selection of printed microlenses for this application was based on the greater coupling efficiency and wavelength independence of refractive lenslets compared to diffractive ones, the high lenslet speeds required ($f/\# \approx 1\text{--}2$), and the greater thermal durability of optical epoxy compared to the photoresist used in photolithographically fabricated refractive lenslets.

7.2.3. Microlens on Fiber: Increasing Acceptance Angle

Increasing the angle of acceptance of light into an optical fiber can relieve the sensitivity of alignment of laser (edge emitting or VCSEL) to an optical fiber in telecom/datacom transmitters, thereby reducing manufacturing costs. Micro-lenses can be printed with differing radii of curvature onto the tips of multimode optical fibers so as to increase their acceptance angles of light from diode laser sources by at least a factor of three (48). In the multimode-fiber case the outer edge of the fiber cladding defines the diameter of the printed lens, so alignment of printing axis to fiber tip is not a critical issue, and the radius of curvature may be increased (within limits determined by surface

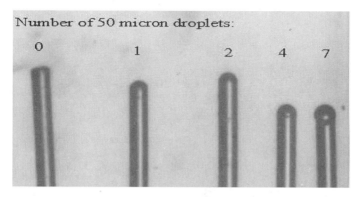

FIGURE 23 Microlenses printed onto tips of 100-μm-diameter optical fibers with indicated numbers of 50-μm droplets.

tension) by increasing the number of droplets of deposited optical fluid, as exemplified in Fig. 23.

To achieve similar acceptance-angle increases in single-mode fibers is a more challenging proposition, because it requires placement of a much smaller, faster microlens at the center of the fiber tip (49). An example of a hemispherical microlens ink-jet printed onto the end of a single-mode fiber is given in Fig. 24, where the lenslet diameter of 70 μm was achieved by depositing and UV-curing one 50-μm-diameter droplet of optical epoxy, after applying a low-wet coating to the tip of the fiber. Alignment of the microlens to

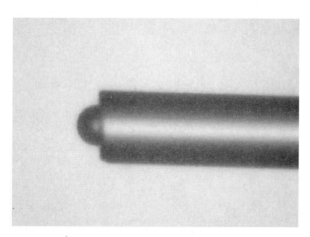

FIGURE 24 70-μm-diameter microlens printed onto the tip of a single-mode optical fiber, cladding outside diameter of 125 μm.

FIGURE 25   Profile view of 65-μm-diameter hemispherical microlenses printed onto the tips of single-mode fibers within a fiber array.

the fiber core within about 5 μm was achieved with a visual targeting system. Printing lenses on single-mode fibers can be easily extended to fiber arrays, as shown in Fig. 25.

7.2.4. Microlens on Fiber: Collimation

Collimation of the output beam of a single-mode optical fiber may also be achieved by printing a microlens onto its tip, but the geometry required is quite different from that needed to increase fiber numerical aperture. Collimation requires a much larger microlens, which is offset longitudinally from the tip by the lenslet focal length. To achieve collimation with a printed microlens, a collet may be attached to the end of the fiber to obtain the requisite geometry, as illustrated schematically in Fig. 26. The ID and OD of these 5-mm-long collets matched the fiber OD and targeted lenslet diameter, respectively. After

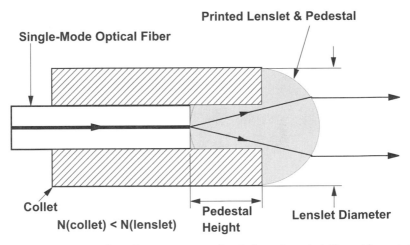

FIGURE 26   Geometry for collimating output of a single-mode optical fiber with a printed microlens.

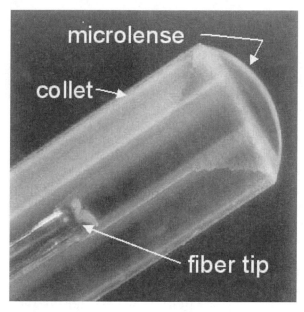

FIGURE 27   Microlens printed into and onto quartz collet attached to a single-mode optical fiber, to collimate output of fiber.

mounting this assembly vertically and aligning it to the print axis, 50-μm droplets of optical epoxy were printed into the collet to fill it to the top and build a convex lenslet surface of optimal radius of curvature on the top. As in the case of multimode-fiber printing, the outside edge of the collet defined the printed microlens diameter, so the radius of curvature could be varied over a significant range by varying the total number of droplets of optical material. A photograph of a fiber (1,550-nm wavelength) with collet and printed microlens is shown in Fig. 27.

7.2.5. GRIN Lenses

Lenses fabricated with a gradient index of refraction (GRIN) built into the material have long provided performance advantages over homogeneous lenses in two geometrical configurations for specialized applications. Cylindrical lenses having radial indexes of refraction gradients (RGRIN), with diameters ranging from about 60 μm in gradient-index, multimode optical fibers (50) to about 10 mm in gradient index rod lenses (51), are currently fabricated by ion interdiffusion in glass rods and are widely used in light collimation applications. Similarly, plano/convex macrolenses with axial index of refraction gradients (AGRIN), currently fabricated by stacking, fusing, and coring glass plates having differing indexes of refraction, then machining a hemispherical surface (52), can produce many reductions in focal spot sizes that homogeneous lenses of the same geometry cannot (53).

Ink-jet printing would allow for inexpensive fabrication of gradient index microlenses of several high-performance configurations (54). For example, ray-trace modeling has indicated that an axial index gradient of only 0.01 in a hemispherical lenslet with 50-μm height could potentially provide a 50-fold reduction in focused spot size over a homogeneous lenslet of identical geometry.

The fabrication process is illustrated conceptually in Fig. 28. It utilizes a dual print head system to deposit sequentially two optical epoxies of differing refractive index at the same location, in order to build an index gradient in the vertical direction. The relative volumes of the two materials and the time allowed for interdiffusion prior to solidification are process parameters that may be adjusted to maximize the axial component of the index gradient and its smoothness. The concept of this approach was tested by using two versions of the same optical material: one with and one without a fluorescein dopant. The undoped material was deposited first, then the printhead was translated to enable deposition of the second, fluorescing material at the same location. After

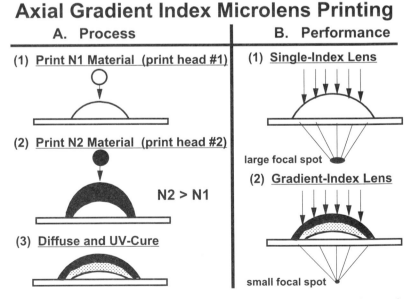

FIGURE 28 Proposed process and desired performance for printed axial gradient index of refraction microlenses.

FIGURE 29   Simulated 300-μm AGRIN microlens, printed by depositing fluorescing optical epoxy onto non-fluorescing epoxy.

an interdiffusion time of 1 minute, the composite structure was cured by UV irradiation. The axial (bottom-to-top) fluorescein gradient created in this hemispherical microlens can be seen under UV illumination in the profile photograph of Fig. 29, where the color changes uniformly from dark to light from the substrate plane to the apex of the lenslet.

7.2.6. Optical Sensors

An application of ink-jet technology currently under development is fabrication of multifunctional fiber-optic biochemical sensors, with potential use in clinical diagnosis (55), manufacturing process control, environmental monitoring, etc. The UV-curing optical epoxies used for microlenses can be adjusted to provide enhanced porosity and doped with chemical indicators. These can then be printed as sensor array elements onto detection surfaces, such as the tips of imaging fiber bundles, providing a sensor configuration as exemplified by Fig. 30.

Biochemical fiber-optic sensors in use today typically consist of an indicator chemistry attached to a single fiber, where the indicator chemistry is designed to change its optical properties (e.g., fluorescence or absorption) quantitatively in response to a target ligand under illumination of a suitable wavelength. The absorption or re-emission of the illuminating radiation in the form of fluorescence is monitored via photosensitive detectors, and a separate sensor is required for each target ligand. Multifunctional array sensors have been fabricated previously with fiber imaging bundles using a series of steps to "grow" sequentially indicator elements by masking the end of the fiber and UV

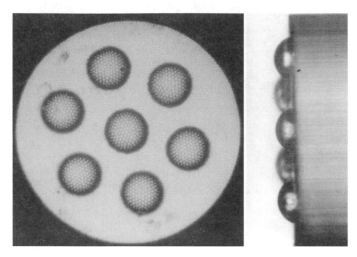

FIGURE 30   Array of 80-μm-diameter indicator elements printed onto 480-μm-diameter fiber-optic bundle.

curing each element out of different polymeric solutions (56), but poor uniformity and reproducibility of indicator element geometries have required calibration of each fiber and have limited their use. Ink-jet printing technology would allow the reproducible fabrication of multifunctional fiber-optic sensors consisting of uniformly sized sensor elements of different indicator chemistries on the same optical fiber bundle. Utilizing multiple printheads with differing indicator chemistries, such multifunctional fiber-optic array sensors could be manufactured with ink-jet printing technology at very high throughput rates and low materials costs.

7.3. DISPLAY MATERIALS

With the growth of the flat panel display market, manufacturing efforts are under way on a variety of display types that are relatively new to the industry. In addition, mature display types are being adapted to larger sizes, higher resolutions, multiple colors, larger volumes, lower cost, and/or higher frame rates. Because ink-jet printing is an additive process and high-quality phosphors are expensive, it offers an economy in the manufacture for phosphor screens that conventional slurry photolithography or screen-printing techniques cannot match, even when they use waste-recovery processes. Other materials and processes in phosphor-based flat panel displays, including conductors and sealants, may benefit from ink-jet deposition processes (57).

For color light-emitting polymer (LEP) displays, lithographic, and screen-printing methods are not practicable due to the sensitivity of the polymers to solvents and aqueous solutions, so development efforts by a number of organizations have focused on ink-jet deposition for the three colors of LEP (58,59). As with phosphor-based displays, the use of ink-jet dispensing for other materials and processes, such as the hole injection layer and conductors for the drive circuitry, is being evaluated. Because light-emitting polymers can be used as both sources and detectors, they also have potential for use in microsensors.

7.3.1. Phosphors

As a demonstration of the feasibility of printing phosphors for displays using ink-jet technology, dispersions of micron-sized phosphors (Superior Micro-Powders, Albuquerque, NM) (60) in "inks" suitable for dispensing from an ink-jet device were formulated as follows. Phosphor powders were prepared using either a ceramic media mill or a high-intensity ultrasonic processor. A low-molecular-weight acrylic polymer was used as a binder, and a mixture of organic solvents was used to modify the rheology of the dispersions. The results obtained for Zn_2SiO_4 : Mn are typical and are discussed here.

The phosphor inks were dispensed from demand-mode ink-jet device with an orifice diameter of approximately $50\,\mu m$. Arrays of spots and lines were printed onto glass substrates. After printing, the binder was burned off in an oven, leaving only the phosphor on the surface. The SEMs of Fig. 31, Fig. 32, and Fig. 33 illustrate the resulting patterns. These printed features are smaller than pixels currently used in many phosphor displays, and the high density and uniformity of particles in the pixels, needed for efficient cathodolumines-cence, are apparent.

7.3.2. Light-Emitting Polymers

Light-emitting polymers are a subset of a broad class of conjugated polymers. To construct practical, active devices with these materials, a uniform layer of approximately $1\,\mu m$ must be created in a structure, and the structure must create an electric field across the light-emitting polymer layer. Whether it is deposited in a spin-coating process or by ink-jet deposition, the polymer is usually suspended in low concentrations (0.5–2% by volume) in a volatile organic solvent. After deposition, the solvent is driven off and the polymer film is left behind. Difficulties with optimizing the solvent for solubility, jetting characteristics, and pattern formation have been reported by some investiga-tors (61). These are inherent difficulties of ink-jet systems, including ink-on-

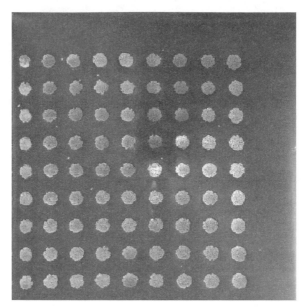

FIGURE 31 SEM of 90-μm spots of ink-jet printed Zn_2SiO_4 : Mn phosphor.

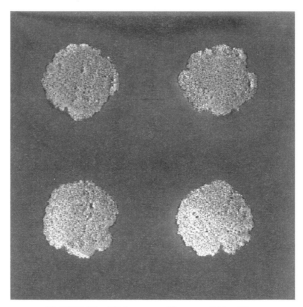

FIGURE 32 Higher magnification of Fig. 31.

FIGURE 33   SEM of 90-µm line of ink-jet printed Zn_2SiO_4 : Mn phosphor.

paper printers. Figure 34 illustrates the spot quality and resolution that can be obtained by ink-jet printing of light-emitting polymers.

7.3.3. Spacers and Adhesives

The high-temperature demand-mode ink-jet process used in printing UV-curing polymer microlenses can be used to create highly controlled spacers in flat panel displays. Figure 35 shows an example of printed spacer bumps that would meet the physical- and thermal (in excess of 200 °C) -durability requirements for flat panel displays. Bumps as small as 25 µm in diameter and 10 µm high can be created, and bumps this size or larger would span the requirements for most spacers in displays.

For spacers, the key parameter to control is the height of the deposited droplets. Height is determined by droplet volume and the degree of spreading

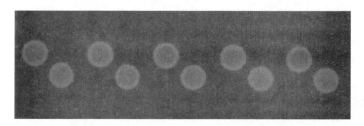

FIGURE 34   75-µm-diameter spots of light-emitting polymer (fluorescing) printed onto glass.

FIGURE 35   Array of 95-μm diameter, 34-μm high, printed spacer bumps, in substrate plane (top) and in profile (bottom).

which occurs on the substrate prior to solidification and curing. Volumetric control for demand-mode ink-jet printing is usually better than 2% (range) for well-behaved fluids. The spreading of deposited material may also be controlled by the use of low-wetting coatings on the target substrates to provide specific contact angles and, additionally, by varying substrate temperature to adjust fluid viscosity prior to solidification.

The printing of adhesives for sealing and bonding of LCD cells can be accomplished by ink-jet printing. Line patterns are printed from data files and aligned to substrate fiducials. The width and height of the cross-section of each printed line is determined by the specified spacings of the deposition sites and number of droplets deposited per site, as seen in Fig. 36.

7.4. Electronic Passives

Not only does the density of circuits on a single chip in an integrated circuit continue to increase, but also multiple functions that used to be implemented onto multiple ICs are now being integrated onto a single IC in many applications. In these cases, discrete passive elements (resistors, capacitors, and inductors) can dominate the component count, take up most of the circuit board real estate, and limit the performance of the device. Direct write of

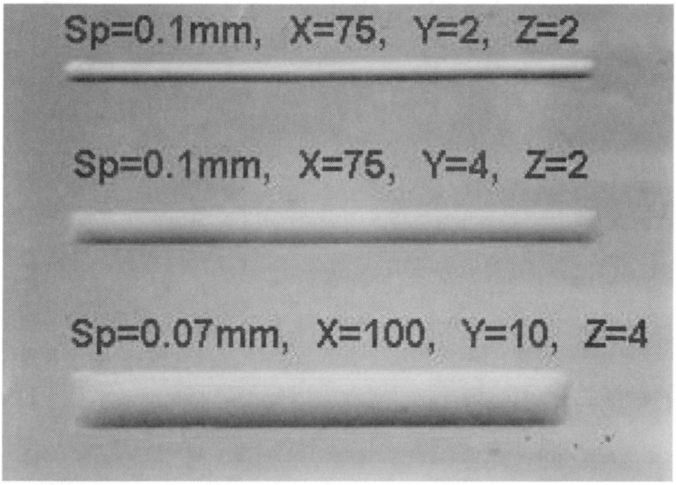

FIGURE 36   Printed thermoset epoxy lines (top is 300-μm wide) made be deposition at 125 °C of 60-μm drops; site spacings (Sp), number of sites in horizontal (X) and vertical (Y) directions, and number of droplets per site (Z) are as indicated.

passives has the potential for decreasing the size and cost of electronic assemblies and, if the passives are embedded, increasing the performance. A number of technologies, including ink-jet technology, are being explored to direct write passive elements.

7.4.1. Resistors

Under a completed National Center for Manufacturing Science research project (62), a UV-curable resistor formulation containing carbon nanotubes was jetted onto a test vehicle into shapes approximating commercial available resistors. An example of the results is shown in Fig. 37, where the resistor dimensions are $4 \times 2.5 \times 0.2$ mm. Thermal cycling data were extremely good for these resistors, due to a Tg of 180 °C for the formulation. However, keeping the nanotubes dispersed was difficult, resulting in standard deviations on the order of 30–40%. Printing these relatively large resistors allowed a pre-existing test vehicle to be employed, but the potential size benefit of a direct-written resistor is lost and throughput is penalized compared to designing the resistor geometry to fit the deposition process.

In an ongoing National Center for Manufacturing Science research project, resistive polymer solutions (aqueous and organic solvent based) are being dispensed to form imbedded resistors on the inner layers of multilayer circuit

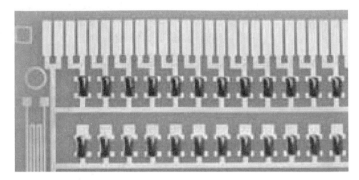

FIGURE 37   0402 resistors printed using MicroJet Technology; filled, UV curing epoxy.

boards. Figure 38 shows one of four test vehicles on an $18'' \times 12''$ core sheet. Resistors ranging from 100 Ω to several MΩ have been created using materials with resistivities as low as 200 Ω/sq. As can be seen in Fig. 38, the size of the smallest resistors printed is an order of magnitude smaller in dimensions than those of Fig. 37 (a requirement for an embedded passive).

7.4.2. Capacitors and Inductors

Capacitors and inductors can be formed onto surfaces (not necessarily planar) using ink-jet printing methods by creating local three-dimensional structures. Figure 39 and Fig. 40 illustrate schematically one method of accomplishing this. For a capacitor, the bottom electrode, dielectric, and top electrode layers are laid down successively, and this process could be repeated to form multilayer capacitors. Both the area and the thickness of the dielectric could be varied to select the value of the capacitance.

For an inductor, a center electrode, ferrite layer, and conductor coil are printed. The inductance would be varied by changing the number of turns in the printed coil.

The difficulty in reduction to practice from these rather straightforward concepts is in the materials required to make capacitors and inductors of practical value. Solder could be jetted as the conductor, but it would only spread into a thin layer on a material that it would wet—in other words, another metal, which would obviate the need for solder. Both organometallic and metal nanoparticle (e.g., silver or silver palladium) solutions can be jetted, but postprocessing temperatures are usually high, with the notable exception of ink-jetable organometallics (gold, copper, palladium, silver) developed at SRI (63). Finally, most high-capacitance (e.g., $Ba_{1-x}Ca_xTi_{1-y}Zr_yO_3$) and high-

FIGURE 38 Embedded conductive polymer resistors printed using ink-jet technology, <200 Ω/sq, ~1 mm long.

inductance materials (e.g., nickel–zinc or manganese–zinc ferrite powders) are ceramics that are sintered at high temperature.

Despite these difficulties, some progress in ink-jet formation of inductors and capacitors has been made. Figure 41 illustrates 250-μm-wide silver nanoparticle lines ink-jet printed onto a ferrite nanoparticle layer, which was also ink-jet printed. Because the conductor lines are parallel in this configuration, they would have to mate with crossover conductor lines in a layer below the ferrite in order to form a spiral conductor path.

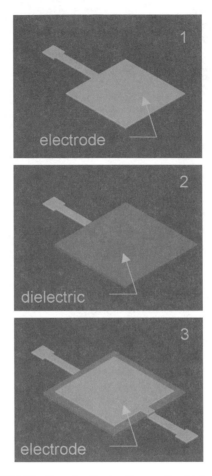

FIGURE 39   Steps for direct-write capacitor fabrication.

7.5. BATTERIES

Because of the ability for some direct-write methods, such as ink-jet, to print onto nonplanar surfaces, recently there has been interest in creating batteries that conform to the shape of the system in which they are employed: credit card GPS, remotely piloted airplane structures, automobile bodies, etc. DARPA is funding direct-write battery development through its Mesoscopic Integrated Conformal Electronics (MICE) program.

Batteries could be printed using ink-jet deposition as shown in Fig. 42. First, an organometallic is printed to form the metal current collectors. Second,

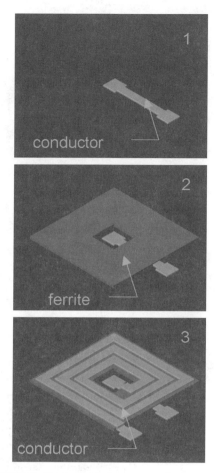

FIGURE 40   Steps for direct-write of inductor.

$LiCoO_2$ ink is printed onto the cathode conductor and a carbon-filled polymer is printed onto the anode. The electrolyte is then printed, covering both anodes and cathodes. The capacity of the battery could be varied by changing the total area printed, and unique shapes could be printed to conform to the features on the substrate.

7.6. BIOACTIVE MATERIALS

Ink-jet printing of bioactive materials has received a great deal of attention in the past two decades because of the rapidly expanding knowledge and

FIGURE 41   250-μm with silver nanoparticle lines ink-jet printed onto ferrite nanoparticle layer
(also ink-jet printed).

capabilities in molecular biology. The key drivers behind the interest in ink-jet
dispensing are the minimal use of rare fluids (both sample and diagnostic
reagents) and parallel processing of a large number of tests, which drives assay
size down and density up.

7.6.1. Immunodiagnostics, Antibody–Antigen Interactions

Early development in ink-jet printing of bioactive fluids (64) centered on
making patterns of antibodies on membrane materials, typically nitrocellulose,
that bound the antibody for use in an assay. The pattern was used as a human
readable display for the assay. Examples include a prototype blood-typing test
shown in Fig. 43. Here, four blood-typing reagents have been printed, using
demand-mode ink-jet, into the characters A, B, and + (the plus contains both
the control and RH positive antibodies), and have been exposed to AB+ blood.
Another example is given in Fig. 44 which shows Abbott's TestPack™ product
line. Here two antibodies (typically, βhCG and a control) are printed onto
nitrocellulose using a two-fluid continuous ink-jet printing system. Over 500
million of these diagnostic test strips have been manufactured to date.

In the early 1990s, with the goals of increasing the number of diagnostic
tests that could be conducted in parallel and of increasing the sensitivity of the
assay by minimizing the amount of analyte bound to the antibody (65),
Boehringer Mannheim (now Boehringer-Roche) Diagnostics developed their
MicroSpot™ system. As many as 196 distinct reactions sites (i.e., spots) would
fit into their disposable reaction well, shown in Fig. 45, and be imaged using a

FIGURE 42 Schematic of direct-write battery process.

fluorescence confocal scanning microscope. The initial pilot line used ten separate ink-jet deposition stations that could deposit ten fluids. Each fluid was printed into multiple spots to provide redundancy, and a real-time inspections system imaged the printed dots using a secondary fluorifor. Figure 46 illustrates the results obtained from two different immunoassays in the MicroSpot™ format (66). Transition of each printing station from one to ten fluids, resulting in a 100-fluid pilot line, was under way in 1999 when the line was shut down.

FIGURE 43 Four-antibody blood-typing test printed using demand mode ink-jet technology.

FIGURE 44 Two-antibody diagnostic assay (pregnancy, Abbott's TestPack™) printed using continuous ink-jet technology.

FIGURE 45 Disposable diagnostic test "well" that can contain as many as 100 individual antibodies (tests) printed using ink-jet technology.

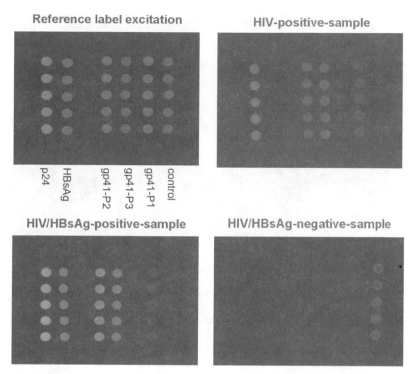

FIGURE 46   Example of MicroSpot™ immunoassay results, obtained using ink-jet deposited spots ~100 μm in diameter. (Image courtesy of Boehringer-Roche Diagnostics.)

7.6.2. DNA and Peptide Arrays

Although Affymetrix's light-activated fabrication method (67) and pin transfer using quill pens have been the most prevalent DNA array fabrication techniques, ink-jet printing methods have been used by a number of organizations, both for synthesis and for deposition of oligonucleotides in microarray format. Deposition of oligonucleotides that are synthesized and verified off-line has been accomplished by using commercially available six-color thermal ink-jet printheads (68); conventional fluid robots that have been modified to hold 4, 8, and possibly up to 96 individual glass capillary piezoelectric demand-mode jetting devices (69); and custom piezoelectric demand-mode array printheads (70). The chief difficulty in deposition of oligonucleotides is the number of fluids that must be dispensed. For very specific genetic resequencing applications (i.e., looking for known sequences or mutations), the number of oligonucleotides that need to be deposited can be as low as ten, and usually would be less than 100. Resequencing applications include clinical diagnostics,

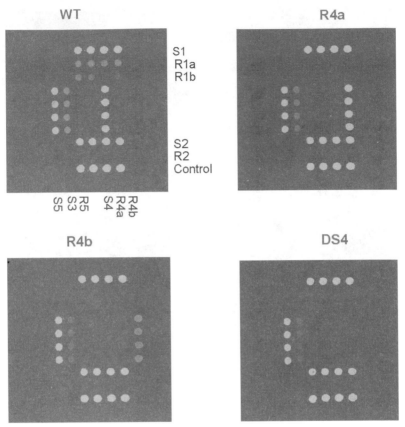

FIGURE 47   Example of MicroSpot™ DNA test for drug-resistant mTB results. Ink-jet deposited spots ~100 µm in diameter. (Image courtesy of Boehringer Roche Diagnostics.)

SNP (single nucleotide polymorphism) detection, point mutation detection, etc. Resequencing assays fabricated using ink-jet deposition of oligonucleotides have been demonstrated for drug resistant *mTB*, as illustrated in Fig. 47.

Synthesis of DNA arrays using ink-jet technology greatly decreases the number of different fluids required. Only the four constituent bases (A, G, C, T) of DNA, plus an activator, are jetted, as illustrated in Fig. 48. The complexity of multistep chemical synthesis in an anhydrous environment is an added problem, but a number of investigators have overcome this difficulty (71,72), and DNA arrays manufactured in this way should be available in the near future.

Peptide arrays for drug and expression screening studies (73) could be formed by ink-jet deposition of presynthesized peptides in a similar manner as DNA and antibody arrays made by deposition, but this has not been reported.

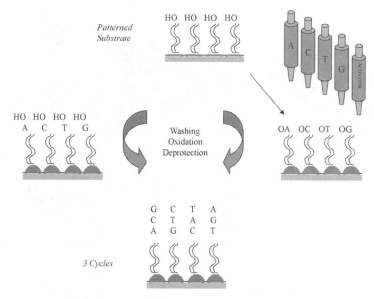

FIGURE 48   Schematic of ink-jet-based DNA array synthesis.

Synthesis of peptide arrays using ink jet can be accomplished in a manner similar to DNA arrays, except that there are 20 naturally occurring amino acids, making the dispensing system more complex, as illustrated in Fig. 49. Multiple companies are pursuing this approach, but have not yet reported on their results.

8. COMMERCIAL SYSTEMS

In many non-ink applications of ink-jet printing technology, the printing system is viewed as a proprietary tool used to manufacture a high value-added product, and part of the value usually includes the intellectual property associated with the ink-jet-based fabrication method. Thus, the availability of commercial ink-jet-based rapid prototyping equipment is limited compared to level of activity in applying ink-jet technology to rapid prototyping.

Commercially available ink-jet-based printing systems that are appropriate to rapid prototyping applications are explained in the following sections and are divided into two categories. In *Products*, systems that are marketed to specific product areas are given. These systems will be more highly developed for their specific target applications, but will also be much less flexible or adaptable to other applications of ink-jet dispensing. Conventional ink-jet printing hardware would also fall into this category.

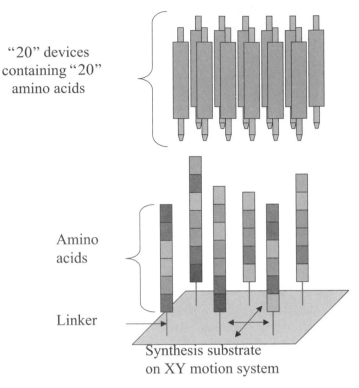

"20" devices
containing "20"
amino acids

Amino
acids

Linker

Synthesis substrate
on XY motion system

FIGURE 49   Schematic of peptide synthesis using ink-jet technology.

Research Systems discusses ink-jet-based hardware that is sold as configurable and/or appropriate to multiple applications. Because they are designed for flexibility and adaptability, they are in general less tailored to any specific application.

In addition to commercial systems, components and subsystems can be purchased from a number of companies for integration into a custom ink-jet printing platform. This includes conventional ink-jet printheads; however, a detailed discussion of such components and subsystems is beyond the scope of this chapter.

8.1. PRODUCTS

8.1.1. Free-Form Fabrication

The ModelMaker II by Sanders Prototype, Inc. (37) uses a demand-mode piezoelectric array printhead dispensing a thermoplastic material (90–113 °C melting point) to build solid models. It also dispenses a support material, also a thermoplastic (54–76 °C), to allow it to construct overhanging structures. The minimum feature size it can achieve is 250 µm, using a 75-µm drop size. The build layer thickness (i.e., the thickness of the slice through the virtual object printed in one pass) is 13–76 µm. The build envelope, or size of the largest part that can be constructed, is $30 \times 15 \times 22$ cm (x, y, and z).

The 3D Systems Thermojet Printer (38) also uses a demand-mode piezo-electric array printhead dispensing a thermoplastic material to build solid models. Printing resolutions (actual addressability) of 300, 400, and 600 dots per inch (drop spacings of 85, 63, and 42 µm) are claimed. The maximum model size $25 \times 19 \times 20$ cm.

The Objet Quadra, recently developed by Objet Geometries Ltd., prints a photopolymer at 600 dots per inch in x and y, and uses a 20-µm layer thickness. The photopolymer is cured during printing so no postprocessing is required. A support material is also printed where required. The maximum model size is $27 \times 30 \times 20$ cm.

Generis GmbH (74) offers a system based on ink-jet dispensing of binder fluid onto particles, similar to the technology developed by MIT (8), except the binder is not burned off after the part is printed. The build envelope is $150 \times 75 \times 75$ cm. They also have a system that prints a water-soluble wax that is ink-jet printed to form the boundary of a poured (in layers) wax part.

8.1.2. Biomedical Diagnostics and Research

Packard Bioscience (75) markets a fluid-handling robot, the BioChip Arrayer, that uses four demand-mode piezoelectric ink-jet devices to aspirate as little as 200 nl into each device from a standard microwell plate, and dispense 350 pl (87-µm diameter) drops. It can dispense onto 5 $1'' \times 3''$ (25×76 mm) glass slides with a 10-µm resolution. Packard has recently announced the second generation of their BioChip Arrayer, but the details were not available as of this writing.

8.2. RESEARCH SYSTEMS

MicroFab Technologies, Inc. (76) manufactures and sells a configurable ink-jet-based printing platform called Jetlab, shown in Fig. 50. Printheads capable of dispensing polymers, solders, or bioactive fluids can be mounted and controlled on this platform. Up to twelve channels (multiplexed) can be independently controlled. Printed areas of 15×15 cm can be addressed at accuracies of 3–10 µm and at rate up to 20 kHz onto a temperature-controlled platen. The system is enclosed to allow for environmental control. Vision

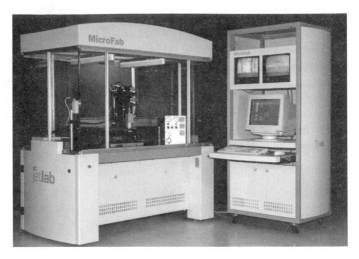

FIGURE 50   MicroFab's Jetlab, a configurable ink-jet-printing platform.

capabilities include substrate/drop alignment, drop-formation setup, and printed pattern inspection.

Microdrop GmbH (77) manufactures and sells a tabletop-dispensing system that can control up to 8 individual dispensers and a 20- × 20-cm *x-y* stage. A vision system allows for observation and setup of the drop-formation process.

Gesim GmbH (78) markets a fluid-handling system, similar to the Packard system (aspirate and dispense), that is more of a research system in its characteristics. It can mount up to 8 single-channel demand-mode piezo-electric dispensers, and dispense onto 30 glass slides (25 mm×75 mm) mounted on the platen. Both 100-pl and 500-pl (60-μm or 100-μm diameters, respectively) drop sizes are available. A vision system allows for observation and setup of the drop-formation process.

9. FUTURE TRENDS

Ink-jet printing technology has become the dominant technology for the home and SOHO (small office/home office) market because of its cost/performance capabilities, the most important of which is color printing of images. However, current and future developments in the conventional ink-jet market are and will be highly targeted to consumer products, and these developments will not directly affect the applicability of ink-jet printing technology to rapid proto-typing applications, where the "printer" is a manufacturing tool.

Developments in two non-ink applications of ink-jet printing are likely to have significant impact on the overall capabilities and availability of ink-jet-based rapid prototyping equipment in the future. The explosion in genetic information from the Human Genome Project has created an ever growing demand for DNA microarrays for gene expression studies, single nucleotide polymorphism (SNP) detection, clinical diagnostics, and other genetic based studies (79). To date, pin transfer methods have dominated the research and low-volume commercial markets for DNA microarrays, and Affymetrix's DNA microarrays fabricated using light-activated synthesis (80) have been widely used by drug companies in their research. However, as the market for DNA microarrays has become more clearly defined, companies such as Agilent and Motorola have begun to develop high-volume DNA microarray production capabilities using ink-jet printing methods. As these and other companies expand their use of ink-jet printing technology for DNA microarrays, standardized and commercially available manufacturing equipment using ink-jet dispensing should become more capable and more generally available. This equipment should be more suitable for, or adaptable to, other rapid prototyping applications than printheads and printers developed for ink deposition.

In the display market, there is a broadly held consensus that, in the future, light-emitting polymer-based displays (81) will displace CRT and LCD displays in many or most existing display products, and will enable a wide range of display products not technically or economically viable today. There is also a broad acceptance that ink-jet printing technology is the most practical method for manufacturing multicolor light-emitting polymer-based displays, and many display manufacturers (Philips, Samsung, Sieko-Epson, Siemens, etc.) are developing ink-jet-based manufacturing methods and equipment. Again, this should result in more capable and generally available ink-jet dispensing for use in standardized and commercially available manufacturing equipment. And this equipment will be more suitable for, or adaptable to, other rapid prototyping applications than are printheads and printers developed for ink deposition.

10. SUMMARY

The ability of ink-jet printing technology to dispense, in a controlled manner, a wide range of materials of interest to rapid prototyping applications has been demonstrated. These materials include optical polymers, solders, thermoplastics, light-emitting polymers, biologically active fluids, and precursors for chemical synthesis. In addition to the wide range of suitable materials, the inherently data-driven nature of ink-jet printing technology makes it highly suited for rapid prototyping applications. Using ink-jet-based methods, commercialization of products is in progress for DNA microarrays, color displays, electronics assembly, and photonic elements. In the future, both the number and type of products fabricated using ink-jet technology should increase, and the availability and capabilities of ink-jet-based prototyping and production tools should expand.

REFERENCES

1. Lewis, A. (1968). U.S. patent.
2. Savart, F. (1833). Memoire sur la constitution des veines liquides lancees par des orifices circulaires en mince paroi. *Annales de Chimie et de Physique* 53, 337–386.
3. Rayleigh, J. W. S. (1879). On the instability of jets. *Proc. London Math Soc.* 10, 4, 4–13.
4. Rayleigh, J. W. S. (1879). On the capillary phenomena of jets. *Proc. Roy. Soc.* 29, 71.
5. Rayleigh, J. W. S. (1892). On the instability of a cylinder of viscous liquid under capillary force. *Phil. Mag.* 34, 145–154.
6. Chaudhary, K. C., Redekopp, I. G., and Maxworthy, T. (1979). The non-linear capillary instability of a liquid jet. *Journ. Fluid Mech.* 96, part II, 257–312.
7. Pimbley, W. T. (1984). Drop formation from a liquid jet: a linear one-dimensional analysis considered as a boundary value problem. *IBM Journ. Res. Dev.* 29, 148–156.
8. Yoo, J., Cima, M., Sachs, E., and Suresh, S. (1995). Fabrication and microstructural control of advanced ceramic components by three dimensional printing. *Ceram. Eng. Sci. Proc.* 16, 5, 755–762.
9. Hansell, C. W. (1950). Jet sprayer actuated by supersonic waves. U.S. patent 2,512,743.
10. Bogy, D. B. and Talke, F. E. (1984). Experimental and theoretical study of wave propagation phenomena in drop-on-demand ink jet devices. *IBM Journ. Res. Develop.* 29, 314–321.
11. Dijksman, J. F. (1984). Hydrodynamics of small tubular pumps. *Journ. Fluid Mech.* 139, 173–191.
12. Adams, R. L. and Roy, J. (1986). A one dimensional numerical model of a drop-on-demand ink jet. *J. of Appl. Mech.* 53, 193–197.
13. Wallace, D. B. (1989). A method of characteristics model of a drop on demand ink jet device using an integral method drop formation model. ASME publication 89-WA/FE-4.
14. Pies, J. R., Wallace, D. B., and Hayes, D. J. (1993). High density ink jet printhead. U.S. patent 5,235,352.
15. Roy, J., and Moore, J. S. (1992). Drop-on-demand ink jet print head. U.S. patent 5,087,930.
16. Hoisington, P. A., Schaffer, R. R., and Fischbeck, K. H. (1989). Ink jet array. U.S. patent 4,835,554.
17. Usui, M., and Katakura, T. (1996). Actuator for an ink jet print head of the layered type with offset linear arrays of pressure generating chamber. U.S. patent 6,033,058.
18. Elrod, S. A., Hadimioglu, B., Khuri-Yakub, B. T., Rawson, E. G., Richley, E., and Quate, C. F. (1989). Nozzleless droplet formation with focused acoustic beams. *J. Appl. Phys.* 65, 9, 3441–3447.
19. Oeftering, R. C. (1994). Directional electrostatic accretion process employing acoustic droplet formation. U.S. patent 5,520,715.
20. Aden, J. S., Bohorquez, J. H., Collins, D. M., Crook, M. D., Garcia, A., and Hess, U. E. (1994). The third generation HP thermal inkjet printhead. *Hewlett-Packard Journal* 45, 1, 41–45.

21. Waldvogel, J. M., Poulikakos, D., Wallace, D. B., and Marusak, R. M. (1996). Transport phenomena in picoliter size solder droplet dispensing on a composite substrate. *J. of Heat Transfer* **118**, 148–156.

22. Deegan, R. D., Bakajin, O., DuPont, T. F., Huber, G., Nagel, S. R., and Witten, T. A. (1997). Capillary flow as the cause of ring stains from dried liquid drops. *Nature* **389**, 827–829.

23. Nguyen, L. T. (1996). Advanced packaging needs for the year 2000. Gorham/Intertech's *Microelectronic Packaging into the 21st Century* (Atlanta).

24. 1998 National Electronics Manufacturers Initiative Roadmap (1998) (*www.nemi.com*).

25. Tuantranont, A., Bright, V. M., Zhang, W., Zhang, J., and Lee, Y. C. (1999). Self-aligned assembly of microlens arrays with micromirrors. *Proc. SPIE-Int. Soc. for Opt. Eng.* **3878** Micromachining and Microfabrication (Santa Clara) 90–100.

26. van Veen, N., and Schwarzbach, D. (1999). Solderjetting, a software driven technology for maskless waferbumping. *Proceedings, IMAPS International Symposium on Microelectronics* (Chicago) 154–159.

27. Argento, C. W., Flynn, T., and Demers, C. (1999). Next generation solder jetted wafer bumping for very fine pitch flip chip technology applications and beyond. *Proceedings, IMAPS International Symposium on Microelectronics* (Chicago) 160–165.

28. Godin, R., Pearson, S., and Lasky, R. (1997). A novel process for solder deposition. *SMT Magazine* **1**, 66–68.

29. Muntz, E. P., Orme, M., Pham-Van-Diep, G., and Godin, R. (1997). An analysis of precision, fly-through solder jet printing for DCA components. *Proc. ISHM 1997* (Philadelphia), 671–680.

30. Priest, J., Jacobs, E., Smith, C., DuBois, P., Holt, B., and Hammerschlag, B. (1994). Liquid metal-jetting technology: Application issues for hybrid technology. *Int. J. Microcir. and Elect. Pack.* **15**, 3, 219–227.

31. Hendricks, C. (1982). Inertial confinement fusion targets. *Proc. of the 2nd International Colloquium on Drops and Bubbles.* NASA-CR-168848, JPL 82-7. 88–93.

32. Smith, T. M., and Winstead, R. E. (1995). Electrodynamic pump for dispensing molten solder. U.S. patent 5,377,961.

33. Schiesser, T., Menard, E., Smith, T., and Akin, J. (1994). Microdynamic solder pump: drop on demand eutectic SnPb solder dispensing device. *Proc. Surf. Mount Int.* (San Jose, CA) 501–509.

34. Hayes, D. J., Wallace, D. B., Boldman, M. T., and Marusak, R. M. (1993). Picoliter solder droplet dispensing. *Microcir. and Elect. Pack.* **16**, 3, 173–180.

35. Hayes, D. J., Cox, W. R., and Grove, M. E. (1998). Microjet printing of polymers and solder for electronics manufacturing. *J. Elect. Manuf.* **8**, 3&4, 209–216.

36. Wallace, D. B. (1995). Method and apparatus for forming microdroplets of liquids at elevated temperatures. U.S. patent 5,415,679.

37. Sanders Prototype, Inc. (2000) (*www.sanders-prototype.com*).

38. 3D System (2000), (*www.3dsystems.com.*)

39. Cox, W. R., Hayes, D. J., Chen, T., and Ussery, D. W. (1995). Fabrication of micro-optics by micro-jet printing. *SPIE Proceedings* **2383**, 110–115.

40. Cox, W. R., Hayes, D. J., Chen, T., Trost, H-J., Grove, M. E., Hoenigman, R. F., and MacFarlane, D. L. (1997). Low cost optical interconnects by micro-jet printing. *IMAPS Int. J. Microcirc. & Elect. Pack.* **20**, 2, 89–95.

41. Daly, D., Stevens, R. F., Hutley, M. C., and Davies, N. (1990). The manufacture of microlenses by melting photo-resist. *Meas. Sci. Technol.* **1**, 759–766.

42. Cox, W. R., Chen, T., Ussery, D., Hayes, D. J., Tatum, J. A., and MacFarlane, D. L. (1996). Microjetted lenslet-tipped fibers. *Opt. Commun.* **123**, 492–496.

43. Baukens, V., Goulet, A., Thienpont, H., Veretennicoff, I., Cox, W. R., and Guan, C. (1998). GRIN-lens based optical interconnection systems for planes of micro-emitters and detectors: Microlens arrays improve transmission efficiency. *Proc. OSA Diffractive Optics and Micro-Optics Topical Meeting* (Kailua-Kona, HW).

44. Ishii, Y., Koike, S., Arai, Y., and Ando, Y. (2000). Hybrid integration of polymer microlens with VCSEL using drop-on-demand technique. *In* "Optoelectronic Interconnects VII; Photonics and Integration II" (M. R. Feldman, R. L. Li, W. B. Matkin, S. Tan, Eds.), *Proc. SPIE* **3952**, 364–374.

45. Cox, W. R., Chen, T., Guan, C., Hayes, D. J., Hoenigman, R. E., Teipen, B. T., and MacFarlane, D. L. (1998). Micro-jet printing of refractive microlenses. *Proc. OSA Diffractive Optics and Micro-optics Topical Meeting* (Kailua-Kona HW) paper #I-00002.

46. Cox, W. R., Chen, T., Ussery, D. W., Hayes, D. J., Hoenigman, R. F., MacFarlane, D. L., and Rabonivich, E. (1996). Microjet printing of anamorphic microlens arrays. *Proc. SPIE* **2687**, 89–98.

47. Liu, Y., Strzelecka, E. M., Nohava, J., Kalweit, E., Chanhvongsak, H., Marta, T., Skogman, D., Gieske, J., Hibbs-Brenner, M. K., Rieve, J., Ekman, J., Chandramani, P., Kiamilev, F. E., Fokken G., Vickberg, M., Gilbert, B. K., Christensen, M. P., Milojkovic, P., and Haney, M. W. (2000). Component technology and system demonstration using smart pixel arrays based on integration of VCSEL/photodetector arrays and Si ASICs. *SPIE Photonics West Symposia.* Paper #CR 76-05, January 25, 2000.

48. Cox, W. R., Chen, T., Ussery, D., Hayes, D. J., Tatum, J. A., and MacFarlane, D. L. (1996). Microjetted lenslet-tipped fibers. *Opt. Commun.* **123**, 492–496.

49. Cox, W. R., Guan, C., Hayes, D. J., and Wallace, S. B. (2000). Microjet printing of micro-optical interconnects. *International Conference on High Density Interconnect and Systems Packaging, HDI 2000,* paper 60720007.

50. Hecht, J. (1999). Fiberoptic communications: an optoelectronics driver. *Laser Focus World,* January, 143–151.

51. Towe, T., and Cai, S. (1999). Gradient index lenses make light work of beam directing. *Laser Focus World,* October, 93–96.

52. Manhart, P. K., and Blankenbecler, R. (1997). Fundamentals of macro axial gradient index optical design and engineering. *Opt. Eng.,* **36**, 6, 1607.

53. Pagano, R. J., Perkins, K., and Manhart, P. K. (1995). Axial gradient-index lenses come of age. *Laser Focus World,* May, 191–196.

54. Cox, W. R., Guan, C., and Hayes, D. J. (2000). Microjet printing of micro-optical interconnects and sensors. *SPIE Photonics West 2000,* paper 3952B-56.

55. Coleston Jr., B. W., Gutierrez, D. M., Everett, M. J., Brown, S. B., Langry, K. C., Cox, W. R., Johnson P. W., and Roe, J. N. (2000). Intraoral fiber-optic-based diagnostics for periodontal disease. *In* "Biomedical Diagnostic, Guidance, and Surgical Assist Systems II" (T. Vo-Dinh, W. Grundfest, and D. Benaron, Ed.), *Proc. SPIE* **3911**, 2–9.

56. Walt, D. R., and Bronk, K. S. (1994). Thin-film fiber optic sensor array and apparatus for concurrent viewing and chemical sensing of a sample. U.S. patent 5,389,741.

57. Cox, W. R., Chen, T., Ussery, D., Hayes, D. J., Tatum, J. A., and MacFarlane, D. L. (1996). Microjetted lenslet-tipped fibers. *Opt. Commun.* **123**, 492–496.

58. Bharathan, J., and Yang, Y. (1998). Polymer electroluminescent devices processed by inkjet printing: I. Polymer light-emitting logo. *Appl. Phys. Lett.* **72**, 21, 2660–2662.

59. Hebner, T. B., Wu, C. C., Marcy, D., Lu, M. H., and Sturm, J. C. (1998). Ink-jet printing of doped polymers for organic light emitting devices. *Appl. Phys. Lett.* **72**, 5, 519–521.

60. Grove, M., Hayes, D., Cox, R., Wallace, D., Caruso, J., Hampden-Smith, M., Kodas, T., Kunze, K., Ludviksson, A., Pennino, S., and Skamser, D. (1999). Color flat panel manufacturing using ink jet technology. *Proceedings, Display Works '99* (San Jose, CA) 131–134.

61. Miyashita, S., Kanbe, S., Kobayashi, H., Seki, S., Kiguchi, H., Morii, K., Shimoda, T., Friend, R. H., Burroughes, J. H., and Towns, C. R. (1999). Patterning of light-emitting poymers for full color displays by ink-jetting. *Proceedings, The 1999 International Conference on Display Phosphors and Electroluminescence.*

62. Cox, W. R. (1999). Microjet printing of resistor material formulations. *In* "Final Report: Polymer Thick Film Electronic Component Materials," CRADA, National Center for Manufacturing Sciences.

63. Sharma, S. K. (1999) *(personal communication)*.

64. Hayes, D. J., Wallace, D. B., VerLee, D., and Houseman, K. (1989). Apparatus and process for reagent fluid dispensing and printing. U.S. patent 4,877,745.

65. Ekins, R. (1994). Immunoassay: recent developments and future directions. *Nucl. Med. Biol.* **21**, 3, 495–521.

66. Eichenlaub, U., Berger, B., Finckh, P., Karl, J., Hornauer, H., Ehrlich-Weinreich, G., Weindel, K., Lenz, H., Sluka, P., and Ekins, R. (1998). Microspot—A highly integrated ligand binding assay technology. *Proceedings, Second International Conference on Microreaction Technology,* New Orleans, LA (W. Ehrfeld, I. H. Renard and R. S. Wegeng, Ed.), The American Institute of Chemical Engineers, New York, 134–138.

67. Chee, M., Yang, R., Hubbell, E., Berno, A., Huang, X. C., Stern, D., Winkler, J., Lockhart, D. J., Morris, M. S., and Fodor, S. A. (1996). Accessing genetic information with high-density DNA arrays. *Science* **274**, 610–614.

68. Okamoto, T., Suzuki, T., and Yamamoto, N. (2000). Microarray fabrication with covalent attachment of DNA using bubble jet technology. *Nat. Biotech.* Volume 18, 4, 438–441.

69. James, P., and Papen, R. (1998). A new innovation in robotic liquid handling. *Drug discovery Today* **3**, 9, 429–430.

70. Cooley, P. W., Hinson, D., Trost, H-J., Antohe, B., and Wallace, D. B. (2001) Ink-jet deposited microspot arrays of DNA and other bioactive molecules. *In* "Methods in Molecular Biology, Vol. 170: DNA Arrays, Methods, and Protocols" (J. B. Rampal, Ed.), Humana, Totowana, NJ.

71. Blanchard, A. P., Kaiser, R. J., and Hood, L. E. (1996). High density oligonucleotide arrays. *Biosens. Bioelectron.* **11**, 687–690.

72. Brennan, T. M. (1999). Apparatus for diverse chemical synthesis using two-dimensional array. U.S. patent 6,00,1311.

73. Frank, R., Overwin, H. (1996). Spot synthesis: Epitope analysis with arrays of synthetic peptides prepared on cellulose membranes. *Meth. Mol. Biol.* **66**, 149–169.

74. Generis GmbH. *(www.generis.de)*.

75. Packard Bioscience *(www.packardbioscience.com)*.

76. MicroFab Technologies, Inc. *(www.microfab.com)*.

77. Microdrop GmbH. *(www.microdrop.de)*.

78. Gesim GmbH. *(www.gesim.de)*.

79. Aschheim, K. (2000). Gene detection by array. *Nat. Biotechnol.* 18, 1129.

80. Fodor, S. P. A., Read, J. L., Pirrung, M. C., Stryer, L., Tsai Lu, A., and Solas, D. (1991). Light-directed, spatially addressable parallel chemical synthesis. *Science* **251**, 767–773.

81. Philips readies the lines for LEP-based display. *Electronics Times* June 29 (1999).

CHAPTER 8

Micropen Printing of Electronic Components

PAUL G. CLEM, NELSON S. BELL, GEOFF L. BRENNECKA, AND DUANE B. DIMOS

Sandia National Laboratories, Albuquerque, New Mexico

BRUCE H. KING

Optomec, Inc., Albuquerque, New Mexico

1. INTRODUCTION

The need for fabrication of high-density hybrid microelectronics increasingly requires diverse materials integration, rapid prototyping, and use of nontraditional substrates, which complicates the use of methods such as screen printing and lithography. One promising approach to these challenges is CAD/CAM controlled direct writing of fluid slurries, by such methods as Micropen™ deposition and "drop on demand" inkjet printing. Such direct write approaches currently are capable of 50-μm resolution printing of a breadth of materials, including conductors, high-value resistors, dielectrics, magnetic materials, and chemically sensitive elements. In particular, micropen deposition enables printing of multilayer material structures on nonplanar substrates, enabling high-density circuitry with integrated passive components. This chapter presents the technologies instrumental to such material integration by the Micropen, including equipment operation, slurry preparation, deposition constraints, and cofiring of multimaterial integrated passives. Application of

Direct-Write Technologies for Rapid Prototyping Applications

these principles will be shown to be instrumental for four example device elements: high-precision resistors, high-capacitance density dielectrics, integrated inductor coils, and chemical sensors.

2. THE MICROPEN

The Micropen™ system (OhmCraft, Inc., Honeoye Falls, NY) is a computer automated device for precision printing of liquids or particulate slurries (1), illustrated in Fig. 1. The system uses a computer-driven x-y stage for printing, where the print pattern is defined by a CAD instruction file. The AutoCAD™ print file can be easily modified, which permits on-line design changes, in contrast to screen printing, where a new screen is required for a pattern change. The cross-sectional area of printed features is determined by the nozzle dimensions and slurry pump rate. The Micropen can use nozzle sizes (Fig. 2) from 2 to 100 mils (50–2,500 μm) to obtain different print geometries, such as fine line traces, or filled dielectric regions. The finest nozzle for high-definition patterns has an inner diameter of 1 mil and an outer diameter of 2 mils, and requires well-dispersed suspensions of very fine particle size to print through this tip. The slurry is delivered to the print head by a pump block, which uses two internal chambers to provide smooth, continuous delivery of slurry. Slurries are easily loaded into a syringe which screws into the pump block

FIGURE 1   Photograph of the Micropen printhead during a printing operation.

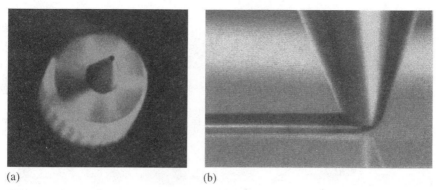

(a)   (b)

FIGURE 2   Views of Micropen tip (a) looking down inner diameter, and (b) in a print operation.
Reprinted with permission of Ohmcraft, Inc., Honeoye Falls, New York.

assembly. A key to obtaining uniform and reproducible processing is elimination of air bubbles in the slurry which is accomplished by centrifuging the syringe and bleeding air from the pump block.

The Micropen uses force feedback control on the pen tip to control printing, and maintain a constant force on the pen tip, enabling conformal writing. A schematic of the Micropen print process is shown in Fig. 3. Feedback control is achieved by balancing the upward force on the pen due to the extruding slurry

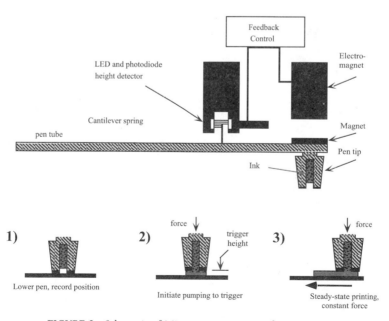

FIGURE 3   Schematic of Micropen apparatus, and writing start-up.

and a downward force applied by an electromagnet. This control leads both to excellent control of the print thickness, and the ability to print over variations in the topography (i.e., height) of the workpiece. The system is also inherently capable of laying down multiple materials in a single layer, which cannot be done with conventional tape casting techniques. However, because the workpiece needs to support the force of the pen, underlying slurry layers must be dried before printing on top of them. Therefore, the system is not directly suitable for continuous processing of multimaterial structures, in contrast to other rapid prototyping techniques, but it is ideally suited for step and repeat procedures, which are appropriate for multilayer, multimaterial electronic components. Other pen-dispense systems are produced by Asymtek and Sciperio, with differing degrees of resolution, vision control systems, processing flexibility, and materials options.

3. RHEOLOGICAL CHARACTERISTICS OF THICK-FILM PASTES

In the production of thick-film pastes, the flow response of the suspension should be tailored to the production method and the desired resolution of the product. Screen-printing formulations are designed to achieve conductor resolutions $\sim$125 µm and dielectric resolutions of 230–250 µm with high reproducibility (2–6). Below the optimal viscosity region for printing, features tend to spread on the substrate and lose resolution. Above the optimal region, there is difficulty in separating the screen from the substrate. The Micropen uses an extrusional method for depositing material, and there are unique writing parameters for each pen tip size, cross-section, writing speed, and paste material. There are several common characteristics in the pastes used in screen printing and the Micropen, and many pastes developed for screen printing are suitable for Micropen deposition. The following section is intended to provide a background in the formation of thick film pastes, and how rheological properties can be tailored to achieve component resolution. A theoretical and phenomenological understanding of slurry rheology enables control of slurry flow reliability and ultimately the final resolution and morphology of printed components.

3.1. PASTE RHEOLOGY

The ingredients of printing pastes include the powder material, solvent, dispersant, viscosity modifiers (i.e., binder), wetting agents, and drying-rate control agents. All these components interact to affect the forces between particles and the solvent with a resultant effect on flow response. In general, it is desired to form a paste with a high solid loading, time-stable rheology, and a homogeneous material distribution. Processing aids are chosen to achieve this state. The first consideration in developing a custom paste for Micropen deposition is finding a method for dispersion of the solid particles. All particles (of the same material) will be attracted to each other through van der Waals interactions (7,8). These forces tend to form loose aggregates of particles with a characteristic fractal structure (9,10), and based on the magnitude of the attractive force, the stress required to make particles flow past each other can prevent printing operations. Clogging of the printing tip is clearly impacted by particle aggregation. Additionally, these aggregates will tend to maintain or reform the aggregated state after printing, and the resultant microstructure will be porous and/or experience significant shrinkage during firing. Percolation theory for the formation of gelled particle networks states that the minimum concentration of spherical solid particles to form an elastic network is ~16 volume % (11). Typical thick-film pastes for screen-printing operations have approximately 40 volume % particulate, and achieving a free-flowing dispersion at this concentration requires a barrier to particle agglomeration.

Dispersant choice for any system requires consideration of the nature of both the solid and solvent phases. In particular, the solvent polarity dictates the applicability of each type of stabilization. In polar solvents, for example, the dissolution of ions is generally supported and particles can have a significant electrostatic charge at the solid-solution interface. If the charge at the surface generates sufficient electrical potential, the resulting "cloud" of counterions will create an osmotic pressure sufficient to prevent particles from approaching one another closely enough to become agglomerated (12). In the case of many thick-film pastes, however, the solvent is generally nonpolar and electrostatic forces are not easily generated. These systems utilize the adsorption of a polymer on the surface to create an entropic barrier to agglomeration (12). The general mechanisms for each type of dispersion are well known, models exist for the calculation of the forces, and they have been measured directly by methods such as the surface forces apparatus and the colloidal probe technique in the atomic force microscope (13). However, the quantitative prediction of the effectiveness and efficiency of a dispersant has not been achieved.

The first and most time-consuming effort in developing a paste is the determination of the optimal dispersant for the solvent system. The dispersion of the solid particles depends on the adsorption of the polymer to the solid, and on the interaction range of the extended layers. The polymer extension relates to the solvency or chemical interaction between the solvent and polymer. This is complicated in thick-film pastes because the solvent is generally composed of more than one solvent component. Also, the drying-rate modifiers or humectant phases are generally higher molecular weight solvents with low vapor pressure, and the solvency parameters of the dispersant can be non-uniform. Wetting agents and viscosity modifiers must also be chemically compatible with the dispersant. The number of variables can become quite large when developing thick-film pastes, and a pragmatic approach is usually adopted by testing a number of sample compositions. However, it is vital to achieve the dispersed state in the beginning in order to maximize solids loading, deposit high-resolution features, and maintain material uniformity.

In all types of suspension-forming techniques of particulate materials, the rheological properties of the concentrated suspension play a key role in controlling the shape-forming behavior and optimizing the properties of the green body. These properties are measured using instruments (viscometers or rheometers) that apply a shear stress to a thin layer of the suspension under various shear rates and determine the coefficient known as the viscosity (14), with

$$\tau_{xy} = \eta \dot{\gamma}_{xy} \tag{1}$$

where τ_{xy} represents the shear stress, η represents the viscosity and $\dot{\gamma}_{xy}$ is the shear rate. Fundamentally, the rheological properties of concentrated colloidal suspensions are determined by interplay of thermodynamic and fluid mechanical interactions. This means there is dependence of the rheological response on particle interactions, including Brownian motion, and the suspension structure (i.e., the spatial particle distribution in the liquid). With particles in the colloidal-size range (at least one dimension < 1 μm), the range and magnitude of the interparticle forces have a profound influence on the suspension structure and, hence, the rheological behavior (15–17). Both fluid mechanical interactions and interparticle forces are strongly dependent on the average separation distance between suspended particles. Due to this, the rheological behavior of concentrated suspensions is a strong function of solids concentration, particle size and shape, and the range and magnitude of the interparticle forces. Following are several important features that relate to the colloidal processing of ceramics and metals from solution, particularly the rheological behavior (viscosity versus shear rate) of suspensions in different solvent/dispersant formulations.

Concentrated, stable suspensions have a particle structure dominated by the interaction energy due to forces between particles and the thermal energy in the system. This tends to form a random particle structure with low packing density within the constraint of the total solids loading. Under the influence of shear stress during a measurement, these suspensions typically display a decreasing viscosity with shear rate ($\dot{\gamma}$) or shear thinning response because of a perturbation of the suspension structure (18). The viscous forces resulting from solvent motion affect the suspension structure by aligning particles in shear bands, and the shear stress to cause particles to flow is lowered. At very high shear rates, these viscous forces dominate and a plateau in viscosity measures the resistance to flow of a suspension with a completely shear-controlled structure. Both the degree of shear thinning and the viscosity at high shear rates increase with an increasing volume fraction of solids.

Based on this conceptual model regarding the effect of shear on the suspension structure, it is expected that models containing low, η_0, and high η_∞, shear viscosity plateaus with a shear-thinning region in between should describe the steady shear behavior of colloidally stable, concentrated suspensions. The Cross equation (19,20) is a model of this type.

$$\frac{\eta - \eta_\infty}{\eta_0 - \eta_\infty} = \frac{1}{1 + \left(\dfrac{\dot{\gamma}}{\dot{\gamma}_c}\right)^m} \tag{2}$$

Here, $\dot{\gamma}_c$ is a critical shear rate related to the Peclet number. This model applies to a suspension of the "hard sphere" type, in which there are no long range interactions and the particles interact with each other only when forced into contact. Often the pastes used in thick-film applications will behave according to the Herschel-Bulkley model (21,22).

$$\tau = \tau_0 + k\dot{\gamma}^n \tag{3}$$

The parameter τ_0 is known as the yield stress of the suspension, and its value relates to the attractive potential between particles and the solids loading through the average number of neighboring particle contacts. A stress of this magnitude must be applied to initiate flow, which thereafter goes through the process of particle restructuring and shear-thinning viscosity. This behavior is characterized by a linear relationship between $\log (\tau - \tau_0)$ and $\log \dot{\gamma}$ with a slope n differing from unity (14).

For a dispersed suspension, the Kreiger-Dougherty law relates the volume fraction dependence of viscosity (19).

$$\eta_{rel} = \left(1 - \frac{\phi}{\phi_M}\right)^{K\phi_M} \tag{4}$$

Each powder has a maximum packing fraction ϕ_M based on the distribution of particle sizes and their shapes. For uniform spheres in a random close-packing arrangement, ϕ_M is predicted to be 63.9 volume %. Under the shear induced ordering of the high shear plateau, the maximum packing fraction is typically higher, reaching 71 volume %. As the particle shape differs from the ideal spherical shape, the maximum packing fraction will decrease. This relationship states that the regions of uniform structure (the low and high shear rate plateaus) will increase in viscosity with solids fraction until they asymptotically approach the maximum packing limit. Solids loading can have an exponential effect on the low and high shear rate viscosity, and is an important parameter in tailoring rheological response. Particle shape is also important, as anisotropic particles display greater changes in shear-thinning response.

An example of the effect of volume fraction and dispersion on viscosity is shown in Fig. 4. This dispersion of silver spheres in dimethylacetamide shows a logarithmic increase in viscosity with shear rate as solids loading is increased. The dispersant used in this system is a triblock copolymer of polyethylene oxide and polypropylene oxide that adsorbs to the solids via the insolubility of the PPO block in the solvent. As solids loading is increased, the particles are forced into close proximity, and flow requires more shear-stress. This dispersant is not optimal because it does not bond strongly to the silver surface. At the high solid content, particles are agglomerating into larger flow units which require more stress to shear and flow. In the shear rate range studied here, the Newtonian plateaus of constant viscosity are not achieved. This occurs when the interparticle attractive forces are long range and continue to affect the particle structure so that increasing shear rates continue to develop the fluid-controlled particle structure in the suspension.

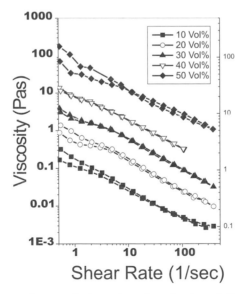

FIGURE 4 Viscosity vs. shear rate for increasing volume fractions of silver spheres in dimethyl-acetamide and Pluronic F87 as dispersant. Power law behavior is exhibited between shear rates of ~3 and 200 sec^{-1}. At low shear rates, the formation of the random structure leads to hysteresis between rising and falling shear rate measurements.

Characteristics that are typically measured for a paste composition are the yield stress to initiate flow, the viscosity as a function of shear rate, and the compressive yield stress. The yield stress has an obvious relation to the printing process as it influences both the pressure to initiate flow and the flow profile during extrusion. The shear-rate dependence of viscosity relates to the printing process in two ways. During transfer it obviously affects the passage through the pen tip. However, it also plays a role in the final deposition of the extrudate, because once printed, the paste will try to conform to the constraints of surface energy with the substrate, and the low shear viscosity is the resistance coefficient during this process. A high value of the low shear viscosity will cause a paste to retain the printed profile. This is advantageous for retaining deposition, but can create surface roughness if the paste cannot flow under the influence of gravity. Finally, the compressive yield stress is a parameter that relates to the stress needed to cause the particle network to densify. Particle migration and the development of gradient-solids concentration occurs during the deposition process and can become a concern for high-resolution printing. It is important to control the printing process so that the paste does not experience this stress in order to prevent the formation of clogs or a nonuniform material deposit.

3.2. PASTE DEPOSITION

The Micropen can be made to produce similar print results for pastes with a wide range of rheological behavior, but an understanding of paste behavior during extrusion is essential for reliable, reproducible results. During deposition, the paste experiences varying shear stresses and rates which affect printing parameters. Exact calculation requires simulation of the die dimensions and solution of the Navier-Stokes equations for the extrudate, but in this study we present more analytical expressions and fundamental predictions of particle behavior during extrusion.

The pressure required to extrude a paste relates to the yield and flow properties of the paste as well as to the constraints of the die dimensions. The theory for flow in straight capillaries has been extensively developed. Figure 5 illustrates the forces involved in flow in a straight capillary. The pressure required must overcome the frictional stresses at the wall and the capillary force of the die. The extrusion pressure (P) for an extrusion (writing) speed v is correlated by the Benbow equation (23).

$$P = 2\tau_y \ln\left(\frac{R_0}{R}\right) = 4(\tau_y + \beta v)\frac{L}{2R} \qquad (5)$$

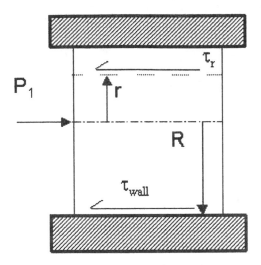

FIGURE 5   Stress diagram for pipe flow.

This equation is derived as the entrance pressure drop to flow from a capillary of size R_0 to a capillary of size R, plus the pressure drop in a capillary of length L, with v as the average velocity in the capillary, and β a constant. This equation introduces an entrance yield stress (τ_0) postulated to have a linear relationship to extrusion speed, α where τ_{y0} is the entrance yield stress at zero extrusion speed.

$$\tau_y = \tau_{y0} + \alpha v \qquad (6)$$

In Shouche et al. (24), die-entry stresses were found to be one to two orders of magnitude larger than the Bingham yield stress, and increased nearly linearly with extrusion speed. The entrance yield stress and extrusion speed was found to have an exponentially increasing relationship with the volume fraction of the solid. The increase became significant at 30 volume % and above. Thick-film inks tend to have a solid loading of ~40 volume %. Considering the small size of the die in the Micropen, it is not difficult to apply pressures of these magnitudes to the paste, but the shear stress profile and flow in the paste will be affected. For very fine tips, these issues become important with respect to forming clogs in the tip.

The shear stress profile for a Newtonian fluid in a capillary has a null value at the die axis, and a linear relationship to a wall stress τ_w.

$$\tau = \tau_w \frac{r}{R} \qquad (7)$$

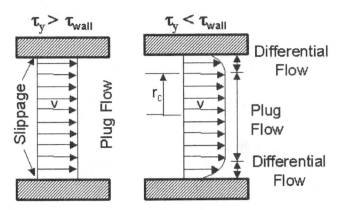

FIGURE 6   Types of flow profiles—plug flow with wall slip or differential flow.

where R is the capillary radius and r is the displacement from the center. For a particle suspension with a yield stress, the stress profile is expected to produce either of two profiles shown in Fig. 6. If the wall stress is lower than the suspension yield stress, plug flow with wall slip is predicted. For those cases where wall stress is higher than the yield stress, differential flow at the periphery is predicted with plug flow in the center. In straight capillaries, wall stress is predicted from the pressure drop ΔP and the capillary length L.

$$\tau_w = \frac{R\Delta P}{2(L + NR)} \tag{8}$$

This expression $2NR$ is the end-effects correction, where N is the equivalent length associated with the end correction (25). It shows the general relationship for generating the wall stress. For fine pen tip sizes, the linear relationship of the shear stress profile suggests that high wall stresses are not likely, and plug flow with wall slip is the expected flow profile in Micropen tips.

When highly loaded particle suspensions are forced through fine capillaries, the volume fraction of solids across the capillary is not expected to remain uniform. Model systems have been examined *in situ* by NMR (26) and particles tend to migrate to the region of plug flow in the center of the capillary. This raises the volume fraction in the center and creates a lubricating layer of increased liquid vehicle at the die wall. Such particle migration is believed to result from shear induced diffusion as described by Leighton and Acrivos (27). In nonuniform shear fields, particles are predicted to migrate from regions of high shear stress to lower shear stress. These effects have been observed in a number of studies (28–30). The diffusion coefficient of particles relates to the parameter $a^2\dot{\gamma}$, where a is the particle radius and $\dot{\gamma}$ is the applied shear rate.

Particle migration is particularly important in the extrusion of suspensions with high solids content, and the migration phenomenon has significant impact on the formation of flow instabilities and wall slip (31–34).

Studies of extrusion of silicon carbide particles in melted wax (injection molding) show a fluid layer on the exterior of the extrudate which correlates well with predictions of the thickness of the fluid layer caused by flow (35). The lubricating layer of particle-depleted solvent is often a desirable extrusion condition. Unstable flow is associated with the formation of particle aggregates and filtration of the vehicle. The preferential extrusion of solvent increases the pressure required to resume flow. Suwardie et al. found that flow instabilities associated with filtration occur when filtration viscosity is comparable to the bulk viscosity of the suspension flowing in the capillary (35). Yaras et al. described the formation of "mats" as related to the presence of unstable extrusion pressure, and above a critical apparent shear rate, these instabilities are eliminated (31).

Controlling clog formation by controlling the interparticle forces is an issue that has not been extensively studied, because most fundamental studies consider particles as hard spheres with little interparticle interaction. Ramachandran et al. have presented interesting studies on the role of hydrodynamic forces at pore entrances on the formation of pore-bridging structures (36,37). Even in systems where colloidal forces are predicted to stabilize particles and prevent particles from agglomerating at the pore surface, the hydrodynamic forces at the entrance are strong enough to cause bridging networks to form. These studies were performed with particles and pores with an aspect ratio of 3:4 (meaning 3:4 particles are required to bridge the average pore) and though this is far from the experimental situation in the Micropen, it is important to note that these hydrodynamic forces are limiting the deposition process.

This all holds in a straight capillary, but the Micropen tips have an approximately 15° angle, so they will undergo continuous compression of the particles, as well as increasing acceleration of the flow field. There is significant potential to develop either a dilatent, or solid aggregate (through hydrodynamic agglomeration) at the plug flow region, and clog the tip. Experimental studies of clogging of the Micropen tips show an approximate relationship between particle size and tip size differing by approximately two orders of magnitude. Particles need to be approximately 1/100th the size of the interior diameter of the tip in order for stable flow to be observed. The 1-mil tips have only been successfully printed using a paste of silver or gold particles with an average diameter of 100 nm.

Elimination of tip clogging requires prevention of particle migration, and prevention of filter pressing. From the studies on capillary flow, flow instabilities are eliminated by increasing the matrix viscosity, increasing the shear stress at the wall, increasing die diameter, and by using smaller, dispersed particles. Using a liquid vehicle with higher viscosity will tend to create a shear rate response curve that is Newtonian, and this response is desired for those situations where the printed material should flow together such as in resistor pads. The elimination of solvent filtration is also inhibited by using high-molecular-weight polymers to increase the solvent viscosity. These agents thicken the solvent and reduce the diffusion through the particle network. If the thickening agents also interact with the particles, they can have the additional effect of reducing the permeability of the particle network and retarding solvent filtration. Thickening agents are suited for applications where high yield stress is desired to maintain the definition of printed materials.

3.3. SURFACE ENERGY EFFECTS

In fine-resolution printing, the effects of surface energy and capillarity will play a role in the structures that can be built by either screen printing or Micropen deposition. Surface energy between the liquid and substrate is important because it affects both resolution and adhesion. A liquid that wets well will spread and thereby limit resolution. A nonwetting liquid will bead, which can cause lines to become nonconnected and give little area for adhesion with the substrate. Adhesion cannot occur without wetting. All printed lines tend to spread beyond the printing dimensions based on surface energy. This limits both line pitch and line width for deposition.

Liang et al. have examined the role of surface-energy effects on resolution in the screen-printing process and their results are expected to be similar to that of the Micropen (38). The surface energy of the paste is usually assumed to be similar to that of the liquid vehicle, and to the authors' knowledge, the role of surface forces on the spreading of a lubricated particle bed has not been considered to date. We therefore make the same assumption and treat a particle suspension as a continuum material with respect to surface energy and flow response. The interaction between a flat surface and a liquid drop is characterized by the contact angle as shown in Fig. 7. The relationship of surface energy to contact angle has long been treated by Young's equation (39).

$$\gamma_{sv} = \gamma_{sl} + \gamma_{lv} \cos \theta \qquad (9)$$

A zero contact angle occurs when the liquid perfectly wets the surface and spreads to coat as much area as possible. *A liquid will only wet a solid if its surface energy is lower than that of the solid.* The substrate surface energy must be higher than that of the paste used in order for wetting to occur. Optimal printing would occur if the contact angle could be tailored to be 90°. Many thick-film inks are formulated in the solvent terpineol with ethyl cellulose

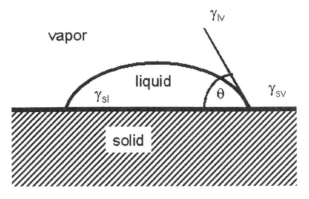

FIGURE 7   Bead contact angle representation.

polymer as a thickening agent. Liang et al. report surface tensions at 30–50 dynes/cm for this solvent-thickener mixture (38).

General effects are seen based on the surface energy of the substrate. Polymeric or epoxy substrates are low-surface-energy solids, and spreading is not usually that serious an issue. High-energy substrates like ceramics and metals are much more subject to spreading phenomena. Liang et al. report that screen-printing of lines with 100-micron widths at a pitch of 100 microns was not achievable on ceramics and glass substrates for a solvent of terpineol and ethyl cellulose binder. Also, lines of 150-micron widths spread to 190-micron widths. For ceramic substrates, high-viscosity pastes are recommended.

As shown in Young's equation (39), there are two obvious methods to prevent excessive spreading of a paste on substrates: increase surface tension of the liquid or lower the surface tension of the solid. In high-energy substrates, modifications in the liquid surface tension are generally inadequate to affect resolution issues. A better approach is to modify the surface with a pre-treatment or to use a high-viscosity ink. The Micropen has the advantage of an easily printing high-viscosity ink; with screen printing, adhesion to the screen can introduce roughness or poor printing. Using a high-viscosity paste with Micropen deposition will prevent spreading of lines, allow for high-definition printing, allow for controlled height, and achieve resolutions of 100 microns.

3.4. Printing Resolution

The Micropen can generally use any thick-film paste appropriate for screen printing, such as those available commercially from DuPont and Ferro (40,41).

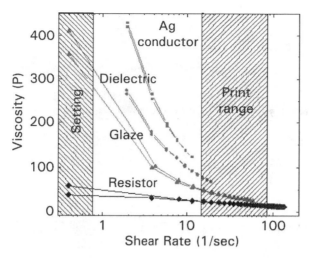

FIGURE 8   Viscosity vs. shear rate behavior for the electronic thick films used. Viscosity was characterized using a Brookfield viscometer with a cone and plate attachment.

Additionally, many novel materials may be integrated as printable slurries by dispersion of powders in a solvent, or even in a solvent containing molecular precursors to a desired matrix phase. Ideal slurries are free of agglomerated particles, have an intermediate viscosity value, and display reliable flow behavior, or rheology. For example, three commercial slurries, a crossover dielectric (Ferro 10-38N), a silver conductor (Ferro 1039), and a RuO_2-based resistor (Ferro 87-102) display tailored rheology for different device elements (40). Viscosities as a function of shear rate for these pastes are shown in Fig. 8, demonstrating that all three thick-film pastes are shear thinning to different degrees. While the viscosities vary greatly at low shear rates, the viscosities tend to converge at higher shear rates typical of the pen printing process. Due to this trend, optimal print parameters need not vary greatly for different thick-film formulations. However, differences in the slurry viscosity at low shear rates, which corresponds to settling conditions, enable either high-definition patterns or smooth filled regions. For example, the most viscous slurry (Ag conductor) leads to sharply defined traces in a filled region (Fig. 9), while the more fluid resistor paste flows during settling to produce smoother, more homogeneous films in filled areas.

In practice, print resolution and morphology are controlled by the paste rheology as described, where shear thinning may be desirable for metal or resistor trace lines, while more Newtonian viscosity may be desired for filled and smooth features, such as capacitors, precision resistors, and RF compo-

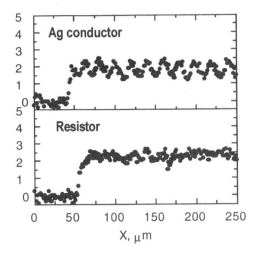

FIGURE 9   Laser profilometry cross-section of filled region using both highly and less shear-thinning thick-film slurries.

nents. An additional constraint on print size is the presence of aggregates in pastes, which limit the ability of the paste to flow through a 1-mil (25 μm) pen tip. This may be avoided in part by good dispersion chemistry, and three-roll milling of pastes to break up any such aggregates. Minimum feature sizes and the minimum spacing between components are also influenced by bead spreading. Electronic components at too fine a pitch may be subject to shorting or electromigration, especially for silver parts in humid environments. Additionally, significant bead spreading of one layer leads to problems during the deposition of subsequent layers, resulting in gross defects within the finished components. Several slurry types have been studied at Sandia National Laboratories to gain an understanding of solvent and rheology influence on bead shape. This information drives both new slurry development and materials processing for high-reliability components.

Four slurries were investigated in one particular case study (42): two commercial high-temperature thick-film pastes and two polymer thick-film (PTF) inks. Two DuPont postfire compositions, 5715, a gold conductor paste, and 1731, a ruthenium oxide-based resistor paste were all printed on dense alumina substrates. The other two commercial pastes were curable polymer thick-film (PTF) pastes, a Minico (Emerson and Cumming) M-4100 silver conductor paste, and an Asahi (Multicore Solders Inc.) TU-1k carbon resistor paste. Both of these pastes consist of solids (either silver or carbon) loaded in an epoxy carrier, designed to be printed on FR4 epoxy board and subsequently cured.

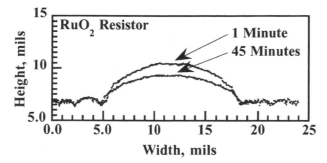

FIGURE 10 Settling profile of a resistor paste bead at 1 and 45 minutes after printing.

The slumping behavior of beads printed from each paste was analyzed by depositing a bead of each material with a 10-mil pen tip, and characterizing the shape of the bead with a CyberScan Cobra laser profilometer. The postfire inks were deposited onto alumina substrates, while the PTF inks were deposited onto FR-4 epoxy board. Profiles similar to the profile shown for the ruthenium oxide resistor paste in Fig. 10 were obtained for all of the pastes printed in the Micropen. Figure 11 plots the width and the cross-sectional area of 4 different material traces as a function of time after printing. For the 4 slurries, minimal spreading was observed after the first minute after printing. However, a significant change in the cross-sectional area of each bead was noted over the course of an hour, attributed to drying shrinkage. In practice, such traces may be printed and dried quickly by radiant or convective heating, before application of subsequent layers. One possible constraint is development of large tensile stresses associated with slurry solvent loss. In particular, aqueous

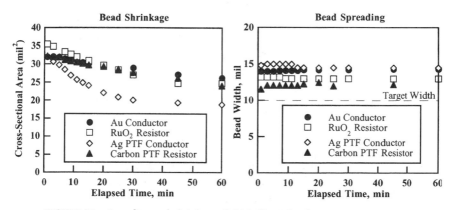

FIGURE 11 Spreading and shrinkage of thick-film inks after micropen printing.

and high-vapor-pressure solvents may dry too quickly, leading to brittle or cracked thick-film components. Approaches effective in mitigating this include use of high solid volume fraction slurries, low-volatility solvents (i.e., terpineol, dodecanol), multisolvent systems, and multiple application and drying cycles to build up thicker (>50 μm) structures.

3.5. DIELECTRIC PASTE DEVELOPMENT AND PROCESSING

Dielectrics for low-temperature, conformal write applications come in essentially two categories: (1) high-permittivity dielectrics for relatively large-value capacitors, and (2) microwave dielectrics. The first class is characterized primarily by the need for dielectric constants >300. The second class is characterized by dielectric loss values < 0.005, and temperature coefficients of capacitance (TCC) below 300 ppm/°C. The processing requirements for the two classes are somewhat different. Primarily, to achieve high dielectric constant values, densification at low temperature is required. To achieve low loss, elimination of impurities such as hydroxyl ions, carbon, and other mobile species is necessary. In our work, deposition routes to both types of dielectrics have been developed for use with a Micropen approach.

Barium titanate ($BaTiO_3$) has long been used as a high-k', low-loss, low-frequency dielectric. The Electronics Industries Association classifies capacitors depending upon the percentage change in capacitance over a given temperature range. In particular, several $BaTiO_3$ compositions satisfy the X7R designation: ±15% capacitance from −55 °C to 125 °C (43). Six commercial powders from two different manufacturers (TAM 262L, 422H, 292N, and Degussa 302L, 402H, 342N)—all of which were reported to have X7R characteristics—were prepared for Micropen deposition. Capacitance results for bulk samples of one powder (TAM 262L) are shown in Fig. 12, measured at three different frequencies.

In formulating printable high-K' pastes, several conditions were considered. The highest possible solids content was desired, as this leads to higher fired densities and less shrinkage. For printing of smooth capacitors, the pastes needed to be stable, near-Newtonian fluids. Several dispersants have been studied for the dispersion and stabilization of dielectric oxide powders, but none identified were better than oxidized Menhaden Fish Oil (MFO, R. A. Mistler) (44). Qualitative rheological stabilities of high-solids pastes were generally very similar to stabilities of low-solids (10 vol%) slurries with otherwise identical composition. Because results were easier to obtain for less viscous fluids than for the final pastes, slurries containing 10 vol% solids

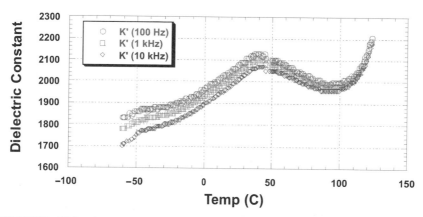

FIGURE 12   Dielectric constant vs. temperature at 100 Hz, 1 kHz, and 10 kHz for a bulk sample of TAM 262L, fired at 1,140 °C.

were studied before the high-solids pastes were prepared. To study the effect of MFO upon the stability of the slurries, 0 to 8% of MFO was added to the slurries, ball milled overnight, and tested (Fig. 13). The low-solids slurries displayed shear thinning until MFO concentration reached a point where particles were sufficiently coated with MFO. Figure 13 demonstrates this behavior for a slurry based on a powder with an extremely high effective surface area; most powders studied stabilized with 2–3% MFO.

High-solids pastes were prepared by hand-mixing appropriate amounts of powder, MFO, terpineol, and 1% (by powder mass) ethyl cellulose as a binder, then milling the paste for 3–4 cycles on a three-roll (shear) mill to break up agglomerates. Aging studies performed (44) showed that paste stability increased slightly with time as the MFO became better dispersed. In order to eliminate such aging, the MFO was added to the powder and ball milled in ethanol overnight, after which the ethanol was evaporated in a fume hood. The predispersed powder was then mixed with appropriate amounts of terpineol and ethyl cellulose and three-roll milled. Rheology results of 35-, 40-, and 45-volume % solids pastes prepared from TAM 262L powder demonstrate that both viscosity and shear-thinning behavior increase with solids content (Fig. 14). For this particular powder, a 40-volume % paste was used for printing. For other powders, especially those with smaller particle sizes, pastes were prepared with solids loadings as high as 60 volume %. Even for the best of powders, however, pastes with more than 50-volume % solids proved problematic when printing due to high viscosities, high degrees of shear thinning, and possible agglomeration.

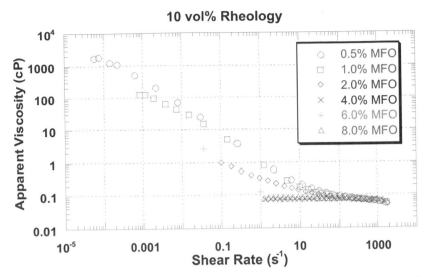

FIGURE 13 Low-solids slurry stability studies: at low amounts of MFO, the slurry is shear thinning, but as the MFO coats the particles, it behaves in a much more Newtonian manner. This particular powder had an extremely high effective surface area; most powders stabilized around 2–3% MFO.

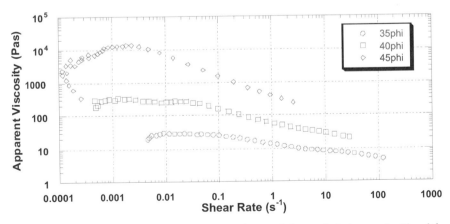

FIGURE 14 Paste rheology: viscosity and shear thinning behavior both increased with solids content for different solids loadings (35, 40, and 45 vol%) of a typical dielectric paste.

4. PROTOTYPING OF COMPONENTS FROM COMMERCIAL SLURRIES

The Micropen system was used to prototype several integrated passive devices using both Sandia-prepared pastes and commercially available slurries, whose viscosity was tailored by the commercial supplier for screen-printing applications (Fig. 8). In this case, the Ag metallization and crossover dielectric pastes displayed a high degree of shear thinning, while the resistor pastes displayed a more Newtonian viscosity. As discussed, the high value of the low-shear-rate viscosity in the silver paste maintains line definition after deposition, and the approximately constant viscosity of the resistor paste allows for separate printed lines to diffuse into each other and form a homogeneous region of constant thickness. Designs were chosen to form a variety of useful electronic components, but the successful implementation of these designs required consideration of material fabrication issues.

Two issues that are critical to the fabrication of multilayer, multimaterial ceramic components are sintering compatibility and compositional stability of adjacent materials. Sintering compatibility is important due to defects caused by differential shrinkage rates in multilayered cofired structures. To assess the cofireability of various thick films, we characterize the sintering behavior of individual thick-film layers prepared with the Micropen using *in situ* measurements of single-layer tapes, and bilayer composites. Combining these data enables a materials library to predict mechanical cofiring compatibility, where differential shrinkage rates may be correlated with overall shrinkage mismatch, and in turn be used to predict degree of warping or cambering (45). A second important issue is the chemical reaction or diffusion between dissimilar materials (46). While some reaction between layers is useful to promote adhesion, extensive reaction needs to be avoided, especially for highly composition-dependent components such as ruthenate-based resistors. Reaction rates between materials, which may be evaluated using hot-stage X-ray diffraction techniques, often determine whether various material systems can be integrated (46). In particular, use of commercial slurries free of glass frits minimizes difficulty in glass interdiffusion between adjacent layers.

For production of multilayer capacitors, a major issue is development of compatible electrodes and dielectric materials. A dominant cost associated with multilayer capacitors is the use of expensive metal electrodes, such as Pt or Pd for pure $BaTiO_3$, which must be fired at temperatures upwards of 1,300 °C. Adding dopants to $BaTiO_3$ not only serves to flatten capacitance peaks, but also can lower the firing temperature to regimes where Ag/Pd solid solutions may be used. Because silver is much less expensive than both Pd and Pt, pure or high-Ag electrodes are desirable. The melting point of the alloy increases continuously from 961 °C for pure Ag to 1,552 °C for pure Pd, making this system attractive for the optimization of low temperature versus high cost. In order to use a 70/30 Ag/Pd electrode, the dielectric must sinter below 1,150 °C; for pure Ag electrodes, the firing temperature must be pushed down to nearly 900 °C. Additionally, cofiring stresses must be minimized to obtain flat structures, as is discussed next.

4.1. MULTILAYER MULTIMATERIAL INTEGRATED CERAMIC COMPONENTS

A wide range of integrated, multilayer components have been fabricated by the Micropen for low-temperature ceramic cofire (47), and ultralow-temperature materials processing (< 400 °C). Figure 15a shows an R-C band reject filter that was Micropen fabricated from commercial slurries, based on the schematic circuit layout in Fig. 15b. The capacitors shown are parallel-plate capacitors with a single dielectric layer. The impedance response is shown in Fig. 15c, which indicates a 27-dB suppression at the reject frequency, which is given by $f_C = (\pi RC)/2$. The sharpness of the attenuation is given by achieving well-matched values for the capacitors and resistors.

The capability of the Micropen to accommodate various topographies has been utilized to build multiturn voltage transformer structures, as shown in Fig. 16a. The transformer was built by interleaving dielectric layers and Ag conductor traces. It was fabricated with an outer winding of 6 turns and an inner winding of 3 turns so that the device can be used for 2 : 1 and 1 : 2 voltage

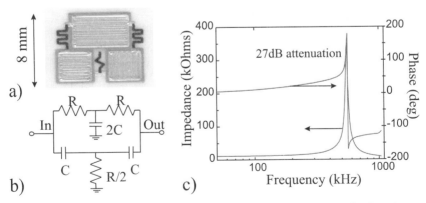

FIGURE 15 a) Photograph of 4-layer band reject filter fabricated with the Micropen, b) schematic diagram of band reject filter, c) impedance response for a typical Micropen band reject filter.

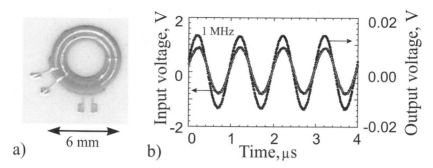

FIGURE 16 a) Picture of multiturn voltage transformer fabricated with independent Ag spiral windings (6 outer turns, 3 inner turns) and crossover dielectric, b) plot of input vs. output voltage for the transformer.

conversion. This construction relies on the ability of the Micropen to conformally print at a range of heights, and would be extremely difficult to fabricate using standard tape/screen-printing methods. This design works well, but the conversion efficiency (Fig. 16b) of this prototype is relatively poor because a low-permeability dielectric was used for the prototyping rather than a high-permeability ferrite. Development of such high-permeability LTCC ferrites is a topic of current research. A flat, 8-turn solenoid (L ~ 1 µH @ 1 MHz), and a multitap voltage divider were also fabricated, which are shown in Fig. 17a and Fig. 17b, respectively. These devices have all been printed on alumina substrates. However, the process has also been used to fabricate components on LTCC tapes and on Mylar and Teflon substrates that allow free-standing components to be built.

The ability to print low-viscosity liquids has been used by Fan, Reed, and Brinker at Sandia to print even chemically sensitive self-assembling nano-

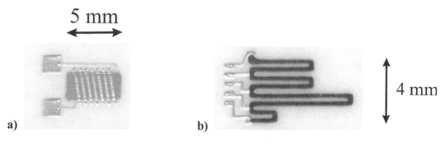

FIGURE 17 a) Flat solenoid fabricated by connecting lower half windings and upper half windings around thick-film dielectric (L ~ 1 mH @ 1 MHz), b) multitap voltage divider for 10:1, 7:1, 4:1, and 2:1 voltage division. A 10 kW/sq resistor with Ag electrodes is used.

structures using the Micropen (48). In this approach, sol-gel precursors are gelled around self-assembled surfactant micelles to form silica frames around cubic or hexagonal mesophases with pore diameters on the order of 100 Å, through a process called evaporation-induced self-assembly (EISA) (49). These silica mesophases themselves are useful as low K' dielectrics ($K' \sim 1.6$), or the pore channels may then be filled with functional molecules such as environment-sensitive fluorescing dyes, pH-sensitive dyes, or biologically sensitive moieties. The large, 400–1,000-m^2/g surface areas possible from such mesophases (48,49) are also of interest for a variety of atmospheric sensing applications. Micropen deposition of such materials in an additive method has been a versatile tool, used to pattern such devices as pH sensors (48) by direct-writing self-assembling nanophases with the same ease as printing commercial dielectrics and conductors.

4.2. SINGLE-LAYER AND MULTILAYER CAPACITOR DEPOSITION

Printing single-layer capacitor samples from commercial powders is a relatively straightforward process once rheology, area filling, and Micropen print parameters are refined. In particular, the AutoCAD software provided with the Micropen allows variable overlap and different fill designs for area fills, such as capacitors. Control of this overlap parameter has a strong impact on final capacitor smoothness, and control of the print cross-section within the MicroCAM print software allows linear control of film thickness (46). For testing, 8×8 arrays of 6.5-mm-square bottom electrode pads were first printed directly on alumina substrates, and dried for 10 min. at 100 °C prior to printing the dielectric layer. The 5.25-mm-square dielectric pads were also printed, then dried for 10 min. at 100 °C. The top electrodes were 3.75 mm square and were dried under the same conditions as the previous layers to develop parallel plate capacitors, as shown in Fig. 18. While sintering temperatures varied, samples were ramped up at 1 °C/min. to 600 °C and held for 2 hours to burn out any organic processing aids, then ramped at 2 °C/min. to the final firing temperature and held for 2 hours. After furnace cooling, the samples were removed and tested.

Cofiring presents not only chemical difficulties, but mechanical ones as well. Differences in thermal expansion between the dielectric and the electrodes lead to stresses at the interfaces. If strong enough, these stresses may lead to partial electrode separation or, in extreme cases, complete deadhesion of the electrode. Poor electrode connection both decreases the effective area of the capacitor, and dilutes effective capacitance, as air gaps serve as low-value capacitors in series with the real dielectric.

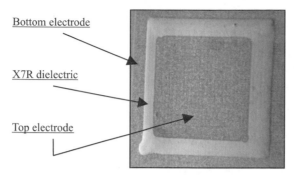

Bottom electrode

X7R dielectric

Top electrode

FIGURE 18   Photograph of direct-written parallel plate capacitor.

To understand origins of de-adhesion, thermal-shrinkage studies were conducted on unconstrained samples of Pt electrode and two low-fire dielectrics, TAM 262L and Degussa 302L. Samples were printed on Teflon, then released. The results (Fig. 19) show that the Pt expands initially, then shrinks at a much higher rate and lower temperature than do the dielectrics, suggesting there would be a high level of tensile stress in the fired capacitors. Evidence of such stress was abundant in several of the capacitor samples printed on Al_2O_3 substrates. The most common problems were curling of the top electrode and complete separation from the dielectric, generally when the firing temperature was higher than recommended for the dielectric used. The bottom electrode, due to the added constraint of the substrate, did not present as many visible problems. A possible remedy to such cofiring problems may be use of Ag/Pd-coated $BaTiO_3$ ceramic powders, such as those available from Superior MicroPowders. Such materials show promise for decreased shrinkage mismatch, since the $BaTiO_3$ powder core would be expected to better match the sintering behavior of the dielectric.

Lower-than-expected dielectric constant measurements can be caused by either incompletely densified dielectrics, or by poor top or bottom electrode contact, as was described. In another method of investigating such electrode effects, samples were printed with only one or neither electrode; after firing, Cr/Au electrodes were sputtered on, ensuring coherent electrode contact. Sputtered electrodes had a negligible effect upon dense samples with good electrode adhesion. Porous samples and those with poor electrode wetting, however, showed clear increases in capacitance with sputtered electrodes. Correlating SEM images with the sputtered-electrode and printed-electrode dielectric measurements, it was determined that thermal expansion stresses generally do not lead to serious problems as long as the electrode properly wets

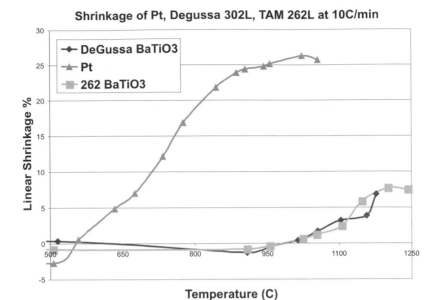

FIGURE 19 Unconstrained shrinkage measurements: plot of linear shrinkage as a function of temperature for printed samples of Pt electrode and two different X7R BaTiO₃ dielectrics.

the dielectric surface, and printed layers are free of cracks, voids, and other significant defects.

Figure 20 compiles the effects of firing temperature and electrode selection on each of the powders tested, with dotted lines connecting each of two manufacturer's high- ($\sim1,300\,^{\circ}$C) and low-fire ($\sim1,050\,^{\circ}$C) compositions. Ag electrodes were used up to $850\,^{\circ}$C, but generally displayed poor adhesion to the dielectric. Despite having similar wetting issues, 70/30 (Ag/Pd) electrodes performed better than pure silver. Platinum electrodes gave the best values for each temperature and powder, even at lower temperatures. Sputter-deposited electrodes produced almost identical K' values for the higher temperature, denser dielectrics. For porous samples fired at lower temperatures, however, measurements on sputtered top electrodes produced incorrect, anomalously high apparent K' values; this error arises because the sputtered electrodes fill open porosity and decrease the actual thickness of the dielectric from the typical assumption $C = \varepsilon_0 K' A / t$, where t is the total thickness of the (porous) dielectric. Aside from this anomaly, for both high- and low-fire dielectrics, capacitors printed with the Micropen were able to reach dielectric constants of

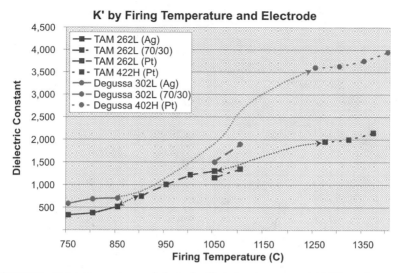

FIGURE 20 Firing temperature and electrode effects for six commercial X7R powders: average dielectric constant results for each of the six tested dielectrics with appropriate electrodes vs. firing temperatures. Dotted lines connect each manufacturer's low- and high-fire compositions.

over 90% (98%+ for the Degussa 402H) of the manufacturer-quoted values for bulk ceramics, an encouraging result for slurry-based Micropen processing.

4.3. MULTILAYER CAPACITORS

One of the most attractive features of the Micropen is its ability to write multiple layers quickly and reliably. Changing the CAD file only slightly and leaving all other parameters equal, multilayer capacitors (MLCs) with up to 7 layers of dielectric were printed and tested on both Al_2O_3 substrates and Teflon sheets. A schematic cross-section is shown in Fig. 21. Those samples that did not electrically short produced dielectric constant values identical to the single-layer capacitors, and displayed capacitance densities as high as $0.18 \, \mu F/cm^2$ ($K' > 1,000$, 950 °C firing, 3 dielectric layers). Shorted capacitors, however, were a serious problem, with only about 25% yield of 7-layer printed MLCs surviving the printing and firing processes. Increasing the number of layers for vertically configured MLCs led to a higher number of short circuits, primarily in the locations indicated in Fig. 21. Proper design changes are expected to minimize this localized defect.

The Micropen is generally a viable, flexible alternative to screen printing for the production of multilayer, multimaterial thick-film components. Direct-

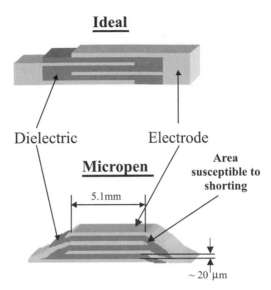

FIGURE 21   Schematic cross-sections of an ideally shaped MLC, and one printed by the Micropen.

written samples printed with the Micropen achieved capacitance values as high as 98% of that of bulk samples fired under identical conditions. Because it is especially well suited for rapid prototyping and small-lot production, the Micropen was a valuable tool for efficient, comprehensive studies of a variety of X7R dielectrics. The total time required to go from dielectric powder to tested samples was less than 2 days (including powder predispersion and capacitor firing), with the printing procedure requiring only minutes for each layer. Most importantly, pattern alterations can be instituted immediately rather than over the course of several days, making pattern and parameter optimization a remarkably swift process.

5. SUMMARY

We have described a direct-write approach for fabricating highly integrated, multilayer components using a Micropen to deposit slurries in precise patterns. With this technique, components are constructed layer by layer, simplifying fabrication. The quality of print features depends on slurry viscosity; however, the Micropen can accommodate a wide range of slurry viscosities corresponding to most commercial thick-film pastes. The direct-write approach provides the ability to fabricate multifunctional, multimaterial integrated ceramic components (MMICCs) in an agile way with rapid turnaround, and has been used to fabricate devices such as integrated RC filters, multilayer voltage transformers, and other passive components. The ability of direct writing to rapidly (≤ 0.75 m/s) and conformally deposit electronic components on a variety of substrates is a considerable advantage of pen dispensing over screen printing or other methods. Vision pattern recognition, *in situ* thermal processing, and other diagnostics promise to lead to future generations of pen-dispense systems by Ohmcraft and Sciperio, enabling yet greater flexibility in component deposition and control.

Several areas of research are still desirable for fuller utilization of Micropen and other pen-dispense direct-write methods. Common to all of these are the need for complete materials densification at low temperature, the ability to hold high tolerances in line dimension, and the development of pastes with a long shelf life. Regarding high line tolerance, two active areas of research include finer (5–25 µm) pen designs, and capillary force control for pastes to enable highly nonwetting or wetting line traces. Overall, with new levels of pen tip sophistication, vision control systems, agile materials delivery, and ultralow temperature (150–200 °C) materials, pen-dispense direct write appears to be a most promising route toward high-speed, multimaterial electronics fabrication.

ACKNOWLEDGEMENTS

We would like to thank Pin Yang, Walter Olson, and Sherry Morissette for excellent technical support and assistance in this work. Sandia is a multiprogram laboratory operated by Sandia Corporation, a Lockheed Martin Company, for the United States Department of Energy under contract DE-AC04-94-AL85000.

REFERENCES

1. C. E. Drumheller, "Dynamic Pen Control in a Synchronous Positive Displacement CAD/CAM Thick Film Writing System," Intl. J. Hybrid Microelectronics, 2 (ISHM '82 Proceedings, Reno, NV), pp. 449–453 (1982); W. M. Mathias, "Direct Writing Applications Update," Hybrid Circuit Technology, June 1986, Lake Publishing Corp., Libertyville, IL; www.ohmcraft.com/micropen/index.html.
2. H. Baudry and F. Franconville, Rheology and printing of high definition thick-film inks. Intl. J. Hybrid Microelectronics, December 1982, pp. 15–23, (1982).
3. D. E. Riemer, "Analytical Engineering Model of the Screen Printing Process: Part I," Solid State Technology, August 1988, 107–111 (1988).
4. D. E. Riemer, "Analytical Engineering Model of the Screen Printing Process, Part II," Solid State Technology, Sept. 1988, 85–90 (1988).

5. J. A. Owczarek and F. L. Howard, "A Study of the Off-Contact Screen Printing Process—Part 1: Model of the Printing Process and Some Results Derived From Experiments," IEEE Trans. On Components, Hybrids, and Manufacturing Technology, 13 [2] June 1990, 358–367 (1990).

6. J. A. Owczarek and F. L. Howard, "A Study of the Off-Contact Screen Printing Process—Part 2: Analysis of the Model of the Printing Process," IEEE Trans. On Components, Hybrids, and Manufacturing Technology, 13 [2] June 1990, 368–375 (1990).

7. J. Israelachvili, Intermolecular and Surface Forces, Academic Press, New York, 1995.

8. H. C. Hamaker, "The London-van der Waals Attraction between Spherical Particles," Physica, 4, 1058–1072 (1937).

9. D. Weitz and M. Olivera, "Fractal Structure Formed by Kinetic Aggregation of Aqueous Gold Colloids," Phys. Rev. Lett., 52, 1433–1436 (1984).

10. D. A. Schaefer, J. E. Martin, P. Wiltzius, and D. S. Cannel, "Fractal Geometry of Colloidal Aggregates," Phys. Rev. Lett., 52, 2371–2374 (1984).

11. M. Sahimi, Applications of Percolation Theory, Taylor and Francis, London, 1994.

12. R. J. Hunter, Foundations of Colloid Science, Vol. 1, Clarendon Press, Oxford, 1995.

13. W. A. Ducker, T. J. Senden, and R. M. Pashley, "Direct Measurement of Interparticle Forces Using an Atomic Force Microscope," Nature (London), 353, 239–241 (1991).

14. R. J. Hunter, Foundations of Colloid Science, Vol. 2, Clarendon Press, Oxford, 1995.

15. W. B. Russel, "Review of the Role of Colloidal Forces in the Rheology of Suspensions," J. Rheol., 24, 287–317 (1980).

16. W. B. Russel, "Concentrated Colloidal Dispersions," MRS Bulletin, August 1991, pp. 27–31.

17. R. L. Hoffman, "Interrelationships of Particle Structure and Flow in Concentrated Suspensions," MRS Bulletin, August 1991, pp. 32–37.

18. L. Bergstrom, "Shear Thinning and Shear Thickening of Concentrated Ceramic Suspensions," Colloids Surf. A: Physiochem. Eng. Aspects, 133, 151–155 (1998).

19. I. M. Krieger and M. Dougherty, "A Mechanism for Non-Newtonian Flow in Suspensions of Rigid Spheres," Trans. Soc. Rheol., 3, 137–152 (1959).

20. M. M. Cross, "Rheology of Non-Newtonian Fluids: A New Flow Equation for Pseudoplastic Systems," J. Colloid Interface Sci., 20, 417–437 (1965).

21. D. Draper, et al., "A Comparison of Paste Rheology and Extrudate Strength with Respect to Binder Formulation and Forming Technologies," J. Mater. Proc. Technol., 92–93, 141–146 (1999).

22. E. Mitsoulis, S. S. Abdali, and N. C. Markatos, "Flow Simulation of Herschel-Bulkley Fluids Through Extrusion Dies," Can. J. Chem. Eng., 71(#1), 147–160 (1993).

23. J. J. Benbow, E. W. Oxley, and J. Bridgwater, "The Extrusion Mechanics of Pastes—The Influence of Paste Formulation on Extrusion Parameters," Chem. Eng. Sci., 42, 2151–2162 (1987).

24. S. V. Shouche, D. K. Chokappa, V. M. Naik, and D. V. Khakhar, "Effect of Particulate Solids on the Rheology of a Lyotropic Gel Medium," J. Rheol., 38(6), 1871–1884 (1994).

25. E. B. Bagley, "End Corrections in the Capillary Flow of Polyethylene," J. Appl. Phys., 28, 193 (1957).

26. S. A. Altobelli, E. Fukushima, and L. A. Mondy, "Nuclear Magnetic Resonance Imaging of Particle Migration in Suspensions Undergoing Extrusion," J. Rheol., 41(5), 1105–1115 (1997).

27. D. Leighton and A. Acrivos, "The Shear-Induced Migration of Particles in Concentrated Suspensions," J. Fluid Mech., 181, 415–439 (1987).

28. F. Gadala-Maria and A. Acrivos, "Shear-Induced Structure in a Concentrated Suspension of Solids Spheres," J. Rheol., 24, 799–811 (1980).

29. J. R. Abbott, N. Tetlow, A. L. Graham, S. A. Altobelli, E. Fukushima, L. A. Mondy, and T. S. Stephens, "Experimental Observations of Particle Migration in Concentrated Suspensions: Couette Flow," J. Rheol., 35(5), 773–795 (1991).

30. R. E. Hampton, A. A. Mammoli, A. L. Graham, N. Tetlow, and S. A. Altobelli, "Migration of Particles Undergoing Pressure-Driven Flow in a Circular Conduit," J. Rheol., 41(3), 621–640 (1997).

31. P. Yaras, D. M. Kalyon, and U. Yilmazer, "Flow Instabilities in Capillary Flow of Concentrated Suspensions," Rheol. Acta, 33, 48–59 (1994).

32. U. Yilmazer and D. M. Kalyon, "Slip Effects in Capillary and Parallel Disk Torsional Flows if Highly Filled Suspensions," J. Rheol., 33, 1197–1212 (1989).

33. S. C. Jana, B. Kapoor, and A. Acrivos, "Apparent Wall Slip Velocity Coefficients in Concentrated Suspensions of Noncolloidal Particles," J. Rheol., 39, 1123–1132 (1995).

34. H. A. Barnes, "A Review of the Slip (Wall Depletion) of Polymer Solutions, Emulsions and Particle Suspensions in Viscometers: Its Cause, Character, and Cure," J. Non-Newtonian Fluid Mech., 56, 221–251 (1995).

35. H. Suwardie, R. Yazici, D. M. Kalyon, S. Kovenklioglu, "Capillary Flow Behavior of Micro-crystalline Wax and Silicon Carbide Suspension," J. Mat. Sci., 33, 5059–5067 (1998).

36. V. Ramachandran and H. S. Fogler, "Plugging by Hydrodynamic Bridging During Flow of Stable Colloidal Particles Within Cylindrical Pores," J. Fluid Mech., 385, 129–156 (1999).

37. V. Ramachandran, V. Venkatesan, G. Tryggvason, and H. S. Fogler, "Low Reynolds Number Interactions Between Colloidal Particles Near the Entrance to a Cylindrical Pore," J. Colloid Interface Sci., 229, 311–322 (2000).

38. T. X. Liang, W. Z. Sun, L.-D. Wang, Y. H. Wang, and H.-D. Li, "Effect of Surface Energies on Screen Printing Resolution," IEEE Trans. On Components, Packaging, and Manufacturing Tech.—Part B, 19, 423–426 (1996).

39. C. J. van Oss, Interfacial Forces in Aqueous Media, Marcel Dekker, Inc., New York, 1994.

40. D. Dimos, P. Yang, T. J. Garino, M. V. Raymond, and M. A. Rodriguez, "Direct Write Fabrication of Integrated, Multilayer Ceramic Components," pp. 33–40 in 8th Solid Freeform Fabrication Symposium Proceedings, Austin, TX, University of Texas at Austin (1997).

41. D. Dimos and P. Yang, "Direct Write Technique Simplifies Fabrication of Thick Film Hybrids," Ceramic Industry, p. 28, December 1997.

42. B. H. King, S. L. Morissette, H. Denham, J. Cesarano, and D. Dimos, "Influence of Rheology on Deposition Behavior of Ceramic Pastes in Direct Fabrication Systems," in 9th Solid Freeform Fabrication Symposium Proceedings, Austin, TX, University of Texas at Austin (1998).

43. A. J. Moulson and J. M. Herbert, Electroceramics, Chapman and Hall, London, 1990, pp. 241–262.

44. S. L. Morissette, J. A. Lewis, P. G. Clem, J. Cesarano, and D. B. Dimos, "Direct Write Fabrication of Pb(Nb,Zr,Ti)O$_3$ Devices: Influence of Paste Rheology on Print Morphology and Component Properties," J. Am. Ceram. Soc. (in press, 2000).

45. T. J. Garino, "Consolidation of Ceramic Thick Films," in Characterization of Ceramics, ed. R. E. Loehman, Butterworth-Heinemann, Boston, 1993, pp. 63–76.

46. P. Yang, D. Dimos, M. A. Rodriguez, R. F. Huang, S. Dai, and D. Wilcox, "Direct Write Precision Resistors for Ceramic Packages," Solid Freeform and Additive Fabrication, MRS Symp. Proc. 542, MRS, Pittsburgh, 1999, pp. 159–164.

47. B. H. King, D. Dimos, P. Yang, and S. L. Morissette, "Direct-Write Fabrication of Integrated, Multilayer Ceramic Components," J. Electroceramics 3(#2) 173–178 (1999).

48. H. Fan, et al., "Rapid Prototyping of Paterned Functional Nanostructuring," Nature 405, 56–60 (2000).

49. C. J. Brinker, Y. Lu, A. Sellinger, and H. Fan, "Evaporation-Induced Self-Assembly: Nano-structures Made Easy," Adv. Mater., 11, 579–585 (1999).

CHAPTER 9

Direct-Write Thermal Spraying of Multilayer Electronics and Sensor Structures

SANSAY SAMPATH, JON LONGTIN, RICHARD GAMBINO
AND HERBERT HERMAN
Center for Thermal Spray Research, State University of New York, Stony Brook, New York

ROBERT GREENLAW
Integrated Coating Solutions, Inc., Huntington Beach, California

ELLEN TORMEY
Sarnoff Corporation, Princeton, New Jersey

1. Introduction
2. Process Description
3. Materials and Microstructural Characteristics
4. Multilayer Electronic Circuits and Sensors by Thermal Spray
5. Fine Feature Deposition by Direct-Write Thermal Spray
6. Summary

1. INTRODUCTION

Rapid prototyping and manufacture of electronic components have received considerable recent interest, with applications emerging in the areas of large substrate ceramic multichip modules, thick-film sensors, and microwave electronics. Application-specific circuit fabrication requires considerable tooling costs and, in general, is not economical for low-volume manufacturing. There is a pressing need to develop methods with which to rapidly and directly translate CAD-based circuit design onto prototype components and, ultimately, for manufacturing. Direct write methods provide an efficient and environmentally conscious means of additive fabrication of electronic multilayers on conformal substrates. Thermal spraying is introduced here as a method for additive fabrication of multilayers through direct write approaches. This effort has recently been initiated at the Center for Thermal Spray Research at Stony Brook and offers considerable promise in the arena of direct write conformal electronics, sensors, and sensor array concepts. Preliminary assessment of the technology, the materials, and various application concepts are presented in this chapter.

Thermal spray is a directed spray process, in which material, generally in molten form, is accelerated to high velocities, impinging upon a substrate, where a dense and strongly adhered deposit is formed by rapid solidification. Material is typically injected in the form of a powder, wire, or rod into a high-velocity combustion or thermal plasma flame, which imparts thermal and kinetic energy to the particles. By controlling the plume characteristics and material state (e.g., molten, softened), it is possible to deposit a wide range of materials (metals, ceramics, polymers, and combinations thereof) onto virtually any substrate in various conformal shapes. The ability to melt, rapidly solidify, and consolidate introduces the possibility of the synthesizing useful deposits at or near ambient temperature. In the case of metals, the particles can be deposited in solid or semi-solid state. In the case of ceramic deposits, it is generally necessary to bring the particles to well above the melting point, which is achieved by either a combustion flame or a thermal plasma arc.

The deposit is built up by successive impingement of droplets, which yield flattened, solidified platelets, referred to as "splats." The deposit microstructure and, thus, properties, aside from being dependent on the spray material, rely on the processing parameters, which are numerous and complex. In recent years, through concerted, integrated efforts of the Center and other organizations, significant fundamental understanding of the process has been achieved, allowing for an enhanced control of the process.

1.1. APPLICATIONS

Thermal spray coatings (i.e., thick films > 5 micrometers) are crucial for the economical, safe, and efficient operation of many engineering components. Numerous industries, in recognition of thermal spray's versatility and inherent economics, have introduced the technology into the manufacturing environment. The technology has emerged as an innovative and unique means for processing and synthesizing of high-performance materials. The main advantages of the process are (1) versatility with respect to feed materials (metals, ceramics, and polymers in the form of wire, rod, or powder); (2) capacity to form barrier and functional coatings on a wide range of substrates; (3) ability to create free-standing structures for net-shape manufacturing of high-performance ceramics, composites, and functionally graded materials; and (4) rapid solidification synthesis of specialized materials. Opportunities exist for many novel applications in advanced materials synthesis and deposition. The technology is rapidly becoming the process of choice for the synthesis of advanced functional surfaces, such as electrical conductors, magnetic components, dielectrics, ferrites, bio-active materials, and solid-oxide fuel cells. Thermal spray offers advantages for manufacture of deposits on large area substrates and for the creation of complex conformal functional devices and systems. A more complete overview is given in a recently published MRS Bulletin Issue (1).

Traditional thick-film-based mesoscale electronics fabrication processes involve screen printing of pastes layer-by-layer followed by a thermal exposure (firing) cycle. The thermal exposures can be as high as 800 °C for low temperature co-fired ceramic modules and up to 1,400 °C for high-temperature co-fired ceramic modules. This temperature exposure severely restricts the substrate and material selection and also limits the size and conformality of substrates. Future thick-film concepts in sensors and electronics require embedding passive components onto actual engineering systems and pertains to conformal geometries and on-site application. It is believed that, once developed, thermal spray will play an important role in these applications.

1.1.1. Direct Write Electronics

The virtues and unique advantages of thermal spray with respect to direct write electronics fabrication and related processes are:

- High throughput manufacturing as well as high-speed direct writing capability
- *In situ* application of metals, ceramics, polymers, or any combination of these materials, both without thermal treatment or curing, and as an incorporation of mixed or graded layers
- Useful materials properties in the as-deposited state
- Cost-effective, efficient processability in virtually any environment
- Limited thermal input during processing, allowing for deposition onto a variety of substrates
- Adaptable to flexible manufacturing concepts; e.g., lean manufacturing
- Robotics-capable for difficult-to-access and severe environments (site applicability using portable tools)
- Readily available for customizing special sensor systems (i.e., prototyping)

- Green technology vis-à-vis plating, lithography, etc.
- Can apply on a wide range of substrates and conformal shapes
- High-aspect ratio conductors and capability for vias production
- Rapidly translatable development to manufacturing (using existing infrastructure).

Thermal spray methods offer means to produce blanket deposits of films and coatings as well as the ability to produce patches, lines, and vias. Multi-layers can be produced on plastic, metals, and ceramics substrates, both planar and conformal. Embedded functional electronics or sensors can be overcoated with protective coating, allowing applications in harsh environments. Such embedded harsh environment sensors can be used for condition-based maintenance of engineering components.

Though limited published accounts exist of the use of thermal spray for electronic applications, there has been some rudimentary consideration of using the technology for the production of electronics components, sensors, ceramic superconductors, waveguide components, insulated metal substrates, various magnetic deposits, photochemical coatings, etc. There has been only moderate success to achieve high-quality functional multilayers by thermal spray. This can be attributed to several deficiencies, among which are lack of fundamental understanding of the process and the ensuing process–materials–property relationships; absence of diagnostic tools to evaluate and to optimize the highly dynamic processes; insufficient process control; and limited personnel expertise with both materials knowledge and advanced processing.

The capabilities of thermal spray technology, even as recently as five years ago, were deficient for meeting direct write requirements. However, a number of important changes have occurred: cost-driven application developments in the automotive industry (e.g., electrical applications), availability of sophisticated, affordable diagnostic tools, enhanced process control and reliability, and, finally, improved fundamental understanding through integrated, interdisciplinary research. Our understanding of the process and the ability to control microstructures now offers unique opportunities with which to synthesize functional surfaces of a variety of complex materials systems. Actually, off-the-shelf current technology remains restricted in its capacity to satisfy the needs of direct write. The present limitation, however, is a classic case of a technology on the verge of a needs-driven upheaval. Thermal spray, therefore, represents a technological disruptive technology.

Thermal spray techniques offer new opportunities for hybrid microelectronics, sensors, superconductors, insulated metal substrates, and other applications, including:

- Embedded sensors in coatings and structural parts for condition-based maintenance engineering, processing, and system monitoring (2),

- Various thick-film materials and sensors, including antennas, magneto-resistive sensors, thermistors, thermocouples, gas sensors, etc.,
- High-frequency inductive components—Inductors, transformers, and antennas, using low-loss ferrites and sprayed windings, interconnects, and dielectrics.

The implications of developing, enhancing, and utilizing this technology in electronics can be far reaching and can impact hybrid electronics, large ceramic substrates, conformal devices, and sensor systems. The developments also can introduce new opportunities in "high definition" thermal spraying, which in itself can broadly impact the general engineering community.

2. PROCESS DESCRIPTION

Figure 1 provides a schematic overview of the thermal spray process. The feedstock material is typically in the form of powder or wire, although ceramic rods and liquid feedstock approaches have also been used in some cases. There are various means to produce thermal spray plumes using either combustion or plasma sources. Several commercial spray devices are available based on these basic methods providing variations with process conditions and, to some extent, allow differentiation. The currently available thermal spray techniques are briefly described in the following sections. More detailed descriptions are available in the literature (3–6).

2.1. HIGH-VELOCITY OXY-FUEL SPRAYING (HVOF)

HVOF is a variation on combustion spraying and has had a dramatic influence on the field of thermal spray. (Figure 2 provides a schematic of a basic HVOF

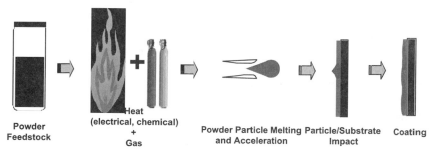

| Powder Feedstock | Heat (electrical, chemical) + Gas | Powder Particle Melting and Acceleration | Particle/Substrate Impact | Coating |

FIGURE 1   Schematic overview of thermal spray deposition process.

High Velocity Oxy-Fuel Process

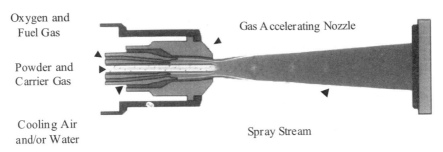

Oxygen and
Fuel Gas

Gas Accelerating Nozzle

Powder and
Carrier Gas

Cooling Air
and/or Water

Spray Stream

FIGURE 2   Schematic illustration of the high-velocity oxy-fuel spray process.

spray torch.) This technique is based on special torch designs, in which a compressed flame undergoes free-expansion upon exiting the torch nozzle, thereby experiencing dramatic gas acceleration, to over Mach 4. By properly injecting the feedstock powder axially from the rear of the torch concentrically with the flame, the particles can achieve supersonic values. Therefore, upon impacting the substrate, the particles spread out very thinly, and bond well to the substrate and to all other splats in its vicinity, yielding a well-adhered, dense coating that is comparable if not superior to plasma-sprayed coatings. It should be noted, however, that the powder particles are limited in the temperature they can achieve due to the relatively low-temperature combustion flame. It is, therefore, not currently straightforward to process high-temperature ceramics using this technique (e.g., zirconia-based systems). However, of particular importance is HVOF's ability to create dense deposits of alumina, spinel, etc.

2.2. PLASMA SPRAY

Plasma spray operates on direct current, which sustains a stable, nontransferred electric arc between a cathode and an annular water-cooled copper anode (4,6). A schematic of a generic DC thermal plasma torch is shown in Fig. 3. A plasma gas (generally, argon or another inert gas, complemented by a few percent of an enthalpy enhancing gas, such as hydrogen) is introduced at the back of the gun interior, swirls in a vortex, then exits out of the front of the anode nozzle. The electric arc from the cathode to the anode completes the circuit, generally on the outer face of the latter, forming an exiting plasma flame, which axially rotates due to the vortex momentum of the plasma gas.

Plasma Spray Process

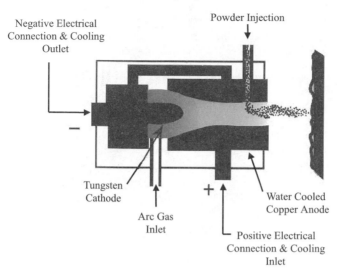

FIGURE 3   Schematic illustration of the DC thermal plasma spray process.

The temperature of the plasma just outside of the nozzle exit is effectively in excess of 15,000 K for a typical DC torch operating at 40 kW. The plasma temperature drops off rapidly from the exit of the anode, and, therefore, the powder to be processed is introduced at this hottest part of the flame. The powder particles, approximately 40 micrometers in diameter, are accelerated and melted in the flame on their high speed (100–300 m/sec) path to the substrate, where they impact and undergo rapid solidification (10^6 K/sec). Depending on the torch design and the powder particle-size distribution, plasma spray can form deposits of greater than 5 micrometers of a wide range of materials, including nickel and ferrous alloys, and refractory ceramics, such as alumina and zirconia-based ceramics.

2.3. COLD SPRAY DEPOSITION

Cold spray deposition and related solid state kinetic energy processes, not strictly "thermal spray," are a new family of spray devices (7). These systems, through special convergent-divergent nozzles, use continuous gas pressure to accelerate ductile materials to supersonic velocities to impact onto virtually any substrate (e.g., plastic, glass, ceramic, metal), where an unusually high adhesive bond is achieved. This unique process can gain a fully dense deposit, for example, of copper and silver, to achieve high electrical conductivity. We have found that cold spray deposits achieve copper conductivity of better than 80% bulk. It is clear that direct write can be readily facilitated by cold spray methods.

3. MATERIALS AND MICROSTRUCTURAL CHARACTERISTICS

Thermal spray offers the ability to deposit virtually any material that can be softened or melted. This encompasses metals, ceramics, polymers, and combinations thereof to produce composites and functionally graded materials (FGMs). Limited applications for thermal spray have been found in the area of *functional materials*. A wide range of materials is used commercially for protective coating. Typically, materials are injected in powder form into the spray plume. The particle sizes range from 5 to 100 micrometers depending on the process and materials considerations. In order to obtain a high-quality deposit it is critical to carefully control the feedstock characteristics. Numerous issues relative to feedstock need to be considered, especially as related to direct write technologies. These include particle size and distribution, the ability to feed fine particles, morphology, flow characteristics, and chemistry (8).

Specific to the deposited material is the occurrence of defects and interfaces associated with the splat-based buildup of the deposit and rapid solidification. In general, the microstructures show a layered anisotropy arising from a sequential buildup of layers. Metallic layers can have high process-induced residual stresses while the ceramic material generally contain microcracks resulting from relief of residual stresses. In the case of complex ceramics, there is possibility of preferential vaporization of components leading to variances in stoichiometry and, therefore, phases. These effects can be compensated for by appropriately manipulating powder feedstock (e.g., $YBa_2Cu_3O_7$ superconductors, quasicrystals, La-doped $SrMnO_3$ (9)). Additionally, deposit surface roughness can be high but controllable through particle-size selection and process parameterization. Finally, metals can oxidize in-flight during spraying under ambient conditions, leading to functionality of the deposit.

A typical thermal spray microstructure is composed of an array of cohesively bonded micrometer-sized splats, which are the result of individual particle impingements and subsequent accumulation to produce the deposit. Figure 4 shows a micrograph of a single splat produced by thermal spray and the deposit produced by successive splats (10).

The splats form by impact and spreading followed by rapid solidification. The splat-based layered microstructure leads to an anisotropic microstructure, which has clear implications on properties. The properties of the deposit

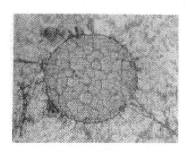
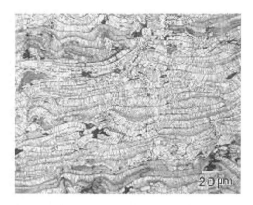

FIGURE 4   Typical example of a thermal spray single splat and deposit cross-section.

depend on a number of factors, principally affected by particle impact conditions (e.g., velocity, temperature), substrate conditions (e.g., roughness, temperature, chemistry) and ambient environment. Control of these variables is key to achieving requisite microstructural quality (10,11).

The advent of high-velocity deposition processes in recent years has considerably expanded the quality and utility of thermal sprayed deposits. Figure 5 shows the microstructure of an alumina dielectric deposited using HVOF oxy-fuel spray processes. HVOF deposits are dense and smooth and have electrical properties that are acceptable for component applications. Other examples of microstructures of deposited electronic materials are shown in Fig. 6. These include a capacitor ($BaTiO_3$) and a magnetoresistive metallic alloy (Permalloy Ni–Fe). The properties of these deposits will be discussed here.

Novel composite and graded microstructures can also be readily fabricated through dual powder feed injection techniques. A micrograph of a thermal sprayed functionally graded material (FGM) is illustrated in Fig. 7. Details related to FGM fabrication by thermal spray processes can be found in a 1995 article by Sampath et al. (12).

Polymers and polymer composites based on thermoplastics can also be produced by thermal spray methods. Polyethylenes, nylons, Teflon-based copolymers can all be thermal sprayed (13).

Table 1 summarizes the properties of various electronic materials fabricated using thermal spray methods over the last year as part of our direct write initiative (14). Also included in the table are typical results that can be observed in bulk materials of equivalent compositions.

Several important observations can be noted from Table 1. These are discussed in detail here in the section on materials development for thermal spray direct write technology.

Log#   1   30.0 KV   Notes:hvall   8/20/1999   04:36   P.M
Mag=   200X   108.860uM   1.0:0

FIGURE 5   HVOF coating cross-section of alumina.

3.1. CONDUCTORS

Conductors of Ag and Cu were produced by HVOF and plasma spray processes. In addition, results were also compared with that of cold-spray-deposited conductor materials. The conventional plasma and HVOF processes resulted in rather rough [5–7 micrometer surface roughness (R_A)] deposits on HVOF alumina with sheet resistances (R_S) of 3 to 5 mΩ/sq depending on conductor thickness (21–33 micrometers). We were able to reduce the R_A of the HVOF copper to $\sim$3 micrometer by using a finer starting powder, but the sheet resistance of these conductors was high (15–17 mΩ/sq) for an average thickness of 17 μm. Partial oxidation of Cu in the HVOF oxygen flame probably contributed to the high resistance.

Table 2A and B provides the sheet resistance and resistivity of the various sprayed materials. In general, the coatings were produced under ambient atmospheric environment, although inert shroud were also investigated. The typical resistivity values for the best plasma and solid-state-deposited materials are approximately 2–3X (i.e., times bulk) for Ag and 4–6X for copper. These

TABLE 1 Summary of Materials Property Results

Component	Materials/processes	Properties in as-sprayed state	Typical bulk
Base Dielectric	Spinel or alumina		
Dielectric constant (K)	HVOF and plasma spray	8–9	~ 8
Loss tangent (tan δ)		0.005 (10 kHz); ≤ 0.007 (1.5 GHz)	< 0.005
Surface roughness (μm)		1.6	
Breakdown voltage (V/mil)		350	
Conductor Traces	Copper and silver		
Resistivity (μΩ-cm)	Plasma and cold spray	~ 4.5 (Ag) to ~ 6.2 (Cu)	Bulk copper ~ 1.8
Surface roughness (μm)		~ 1.1–1.2	Bulk silver ~ 1.6
Linewidth (microns)		200 microns	Thin-film conductors
Thickness (μm)		25–30	~ 4–5
Resistors	$NiCr/Al_2O_3$; $NiAl/Al_2O_3$	HVOF $NiCr/Al_2O_3$	NA
Sheet resistance (Ω/sq)	HVOF and plasma	17 Ω/sq to 54 KΩ/sq	
TCR (ppm/°C)		175–300 at 10 Hz	
Surface roughness (μm)		~ 1.8	
Capacitors	$BaTiO_3$ and BST (68/32)		
Dielectric constant (K)	HVOF and plasma spray	120–190 (as-sprayed)	500 to 10,000
Loss tangent (tan δ)		~ 0.01	0.005 to 0.01
TCC (ppm/°C)		< 500 ppm/°C	
Surface roughness (μm)		~ 2–3	
Ferrites	$Mn_{0.27}Zn_{0.26}Fe_{2.47}O_4$		Bulk fired crystals
Sat. magnetization (Gauss)	$Mn_{0.35}Zn_{0.17}Fe_{2.48}O_4$	3,500–4,000	4,000–5,000
Coercivity (Oersteds)	HVOF and plasma spray	50–70	2
Resistivity (Ω-cm)		70	200
Permalloy	Ni–Fe		
Sat. magnetization (Gauss)	Air plasma spray	7,000	> 7,000
Coercivity (Oersteds)	Low pressure plasma spray	5 (in low pressure 2)	1
Resistivity (Ω-cm)		130	50

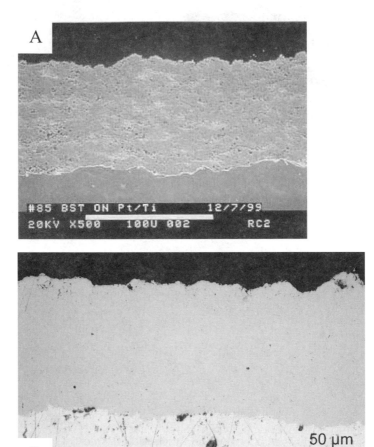

FIGURE 6 Cross-section microstructures of plasma sprayed Ba(Sr)TiO₃ (A) and Ni–Fe Permalloy (B).

were produced under ambient conditions and clearly there is considerable room for improvement. However, these conductor lines are useful for a number of electronic applications at their current levels.

Figure 8 shows the relationship between resistivity and line thickness for solid-state-deposited Ag lines. Nominally these lines were of similar width in the range of 1.5–2.0 mm and of similar surface roughness. The deposits were produced on 96% pure alumina substrates. The reduction in resistivity as a function of thickness may be attributed to roughness effects, measurement discrepancies, and interface adhesion.

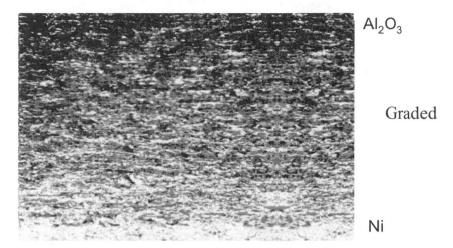

Al₂O₃

Graded

Ni

FIGURE 7   Cross-section microstructure of a functionally graded deposit (~1 mm thick).

Figure 9 shows the microstructure of plasma-spray-deposited Cu and solid-state-deposited Ag samples. The deposits are characterized by their high density. The deposit was produced using advanced nozzle designs that enhanced the deposit quality through proprietary parametric modifications which are not described here.

Preliminary results on plasma spray deposition of Ag using powder is very promising. In open atmosphere, under conditions similar to Cu deposition, a sheet resistance of 2.7X bulk was achieved for Ag patches. New powder morphologies (Ag) and compositions (Ag/Pd) will provide further improvements. These are being investigated for an optimized process.

3.2. DIELECTRICS

A wide range of dielectrics was evaluated as part of a screening study on dielectrics. They include alumina, alumina-titania, magnesium aluminate spinel, yttria, and partially stabilized zirconia. Both plasma and HVOF processing were conducted to examine the characteristics of the thermal sprayed dielectrics. Table 3 provides a summary of the results.

Considerable development effort was carried out on $MgAl_2O_4$ spinel material for dielectric applications. The choice of spinel over alumina was based on the formation of a metastable gamma-phase in plasma sprayed alumina under rapid solidification conditions. Gamma-alumina is hygroscopic and can play a deleterious role in the dielectric properties of the material. Figure 10 shows the cross-section of plasma and HVOF sprayed spinel

TABLE 2(A) Characteristics of Plasma Sprayed Copper[a] Conductor Lines

Process	Avg. line width (mm)	Avg. line thickness (micrometer)	R_s (mOhm/Sq.)	ρ (μOhm-cm)
Conv. plasma w/o shroud	0.698	21.7	8.54	18.5
Conv. plasma w shroud	0.627	18.1	6.49	11.74
Mini-plasma w shroud	0.287	32.3	1.92	6.20
Mini-plasma w/o shroud	0.198	57.6	1.44	8.32

[a] Resistivity of bulk copper is 1.77 μOhm-cm.

TABLE 2(B) Characteristics of Solid-State-Deposited Ag Lines

Line ID	Avg. thickness (micrometer)	Avg. surface roughness (R_A) (micrometer)	Avg. width (mm)	Line length (mm)	No. of squares	Sheet resistance (R_s) (mΩ/sq)
1	33.3	2.2	2.4	33.7	14.2	1.03
2	33.1	2.0	2.4	33.3	14.0	0.84
3	67.9	1.9	3.2	32.5	10.2	0.29
4	63.5	1.9	3.2	32.5	10.2	0.28

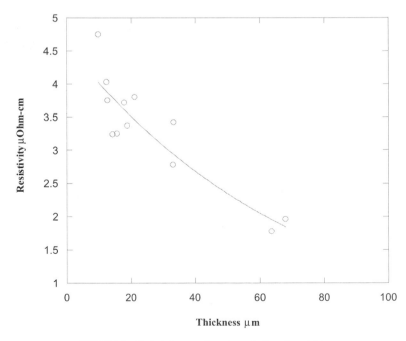

FIGURE 8   Resistivity as a function of silver line thickness.

a

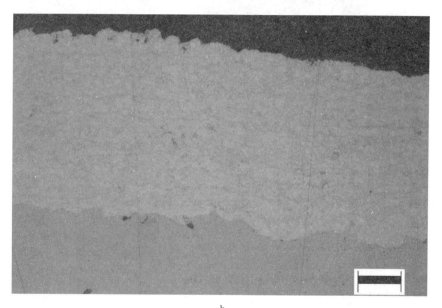

b

FIGURE 9   Microstructures of (a) solid-state-deposited silver and (b) plasma sprayed copper.

TABLE 3 Properties of Thermal Sprayed Dielectrics

Material	Powder source	Spray process	Thickness (mm)	Substrate	Diel. prop @ 10 KHz	Diel. prop @ 1 MHz
E-TFE	Du Pont	Combustion	0.910	Free standing	K ~ 2.29 tan δ 0.0006	K~ 2.29 tan δ 0.0002
Al_2O_3	Plasmtex	Plasma	0.8	Free standing	K ~ 9.2 tan δ 0.010	
Al_2O_3	Praxair	HVOF	~ 1 mm	Free standing		K ~ 7.87 tan δ 0.0008
Al_2O_3-13 TiO_2	Praxair	HVOF	~ 1 mm	Free standing		K ~ 8.46 tan δ 0.005
$MgAl_2O_4$ Spinel	Norton St. Gobain	HVOF	~ 1 mm	Free standing		K ~ 6.67 tan δ 0.13
$MgAl_2O_4$ Spinel	Norton St. Gobain	Plasma	0.76	Free standing	K ~ 8.6 tan δ 0.015	
$MgAl_2O_4$ Spinel	Norton St. Gobain	Plasma	0.072	Ti	K ~ 8.1 tan δ 0.005	
Ta_2O_5 99.99% Pure	CERAC optical grade	Plasma	0.0983	Ag/Pd on Ti	K ~ 121.14 tan δ 0.706	K ~ 55.49 tan δ 0.136
Y_2O_3	Japanese Abrasive	Plasma	0.1330	Ag/Pd on Ti	K ~ 11.83 tan δ 0.002	K ~ 11.75 tan δ 0.004
ZrO_2-$7Y_2O_3$ High purity	Metco AE7592 HOSP	Plasma	~ 1 mm	Free standing		K ~ 15 tan δ 0.0009
ZrO_2-$7Y_2O_3$ High purity	Metco-HOSP (−325 mesh)	Plasma	0.0946	Ag/Pd on Ti	K ~ 29.47 tan δ 0.008	K ~ 28.82 tan δ 0.007

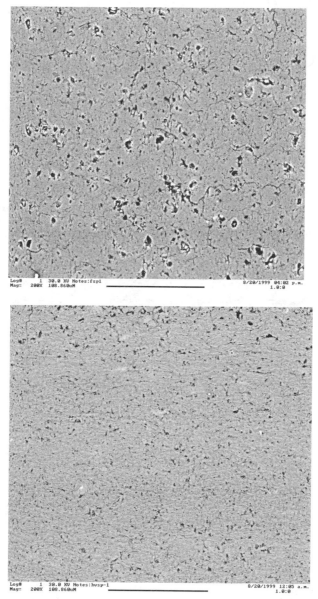

FIGURE 10 Cross-section microstructures of spinel dielectrics plasma sprayed (a) and HVOF sprayed (b).

TABLE 4   Properties of HVOF and Plasma Sprayed Spinel (MgAl$_2$O$_4$) Coatings on Ti Substrates

Spray process	Median particle size (μm)	Metal substrate	Thickness (μm)	R$_A$ (μm)	Dielectric properties K/tan δ @ 10 kHz	XRD data
Plasma	39–42	Grit-blasted Ti	72	6.8	8.1/0.005	Cubic spinel
HVOF	26	Etched Ti	20	3.5		
HVOF	21	Etched Ti	35	2.6	6.5/0.009	Cubic spinel
HVOF	21	Etched Ti	80	2.6	7.0/0.008; BDV ~ 350 V/mil	Cubic spinel
HVOF	6.3	Etched Ti	15	1.6	7.1/0.005	Cubic spinel
HVOF	10	Etched Ti	62	1.7		
HVOF	10	NiAl on etched Ti	60	1.7		

deposits. The HVOF deposits show a higher density, yet layered morphology compared to plasma sprayed materials. The spinel material is observed to have a dielectric constant of ~ 6–8 in the as-deposited state with tan δ of < 0.005 (see Table 4). This low loss is maintained over a range of frequency from 100 mHz to 5 GHz. The dielectric roughness in the as-sprayed state is between 1 and 2 micrometers.

3.2.1. Spinel Dielectric High-Frequency Measurements

Ring resonators were produced using conventional thin-film metallization (Ti/Pt/Au) on 175-micrometer- and 625-micrometer-thick HVOF spinel on Ti substrates. Preliminary measurements on the 175-micrometer-thick sample indicated that the fundamental frequency is ~ 4 GHz and the Q (quality factor—energy stored/energy dissipated per cycle) of the resonator is reasonably high. From these data, a relative dielectric constant of ~ 8 was calculated at 4 GHz. The associated loss tangent was 0.005. Ring resonator measurements on another sample (with a nonoptimal parameters) gave a relative dielectric constant of ~ 9, calculated at 1.5 GHz, with a loss tangent of 0.007.

   Measurements were also made on the ring resonators produced on 500-micrometer-thick spinel base dielectric deposited by thermal spray on a Ti substrate. Well-defined resonances were observed (Fig. 11). From the quality factor of the first resonance curve, a loss tangent of 0.0095 was calculated at 2.1 GHz. The dielectric constant was ~ 8, consistent with previous measurements. The loss tangent value includes the effect of the spinel surface roughness; R$_A$ = 1.85 micrometers. Note that for a 175-micrometer-thick specimen

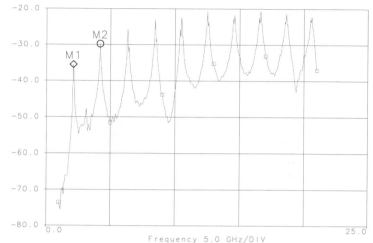

M1 Frequency=2.10000000 value=-35.5099808
M2-M1 dFrequency=2.10000000 dValue=5.73008036

FIGURE 11 Typical ring resonator response. Resonator radius = 0.400″, resonator line width = 0.010″, spinel substrate thickness = 0.020″, bandwidth = 50 MHz.

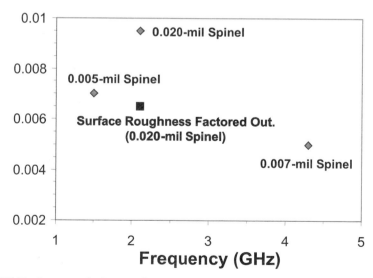

FIGURE 12 Summary of microwave loss tangent measurements to date on thermal spray spinel-on-Ti substrate. Effect of surface roughness is not factored out, except for the point shown as a rectangle illustrating the deleterious effect of surface roughness.

having a surface roughness of $R_A = 1.05$ micrometers, the loss tangent is 0.005. The effect of surface roughness has been modeled to extract the material loss tangent. Removing the effect of the surface roughness, the material loss tangent is 0.0065 (see Fig. 12). Based on these measurements, it is concluded that thermal sprayed spinel is a good base-dielectric material at microwave frequencies.

3.3. RESISTORS

Synthesis of resistors remains a challenging task for thermal spray direct write technology. Conventional thick-film resistors such as RuO_2 sublimes during thermal spraying and, as such, quality deposits could not be produced. However, thermal spray does offer a highly flexible approach toward synthesis of metal-ceramic composites that will allow fabrication of tuned resistors. Most of the resistor development effort to date was done with NiCr/Alumina mixtures and HVOF spraying. The NiCr powder (80% Ni/20% Cr composition) was mechanically mixed with an HVOF-sprayable alumina powder prior to spraying. Plasma sprayed resistors on HVOF spinel were also evaluated, but plasma spraying resulted in considerably rougher surfaces than with HVOF. Some data for plasma sprayed NiCr/Alumina resistors, and the results for the

TABLE 5   NiCr/Alumina Resistors on Spinel Coated Titanium

Metal content	No. of resistors measured	Avg. R_A (micrometers)	Avg. thickness (micrometers)	Avg. no. sq.	R_s (Ω/sq)	RT-125 °C TCR (ppm/°C)
Plasma sprayed NiCr/Al$_2$O$_3$ resistors on spinel						
4% NiCr	2	4.1	51.0	2.2	4.5 k	1120
	4			~ 2	> 20 M	
4.5% NiCr	1	4.2	36.1	2.0	4.2	
	5		34–42	~ 2	> 20 M	
5% NiCr	3	5.9	35	1.9	1.44 k	990
	1	6.8	35.3	2.5	5.9 k	2450
	2		~ 35	~ 2	> 20 M	
6% NiCr	3	4.4	39.2	2.3	33	760
HVOF sprayed NiCr/Al$_2$O$_3$ resistors on spinel						
8% NiCr	1	1.9	55.3	1.0	54 k	217
	3	1.9	53.4	~ 1	> 20 M	
10% NiCr	4	1.9	38.2	1.1	88.0	1476
12% NiCr	4	1.9	27.4	1.1	47.6	1823
14% NiCr	4	1.8	25.3	1.0	17.1	2714

TABLE 6   NiCr/Alumina Resistors Terminated with Sprayed Ag on HVOF Spinel

Resistor composition	Resistor thickness (μm) avg./range	Average Ag termination thickness (μm)	Avg. resistor surface roughness R_A (μm)	R_s (ohm/sq) average/range	TCR (room temp. to 125 °C) (ppm/°C) average/range
9% NiCr	42.7			699	303
	39.7–47.1			309–1381	282–339
10% NiCr	55.6			202	204
	52.3–59.0			145–236	183–233
10% NiCr	24.5			1009	140
	21.4–28.2			237–3392	−95–264
10% NiCr	60.8	15.6	1.83	161	191
	57.3–64.4			108–220	133–216
10% NiCr	57.3	13.9	1.82	123	141
	56.5–59.1			95–163	124–181

initial HVOF resistors on spinel—all of which were post-terminated with Ag paint—are summarized in Table 5. Note that there are variations in the results within a group of samples. This is attributed to improper mixing of the composite powder, leading to segregation.

HVOF resistors were also post-terminated with thermal spray deposited Ag or Cu conductors; data for these resistors are summarized in Table 6 and illustrated in Figure 13. Resistors with 2-square geometries (4 mm wide × 8 mm long) were HVOF sprayed onto spinel-coated Ti substrates and

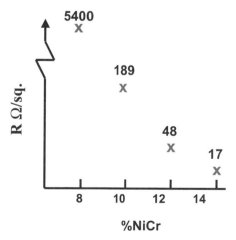

FIGURE 13   Resistors obtained from NiCr-Alumina composite with various volume fractions.

were then terminated with thermal sprayed Ag and/or Cu using metal masks. Some buried resistors were also fabricated and evaluated. The 10% NiCr compositions resulted in $\sim$ 150–300 Ω/sq resistors when sprayed $\sim$ 50 micrometers thick and had a room temperature to 125 °C temperature coefficient of resistance (TCR) $\sim$ 200 ppm/°C.

Figure 13 illustrates the resistances as a function of the metal/ceramic ratio in the composite resistors. It is clear that this concept can be further tuned through development of appropriate powder composites and process developments.

3.4. CAPACITORS

BaTiO$_3$-based capacitor materials were examined under a variety of operating conditions. Both HVOF and plasma spray processes were investigated for BaTiO$_3$ as well as Ba(Sr)TiO$_3$ (referred to as BTO and BST, respectively). Two types of BaTiO$_3$ powders were evaluated. One was a ground material with a median particle size of 20 micrometers and the other was a spray dried and sintered material with median particle size of 18.3 micrometers. Both materials behaved similarly in both HVOF and plasma spraying. BST [Ba$_{0.68}$Sr$_{0.32}$TiO$_3$ with Curie temperature (T$_c$) $\sim$ 5 °C] was evaluated because of its low T$_c$ and also higher intrinsic dielectric constants relative to BaTiO$_3$. The dielectric materials were sprayed onto thin-film platinum or thick-film AgPd-coated Ti substrates and also onto steel substrates. The substrate material did not have a noticeable effect of the dielectric properties of the deposits. Plasma sprayed deposits were evaluated in terms of dielectric properties at 10 kHz, surface roughness using a stylus profilometer (500 micrometers scan length, 10 micrometers/sec scan speed), XRD for phase content, and cross-sectioned and polished for microstructural evaluation. For all of the samples, an Ag paint electrode ($\sim$ 0.38–0.5-inch-square pad) was applied to the top surface of the deposits and dried at 100 °C in order to form a parallel plate capacitor structure for measurement of capacitance and tan δ using fine-tipped probes and an HP 4275A LCR meter (@ 0.2 V, 10 kHz). Dielectric constants were calculated from the measured capacitance and deposit-thickness data for each sample. For a few samples, dielectric measurements were also made with a thermal sprayed Ag top electrode. The dielectric properties of these were comparable to those made on the same samples with Ag paint top electrodes.

These materials yielded the most significant challenge in the sprayed deposits in terms of achieving requisite properties. Several important findings emerged. The deep eutectic nature of the BaTiO$_3$ compositional phase field tends to kinetically suppress the formation of the crystalline perovskite phase, promoting the formation of an amorphous phase. This amorphous phase is typically found at the interface between the substrate and deposit and creates a series capacitive network, thereby reducing the through-thickness dielectric constant.

Figure 14 shows the XRD patterns of several HVOF sprayed $BaTiO_3$ samples with varying degrees of amorphicity (15). The associated dielectric constants illustrate the strong dependence on amorphous phase content. HVOF spraying resulted in rough deposits ($R_A \geq 6$ micrometers), with fairly low dielectric constants (~ 20–100) and dielectric losses < 0.05 at 10 kHz. Polished cross-sections show fairly dense material with microcracks, which presumably form on cooling due to thermal expansion mismatches and/or the 130 °C $BaTiO_3$ phase transition. Despite the microcracks, reasonably high break-down voltages were measured on some of the thicker (100 to 150-micrometer-thick) deposits. XRD analysis of the as-sprayed material shows mostly cubic $BaTiO_3$ and a trace or small percentage of amorphicity. In general, the data for the HVOF sprayed materials indicates that the dielectric constant increases with increasing deposition thickness. This may be the result of formation of a graded deposit, wherein the material initially deposited is amorphous and subsequent layers crystallize as the deposit builds up.

Plasma spraying the $BaTiO_3$ powders resulted in smoother deposits ($R_A = 3$–5 micrometers), which are gray in color (the starting material is an

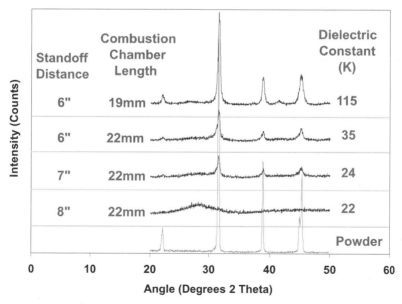

FIGURE 14   Effect of amorphous phase on dielectric constant of HVOF sprayed $BaTiO_3$.

off-white powder) and comprised mostly of cubic $BaTiO_3$ with a moderate amount of amorphicity. Polished cross-sections of the plasma sprayed material show a dense two-phase material also with microcracks. Typical microstructures of the HVOF and plasma sprayed $BaTiO_3$ coatings are shown in polished cross-section SEMs in Fig. 15.

Incorporation of heating cycles during spraying significantly increases the dielectric constants for both $BaTiO_3$ and BST. The highest dielectric constants for both materials were achieved using heating cycles during spraying, which probably enhances crystallization in the deposits, thereby increasing the dielectric constants. The SEM of $BaTiO_3$ plasma sprayed with heating cycles ($K \sim 189$) (Fig. 16(a)) shows relatively little amorphicity compared with those of prior samples with much lower dielectric constants. Figure 16(b) shows a micrograph of a typical BST deposit.

Heat treating both types of coatings at $400°$ to $500°C$ increased the dielectric constants by factors of 4–6, and also increased the percentage of crystalline material with only a trace or no amorphicity remaining in some cases. The changes in dielectric constant with heat treatment for both HVOF and plasma sprayed $BaTiO_3$ coatings on Ti are shown in Fig. 17.

The data for the HVOF sprayed $BaTiO_3$ on Ti samples indicated that as the thickness of the deposits increases, so do the dielectric constants. As was seen for HVOF spraying of BTO, this may be the result of the formation of a "graded deposit," whereby the initial material deposited is highly amorphous and subsequent layers are more crystalline as the deposit builds up. To investigate this hypothesis, two samples were ground down in steps and the dielectric constants and phase content (by XRD) was measured after each grinding step. The results of this experiment are summarized in Table 7. In general, the K values decrease with grinding (as the coatings became thinner) and the XRD analysis displayed slight changes in phase composition. The as-sprayed materials were cubic $BaTiO_3$ with trace amounts of amorphous phase, the percent amorphous phase increasing (by a few percent) as the coatings were ground thinner. There may be other factors that can also contribute to this effect, such as the formation of a low-K oxidized metal layer between the $BaTiO_3$ and substrate.

Some of the plasma sprayed $BaTiO_3$ and BST samples were characterized for temperature coefficient of capacitance (TCC). Capacitance measurements at room temperature (RT) and $125°C$ were used to calculate TCC values, reported in ppm/°C. The $BaTiO_3$ coatings have positive TCC values since the Curie temperature ($130°C$) is above ambient and is in the 400–500-ppm/°C range. The BST coatings have negative TCC values, ranging from $-1,050$ to $-1,600$ ppm/°C, because the Curie temperature is below room temperature ($\sim 0°C$).

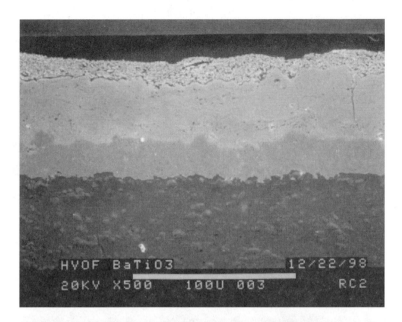

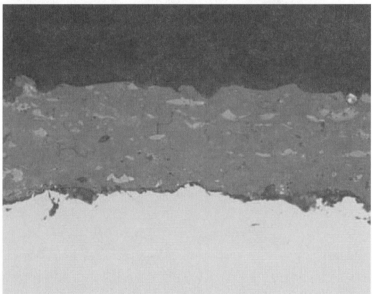

FIGURE 15 SEM micrographs of HVOF and plasma sprayed BaTiO₃. Note: the top figure shows a thick-film Ag electrode layer on the BaTiO₃ deposit.

286

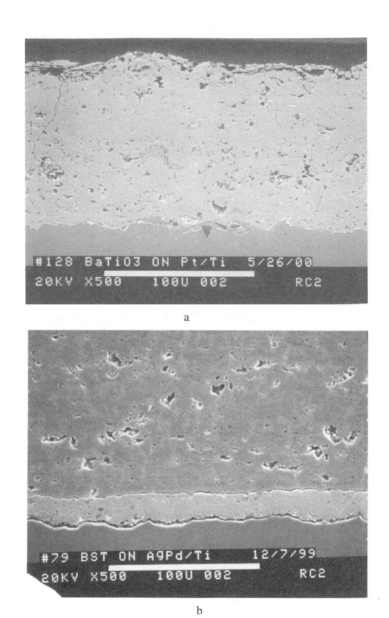

FIGURE 16 Polished cross-section SEM of plasma sprayed BaTiO$_3$ on Pt/Ti substrate (a) and polished cross-section SEM of plasma sprayed BST on AgPd/Ti substrate (b).

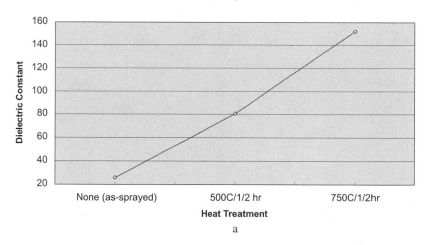

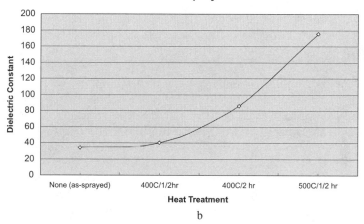

FIGURE 17   The effect of heat treatment on dielectric constant for HVOF sprayed BaTiO₃ (~75 μm thick) (a) and the effect of heat treatment on dielectric constant for plasma sprayed BaTiO₃ (~75 μm thick) (b).

3.4.1. High-Frequency Response of BTO and BST

The capacitances of some of the BaTiO₃ and BST samples were measured over a range of frequencies ranging from 0.1 kHz to 3,000 kHz.

The samples were prepared by drying at 100 °C for 4–6 hours followed by definition of a small parallel-plate capacitor. The metal back of the samples was

TABLE 7   Results of HVOF BaTiO$_3$ Grinding Experiment

Sample ID	Grinding step	Coating thickness (micrometers)	Dielectric const./ loss @ 10 KHz	XRD results
HVOF T.T. BaTiO$_3$ on NiAl/Ti; 3/10/99	As-sprayed	~150	113 tan δ 0.011	Cubic BaTiO$_3$ + trace amorphous
	1st	82	112 tan δ 0.019	Cubic BaTiO$_3$ + trace amorphous
	2nd	38	37 tan δ 0.024	Cubic BaTiO$_3$ + weak amorphous (amount > after 1st grind)
	3rd	27	not measurable- shorted	Cubic BaTiO$_3$ + weak amorphous (amount > after 2nd grind)

smoothed and the samples were attached by Ag epoxy to metal blocks. Areas ranging from 27 mm^2 to 100 mm^2 were also defined by Ag epoxy near an edge on the top of the dielectric. Small wires embedded in the epoxy during curing were soldered to small stand-offs after curing. Larger wires from the stand-offs and the metal blocks formed the leads of the test capacitor. The capacitance of the four samples and four commercial mica capacitors as controls were measured over the 0.1 kHz–5,000-kHz range on a Hewlett Packard model HP4192A impedance analyzer. The measurements, shown in Fig. 18, show a slight reduction in capacitance with increasing frequency. The BST (68/32) sample showed greater variation than the BaTiO$_3$ samples. This behavior of the dielectric constant monotonically decreasing with frequency is similar to that reported for single-layer polycrystalline BaTiO$_3$ (16). The frequency range was limited on the low side by the resolution of the meter for low-value capacitances. Above 5,000 kHz, the measured values of both the samples and the control capacitors increased significantly. The control capacitors remained steady (± 0.005) in the measurement range.

The results address the potential of thermal sprayed capacitors along with concomitant challenges to direct write in terms of low-temperature deposition. The rapid solidification process which is enabling technology for direct write onto low-temperature substrates can adversely affect perovskite phase formation. Improvements in feedstock chemistry and process optimization will yield improvements in as-deposited properties, but it is likely that some form of localized thermal exposure may be required to improve the dielectric constants

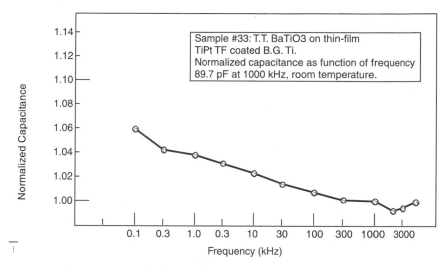

FIGURE 18   Normalized capacitance as a function of frequency for BaTiO$_3$.

of perovskite oxides. Nevertheless, applications can be developed with the currently achieved properties.

3.5. FERRITES AND PERMALLOY

Ferrites are important for the development of magnetic layers for high-frequency circuits. They represent complex chemistries, making it difficult to thermal spray these materials. Mn–Zn-ferrite compositions with different Mn/Zn ratios were selected for high-frequency permeability. Two compositions of ZnO/Fe$_2$O$_3$/MnO were examined; one with a slightly higher zinc content, in the event that some of the zinc was lost from the thermal spraying. The average particle diameter for both compositions was 25 micrometers. Optimization of the deposition parameters for this material was carried out using a conventional plasma spray system as well as two different HVOF systems; the Sulzer-Metco DJ2700 and Praxair HV2000.

The results of a process study are shown in Table 8. An as-deposited microstructure of HVOF sprayed ferrite is shown in Fig. 19. The resistivity is low for the plasma sprayed and HV2000 coatings, probably because of some reduction of trivalent to divalent iron. The ferrite deposited with the HVOF (DJ) torch has the highest resistivity, perhaps because the propane fuel used in the DJ gun is less reducing than the hydrogen fuel for the HV2000. The

TABLE 8   Process Properties for Thermal Sprayed MnZn Ferrite[a]

Ferrite type/process	Thickness (microns)	σ (emu/g)	Sat. Mag Gauss	Hc Oe	ρ (Ω-cm)
High Zn/air plasma	60	51.3	3223	66	2
Low Zn/air plasma	60	64.0	4021	72	2
High Zn/HVOF diamond jet	60	64.9	4080	51	78
High Zn/HVOF HV-2000	250	55.8	3506	177	1

[a] High-Zn composition: $Mn_{0.27}Zn_{0.26}Fe_{2.47}O_4$; Low-Zn composition: $Mn_{0.35}Zn_{0.17}Fe_{2.48}O_4$.

coercivity is also lowest in the HVOF (DJ) sprayed ferrite. The coercivity is increased by porosity and by poor coupling among the grains, so optimizing spray parameters can be expected to further reduce coercivity. Initial measurements on thicker coatings with the DJ torch indicate a resistivity of about 250 Ω-cm, which is about as high as would be expected for this ferrite. The best ferrite properties we have observed so far show that one can prepare cores for

50μm

FIGURE 19   Optical micrograph of the HVOF ferrite deposits.

high-frequency transformers, but the coercivity must be reduced for filters, antennas, and inductors. This can be accomplished with improved powder size, morphology, and enhanced density deposition.

4. MULTILAYER ELECTRONIC CIRCUITS AND SENSORS BY THERMAL SPRAY

4.1. MULTILAYER FERRITE INDUCTOR

One of the capabilities of thermal spray is the ability to fabricate multilayered systems rapidly, with no postfabrication firing required. As a demonstration of this, a multilayer inductor was produced using thermal spray of all of the inductor components: dielectric, ferrite core, and metal conductors. A Mn–Zn ferrite-based inductor with the 10-turn coil was fabricated using plasma spray to form the insulator (spinel) and ferrite (Fig. 20). Complete connectivity in the conductor was achieved. This enabled preliminary evaluation of the inductor. The high-frequency performance of the inductor was evaluated from 40 Hz to 110 MHz. Using a series equivalent circuit model, a plot of the series inductance L_s and series resistance R_s shows a well-defined resonance at 35.2 MHz. The inductance below the resonance is about 165 nH and above the resonance the inductance was 50 nH. The series resistance is a measure of all the losses in the circuit, mainly core losses at high frequency. At 40 Hz R_s is less than 1 ohm, consistent with the DC resistance of the coil. The core loss has a hysteresis term, linear in frequency, and an eddy current term with a frequency-squared dependence. The series resistance increases to about 15 ohms at 110 MHz. By fitting the frequency dependence of Rs to a sum of f and f^2, we find that at 110 MHz most of the loss is from eddy currents, as expected for a ferrite with a resistivity of 1 ohm-cm. The resistance of the ferrite can be increased to 100 ohm-cm, which will significantly decrease the eddy current loss.

The Mn–Zn ferrite composition was selected because of its high permeability at high frequencies. The relatively low hysteresis loss at high frequency is consistent with high initial permeability. The coercivity of the ferrite core is 50 Oe, so in high-power applications approaching saturation the hysteresis loss may become significant. Therefore, for some applications it will still be necessary to find process conditions that produce lower coercivity ferrite cores. The inductor has a Qmax of 143 at 33 MHz.

Future activities include increasing the number of coils with finer line widths, allowing for larger current capability as well as for a fabrication of the flux gate magnetometer.

292

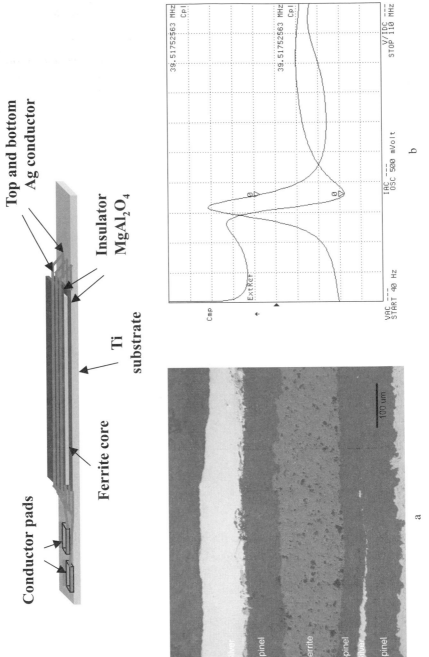

FIGURE 20   Cross-section of Mn–Zn ferrite-based inductor (a) and its behavior (b).

4.2. SENSOR FABRICATION USING DIRECT-WRITE TECHNOLOGY

One of the exciting prospects thermal spray technology offers is the fabrication of integrated, embedded sensors directly onto thermal-spray-coated components. The idea is that one, or *many* sensors could be integrated into a traditional thermal spray coating for "smart" components (2). Because of the great variety in possible materials that can be thermal sprayed, there is a wealth of possibilities for innovative sensor fabrication. The following section discusses the development of a thermistor-based temperature sensor fabricated exclusively with thermal spray technology.

4.2.1. Si-based Thermistor

A silicon resistive sensor system has been designed consisting of a polycrystalline Si coating with Ni ohmic contacts, as shown in Fig. 21. A metal substrate is insulated with $MgAl_2O_4$ spinel insulation. The goal is to determine the TCR of sprayed polycrystalline silicon with this device as well as thermal time constants.

The thickness of the silicon element is approximately 25–30 μm while the nickel contacts are about 45–50 μm. The silicon element was deposited onto the nickel contacts resulting in two interfaces between the two materials. Low

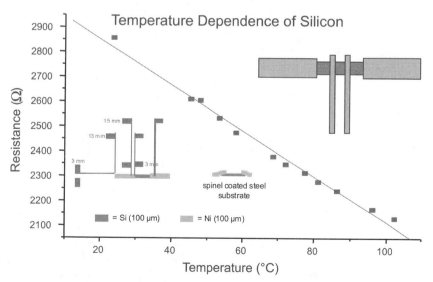

FIGURE 21   Si-based thermistor design and the temperature response of resistance.

resistance contacts have been obtained by spraying Si onto Ni, but we have not yet determined if the contacts are ohmic. We have previously shown that Ni sprayed on bulk polycrystalline Si produces an ohmic contact.

The sensors were tested by soldering copper wires to the two Ni contacts and connecting the sensor to a digital multimeter configured to read resistance. The nominal resistance of tested devices was in the range of 1–10 k which is consistent with commercial thermistor devices. A device was then placed in an electrically heated oven with an ambient (air) atmosphere. The oven temp-erature was increased to a desired temperature, left at that temperature for some time, and then either changed to a second elevated temperature, or allowed to cool to room temperature. The results for one sample are show in Fig. 21.

The device clearly functions as a thermistor-type temperature sensor, as resistance decreases proportionally with temperature. The resistance-versus-temperature plot has some curvature to it, which is typical of semiconductor-based thermistors.

There appears to be significant hysteresis in the preliminary devices tested, particularly for those tests in which the device is left at elevated temperatures for several hours. The nature of this hysteresis is not known at this time though a likely cause is the change in the Ni–Si interface at elevated temperatures. Further work will focus on isolating the nature of hysteresis. One concept currently being explored includes performing the test in an inert environment (e.g., nitrogen or helium) to minimize oxidation at elevated temperatures.

Another interesting aspect of a Si-based device is that it should be highly sensitive to strain as well as temperature. As the substrate temperature increases, it will expand, resulting in a strain-induced resistance change in the thermistor. This may provide for combined strain and temperature measurement by careful sensor design. Further characterization is under way that includes (1) fabricated Si-based thermistors on very low thermal-expansion optical glass (to remove the strain component in the resistance change); and (2) straining the existing gauge at room temperature by applying a force to the substrate material (to remove the temperature component in the resistance change).

5. FINE FEATURE DEPOSITION BY DIRECT-WRITE THERMAL SPRAY

To enable direct write technologies by thermal spray would require the ability to miniaturize the spray stream to enable fine-feature deposition. Current thermal spray devices generally employ nozzles in the several mm regime and are targeted toward large area deposition. As such, no commercial technologies exist for fine feature development. Using recently advanced modeling and diagnostic tools for thermal spray, several new designs for thermal spray systems are being currently investigated. Stony Brook's program emphasizes modified torches with special nozzles, implementing specially designed hardware, with a goal toward achieving a narrowly focused plasma-particle stream. We have demonstrated the feasibility of fine-feature spray (down to 100s of micrometer range). Examples of a ceramic and metallic direct written thermal spray lines are shown in Fig. 22.

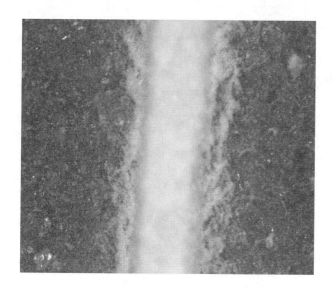

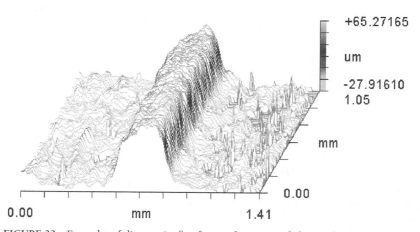

FIGURE 22   Examples of direct write fine feature deposition of alumina by thermal spray.

5.1. ULTRAFAST LASER TRIMMING AND DRILLING OF DIRECT WRITTEN STRUCTURES

One of the many benefits of direct write technologies is the large spectrum of materials that can be deposited and fabricated onto a substrate. To compliment the additive nature of thermal spray, a subtractive process in which sprayed material can be removed at appropriate locations improves the total manufacturing capability of the technology. Laser processing using ultrafast lasers represents an innovative technique for such material removal.

Ultrafast lasers produce laser pulses with durations of femtoseconds or (10^{-15} s) picoseconds (10^{-12} s). At such short time scales, material is removed by direct vaporization in the vicinity of the incident laser pulse. Material occurs by a *photochemical* process, in which chemical bonds are broken directly, rather than a *photothermal* process, in which the material is heated by the laser until it melts and then vaporizes. Photothermal processes dominate laser processing with longer (nanosecond and greater) pulse durations, and can result in extensive thermal damage to the substrate, poor feature quality, and substantial redeposition of ablated material near the irradiated site. Ultrafast laser processing also works equally well for a very wide range of materials, including metals, semiconductors, dielectric materials (polymers, glass, ceramics, liquids), and composite materials. This material versatility is a natural match with that of thermal spray. The following discusses recent developments in ultrafast laser processing of direct written thermal spray structures.

5.1.1. Ultrafast Laser Trimming of Thermal Spray Lines

One important application of ultrafast laser processing is trimming thermal sprayed lines to a desired width. While providing very fast throughput, large versatility in deposited line width, and feature geometry, it is, to date, still technically challenging to fabricate a thermal spray line (TSL) with a width of less than a few hundred microns. Also, in some cases, it is desirable to have sharp edges to sprayed lines. As-deposited thermal spray lines often have a Gaussian distribution of line thickness, with the thickest region at the torch centerline, and tapering off in thickness as one moves away from the center line. The idea is to use the ultrafast laser to trim these "wings" from the sprayed line to reduce the line width as desired. The advantages of this technique include very precise control over the line width, including continuously variable lines, and sharp edge definition.

In the following example, a TSL is trimmed by the laser making multiple passes on both sides of the line. The substrate to be processed is mounted to a 3-axis translation stage with 0.5 micrometer repeatability. The TSL is trimmed by starting from the outside and working toward the center. The thickness of the line is determined by stopping at a prescribed distance from the center line, while the depth of the machining into the TSL and substrate is determined by the stage speed.

An SEM image of a trimmed line is shown in Fig. 23. The material is Ag sprayed onto a Ti substrate. The original line width as sprayed is roughly 500 μm, while the laser-trimmed region is 80–100 micrometers in width, and 200 micrometers in length. For this case, 10 laser cutting paths were used on each side of the line with the laser making two passes over each path. The stage speed was 5 mm/s, and the process proceeds from the outside toward the center line of the line such that the final pass on each side is closest to the center line, which is done to avoid redeposition of material on the trimmed portion of the line. The entire processing time was less than 30 sec.

Feature quality and uniformity are good. Though not obvious, the laser-machined regions cut into the Ti substrate as well as the TSL. This happens because there is no indicator to instruct the laser to stop cutting when the TSL has been completely removed and the substrate is being removed. For this work the stage speed was run slower than needed to guarantee the entire TSL line was removed. To optimize the technique, parameters can be empirically

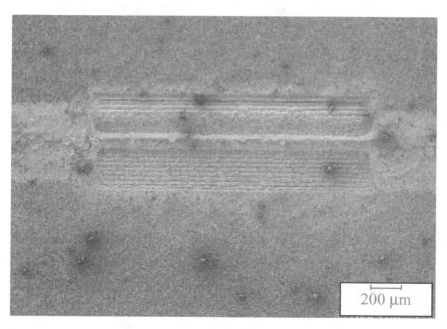

200 μm

FIGURE 23   Ultrafast laser trimming of thermal spray line.

determined to provide sufficient removal of TSL material, while minimizing material removal from the substrate. A better solution—and one being investigated—is to dynamically monitor the laser-processed region to determine when the substrate has been reached. Two competing ideas include (1) monitoring the laser-processed feature using a video camera; and (2) using a fiber-optic spectrometer to analyze the ablated material, shutting off the laser when substrate material begins to be ablated.

Another observation made is that the trimmed lines are not perfectly sharp. In Fig. 24, it can be seen that the spatial profile of the laser beam influences the trimmed line. The tighter the beam is focused (for a smaller spot size), the more pronounced the hourglass shape of the beam becomes. As the beam passes near the TSL, the wider beam near the top will remove a portion of the line. Solutions to this issue may include (1) reducing the beam power so that only the focal region (waist) of the beam has sufficient energy to initiate material removal; (2) removing only the TSL material and minimizing substrate removal (see preceding discussion as well); and/or (3) prescribing a more complicated laser-material path to minimize beam-profile effects that tend to round the tops of the trimmed lines.

5.1.2. Fabricating and Filling Vias

When fabricating electronics by direct write processes, multilayer structures are often encountered in which metal conductor lines from different layers must be connected electrically. Similar to vias used in printed-circuit-board technology, vias can be fabricated in thermal spray structures using ultrafast laser processing. Material at an intersection of two conductors at different levels is removed using the laser until the bottom conductor is exposed. Thermal spray can then be used to refill the via with an electrically conducting material to electrically connect the conductor lines in each layer.

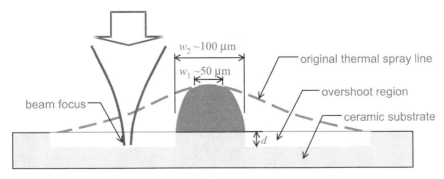

FIGURE 24   Schematic of laser trimming process.

Vias have been fabricated into a thermal-sprayed multilayer structure using the motion control system discussed before. The vias are fabricated in a thermal-sprayed electrical inductor consisting of eight layers: Ti-substrate, bonding layer, ceramic insulator, bottom Ag conductor, ceramic insulator, ferrous inductor material, insulator, and top Ag conductor.

Feature quality and edge definition are very good, as shown in Fig. 25. Note that the perspective view on the right is slightly deeper near the edges. This occurs because the stage cannot accelerate or decelerate infinitely fast, and the stage velocity is slower in this region. Because the laser provides a constant number of pulses per second, the edges receive more pulses per site, resulting in deeper features. This issue is readily addressed by coordinating stage motion with laser pulse control.

The bottoms of the vias are somewhat rough, and this is a current area of research. The same issue as that concerning trimming of thermal spray lines is present with processing the vias, namely it is difficult at this time to accurately control the feature depth. Because the machining processes are very similar, however, improvements made on this issue for laser trimming will also benefit via fabrication.

To complete the electrical connection, the via was backfilled using a subsequent thermal spray pass of pure Ag. The via was then sectioned to determine the quality of the fill. The via section is show in Fig. 26.

As can be seen, the second thermal spray pass provides dense Ag in the sprayed via. Also, note that the Ag re-spray forms a contiguous, dense interface with the original Ag coating (the interface is the thin line in the light region near the top of the figure). Because this particular via was laser micromachined manually, the depth varies considerably, hence the uneven contour of the via cross-section. In this example, there is also a large buildup (not shown) of Ag above the via. Further process optimization is required to minimize this effect.

6. SUMMARY

Using both traditional and hybrid thermal spray methods it is possible to produce multilayer deposits of ceramics and metals having properties required by the electronics industry. Exemplifying this early success are the values of resistivity obtained for thermal-sprayed-copper-versus-bulk values: about 3X. In the case of the ceramic insulator dielectrics, spinel yields excellent in-sulative properties, as does alumina. The same trend can be attributed to ferrites, which display reasonably good values for low-temperature deposition and can clearly be improved with further process optimization. Components, ranging from insulated substrates, capacitors, conductors, resistors, to induc-

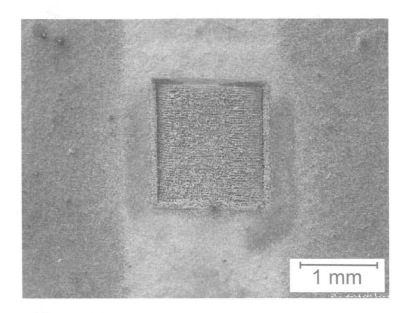

FIGURE 25   Laser-machined vias in multilayer thermal spray structure.

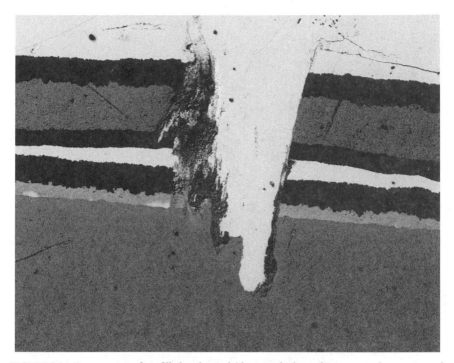

FIGURE 26 Cross-section of via filled with metal (the jagged edge is due to manual movement of the stage during laser drilling, which left an uneven edge).

tors and sensors, can be fabricated using a group of established and novel thermal spray processes. For both feedstock and substrates, versatility relative to materials, processing convenience, and cost efficiency make the family of thermal spray devices particularly suitable for the production of electronic components and circuits.

ACKNOWLEDGMENTS

This project was supported by DARPA/ONR award N000140010654 under the auspices of Dr. W. Warren and Dr. C. Wood. Use of shared facilities supported partially by the MRSEC program of NSF under award DMR 0080021 is duly acknowledged. The authors thank Andrew Dent, Josh Margolies, Ashish Patel, Anand Kulkarni, Bonnie Thompson, Jonathan Gutleber, and Glenn Bancke for their assistance in this project.

REFERENCES

1. *MRS Bulletin*, Guest Editors: S. Sampath and R. McCune, No. 7, **25** (2000) 12–53.
2. M. Fasching, F. Prinz, and L. Weiss, Smart Coatings, *J. Thermal Spray Tech.*, 4(2) (1995) 133.
3. American Welding Society, *Thermal Spraying: Practice, Theory, and Application*, 1985; Library of Congress No: 84-62707.
4. H. Herman, Plasma Sprayed Coatings, *Sci. Am.*, **256** (1988) 113.
5. H. Herman, *MRS Bulletin*, No. 4(13) (1988) 60.
6. D. A. Gerdeman and N. L. Hecht, *Arc Plasma Technology in Materials Science*, Springer, New York, 1972.
7. A. P. Alkhimov, A. N. Papyrin, V. F. Kozarev, N. I. Nesterovich, and M. M. Shushpanov, U.S. Patent No. 5,302,414 (April 12, 1994).
8. H. Herman, Powders for Thermal Spray Technology, *KONA—Powder Science and Tech.*, No. 9 (1991) pp. 187–199.
9. R. A. Neiser, Ph.D. Thesis, State University of New York at Stony Brook (1989).
10. S. Sampath and H. Herman, Rapid Solidification and Microstructure Development during Plasma Spray Deposition, *J. Thermal Spray Tech.*, 5(4) (1996) 445.
11. H. Herman and S. Sampath, in *Metallurgical and Protective Coatings*, ed. K. Stern, Chapman and Hall, New York, 1996, p. 261.
12. S. Sampath, H. Herman, N. Shimoda, and T. Saito, Thermal Spray Processing of FGMs, Invited Review, *MRS Bulletin*, Vol. XX, No. 1 (Jan. 1995) 27.
13. J. Brogan, *MRS Bulletin*, No. 7, **25** (2000), 48.
14. S. Sampath, H. Herman, A. Patel, R. Gambino, R. Greenlaw, and E. Tormey, MRS Symposia Proceedings on Direct Write Technologies, 624 (2000) pp. 181–188. Spring 2000, San Francisco.
15. A. H. Dent, A. Patel, J. Gutleber, E. Tormey, S. Sampath, and H. Herman, HVOF and Plasma Deposition of $BaTiO_3$ and $(Ba,Sr)TiO_3$, *Mat. Sci. and Eng. B.*, in press.
16. Q. X. Jia, Z. Q. Shi, and W. A. Anderson, $BaTiO_3$ Thin Film Capacitors Deposited by R.F. Magnetron Sputtering, *Thin Solid Films*, 209 (1992), 230–239.

Dip-Pen Nanolithography: Direct Writing Soft Structures on the Sub-100-Nanometer-Length Scale

CHAD A. MIRKIN, LINETTE M. DEMERS, SEUNGHUN HONG

Institute for Nanotechnology and Center for Nanofabrication and Molecular Self-Assembly, Northwestern University, Evanston, Illinois

1. INTRODUCTION

As physical processes for generating miniaturized structures increase in resolution, the types of scientific questions one can ask and answer become increasingly refined. Indeed, if one had the capability to control surface architecture on the 1–100-nm-length scale with reasonable speed and accuracy, one could ask and answer some of the most important questions in science and, in the process, develop technologies that could allow for major advances in many areas, including surface science, chemistry, biology, and human health. This length scale, which is exceedingly difficult to control, is the primary length scale of much of chemistry and most of biology. Indeed, chemical and biochemical recognition events are essentially sophisticated examples of pattern recognition processes. Therefore, if one could pattern on this length scale with control over feature size, shape, registration, and composition, one could systematically uncover the secrets of recognition processes involving

Direct-Write Technologies for Rapid Prototyping Applications

extraordinarily complex molecules. A recent invention, Dip-Pen Nanolithography, may provide both access to this type of control over surface architecture and entry into an entirely new realm of structure-versus-function studies available to the chemist, biologist, physicist, and materials scientist.

2. SCANNING PROBE MICROSCOPE METHODS

Since the invention of scanning probe microscopes (SPMs) and the realization that one could manipulate matter in a one-atom-at-a-time fashion, many scientists thought that this elusive goal in controlling surface architecture might be possible. However, although early attempts to develop patterning methodologies from SPMs were able to demonstrate the high-resolution capabilities of these instruments, they were fraught with limitations of speed or with respect to the types of molecules with which one could pattern. Throughout the 1980s and most of the 1990s, SPM surface patterning methods focused on impressive, but inherently slow, serial scanning tunneling microscope (STM) methods that move individual atoms around on a surface under ultrahigh vacuum (UHV) and low-temperature conditions, or by indirect multiple step STM or atomic force microscope (AFM) etching-backfilling methods (1). These approaches are limited because it is difficult to create parallel patterning methods from them, controlling both feature size and the types of molecules that can be patterned within nanoscopic dimensions. In other words, one can utilize a multiple-tip SPM instrument with multiple independent-feedback systems to control parallel etching procedures, but it is difficult, if not impossible, to selectively "fill in" such features with different types of molecules on the sub-100-nm-length scale.

3. DIP-PEN NANOLITHOGRAPHY METHODS

In 1999, we introduced a new direct-write SPM-based lithographic method termed "Dip-Pen Nanolithography" (DPN) to the scientific community (2), shown schematically in Fig. 1. DPN allows one to transport molecular substances to a surface, much like a macroscopic dip-pen transfers ink to paper, but with the resolution of an AFM. Two important observations made DPN possible. First, in studying the dynamics of the capillary effect and water transport from an AFM tip in air, we realized we could use the meniscus, which naturally forms in air between tip and sample, as a nanoscale reaction vessel and "ink" transport medium thereby providing a way of controlling molecular

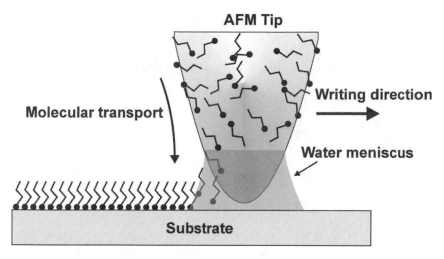

FIGURE 1 Schematic representation of DPN. A water meniscus forms between the AFM tip coated with ODT and the Au substrate. The size of the meniscus, which is controlled by relative humidity, affects the ODT transport rate, the effective tip-substrate contact arc, and DPN resolution. (Reprinted with permission from R. D. Piner, J. Zhu, F. Xu, S. Hong, C. A. Mirkin, *Science* **1999**, *283*, 661. Copyright 1999 American Association for the Advancement of Science.)

transport between an ink-coated AFM tip and substrate. Second, if one used inks that were designed to react with the surface to be patterned, one would have a chemical driving force that would favor the movement of such inks from the AFM tip via the water-filled capillary to the substrate. Moreover, chemisorption of the ink onto the substrate would lead to stable, chemisorbed nanostructures.

Initial DPN demonstrations showed that alkanethiols could be patterned onto gold surfaces and that the feature size and basic transport process could be controlled by regulating humidity and temperature. Hydrophilic and hydrophobic molecules, including biomolecules such as DNA, have now been patterned via the DPN process, and through the development of controlled environment chambers, which will allow one to replace the water meniscus with other solvents (2), it seems likely that it will be possible to extend DPN to a wide range of molecule types and substrates. One feature that sets DPN apart from other high-resolution nanolithography techniques is its nanostructure registration capabilities (3) (Fig. 2). With the DPN process, one uses the same high-resolution tool to "read" and write, so it therefore can generate multiple, chemically pristine nanostructures made of the same or different inks. In addition, they can be aligned with respect to one another with near-perfect precision (less than 5-nm alignment resolution). This capability makes DPN a tremendous customization tool for generating multifunctional

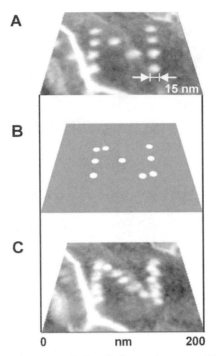

FIGURE 2 Schematic diagram with lateral force microscopy (LFM) images of nanoscale molecular dots showing the essential requirements for patterning and aligning multiple nano-structures via DPN. (A) A pattern of 15-nm-diameter 16-mercaptohexadecanoic acid (MHA) dots on Au(111) imaged by LFM with an MHA coated tip. (B) Anticipated placement of the second set of MHA dots as determined by calculated coordinates based on the positions of the first set of dots. (C) Image after a second pattern of MHA nanodots has been placed within the first set of dots. Note that the white jagged line is an Au(111) grain boundary. (Reprinted with permission from S. Hong, J. Zhu, C. A. Mirkin, *Science* 1999, *286*, 523. Copyright 1999 American Association for the Advancement of Scirnce.)

nanostructures and for making chips with integrated nanostructures composed of different chemical components, including biomolecules such as peptides and DNA. For this reason, DPN could have a major impact in molecular electronics, biodiagnostics, catalysis, and tribology.

With conventional cantilevers, DPN offers 12-nm linewidth and 5-nm spatial resolution and, therefore, rivals other techniques like e-beam lithography for patterning solid substrates. The resolution of DPN will undoubtedly improve through the use of sharper cantilevers, particularly as we further our understanding of the ink transport process. Significantly, DPN has been transformed from a serial to parallel process through the use of eight cantilevers used as a single cantilever in a conventional AFM instrument (4). (See Fig. 3 and the scheme shown in Fig. 4.) In these experiments, it was

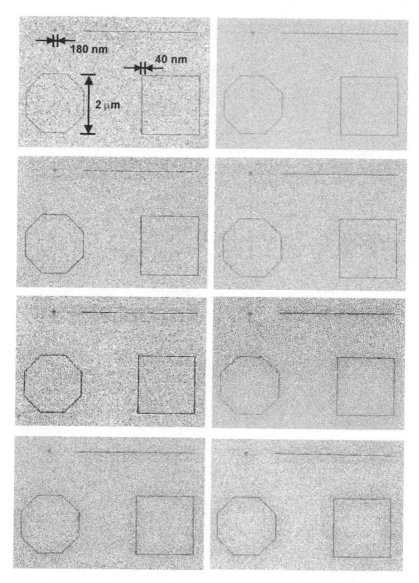

FIGURE 3 LFM images of eight identical patterns generated with one imaging tip and eight writing tips coated with 1-octadecanethiol (ODT) molecule. (Reprinted with permission from S. Hong, C. A. Mirkin, *Science* 2000, *288*, 1808. Copyright 2000 American Association for the Advancement of Science.)

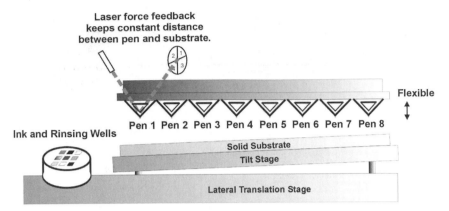

FIGURE 4　Schematic representation of an eight-tip DPN system using modified AFM cantilevers for parallel processing. (Reprinted with permission from S. Hong, C. A. Mirkin, *Science* 2000, *288*, 1808. Copyright 2000 American Association for the Advancement of Science.)

shown that the feature dimensions in a DPN experiment are almost independent of tip-substrate contact force over a two order of magnitude contact force range. Consequently, a bundle of tips can be used in parallel-writing fashion simply by applying a contact force suitable to engage all of the tips in the bundle with the substrate to be patterned. Interestingly, each of these tips also can be engaged individually allowing one to use the multiple-pen system in serial fashion like a nanoplotter; with the nanoplotter mode, one can use DPN to generate customized features composed of multiple inks without the need to change tips during a patterning experiment. Rinsing and inking wells located on the periphery of the sample make DPN an almost totally automated process.

Although DPN is particularly well suited for patterning flat solid substrates with soft matter when combined with wet chemical etching procedures, it also can be used, thus far, to fabricate silicon structures on the greater-than-40-nm-length scale (5) (Fig. 5). Recently, silicon nanostructures were generated via a multistep process that utilizes DPN to generate soft etch-resist structures composed of octadecanethiol on gold-titanium-silicon trilayer structures. The unmodified gold could be selectively etched with a ferri-ferrocycanide-based etching solution leaving behind an alkanethiol capped gold nanostructure. Subsequently, the exposed Ti-coated silicon subsequently could be anisotropically etched with KOH-yielding raised trilayer Au-Ti-Si nanostructures. Finally, the Au and Ti layers can be removed by treatment with aqua regia. This approach, in principle, could be extended to silver- or platinum-coated silicon substrates.

Although DPN is still in its infancy, its development already has led to the accomplishment of several important milestones in nanotechnology, and

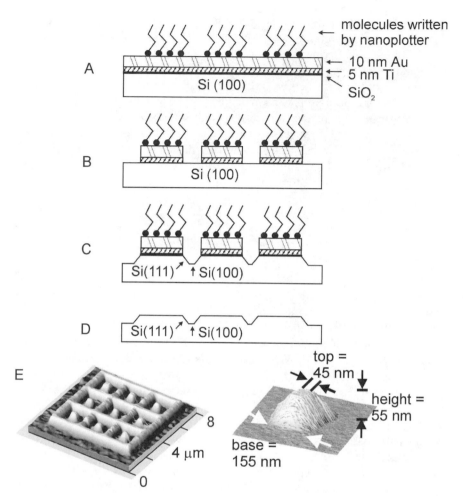

FIGURE 5 A schematic representation of the deposition and multistage etching procedure used to prepare three-dimensional architectures in Au/Ti/Si substrates. (A) Deposition of ODT onto an Au surface of a multilayer substrate using nanoplotter. (B) Selective Au/Ti etching with ferri-ferrocyanide-based etchant followed by passivation of Si surface with HF. (C) Selective Si etching with basic etchant and passivation of Si surface with HF. (D) Removal of residual Au and metal oxides with aqua regia and passivation of Si surface with HF. (E) AFM topography of a nanometer-scale structure prepared according to procedures A–D. (From D. A. Weinberger, S. Hong, C. A. Mirkin, B. W. Wessels, T. B. Higgins, "Combinatorial generation and analysis of nanometer- and micrometer-scale silicon features via 'Dip-Pen' nanolithography and wet chemical etching," *Adv. Mat.* **2000**, *12*, 1600–1603.)

scanning probe lithographies in particular (6). First, DPN is now a user-friendly form of lithography accessible to any group with a conventional AFM. This alone should accelerate its use throughout the scientific community at large and in the chemical and biochemical communities in particular. Second, its direct-write nature sets it apart from all scanning probe lithography alternatives for making soft nanostructures and provides some unique capabilities with respect to nanostructure registration and the chemical complexity one can incorporate into such structures. Third, it offers comparable resolution to e-beam lithography, but it is substantially more general with respect to the types of structures (hard and soft) one can pattern. Finally, it offers parallel writing capabilities with minimal increase in the complexity of the patterning instrumentation.

4. FUTURE ISSUES

Future issues facing the development of DPN involve the advancement of this tool from being an academic curiosity to becoming a major discovery tool for the nanotechnologist. This will be accomplished by the further development of parallel writing capabilities and a greater understanding of the fundamental ink transport processes. DPN, when used in nanoplotter mode, can be used as a very powerful means of doing combinatorial nanotechnology. Essentially, the nanoplotter can be programmed to generate a series of monolayer patterns that vary with respect to composition, feature size, and feature spacing (Fig. 6). These patterns subsequently can be used to study a wide range of chemical or physical processes, including crystallization, catalysis, chemical and biochemical recognition, etching behavior, and surface molecular transport. Essentially, one can determine the type of pattern appropriate for a given application and, because of the dimensions associated with these patterns, be able to use the same scanning probe instrument used to generate such structures to screen thousands of them relatively quickly (within minutes). For this reason, DPN could have a major impact on biotechnology, especially in the development of devices that recognize and detect important biological molecules (e.g., DNA, RNA, and proteins). It is important to note that DPN does not need to compete with conventional microfabrication processes with respect to speed. Indeed, we are not attempting to develop a tool that will displace photolithography. Rather, we are interested in a tool that offers capabilities currently not available through any other type of lithography, and which can significantly complement conventional lithographies. For many applications (e.g., sensor design and nanoelectronics) DPN will be interfaced with architectures with microscopic features generated via the parallel conventional light-driven or stamping processes. For such applications, the speed of DPN is already adequate because

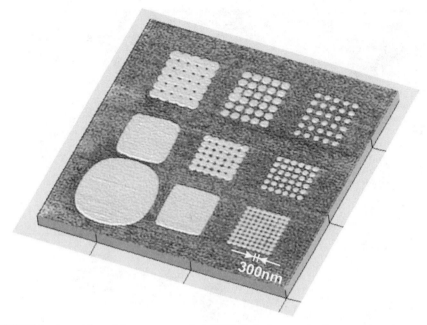

FIGURE 6 A combinatorial array-generated via DPN. Each dot is made of mercaptohexadecanoic acid chemisorbed to a gold substrate.

the surface area that needs to be patterned is quite small (e.g., the functional components of a microelectrode array).

ACKNOWLEDGMENTS

CAM acknowledges the AFOSR and DARPA for support of this work. L. Demers and S. Hong are also acknowledged for their help in preparing most of the figures used in this chapter. Portions of this chapter have appeared in the MRS Bulletin, vol. 26, p. 535.

REFERENCES

1. J. K. Schoer, R. M. Crooks, "Scanning probe lithography: Characterization of scan tunneling microscope-induced patterns in n-alkanethiol self-assembled monolayers," *Langmuir* 1997, *13*, 2323; S. Xu and G. Liu, "Nanometer-scale fabrication by simultaneous nanoshaving and molecular self-assembly," *Langmuir* 1997, *13*, 127.
2. R. D. Piner and C. A. Mirkin, "Effect of water on lateral force microscopy in air," *Langmuir* 1997, *13*, 6864; R. D. Piner, J. Zhu, F. Xu, S. Hong, C. A. Mirkin, "Dip pen nanolithography," *Science* 1999, *283*, 661; S. Hong, J. Zhu, C. A. Mirkin, "A new tool for studying the *in-situ* growth process for self-assembled monolayers under ambient conditions," *Langmuir* 1999, *15*,

7897; C. A. Mirkin, "A DNA-based methodology for preparing nanocluster circuits, arrays, and diagnostic materials," *MRS Bull.* (January) **2000**, 43.

3. S. Hong, J. Zhu, C. A. Mirkin, "Multiple ink nanolithography: Toward a multiple-pen nanoplotter," *Science* **1999**, *286*, 523.

4. S. Hong, C. A. Mirkin, "A nanoplotter with both parallel and serial writing capabilities," *Science* **2000**, *288*, 1808.

5. D. A. Weinberger, S. Hong, C. A. Mirkin, B. W. Wessels, T. B. Higgins, "Combinational generation and analysis of nanometer- and micrometer-scale silicon features via 'dip-pen' nanolithography and wet chemical etching," *Advanced Materials* **2000**, *12*, 1600.

6. C. A. Mirkin, "Programmong the assembly of two- and three-dimensional architectures with DNA and nanoscale inorganic building blocks," *Inorg. Chem.* **2000**, *39*, 2258.

7. Mirkin, C. A., *MRS Bull.* **2001**, *26*, 535–538.

8. Mirkin, C. A., Hong, S., Demers, L. *Chem. Phys. Chem.* **2001**, *2*, 37–39.

9. Demers, L. M., Mirkin, C. A. *Angew. Chem. Int. Ed.* **2001**, *40*, 3069–3071.

10. Demers, L. M., Park, S. J., Taton, T. A., Li, Z., Mirkin, C. A. *Angew. Chem. Int. Ed.* **2001**, *40*, 3071–3073.

11. Ivanisevic, A., Mirkin, C. A. *J. Am. Chem. Soc.*, **2001**, *123*, 7887–7889.

12. Li, Y., Maynor, B. W., Liu, J. *J. Am. Chem. Soc.* **2001**, *123*, 2105–2106.

13. Maynor, B. w., Li, Y., Liu, J. *Langmuir* **2001**, *17*, 2575–2578.

Nanolithography with Electron Beams: Theory and Practice

MARTIN C. PECKERAR, R. BASS, AND K.-W. RHEE

U.S. Naval Research Laboratory, Electronics Sciences and Technology Division, Washington D.C.

C.R.K. MARRIAN

Defense Advanced Research Projects Agency, Electronic Technology Office, Arlington, Virginia

1. INTRODUCTION

An "image" is projected onto a surface. This image is a light intensity pattern or a particle density pattern—some form of radiation incident on the surface. We begin our discussion with this general concept and those elements that are common to any patterning technique. The "image" creates chemical or structural change in the material beneath the surface. Material exposed to incident flux may be "hardened" or "softened" with respect to removal processes (an etch or solvation). The incident image is thus transferred to the underlying material during this removal, as illustrated in Fig. 1. This process, in which an image is formed on a surface and underlying material is subsequently patterned, is called *lithography*.

In nanolithography, the primary focus of this chapter, we form patterns whose critical dimensions may be on the order of 1 nm (10 Å, or 10^{-7} cm). In practice, these features may be somewhat larger (by more than an order of

Direct-Write Technologies for Rapid Prototyping Applications

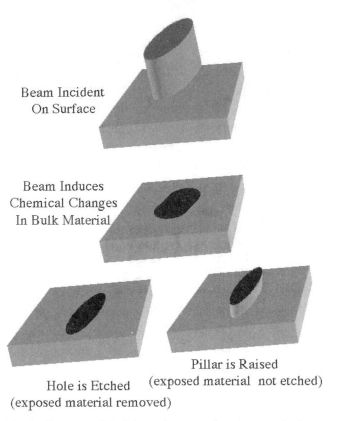

Beam Incident
On Surface

Beam Induces
Chemical Changes
In Bulk Material

Pillar is Raised
(exposed material not etched)

Hole is Etched
(exposed material removed)

FIGURE 1   An illustration of the lithography process from image projection to etch.

magnitude). In any event, we distinguish between nanolithography and microlithography which is more routinely practiced in the chip fabrication industry. An overview of where the microchip industry is moving in lithography is included as Fig. 2. The industries' "best estimates" of minimum resolved feature as a function of time is shown for the various years in which the projections were made.

A few things are evident from this chart. First, there is considerable uncertainty in the development timeline, and estimates change year by year. There is also a trend to estimate more aggressively as the years progress (culminating in the most aggressive International Roadmap Committee (IRC) recommendation of 1999). Finally, note the phrase "Dense Lines (DRAM Half-Pitch)" in the upper right-hand corner. The roadmap committee recognized (and rightly so) that capabilities for different types of feature shape and density lead to significantly different capability specifications. This subject will be dealt with in some detail in this book.

315

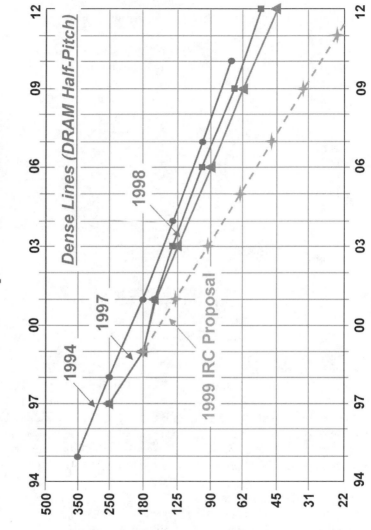

FIGURE 2   An illustration of the lithography process from image projection to etch.

In any event, we see from this figure that industrial planning does not reach levels approaching nanolithography before the year 2006. After 2006, we can anticipate a twofold change in the microelectronics industry. First, optical lithography, which has been the mainstay of chip production over the lifetime of the industry, will at least be augmented by some nonoptical techniques. Second, there will be a broader class of devices for designers to choose from. Specifically, "quantum-effect" devices (such as resonant tunneling structures, single electron transport (SET) devices, etc.) will become available (1). There is a wealth of literature in optical lithographic techniques (2), which will probably be of minimum use in nanostructure patterning technology. This chapter deals with what has become the work-horse of the industry in providing ultimate control in boundary placement—the probe-forming e-beam tool as it is applied to nanostructure patterning.

The problem of nanolithography can be divided into three parts:

1. The creation of the high-resolution areal image on a surface (probe formation and translation).
2. The formation of a "latent" image in some medium beneath the surface (scattering effects).
3. The transfer of the latent image pattern to underlying material.

This chapter deals with the first two of these issues.

Throughout this introduction, we have not used the word "resolution." This, at first, would seem odd in a discussion of image technology. Optical science has provided a number of criteria (such as that of Rayleigh and Sparrow) to define resolution (3). The problem is that these quality factors were designed for something entirely different from what we are attempting in nanolithography. Optical resolution criteria were designed to predict when the human eye would perceive two point light sources as distinct. In nanolithography we are trying to define techniques for creating distinctly isolated features whose boundaries fall at predetermined points on a surface.

An illustration of this concept is shown in Fig. 3. The example worked in this figure is for an $f\backslash1$ lens[1] at unity magnification. Point sources of 0.254 micron wavelength light are assumed. Here, the eye can clearly discern two overlapping "pulses" in the image plane. Now let us assume that the medium in which we wish to print this image has a switchlike response function. That is, any amount of exposure over a critical exposure level leads to full exposure, and any exposure level less than this leads to no exposure. A resist whose critical level is at 0.5 (line A) in this figure will create a single line of exposure

---

[1] The "f number" of an optical system is defined as the ratio of the lens focal length to the effective diameter of the aperture. Here, "effective" refers to the diameter of the incident light beam which, on entering the lens, actually fills the real aperture, or "stop."

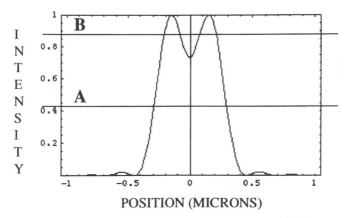

POSITION (MICRONS)

FIGURE 3 The impulse response function for two point sources of light separated by one Rayleigh distance. An incident wavelength of 0.248 microns is assumed. The normalization of the plot is arbitrary.

(no separation) about 0.7 microns wide. If the critical exposure level is greater than 1, there will be no pattern exposed. Critical exposure levels between about 0.75 and 1 (line B) yield variable size "spots" whose boundary separation increases from zero to the correct amount (one Rayleigh separation, or about 1.22λ, where λ is the wavelength of the incident light—in this case about 0.25 microns.

It is important to note that there is only one exposure density that will provide two point exposures separated by a Rayleigh resolution distance. That is the intensity 1 point. By varying exposure intensities, a variety of areal images can be obtained for a given projected pattern. Of course, the switchlike exposure material response curve is usually not obtainable in real materials and any other areal image with variations in exposure flux may not be as severe as the case examined here. But clearly many factors (in addition to the optical quality of the imaging system) determine final feature geometries.

As a result of these considerations, we use minimum feature size as a quality factor. This refers to the smallest critical feature dimension reproducibly obtained by a given process. But even this simple definition is ambiguous. How do we define "critical dimension?" If the side wall of the pattern feature has some slope to it, do we take the top of the feature or the bottom for measurement purposes? Even before we reach true nanometric dimensions, the measurement tool has a response function which is convolved with the actual feature profile data. Some de-convolution is necessary to ascertain boundary position. This de-convolution is not always possible. Sidewall slopes, and the ability of a measurement tool to accurately follow a boundary, depend on the density and shape of the features measured.

Issues of resist performance and its impact on boundary placement are discussed further in Section 4. But the issues of sidewall slope receive somewhat of a cursory treatment, as the full impact of this parameter on nanostructure formation is currently unclear. Research in this area is ongoing. But much of nanolithography is accomplished with "ultrathin" resist (resists thinner than 200 nm). Thus, there is very little change in exposure density through the film thickness. Two-dimensional models seem to work fairly well at present.

2. THE AREAL IMAGE

In direct-write e-beam lithography, a pencil-like beam is used to write the image in a serial fashion. In the later case, an image is formed as a flux-density pattern in a manner similar to optical image projection.

In the case of e-beam lithography, we may make further distinctions. In the past, we might have divided the field into high-energy (>10 KeV) and low-energy (<10 KeV) cases. Now, atom scanning tunneling microscopes (STMs) and atomic force microscopes (AFMs) are being used in nanodevice fabrication. These "proximal probe" techniques provide ultralow energy beams (<1 KeV). Despite their low energy, beams created by proximal probes have been shown to create significant chemical modification of material surfaces and the bulk regions immediately below these surfaces. Even though the single-point serial exposure technique is slow, it may be possible to utilize it for making a small number of critical devices embedded in a larger integrated structure. Because of their potential importance, proximal probe lithography is discussed here separately.

3. CONVENTIONAL PROBE-FORMING E-BEAM TOOLS

In a probe-forming system, a pencil-like beam of charged particles is focused to a circular spot. The spot is moved to the various exposure locations by a combination of magnetic and mechanical controls. For many years, this type of e-beam technology has played an important role in integrated circuit development (4). Initially, attempts were made at writing the patterns directly on production wafers. This process proved too slowly for high-volume manufacturing and the e-beam equipment at the time did not have sufficient reliability to be viable in manufacturing. IBM corporation led the way in personalizing gate arrays (writing the final level of metallization) using a direct-write tool. For limited volume runs, this process is still used today.

The major use of e-beam tooling is in the mask-making area. Here, the e-beam is used to write a negative of the image that is usually used in an optical projection system for mass replication. The mask image is usually reduced by some integer factor on the work piece. This is referred to as "nx" reduction. As the mask volume production requirement is not as high as the wafer production requirement, lower throughput is tolerated.

The simple fact that e-beams have created the smallest lithographically produced features accomplished to date (5) is of particular interest to this discussion. As we shall see, electrons can be accelerated to voltages at which diffraction effects are negligible. Of course, when the beam hits the target substrate, it splays apart and beam–matter interactions determine ultimate resolution. But even these effects can be controlled to some extent. The degree to which this control is possible is a principal subject of this book.

We begin our discussion of conventional e-beam technology with an overview of the system. We refer to the system block diagram shown in Fig. 4. This is very much a bare-bones system snapshot, glossing over many of the critical subsystems necessary to make the tool work. Next, we move through the system from the "top down," beginning with the electron gun.

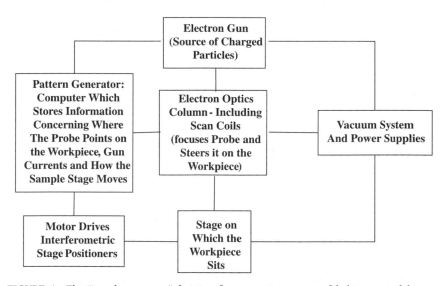

FIGURE 4 The "impulse response" function for two point sources of light separated by one Rayleigh distance. An incident wavelength of 0.248 microns is assumed. The normalization of the plot is arbitrary.

3.1. THE ELECTRON GUN

The gun is the source of electrons—the exposing particle. It also accelerates these electrons to the energy at which they transverse the column. Because the image projected onto the workpiece is a demagnified image of the source, the source therefore helps determine the smallest probe size the system can attain. We begin our discussion of electron sources with some basic concepts and definitions.

Free electrons are created in a vacuum by thermionic processes, field emission processes, or some combination of each (6). In thermionic emission, a solid is heated to the point that electron kinetic energies become high enough to surmount the surface potential barriers which normally confine these electrons to the solid. In field emission, the electric field at the surface is sufficient to lower the surface potential barrier below the mean kinetic energy of free electrons in the solid, and electrons stream out into vacuum. As temperature increases the mean kinetic energy of the free electrons in the solid, emission can take place by a combination of the two processes. This is known as the Schottky emission.

Thermionic emission is characterized by the Richardson equation:

$$J_R = AT_c^2 \exp\left(\frac{-\phi}{kT_c}\right) \tag{1}$$

where A is a material-dependent constant, T_c is the cathode temperature (K), ϕ is the work function of the cathode material, and k is Boltzmann's constant.

Field emission current densities can be described by the Fowler–Nordheim equation:

$$J_{FN} = \frac{k_1|E|^2}{\phi} \exp\left(\frac{-k_2\phi^{3/2}}{|E|}\right) \tag{2}$$

where $|E|$ is the magnitude of the electric field normal to the surface of emission.

Langmuir (7) observed that there is a limit to the amount of current that can be brought to focus on a spot of some area, which is independent of diffraction and aberration conditions. This limit depends on the accelerating potential E_a of the gun, the current density at the emitting cathode J_0, and α, the solid angle subtended by the exit aperture of the electron optical system when viewed from the point of intersection of the sample surface with the optical centerline of the e-beam column. Next, we outline the basic considerations leading to Langmuir's result, and we present this result itself. This approach clarifies some of the basic notation and underlying principles associated with electron optical systems. Also, knowledge of the maximum current density in the image spot

Exit Aperture
Area $= \pi\, r^2 = \pi\, (R\, \alpha)^2$

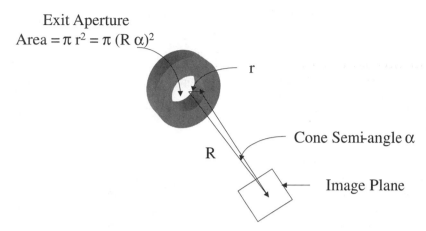

r

Cone Semi-angle α

R

Image Plane

FIGURE 5   Definition of terms used in the brightness calculations.

allows a determination of the exposure speed, which is important in manufacturing facilities.

The Langmuir result is an extension of the brightness conservation theorem of classical optics (8). This theorem states that the image cannot be brighter than the source; and, if there are no absorption losses in the system, the image brightness will equal that of the source. The brightness, β, of a particle source is the current per unit/area per unit of solid angle subtended by the source, $d\Omega$. Referring to Fig. 5, let us take the solid angle to be defined by some pupil aperture of the electron optical system, which we assume to be a circle of radius r.

In that case, the solid angle is defined by a cone whose base is just the area $A_{pupil} = \pi r^2$ and whose height is R, the separation of the aperture from the image plane. Thus, we may write:

$$d\Omega = \frac{A_{pupil}}{R^2} = \pi \frac{r^2}{R^2} \tag{3}$$

If α is small, we have $r = R\alpha$, and:

$$d\Omega = \pi\alpha^2 \tag{4}$$

Given these definitions, the maximum gun brightness is written:

$$\beta_{max} = \frac{J_{max}}{d\Omega} \tag{5}$$

or, alternatively:

$$J_{max} = \pi\beta_{max}\alpha^2 \tag{6}$$

The total number of electrons per second (flux) admitted into the optical system from a source of area A is

$$F = \beta A_o d\Omega_o \tag{7}$$

Similarly, the flux exiting the system is given as

$$F' = \beta A_i d\Omega_i \tag{8}$$

If there is no absorption or reflection in the system:

$$F = F' \tag{9}$$

Now let us consider what happens to the image of the object (the primary source) as it propagates through the optical system. The size, S_i of the image of the source is related to that of the object, S_o, through the magnification factor, M:

$$S_i = MS_o \tag{10}$$

and so the area of the object and image are related as shown:

$$\frac{A_i}{A_o} = M^2 \tag{11}$$

The magnification factor for electron optics is the same as that for light optics:

$$M = \frac{r_i}{r_o} \tag{12}$$

or:

$$M^2 = \left(\frac{r_i}{r_o}\right)^2 \tag{13}$$

where r_i and r_o are the separations of the image and object from the pupil plane. The ratio of solid angles is also related to the magnification factor:

$$d\Omega_o = \frac{A_{pupil}}{r_o^2} \tag{14}$$

and

$$d\Omega_i = \frac{A_{pupil}}{r_i^2} \tag{15}$$

Thus:

$$\frac{d\Omega_i}{d\Omega_o} = \left(\frac{t_o}{r_i}\right)^2 = \frac{1}{M^2} \tag{16}$$

From these results we see that the "optical throughput" of the system, $Ad\Omega$, remains constant through the system. From the consistency of the flux, we have

$$\beta_i A_i d\Omega_i = \beta_o A_o d\Omega_o \qquad (17)$$

Because the $Ad\Omega$ terms are equal, and the total number of electrons entering the system leave the system (i.e., unity transmission factor), the brightness of the image and object are the same.

Given the brightness conservation theorem, Langmuir first evaluated the brightness of the source. He did this by using a simplification of the Fowler–Nordheim approach, in which the energies of electrons in the emitting solid could be expressed as a Maxwellian distribution. The electrons incident on the surface could escape if their incident energy was greater than some critical energy, E_c. Any external field, E_a, applied normally to the emitting surface would lower this critical energy. This determined the current density of the emission. Emission from the surface was taken as isotropic (independent of angle). Thus, the brightness of the source could be obtained by dividing this current density by 2π Sr. The resulting expression is given here:

$$\beta_{source} = \frac{J_0}{\pi}\left(1 + \frac{eE_a}{kT_c}\right) \qquad (18)$$

where e is the charge on the electron and T_c is the cathode temperature. J_0 is, essentially, the Richardson J_c (Eq. 1). E_c is included in this term. As the second term in parentheses is much larger than 1, we approximate:

$$\beta_{source} = \frac{J_0}{\pi}\frac{eE_a}{kT_c}\left(\frac{E}{T}\right) \qquad (19)$$

The same brightness will be obtained for the image. If we move the object closer to the entrance pupil to capture more electrons, the image-magnification factor increases, spreading the image over a larger area, reducing the current density. This maintains the throughput fact as a constant across the optical system.

We can thus evaluate the current density in the image spot by multiplying this brightness by a solid angle. The correct solid angle is that which is defined by the pupil aperture, because no electrons are incident on the image from regions outside this solid angle. As discussed, the pupil solid angle defined at the image spot is $\pi\alpha^2$. Thus the maximum current density in the image spot is, after substitution of constant values,

$$J_{max} = 11{,}600J_0\frac{E_a}{T_c}\alpha^2 \qquad (20)$$

TABLE 1   Critical Parameters for Various Types of Electron-Emitting Sources

Source	T_c (K)	ϕ	J_c	r	ΔE
Tungsten (W) hairpin	2,500–3,000	4.5 eV	1–3 A/cm²	0.01 μm	1–2 eV
Lanthanum hexaboride	1,400–2,000	2.7 eV	20–50 A/cm²	0.005 μm	1 eV
Cold field emission (W)	300	4.6 eV	2×10^5 A/cm²	0.1 μm	0.2–0.4 eV
Hot field emission (W)	1,800	4.6 eV	5×10^6 A/cm²	0.01 μm	0.5–0.7 eV

In computing the available current, one other factor must be described—curvature induced field enhancement. Lines of electric field concentrate around points of high curvature. The magnitude of the field at the surface of the emitter is, in fact, dominated by this curvature. If U is the potential difference between the emitter and the first extraction anode, the electric field is:

$$|E| = \frac{U}{r} \qquad (21)$$

where r is the radius of curvature of the emitting surface. Table 1 allows for evaluation of the maximum spot current densities for a number of common electron emission sources.

The width of the energy distribution of emitted electrons is important because it creates a "chromatic aberration" in the final image. (This will be discussed later.) The energy spreads for the various emitters are shown in the table. Needless to say, there is a continuing drive to reduce the energy spread in the emitted beam. Thus "cold cathode" emitters are actively studied (9–11). The thrust of this work is to use a combination of low (or negative) electron affinity materials in a field-emission configuration (shown in Fig. 6) to reduce the width of the electron energy distribution. Many of these cathodes were designed for high current density operation as well as for small chromatic aberration. However, the stochastic nature of electron–electron interactions at cross-over broaden the distribution. This will be discussed further.

The requisite level of irradiation necessary to achieve a critical exposure level is called the sensitivity, S. For electron-beam irradiation, this is given in Coulombs per square centimeter ($[C]/[cm]^2$). Clearly, sensitivity is a function of incident energy, under this definition. By knowing both the maximum current density possible and the sensitivity, we can work out the flash time (minimum exposure time) per pixel:

$$t_{flash} = \frac{S}{J_m} \qquad (22)$$

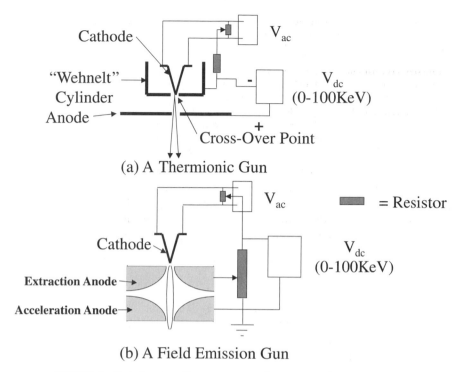

Cathode

"Wehnelt"
Cylinder

Anode

V_{ac}

V_{dc}
(0-100KeV)

Cross-Over Point

(a) A Thermionic Gun

Cathode

V_{ac}

= Resistor

Extraction Anode →

Acceleration Anode →

V_{dc}
(0-100KeV)

(b) A Field Emission Gun

FIGURE 6 Typical gun configurations currently in use in e-beam technology.

Now let us turn our attention to the actual physical construction of the electron gun. We do this with reference to Fig. 6. Here, we see two types of gun: the basic thermionic gun (in which high temperatures are responsible for electron emission) and the field-emission gun (in which high fields assist the emission process). With the field-emission gun, a secondary cross over is obtained, which defines the source brightness.

3.2. LENSES FOR ELECTRON OPTICS

In this section, we consider the optical components that bring the electron beam to focus on the imaging plane. The magnetic lens is simply a wire coil wound as shown in Fig. 7. The "fringing" of the field is responsible for the focusing effect. If the electron trajectory were parallel to the magnetic field lines, there would be no magnetic force. The fringing gives rise to a y component of the magnetic field that in turn gives rise to a component of

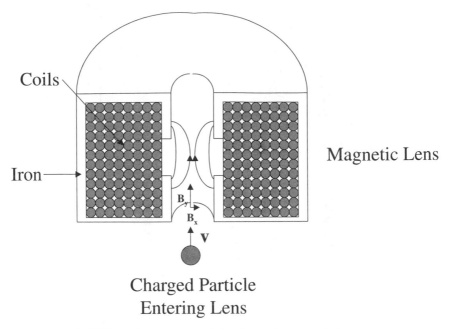

Coils

Iron

Magnetic Lens

B_y

B_x

V

Charged Particle
Entering Lens

FIGURE 7   A typical magnetic lens for an electron optical system.

force into the plane of the paper, as the electron enters the lens with its velocity vector directed along the x axis.

As the electron moves through the fringing field, it acquires a velocity component directed into the plane of the page. The $v_z \times B_x$ force that results forces the electron toward the optical axis of the system. The y component of the magnetic field is stronger as we move off-axis. This drives the electrons entering further off-axis to be pushed more vigorously to the optical axis than are those entering closer to the optical axis. Thus an extended beam will be focused, as in a light optical system.

Just as in a light optical system, the focused beam will suffer aberrations—departures from an ideal, diffraction-limited circular spot. The dominant aberration in most electron optical systems is a result of the fact that the nature of the forces in the magnetic lens does not allow all off-axis rays to be brought to a common focus. This is the charged-particle analogy of spherical aberration. For magnetic lenses, electrons incident on the lens further from the optical axis cross that axis closer to the lens than do those electrons initially closer to the optical axis. In addition, even if the forces were such as to create an ideal focus, that condition would hold only for electrons of some single energy. As there is a naturally occurring electron energy spread there is a spot blurring which is analogous to chromatic aberration. Finally, local charging effects in the lens or in the column distort the spot along a given direction. A line may be well focused in one orientation and out of focus in another orientation as a result of this uniaxial distortion. This behavior is similar to astigmatism in light optics (8).

As a result of these effects, electrons are not brought to a point focus; rather, they form a "spot of least confusion" on the image plane. The degree to which each of these factors influences the diameter of this spot is determined by an aberration coefficient and the size of the final aperture as seen from the image plane. The latter is determined by the cone half-angle, α, of the final aperture as drawn from the image spot. Each aberration makes its contribution to the broadening of the image spot diameter.

In addition to these optical aberration effects, diffraction plays some role in determining spot size. Furthermore, as a result of conservation of gun brightness, there is a geometric probe diameter, d_g, which exists independently of any aberration in the system. To see this, we realize that the total current in the probe is given as:

$$I_p = \frac{\pi}{4} d_g^2 j_p = \frac{\pi^2}{4} \beta d_g^2 \alpha^2 \tag{23}$$

The equality on the right was obtained using Eq. 26. We can solve for d_g^2, yielding:

$$d_g^2 = C_0 \alpha^{-2} \tag{24}$$

where:

$$C_0 = \frac{4 I_p}{\pi^2 \beta} \tag{25}$$

The functional expression for each of these terms and some typical values for the aberration coefficients are given in Table 2. The chromatic aberration coefficient, C_c, is approximately equal to the focal length of the lens. U is the accelerating voltage of the electrons.

All of these individual broadening factors contribute to the final spot size, summing as a sum of squares:

$$d_p^2 = d_g^2 + d_d^2 + d_s^2 + d_c^2 \tag{26}$$

The resulting equation is a polynomial in α which exhibits a minimum depending on the accelerating energy, E, the aberration coefficients, the beam brightness, β, and the probe current, I_p. This is shown in Fig. 8.

TABLE 2   Terms of the Probe Size Calculation

Aberration	Coefficient	Form	Magnitudes
Spherical aberration, d_s	C_s	$0.5\,C_s\alpha^3$	10–100 mm
Chromatic aberration, d_c	C_c	$C_c\left(\dfrac{\Delta E}{E}\right)\alpha$	1–10 mm
Diffraction broadening, d_d	—	$0.6\dfrac{\lambda}{\alpha}$	$\lambda = \dfrac{1.226}{\sqrt{U}}$
Base probe diameter, d_g	C_0	$C_0\alpha^{-2}$	Eq. 25

More recently, this approach has been extended to multiple-lens systems, assuming correlation among the aberrations of the system components (12,13). The new relationship is:

$$d_g^2 = \frac{\lambda^2}{4\alpha^2} + \frac{\alpha_o^6 M^2}{16}\left(C_{s1} + \frac{C_{s2}}{M^4 r^{3/2}}\right)^2 + \frac{\alpha_o^2 M^2}{4}\left(\frac{\Delta E}{E}\right)^2\left(C_{c1} + \frac{C_{c2}}{M^2 r^{3/2}}\right)^2 + M^2 d_0^2$$

$$(27)$$

where C refers to the aberration coefficients (spherical or chromatic) of each lens, M is the magnification factor of the system, and r is the ratio of accelerating voltages in the gun and in the column, respectively. The d_0 term is the size of the

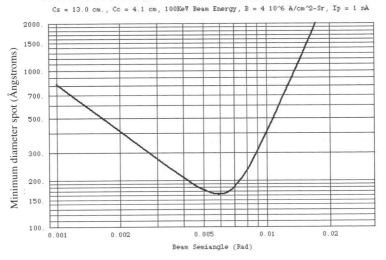

MINIMUM SPOT SIZE AS A FUNCTION OF SEMI-ANGLE

Cs = 13.0 cm., Cc = 4.1 cm, 100KeV Beam Energy, B = 4 10^6 A/cm^2-Sr, Ip = 1 nA

FIGURE 8   Spot size as a function of semiangle for a typical e-beam tool.

electron source. The expression for λ is given in Table 2. The terms α and α_0 are the gun and final aperture semiangles, respectively. This expression can be optimized for both the semiangles and the magnification factor, M.

3.3. STOCHASTIC EFFECTS

The considerations provided in the preceding section parallel discussions of classical geometric optics. Electrons behave more like particles than waves as they pass down an e-beam column. There is a considerable graininess to the beam (which may be present in photon beams in the low-light-level limit). But electrons repel one another, while photons do not. This has considerable impact on the ability of an e-beam system to reproducibly form a minimum size spot. This blurring occurs in a surprising way. If the beam has sufficient current in it, and a large number of electrons appear at crossover at once, there is a well-defined blurring that can be compensated for by re-focusing (changing the lens coil current). Problems occur because in the relatively low current range used in lithography, there is a considerable statistical (i.e., "stochastic") variation in the number of electrons at crossover at a given time. This statistical broadening cannot be eliminated by refocus, as described.

There are two types of stochastic interaction of concern here. First, as the beam propagates through the column, energy redistribution takes place via the Coulomb interaction. Thus the beam thermalizes and the energy distribution of the beam broadens. This manifests itself as an enhancement of chromatic aberration, referred to as the Boersch effect (14). The second effect is the result of the random lateral displacement of particles which naturally occurs as a result of Coulombic repulsion (mainly at crossover). These are referred to as trajectory displacement effects. Both effects are described in some detail in the monograph by Jansen (15).

3.4. WRITING STRATEGIES

As is evident in Fig. 2, the e-beam system is far more than an electron source and focusing optics. Data concerning the shapes to be written must be created by graphics-language computer programs. These data must be further processed in that information on how the shapes are to be written must be appended to the original graphics file. The shapes must be broken up into individual beam flashes called *pixels*. The size and placement of the pixels is known as the *address-structure* of the pattern. Information on how long the beam dwells on a given pixel is expressed in terms of system clock frequency. The resulting machine-control data file will provide a data stream to magnetic deflection coils and beam blankers in the e-beam column. The coils deflect the beam to the addressed pixel and the beam blankers adjust the dwell time for a given addressed pixel. The resulting data package is called a *job file*.

E-beams are severely constrained in the ways they can write a pattern. Furthermore, writing approaches (or strategies) are brand dependent. There are no e-beam tools currently available which can arbitrarily adjust the address structure and clock on a pixel-by-pixel basis. This has dramatic impact on our ability to achieve a given patterning goal. This will become more evident when we discuss beam-matter interaction. Let us take, as an example, one of the e-beam tools currently available for nanostructure development—the "nano-writer" produced by Japan Electronics Corporation (JEOL).

The nanowriter is a vector-scan machine. This is a contrast to the raster-scan machine that is in widespread use in mask-making for the IC industry. In a raster-scan tool, the address of every pixel in a scan field is loaded onto the deflection coils. The beam is blanked or unblanked, depending on the positions of the various features written. In a vector-scan tool, the beam is "vectored" to a shape and the shape is "painted" by addressing all the pixels in a given written shape. Although vector-scanned tools would appear to be much faster than raster-scan tools, such is not necessarily the case. When the beam is moved over a large distance, the scan-coil currents change dramatically. This gives rise to eddy currents which must subside before the beam stabilizes on a target pixel. Also, other machine "overheads" figure in to the write-time determination. For example, the beam must be periodically returned to an alignment fiducial which compensates for "column drift." This drift arises from a variety of causes including local charging effects in the column and instabilities in the electronics.

The "address structure" (center-to-center spacing of the pixels) is fixed through the writing job. As long as each pixel is assigned a different data type designation, the clock (reciprocal dwell time) can be varied on a pixel by pixel basis. This allows considerable flexibility in modulating the dose in every pixel. However there is a limit on the number of "clocks" which can be appended to each pixel. A single line of pixels can be written with up to 16 separate clocks. A feature which exists over some area can be written with 64 clocks.

The nanowriter strives for high resolution. As a result of the brightness-conservation theorem just discussed, minimum spot size and beam current density are related. Thus, changing the beam current will change spot size. It is therefore not practically possible to change gun current to modulate dose. Other tools, like the ETEC MEBES raster-scan-pattern generator work with relatively large spots (on the order of, or greater than, 100 nm). Here, beam current can be changed during writing. The MEBES writing strategy is presented here as it illustrates how complex strategies can be employed to achieve improved edge acuity and boundary placement.

Write strategies in MEBES center around the "multiphase printing" (MPP) option. MPP is broken into two parts: offset-scan voting (OSV) and phased feature pixelization (PFP). In PFP a "superpixel" is created by grouping pixels at the smallest available address size into 2×2 or 4×4 groups. The pattern is exposed with spot and address sizes which are bigger by a factor of two or four (depending on the pixel group size). The pattern is exposed on four separate passes, each pixel written on the same center as it would have been in a normal write using the finest address structure available. The net effect in the $2 \times$ case is to improve edge placement and edge acuity by an averaging process which cancels the effect of random column drift and charging.

The $4 \times$ case is an update of what used to be called virtual addressing. During each pass, one-sixteenth of the superpixel is written. The beam visits the superpixel 4 times in total (with a beam size and address grid magnified by a factor of 4). Thus, only one-fourth of the total number of fine-address pixels are written. This leads to a significant improvement in throughput. The averaging effects of multiple passes in a single superpixel still improves pattern placement. The fact that the beam is dark during a portion of the superpixel write allows for feature resizing and a "firewall" approach to proximity[2] control.

Now let us turn our attention to OSV. Remember that in PFP a single superpixel is exposed four times. In OSV, the first writing phase, the data is written as it would be without OSV. In the next phase the data is displaced within the write scan by one-fourth of the superpixel height and the write scan is also displaced an equal amount in the opposite direction. Thus each of the fine-address pixels is written four times. This again provides an average that improves boundary and feature placement overall.

To conclude, we see that e-beam tools are not infinitely flexible; there are limits on how the beam is sized and on how the energy is deposited on a pre-assigned grid. Studying the limitations of each machine allows us to create effective strategies to optimize boundary placement, feature placement, and edge acuity. These concerns are of a practical nature, because they pertain to limitations created by actual instantiations of the tool. In the following section we cover those limitations that pertain to the basic physics boundary position control created by scattering processes in resist and substrate materials.

4. MATHEMATICAL APPROACHES TO PROXIMITY CONTROL

This section explores the resolution limits of e-beam lithography, by looking at a mathematical optimization technique known as the simplex method, in use since the Second World War. Simplex is known to provide parameter

---

[2] Proximity effect is the exposure of distant pixels by writing in a single pixel. This is an important topic for discussion in this text, and will be discussed in great detail in Section 2.

optimization in light of previously agreed-upon criteria in a mathematically rigorous sense (16). Furthermore, this technique will inform the user of the feasibility of achieving a given feature in a given resist. This technique was applied to the e-beam proximity-effect problem in the early eighties (17). This work was performed on extremely small data sets (<100 pixels), and in one dimension only). This work extends such an approach to data sets including tens of thousands of pixels, whose dimensions are all less than 0.1 μm. Two-dimensional patterns are used and experimental verification of the approach is provided.

For systems in which strong coupling exists among all computational points (as is the case in e-beam proximity-effect correction) the simplex method can be slow. Its order of complexity is higher than that of matrix inversion (greater than third-order complexity). The situation is greatly improved using parallel processors, and parallel codes have been written for this problem. Even without the use of parallel computation, patterns of meaningful dimension can be corrected and strategies for more ambitious data sets can be derived.

In the following sections, we provide some background on previous approaches to the proximity-correction problem, the experimental and theoretical methods employed are described, and results (including those obtained computationally and by experiment) are presented.

4.1. A BRIEF HISTORY OF PROXIMITY CORRECTION

It is well known that energy deposited in one pixel can migrate to pixels far removed from the original point of deposition (18). The total energy absorbed in any one pixel is a convolution of energy deposited in all other pixels. The convolution integral is expressed numerically as follows:

$$\mathbf{M}d_I = d_A \tag{28}$$

Here, $d_A(\mathbf{r})$ is a column vector each of whose entries is the total energy absorbed in a specific pixel; $d_I(\mathbf{r})$ is also a column vector, each of whose entries is the individual dose applied to the ith pixel; $\mathbf{M}$ is an interaction matrix, derived from the point spread function. Each entry, M_{ij} is the fraction of energy deposited in pixel j from a unitary dose applied to pixel i. The M_{ij}s are coupling constants describing the strength of interaction between pixels. It should be noted that, as the range of secondary electrons is large, the interaction matrix is not sparse. Attempts to control proximity by changing pixel-to-pixel doses are called "dose modulation" schemes.

It would seem easy to simply invert Eq. 28. One can solve for the requisite pixel-dose values needed to achieve the desired integrated dose per pixel. In fact, a number of authors have done this (19,20). Even if it were computationally tractable to accomplish this inversion, the results would be incorrect. In order to solve the $(n \times n)$ array of linear equations, "negative doses" are invariably required. It is not possible, physically, to apply such doses. Simply ignoring these doses provides totally erroneous results (21).

One way to deal with the issues of computational complexity and the possibly singular nature of the interaction matrix is to recast Eq. 28 as a problem in optimization. We begin by forming a "cost function":

$$f = (\mathbf{Md_I} - \mathbf{d_A})^{\dagger}(\mathbf{Md_I} - \mathbf{d_A}) \tag{29}$$

Which expands to:

$$f = \mathbf{d_I^{\dagger}M^2d_I} - \mathbf{d_I^{\dagger}Md_A} - \mathbf{d_A^{\dagger}Md_I} + \mathbf{d_I^{\dagger}d_A} \tag{30}$$

(assuming $\mathbf{M^{\dagger} = M}$). We can view this as a quadratic, which we minimize by variation of individual pixel doses (contained in the vector $\mathbf{d}$). Clearly, when $\mathbf{Md_I} = \mathbf{D_A}$, f is zero and we have a solution to the problem. Furthermore, even when no exact solution is possible, this approach provides an optimum solution for the individual pixel doses, in a "least-squares" sense. Also, by viewing the problem this way, one sees the issue of negative doses more clearly—the paraboloid that Eq. 3 defines simply has a minimum in the negative orthant.

But we must still find a way to avoid negative doses. The following approach was outlined in 1995 (21). We must, of course, find a way to constrain the solutions to the real positive numbers. Such constrained optimization can be performed using the Method of Lagrange Multipliers. For the case at hand, we simply add a functional form to Eq. 30 which becomes large as we approach a negative solution. Such a form is called a "regularizer." Any function which is small when the d_is are positive and large when they approach negative values will do. The regularizer we choose is the informational entropy of the dose image. We define this entropy, S, as:

$$S = \sum_i \left(\frac{d_i}{d_T}\right) Ln\left(\frac{d_i}{d_T}\right) \tag{31}$$

where d_i is the individual excel dose, d_T is the summed dose over all the pixels and Ln is the natural logarithm. The sum, i, is over all the pixels. The new cost function becomes:

$$f' = f - \lambda S \tag{32}$$

The value of this approach is that it is iterative, and imaging improvements can be achieved at each cycle of the iteration. We can terminate iteration when the computation exceeds some bound in time. However, many iterates may be needed to achieve complete optimization of an image.

Nonmathematical approaches have been attempted. One of the most popular of these is the GHOST technique (22). Here, the normally unwritten field is written with a broadened, attenuated Gaussian beam. This equalizes the background dose and makes contrast uniform from feature to feature (without regard to size or shape). This approach has been demonstrated to fail for certain classes of shapes (23). Furthermore, as confirmed by the analysis in this chapter, attempts to GHOST write the unexposed field lead to significant contrast degradation.

4.2. DESCRIPTION OF THE SIMPLEX APPROACH

Based on the preceding discussion, it appears that dose modulation should provide the best imaging results if issues in computational complexity could be resolved. From a very basic point of view, it would be interesting if a dose-optimization scheme could be developed which would provide an "exact" solution to the problem in some generally accepted and mathematically rigorous sense. This approach would allow us to study the limits of lithographic processing in a systematic way. It could provide us with a laboratory for studying resist contrast requirements and it would help us avoid futile efforts to achieve unrealistic goals.

The approach we have chosen is known as the simplex method of linear programming. The simplex method solves the following problem: We are given a function which we would like to optimize (i.e., find its maximum or minimum) within certain constraints. While we are free to select a large number of constraints on our own, simplex demands that all optimized variables must be positive. We can state this mathematically. The problem is to find the extrema of a cost function, $c(\mathbf{x})$:

$$c(\mathbf{x}) = \mathbf{w} \cdot \mathbf{x} \tag{33}$$

satisfying the constraints:

$$\sum_{i=1,j=1}^{m,n} M_{ij}x_j > a \tag{34}$$

$$\sum_{i=n-m,j=1}^{n,n} M_{ij}x_j < b \tag{35}$$

and

$$x_i > 0 \quad \text{for all } i \tag{36}$$

This equation set fully models the proximity problem. The individual entries of the vector x represent the applied dose. Pattern information is included in vector w. We demand that in the exposed region of the pattern, the total absorbed dose must exceed some critical value, a. In the unexposed field, total dose must be less than b. Of course, all doses must be positive (Eq. 36). Thus, the minimum contrast of the exposed pattern (modulation factor) is:

$$M_f = \frac{(a-b)}{(a+b)} \tag{37}$$

We have a certain amount of flexibility associated with the cost function. The simplest approach seems to be the one recommended by Carroll (17). In this case, w is a vector of all 1^s.

It should be noted that, from a practical point of view, Eq. 10 incorporates everything needed to specify the resist response. The a and b parameters can be associated with features on the resist development curve. Certainly edge slope and 3-D modeling are valuable, but our ability to "pull a feature out" on development is most closely associated with M_f. It is also true that a relaxation of the contrast requirements can lead to feature size variation across the exposure field. Solving equations 33–36 allows a determination of minimum acceptable contrast to achieve a given feature for a given set of beam conditions. Experiments may modify this number to accommodate feature-size variations across a field.

The simplex equation set can be solved by systematically varying all the dose values and looking for those values which minimize the cost function while satisfying the constraints. Clearly, this would be an enormous task, outside the ability of modern computers. However, highly efficient codes have been developed which allow for rapid solution of meaningful data sets (24). Also, these techniques are amenable to parallel computing (25).

Without undue attention to mechanics, the basic idea of the method is as follows. Equations 34–36 represent planes in some hyperspace whose coordinate axes are the individual pixel doses. Each plane intersects with neighboring planes to form a face of a geometric shape—an n-gon known as a *simplex*. The cost function (Eq. 33) is another plane in the space. By multiplication of Eq. 33 by -1, we convert it to a "greater than" inequality. Thus, the region outside the volume of the resulting simplex is allowed (feasible). The position of the cost plane in hyperspace, as defined by Eq. 33, is set by the right-hand side of this equation. We may lower the cost plane until it just touches the simplex at a single point. This point occurs at a vertex of the simplex and it represents the optimum (lowest value) cost. The coordinates of this vertex represent the solution doses.

The technique (24), represents an efficient way to test the individual vertex points for optimality. It may be that after searching all vertex points, no feasible solution is found. That is to say, the inequalities established by equations 34–36 are inconsistent. The calculation will cease and an error flag is set. This is valuable information, because it indicates that for the particular contrast chosen, no solution is possible. The feature cannot be resolved, and thus the technique represents an important checkpoint for nanolithography.

4.3. CALCULATIONS PERFORMED

A number of shapes were studied to get a feeling for the types of strategies that simplex indicated would be useful for resolving features with critical dimensions at or smaller than 0.1 μ. These included closely spaced rectangles, "box-in-box" patterns, and "bar-in-bar" patterns, and features used in our quantum electronics program (single-electron transistor patterns). Examples of each are shown as Figures 9–12. The writing surface assumed was a GaAs substrate coated with 50 nm of Si_3N_4. The energy deposition profiles were obtained by Monte Carlo analysis which we previously described (26). The point-spread energy deposition thus obtained was convolved with a 6-nm Gaussian round beam to estimate the effect of the electron optical system. Pixel-to-pixel spacing was taken as 30 nm.

Closely spaced rectangles, shown in Fig. 9(a), are predicted to form neck-like bridges, as shown in Fig. 9(b). The simplex-corrected dose file is shown in Fig. 9(c). Many interesting observations can be made. First, the exposed pattern appears to be scored. That is, lines of pixels lying two pixels from the geometric boundaries are unwritten. The three outer corners of each square receive high spike-like doses. These apparently serve the same purpose as anchor additions in optical proximity correction. Energy loss is greatest in regions of high perimeter to enclosed area, such as at corners. This requires extra doses. This is necessary, even for the two contiguous vertices. However, to prevent overexposure of the contact point between the squares, the spikes are offset. Scoring and spiking are also noted in the "box-in-box" cases shown in Fig. 10.

There appears to be a small amount of extra dose on the corner of the lower square, upsetting the symmetry of the structure. This may be a "numerical noise" issue, as the dose is small. Carroll (17), however, points out that rigorous symmetry is not required by the simplex approach. Any pattern that solves the simplex problem, as stated in equations 6–9, yields a reported solution. Symmetry is not part of the cost or the constraint. It is also interesting to note that a pattern rotation of 180° would also yield a valid solution. This is a manifestation of the "ill-posed" nature of the problem, as defined by Hadamard (27): multiple solutions of the problem are possible.

Another interesting strategy is shown in Fig. 11. The center bar appears to be an intense radiative source, which overexposes the outer bars. To accommodate this, the outer bars are thinned in the regions closest to this source.

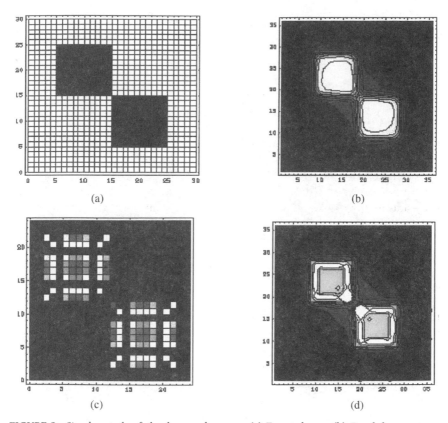

FIGURE 9 Simplex study of closely spaced squares. (a) Target shapes. (b) Equal dose contours for the uncorrected pattern. (c) The simplex-generated dose file. Note scoring and "anchoring" approaches at corners (spikes and offsets) as discussed. (d) Equal dose contours for the corrected pattern.

While no new strategies emerge on correction of the quantum device pattern (Fig. 12), a significant degree of improvement created by the correction is visually evident in comparing Fig. 12(c) to Fig. 12(d).

4.4. EXPERIMENTAL TECHNIQUE

Patterns, as shown in the (a) portions of Figures 1–4, were transformed to point exposures with distance better-point exposures (pixels) at 25 nm. The pixels were coded as paths of length 2.5 nm and 0 width. A JEOL JBX 5DII

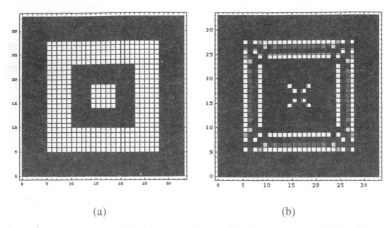

(a)   (b)

FIGURE 10 "Box-in-box" study. (a) Target shapes. (b) Simplex-generated dose file. Note the scoring and anchor strategies employed once again.

e-beam operating at 50 kV was used to expose the patterns; the 0-width paths are converted to point exposures by the JEOL software. Each pixel in the CAD and resized patterns was exposed with the same dose. Simplex data consisting of pixel position and relative intensity was read and then converted to GDSII data by a graphics program (developed by NRL staff) which sorted the pixel into ten different dose values based on the relative intensity of each pixel. The dose of pixel was specified by the data-type variable used by the JEOL to determine the exposure dose.

Because the base dose for the exposure was not known, a dose matrix consisting of 20 different base doses was written. We report on 6 different exposure runs: 2 runs each for e-beam currents of 15, 30, and 50 pA. All 6 runs were written on the same substrate. The sample was evaluated using a LEICA Streroscan 360FE SEM. Each of the 6 runs was evaluated individually. For each pattern type (Simplex, resize and CAD) a dose set was chosen for measurement based on the evaluation of pattern S18X7 (100-nm gap). The base dose having the smallest relative error for this pattern was chosen for measurement of all five patterns.

AZ PN114 (Clariant) negative resist diluted (1 : 2) with AZ1500 Thinner was spun at 3 KRPM onto a GaAs wafer coated with 50 A of chrome to help adhesion. The sample was baked at 100 °C on a vacuum hot plate for 4 minutes, postexposure baked at 98 °C in an oven, and developed in MF-322 for 2 minutes. The resist thickness of a large (50 × 80 μm) pad was determined to be 190 nm.

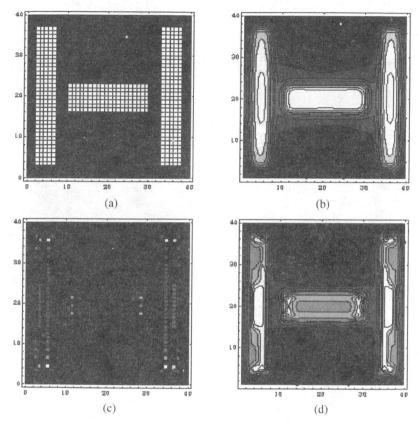

FIGURE 11 "Bar-in-bar" pattern study. (a) Target shapes. (b) Equal dose contours for the uncorrected pattern. (c) The simplex-generated dose file. Note the thinning of the outer bars in regions closest to the center bar. (d) Equal dose contours for the corrected pattern.

4.5. EXPERIMENTAL RESULTS

In this section we concentrate on numerical results obtained from the box-in-box and bar-in-bar experiments. First, consider the box-in-box experiment. The general format of the test feature and the range of subelement dimensions for this part of the study is shown in Fig. 13.

In Fig. 14 we see the results correction boundary placement as compared with a rudimentary feature-sorting algorithm (28).

In fairness, it should be pointed out that the feature-sorting algorithm employed (28) can make use of a much finer address grid, which would

340

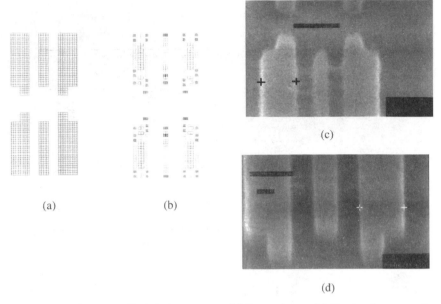

(a) (b)

(c)

(d)

FIGURE 12 "Quantum effect" device structure. (a) Target shapes. (b) Simplex dose correction pattern. (c) SEM images of uncorrected pattern. (See text for exposure details.) (d) SEM images of corrected pattern.

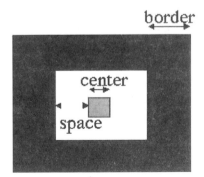

Device Number	Center Dimension (nm)	Space Dimension (nm)	Border Dimension (nm)
1	150	120	210
2	150	120	150
7	90	120	150
8	90	120	210

FIGURE 13 Box-in-box base shape and study details (table).

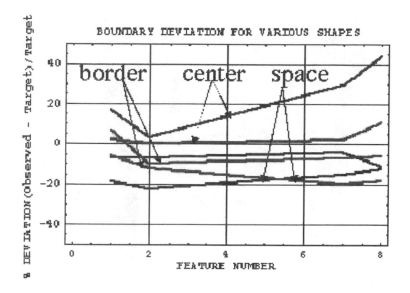

Grid: 30nm

FIGURE 14 Linearity study showing the spread of target deviations over the whole data set described in Fig. 13.

improve boundary placement considerably. However for the number of pixels corrected for, the results cited here serve as a fair comparison.

Figure 6 indicates that the simplex algorithm is particularly good in defining solid shape dimensions. The center square and border boundary placement accuracy was always better than 10% for all features chosen—a fairly linear result. Space control was generally better than 10%, and significantly more linear than feature sorting over the range of features studied.

Figure 15 shows the basic bar-in-bar pattern and the dimension parameter range studied.

Figure 16 shows relative error, defined by the relationship:

$$R_e = \frac{|Target\ Dimension - Dimension\ Obtained|}{Target\ Dimension} \tag{38}$$

for all features used in this study.

Mean relative error for all simplex-corrected features was better than 10%. These results were compared with uncorrected patterns and with a "manual" attempt at feature resizing. In all cases, simplex correction was better than or as

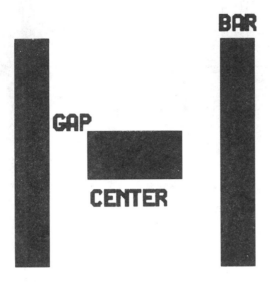

PATTERN	BAR nm	GAP nm	CENTER nm
S6X7	125	250	150
S10X7	125	200	250
S14X7	125	150	350
S18X7	125	100	450
S20X7	125	75	500

FIGURE 15   Bar-in-bar base shape and study details (table).

good as the other methods. Figure 17 shows a cumulative frequency distribution of errors, again supporting these conclusions.

4.6. DISCUSSION

From experimental results, we see that the simplex approach clearly works as a mathematically rigorous proximity-correction approach. It is, however, clearly not suitable for the massive databases representative of very large-scale integrated circuit technology. However, the methods described here are of use in small database projects such as those that are typical of nanostructure pattern development.

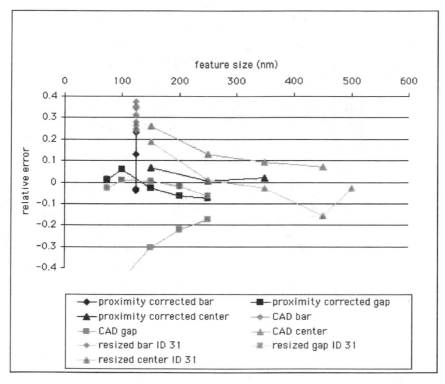

FIGURE 16 Relative error vs. feature size for the various shapes used in this study.

Using simplex algorithms, we may ascertain the limits of our ability to resolve a given feature, without going through weeks of futile experiments. Also, strategies emerge through study of the simplex-corrected pattern that would enable extension to even larger databases by scaling (such as the scoring and anchor-addition approaches described).

Limitations of the approach include the fact that our solutions are 2-D in nature. We do not include dose variation in the depth of the film. However, for the ultrathin resists used in nanolithography, this should not be an issue. Figure 4 indicates relatively steep sidewalls even for resists in the 200-nm-thickness range. The influence of the resist seems to be well modeled through contrast factor, an essential part of the simplex calculation.

Of course, developing an effective "imaging inverse" approach to deal effectively with proximity control is only part of the larger problem of database management in e-beam technology. Because e-beam databases are hierarchical in nature, we are constrained to work at the lowest level of hierarchy accessible

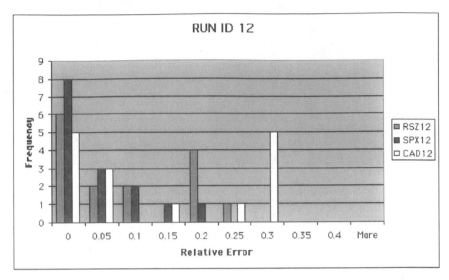

FIGURE 17 Cumulative error distributions for the features recompiled from Fig. 16. Note the distribution of errors for the simplex case has a mean close to zero and a relatively low full-width and half maximum.

through our systems. This may not be at the flattest level we can envision—the blanked/unblanked data pattern of single pixels. Those machines frequently referred to as "nanowriters" have access to this level, as we have demonstrated here. Production-level machines may not support such access. Studies such as those described here may establish the lowest level of accessible hierarchy.

5. SUMMARY AND CONCLUSIONS

Throughout this chapter, we have attempted to present both the practical and theoretical underpinnings of e-beam nanolithography as it pertains to nano-structure science and technology. We have discussed the various subsystems and overall systems architecture of today's e-beam tools. We have shown how these architectures can lead to limitations in our ability to achieve optimum boundary-placement control. We have provided an indication of how future architectures must progress in order to achieve optimization in this area.

We have described a method for studying boundary-placement accuracy limits in e-beam lithography using the simplex algorithm. This approach appears to be significant for small database modeling of nanostructure patterns. It is mathematically rigorous, and it avoids the negative dose problem without sacrificing contrast. It naturally includes resist impact through the contrast factor. While the approach is computationally expensive it is amenable to parallel computing approaches, which are currently being studied in our laboratory.

The discussion of hierarchies which concluded the last section indicates that full-dose modulation approaches may not be practical for the circuits of the future. The solution may ultimately involve a fusion of techniques. For example, the strategies developed here may serve as templates for future feature-sorting approaches, which enable database specification at a much higher level of hierarchy.

ACKNOWLEDGMENTS

The authors would like to acknowledge Dr. F.K. Perkins of NRL, Dr. D. Gerber of Gerber Consulting, Dr. T. Groves of Leica, and Dr. S.-Y. Lee of Auburn University for stimulating discussions concerning this work.

REFERENCES

1. C. Weisbuch, B. Vinter, *Quantum Semiconductor Structures: Fundamentals and Applications*, Academic Press, Boston, MA 1991.
2. W.M. Moreau, *Semiconductor Lithography: Principles, Practices and Materials*, Plenum Press, New York, NY 1991.
3. G.O. Reynolds, J.B. DeVelis, G.B. Parrent, B.J. Thompson, *The New Physical Optics Notebook: Tutorials in Fourier Optics*, SPIE Press, Bellingham, WA 1989.
4. G.R. Brewer, ed., *Electron-Beam Technology In Microelectronic Fabrication*, Academic Press, Boston, MA 1980.
5. W. Chen, H. Ahmed, "Fabrication of Sub-10 nm Structures by Lift-off and by Etching After Electron-Beam Exposure of Poly(methylmethacrylate) Resist on Solid Substrates," J. Vac. Sci. Technol. B **11**(6), 2519(1993).
6. R. Gomer, "Field Emission and Field Ionization," in *American Vacuum Society Classics*, American Institute of Physics, New York City, NY 1993.
7. D. B. Langmuir, "Theoretical Limitations of Cathode Ray Tubes," Proc. IRE **25**, 977(1937).
8. D.C. O'Shea, *Elements of Modern Optical System Design*, Wiley-Interscience, New York, 1985, p. 100.
9. W.K. Lo, G. Parthasarathy, C.W. Lo, D.M. Tannenbaum, H.G. Craighead, M.S. Isaacson, "Titanium Nitride Coated Tungsten Cold Field Emission Sources," J. Vac. Sci. Technol. **14**(6), 3787(1996).
10. M.L. Yu, H.-S. Kim, B.W. Hussey, T.H.P. Chang, W.A. Mackie, "Energy Distributions of Field Emitted Electrons from Carbide Tips and Tungsten Tips with Diamondlike Carbon Coatings," J. Vac. Sci. Technol. **14**(6), 3797(1996).
11. J.E. Schneider, A.W. Baum, G.I. Winograd, R.F.W. Pease, M. McCord, W.E. Spicer, K.A. Costello, V.W. Aebi, "Semiconductor on Glass Photocathodes as High-Performance Sources for Parallel Electron Beam Lithography," J. Vac. Sci. Technol. **14**(6), 3782(1996).

12. L. Veneklasen, in *Physical Aspects of Electron Microscopy*, B. Siegel and D.R. Beaman, eds. Wiley, New York 1975.
13. H. Pearce-Percy, F. Abboud, R. Garcia, M. Mankos, "Optimization of Field Emission Columns for Next-Generation MEBES Systems," J. Vac. Sci. Technol.15(6), 2754(1997).
14. H. Boersch, Zeitschr. Phys. **139** (1954).
15. G.H. Jansen, *Coulomb Interactions in Particle Beams*, Academic Press, Boston, MA 1990.
16. R. Dorfman, P.A. Samuelson, R.M. Solow, *Linear Programming and Economic Analysis*, Dover Publications, NY 1986.
17. A.M. Carroll, "Proximity-Effect Correction with Linear Programming," J. Appl. Phys. **52(1)**, 434(1981).
18. T.H.P. Chang, "Proximity Effect in Electron-Beam Lithography," J. Vac. Sci. Technol. **12**, 1271(1975).
19. M. Parikh, "Corrections to Proximity Effect in Electron Beam Lithography," J. Appl. Phys **50(6)**, 4371(1979).
20. D.P. Kern, "A Novel Approach to Proximity Effect Correction," in *Proc. Symp. on Electron and Ion Beam Science and Technology; Ninth Int'l Conf.*, R. Bakish, Ed., Vol. 80-6, The Electrochemical Soc., Inc. 326(1980).
21. M.C. Peckerar, S. Chang, C.R.K. Marrian, "Proximity Correction Algorithms and a Coprocessor Based on Regularized Optimization. I Description of the Algorithm," J. Vac. Sci. Technol. **13(6)**, 2518(1995).
22. G. Owen, P. Rissman, "Proximity Effect Correction for E-beam Lithography by Equalization of Background Dose," J. Appl. Phys. **54(6)**, 3573(1983).
23. M.C. Peckerar, C.R.K. Marrian, F.K. Perkins, "Feature Contrast in Dose-Equalization Schemes Used for Electron-Beam Proximity Control," J. Vac. Sci. Technol. **13(6)**, 2518(1995).
24. W.H. Press, B.P. Flannery, S.A. Teukolsky, W.T. Vetterling, *Numerical Recipes: The Art of Scientific Computing*, Cambridge University Press, Cambridge (1986).
25. *see:* http://ism.boulder.ibm.com/es/oslv2/features/lib.htm
26. C.R.K. Marrian, F.K. Perkins, D. Park, E.A. Dobisz, M.C. Peckerar, K.-W. Rhee, R. Bass, "Modeling of Electron Elastic and Inelastic Scattering," J. Vac. Sci. Technol. **B 14(6)**, 3864(1996).
27. C.W. Groetsch, *The Theory of Tikhonov Regularization for Fredholm Equations of the First Kind*, Pitman Advanced Publishing Program, Boston, MA 1984.
28. S.-Y. Lee, J. Laddha, "Adaptive Selection of Control Points for Improving Accuracy and Speed of Proximity Effect Correction," J. Vac. Sci. Technol. **B 16(6)**, 3269(1998).

CHAPTER 12

Focused Ion Beams for Direct Writing

KLAUS EDINGER

Institute for Research in Electronics and Applied Physics, University of Maryland, College Park, Maryland

1. Introduction
2. Equipment
3. Ion Solid Interaction
4. Applications
5. Conclusions

1. INTRODUCTION

Over the past thirty years focused ion beam (FIB) systems have evolved into high-performance commercial tools for direct writing microfabrication. In these systems ions are generated by a liquid metal ion source, accelerated to energies in the range of 25 to 150 keV and focused onto the target. When this energetic beam of ions collides with the substrate surface, a variety of effects occur which can be used to characterize or locally modify the target material: (1) secondary ions and electrons are emitted, which are used to generate scanning ion microscopy images; (2) substrate atoms are ejected, resulting in sputtering (removal) of material; and (3) finally the ions are implanted into the material. If the substrate is exposed to a suitable precursor gas, the incident ions might also cause reactions at the surface, leading to a deposition of material through fragmentation of the precursor molecules or to enhanced material removal by gas-assisted etching.

By far the most widely used systems are low-energy (25–50 keV) ion-beam tools for surface micromachining applications utilizing sputtering and ion-beam-induced surface reactions. These systems operate exclusively with

Direct-Write Technologies for Rapid Prototyping Applications

347

gallium ions and achieve current densities up to 10–20 A/cm^2. Depending of the total ion current at the target, which can be selected with a variable beam-limiting aperture, these current densities correspond to beam diameters below 10 nm at low currents (<10 pA) up to several hundred nanometers at maximum currents of 10–50 nA. Low-energy focused ion-beam systems are well-established tools in the semiconductor industry and are very successfully used for applications such as failure analysis (e.g., cross-sectioning of transistors or contacts and transmission electron microscopy sample preparation), prototype-device modification, and lithographic mask repair.

A second type of focused ion-beam system has been developed, operating at beam energies of up to 200 keV, mainly for direct-implantation applications. These systems use alloy liquid metal ion source and can produce a large variety of ions, including Be, B, As, and Si, the main dopants of silicon and III-IV semiconductors. So far, high-energy focused ion-beam systems are used at the research level for prototyping of novel or improved electronic and opto-electronic devices.

In this chapter the main components and principle of operation of a focused ion-beam system will be outlined, followed by discussion of ion solid interactions, with emphasis on the main effects important for micromachining applications—image generation, sputtering, and ion-induced surface reactions. Finally, applications of focused ion-beam direct writing will be reviewed.

2. EQUIPMENT

With respect to their targeted application focused ion-beam systems can be divided into two groups. First, systems for micromachining exclusively use gallium ion sources and operate at ion energies in the range of 25–50 keV. Several hundred of these machines are in operation worldwide, and they are extensively used in the semiconductor industry for applications such as failure analysis, circuit editing, and mask repair, all of which require precise removal and deposition of material. Second, focused ion-beam systems for ion-beam lithography and maskless ion implantation operate at variable voltages of up to 200 kV and can produce a variety of ion species such as the semiconductor dopants Be, B, Si, and As.

Depending on performance and complexity, the cost of a low-energy focused ion-beam system can range from $300,000 to $3,000,000 or more. Currently, focused ion-beam guns and complete systems are manufactured by FEI Company (formerly FEI Co. and Micrion Corp.) and Schlumberger Corporation in the U.S.; Orsay Physics in France and Hitachi, Ltd. and Seiko Instruments in Japan. High-energy systems are at present built by NanoFab

Inc. in the U.S. and in Japan by Eiko Engineering. The cost of these machines is $1–2 million.

The components of a focused ion-beam system can be divided into three parts as schematically shown in Fig. 1: the ion source, the ion optics, and the vacuum chamber with sample stage and auxiliary components. The principal function of the system is to focus an ion source—ideally a point source—onto the sample by means of an electrostatic optical column. The optical column

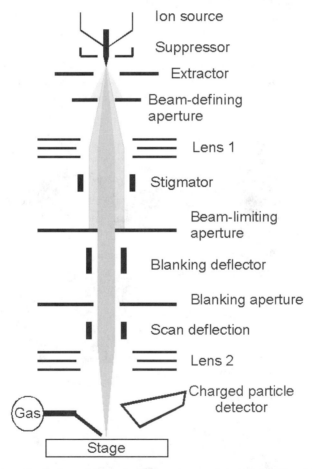

FIGURE 1 Schematic of a focused ion beam (FIB) system consisting of a liquid metal ion source, ion optical column, and sample stage. Additional components include a gas delivery system for ion-induced deposition or gas-assisted etching and an electron gun showering the sample with low-energy electrons to avoid charging of insulating samples.

also includes deflection mechanisms for beam steering and beam blanking (e.g., turning the ion beam "off").

The next section provides a basic overview of the components and functions of a focused ion-beam system. For a more detailed discussion of the topic see for example (1–4) and references therein.

2.1. ION SOURCES

Present-day focused ion-beam systems are equipped exclusively with liquid metal ion sources (LMIS). This source consists of a needle, usually made out of tungsten, with an end radius of about 10 μm, which is covered with a liquid metal film. The needle is attached to a heated filament and a reservoir to provide a supply of liquid metal (see Fig. 2). The needle faces a nearby extracting aperture. If a positive voltage (2–10 kV) is applied to the needle relatively to the extraction electrode the liquid metal is pulled into a sharp cone

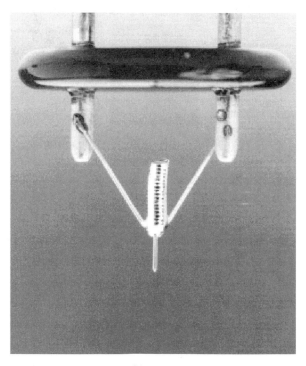

FIGURE 2   Picture of a Gallium liquid metal ion source. Most of the metal is contained in a coil-shaped reservoir, which can be heated by passing a current through the supporting wires.

and the electric field at the apex of the cone becomes high enough to generate positive ions through field evaporation and field ionization. The positive ions are accelerated away from the tip and pass through the hole in the extraction aperture. The emission current is usually controlled by a servo system, which includes an auxiliary electrode (called a suppressor) to adjust the extraction field. Optimal extraction currents are in the range of 1–3 μA. Above this value the energy spread of the ions, which is caused by coulomb repulsion and is typically ~5 eV, will increase rapidly. The energy spread of the ions causes chromatic aberration, which will increase the beam diameter at the focal plane.

Because of its low melting point of 30 °C, low vapor pressure and relatively low-reactivity gallium is the metal of choice for liquid metal ion sources, and commercial Ga sources with a source life of 500–1,500 (μA hours) are usually used in low-energy focused ion-beam systems. For applications such as focused ion-beam lithography and direct implantation, a number of liquid metal ion sources have been developed that offer a wide variety of ion species including Al, As, Au, B, Be, Bi, Co, Cs, Cu, Ge, Er, Fe, In, Li, Ni, Pb, P, Pd, Pr, Pt, Si, Sn, U, and Zn. In order to lower the melting point and to control the reactivity and volatility most of these sources contain metal alloys. For example, PdAs, PdAsB, AuSi, and AuSiBe alloy sources are frequently used to deliver the main dopants for silicon and III-IV semiconductors.

2.2. Ion Column

The ion optical column contains electrostatic lenses to focus the ions emitted from the ion source onto the target and has several deflection electrodes for bean scanning, stigmatism correction, and beam blanking.

2.2.1. Ion Optics

The performance of the optical system in terms of beam diameter and current density at the focal plane is limited by spherical and chromatic aberrations of the lenses and ultimately by the apparent or "virtual" source size. Chromatic aberration is the dominant factor over a large range of typical operation conditions. Both lens aberrations are a function of the acceptance angle of the optical system, which determines how much of the ion current emitted from the ion source into a cone are passing through the lenses. For this reason the optical column contains a beam-limiting aperture with variable size, which limits the acceptance angle and in turn determines the total current and the beam diameter at the focal plane. Over a certain operation range the current density in the focal spot is approximately constant so that increasing the ion current by selecting a larger aperture will result in a corresponding increase in beam diameter. Typical values reported for state-of-the-art focused ion-beam systems are a beam diameter of 5 nm at ~1 pA and ~500 nm for 10 nA of beam current. The spatial current distribution in the focal spot is approximately gaussian. However, especially at high currents, the current distribution farther away from the central peak exhibits long tails. This deviation from the gaussian distribution starts at current densities that are about 2–4 order of magnitude below the peak value, and the current in these so-called "beam tails" decreases exponentially with distance. Although the current in the beam tails is relatively low, the fact that these tails extend for several tens and even hundreds of micrometers can be of concern, for example, in electronic devices, which are very sensitive to gallium contamination.

2.2.2. Deflection

Scanning the ion beam across the sample is accomplished by applying a traverse electric field in the x and y directions using deflection electrodes. Usually an octopole deflector located just before or after the final lens is used for this purpose. The maximum scan field is typically several hundreds of micrometers. However for large field sizes distortions (beam stigmation) at the edges of the scan field can be expected. In addition, relatively high-deflection voltages are required and the positioning accuracy might be limited by electronic noise from the high-voltage, large-bandwidth deflection amplifiers as well as by the bit resolution of the digital-to-analog converter that controls the pattern writing. Therefore, large fields should be used only for imaging and navigation, and typical scan fields for high-resolution writing are about $100 \times 100 \,\mu m$ and below, depending on the required precision. The writing speed is not only limited by the bandwidth of the electronics controlling the deflection electrodes, but ultimately by the time it takes the relatively slow ions to travel through the deflection assembly (the velocity of a 30 keV Ga^+ ion is 2.9×10^5 m/sec). In a digitally scanned system a measure for the maximum scan speed is the dwell time, which refers to the time the beam remains at each point. Typically the minimum dwell time is about 0.1 μsec. Another important feature for direct-writing application is the ability to turn off the beam rapidly during scanning so as to not expose certain areas of the scanned field. This so-called beam blanking is done by deflecting the beam with a separate set of deflection plates so that it does not pass through an underlying blanking aperture.

2.2.3. Mass Filter

High-energy systems are also equipped with a mass separator to filter out unwanted ion species emitted from alloy liquid metal ion sources. This filter generates a crossed electric and magnetic field perpendicular to the optical axes (ExB filter), which deflects the ions according to their velocity, so that only the desired ion passes through a downstream aperture.

2.3. PERIPHERAL

Usually the system is controlled by a computer, which sets the lens voltages for focusing and generates the signals for beam scanning and blanking. Patterns can be written with a predefined shape and ion dose by either drawing them directly onto the computer screen or using various types of bitmap files. In both cases alignment of the pattern with the sample is obtained by acquiring an image in the scanning ion microscope mode and storing it on the computer screen. Because the same apparatus is used for producing the image and the direct writing the placement accuracy is very high and only limited by stage or beam drift. Most commercial systems can also be interfaced with CAM/CAD tools to provide navigational data. The positioning accuracy in this case depends on the precision of the sample stage, which for encoded stages is in the range of 1–2 µm. Optionally, interferometrically controlled stages are available, which allow positioning with an accuracy from ~100 nm down to <10 nm. High-energy focused ion-beam systems for direct implantation are always equipped with these high-precision stages, because imaging the substrate with the ion beam would implant considerable amount of dopants.

The sample stage is housed inside a vacuum chamber which contains other necessary and optional equipment, such as a secondary electron or ion detector that generates the image information for the scanning ion microscopy mode, an electron flood gun, which showers the sample with low-energy electrons to prevent charging of insulating samples, a gas injection system for ion induced surface chemistry, and optical or electron microscopes to assist in navigation and inspection.

3. ION SOLID INTERACTION

As is shown in Fig. 3 an energetic ion that collides with a solid surface causes a variety of interactions, which can be used in some form or another to modify or characterize a material. While this makes focused ion beams versatile tools for micro- and nanofabrication and material analysis, the fact that the interactions are not generally separable may in some cases lead to unwanted side effects that need to be considered.

Implantation. As the ion penetrates into the solid, it loses its energy due to interaction with the electrons and the atoms of the solid. The two rates of

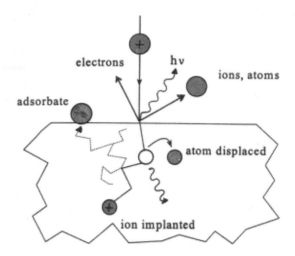

FIGURE 3 Schematic of ion-solid interactions. The incident ion penetrates the solid and is implanted. The energy lost by the ion leads to lattice damage in the bulk, sputtering of atoms (or ions) at the surface, and emission of photons and secondary electrons. In addition, if a precursor gas is adsorbed at the surface, chemical reactions can be induced.

energy loss are known as electronic and nuclear stopping power, and in general electronic stopping is dominant for fast-light ions whereas nuclear stopping is more dominant for slow and heavy ions. While the ion is traveling deeper into the substrate its energy continuously decreases and more and more energy is lost due to the nuclear stopping in binary collisions with the substrate until it finally comes to rest. Because of the scattering caused by the momentum transfer during the binary collisions, the final resting place of the ion is statistically distributed around an average depth called the ion range R_p. The vertical deviation (half width at half maximum) around this depth is called range straggle ΔR_p and the horizontal deviation is called lateral or traverse straggle ΔR_t. The values for these quantities depend on the ion mass and energy and on the properties of the target material (e.g., atomic mass and density). Both range and straggle decrease with lower ion energy and increasing ion mass. Values for a particular material system can be obtained using, for example, a Monte Carlo simulation program (SRIM (5)). Commercial FIB systems for micromachining exclusively use Ga ions in an energy range of 25–50 keV. For these systems the average penetration depth or range R_p is approximately 26 nm (30 keV Ga^+ in silicon). The energy that is transferred from the ion to the substrate during the implantation process is dissipated through several mechanisms as shown.

Damage. If the energy transferred to a substrate atom in a binary collision is high enough, the target atom is displaced from its lattice site and can in turn collide with other substrate atoms leading to a collision cascade. This ion-induced damage can change the electronic properties of the material significantly even for low-ion doses. Higher-ion doses ($>1 \times 10^{14}$ ions/cm^2 for Si (6)) can completely destroy the atomic lattice, rendering the substrate amorphous.

Sputtering and Surface Reactions. Most important for micromachining applications is the case where the collision cascade involves a substrate atom at the surface. If the collision imparts enough energy and momentum to the surface atom to overcome the binding energy (4.7 eV for Si) the atom is ejected. This effect is called sputtering or ion milling. Because it is a physical process and depends only on momentum transfer, essentially any material can be removed by focused ion-beam milling. In addition to this physical sputtering effect, FIB micromachining also utilizes ion-induced chemical reactions at the surface for enhanced material removal and deposition.

Sputtering and ion-beam-induced surface chemistry are by far the most utilized effects in focused ion-beam applications and are therefore discussed in the following chapters in greater detail. However, note that these effects, which are confined to the top surface layer, are always accompanied by ion implantation and ion-induced damage. This might pose a problem for micromachining of materials that are sensitive to damage and implantation such as in electronic or optical applications. In this case the effective minimum achievable size of a micromachined feature is ultimately limited by the size and depth of the implanted area. For 30 keV Ga ions this area is approximately 35 nm deep ($R_p + \Delta R_p$) and 13 nm wide ($2\Delta R_t$).

Bulk Chemical Reactions. In addition to damage to the substrate lattice, the energy deposited by the incident ion can also lead to chemical reactions within the bulk of a compound material. In analogy to optical or electron-beam lithography, this effect can be used for maskless focused ion-beam lithography. Exposing a suitable organic resist with an ion beam either results in cross-linking of molecules, making the resist insoluble in the developer solution (negative resist), or causes chain scission of polymer molecules and the resist becomes soluble (positive resist). Because cross-linking and chain scission are also the major mechanisms in organic electron-beam resists it can be assumed that these chemical reactions are caused by electronic interaction. This has been studied for the positive resist polymethyl methacrylate (PMMA) and a strong correlation of the exposure dose with the electronic stopping power of the ion has been found (7).

In addition to organic resists, focused ion-beam chemical modification of inorganic materials has been demonstrated (8–11). In most of these reports the inorganic materials have been used as resist, where the modified areas act as a masking layer in further processing steps. However, chemical reactions as well as ion implantation and ion damage can also be used to introduce functionality in a one-step process by locally modifying, for example, the electrical, optical, chemical, or mechanical properties of a material.

3.1. PHOTON AND CHARGED PARTICLE EMISSION

While the ion-solid interactions introduced so far are used for modification of a material, the detection of particles and light emitted during ion-beam exposure can be utilized for material characterization and analysis. In analogy to a scanning electron microscopy, secondary electrons or, to a lesser extent, secondary ions emitted while scanning the ion beam over the sample, are used for image generation. As in scanning electron microscopy, the image is obtained by creating a map of the secondary electron or ion intensity as a function of ion-beam position. The image contrast can be attributed to topographic contrast, material contrast, ion channeling, and electrical charging.

Topographic contrast is caused by differences in the orientation of the sample surface with respect to the incident ion beam and to the detector. Because of the limited escape depth of both secondary electrons (1–50 nm) and ions (<1 nm), only electrons or ions generated close to the surface contribute to the detected signal. Increasing the angle between the incident ion beam and the sample surface will deposit more energy into the top surface layer and therefore increase the secondary particle yield. This effect together with the small penetration depth of the primary ion accounts for the very high sensitivity of scanning ion microscopy to surface topography.

If the incident ion beam is closely aligned with one of the low-index crystallographic orientations of the sample, the ion can travel between the lattice atoms without major scattering. Due to this channeling effect the ion will be implanted much deeper into the material and because more energy is dissipated further away from the surface, a reduced secondary electron and ion yield as well as a decrease in sputter yield will result. Because channeling orientations appear dark in the image, images taken at various tilt and azimuthal angles deliver detailed information about the local crystallography of a sample. Applications of this technique include, for example, determination of grain-size distribution in polycrystalline samples (12,13).

Imaging insulating or not electrically grounded materials with an ion or electron beam presents problems because of charge build-up. In the case of a primary ion beam the material charges to positive potentials, because of the implantation of positive ions from the ion beam and the fact that more secondary electrons than secondary ions are emitted. Because the secondary electrons typically have low energy (<20 eV), they are retracted by this potential, and charged areas appear darker. In the examination of integrated circuits this effect is frequently used to distinguish floating from grounded metal lines. In scanning electron microscopy, insulating material generally leads to a negative charge build-up resulting in bright images, which usually have to be coated with a thin conductive layer to obtain high-resolution. Charge build-up from focused ion beams on the other hand can be greatly reduced by simultaneously exposing the material to a broad, low-energy electron beam while detecting the secondary ions. Therefore FIB systems equipped with an electron flood gun are well suited for imaging and micro-machining of insulating samples.

As in scanning electron microscopy, the secondary-electron yield is also a function of material properties such as its work function, which determines the minimum energy for emission of low-energy electrons. Also, the generation of secondary ions very much depends on the chemistry of the material. For example, it is known from secondary ion mass spectroscopy (SIMS) that halide or oxygen atoms have a high yield for negative ionization and alkali atoms produce positive ions. In addition, the ionization yield is strongly influenced by the chemical environment, a condition known as the matrix effect. The detection of secondary ions can be used not only to generate an image of the sample, but also to acquire chemical information through spatially resolved SIMS. In this case an ion-collection optics attached to a mass analyzer (usually a quadruple mass spectrometer) is installed in close proximity to the sample. Secondary ions generated during the raster scan of the primary ion beam are collected and separated according to their mass/charge ratio, which is then used to generate a two-dimensional map of the elemental composition. Lateral and depth resolution of 20 nm has been reported for focused ion beam SIMS (14). However, because of the small volume analyzed and the low ionization yield from the primary gallium ion beam, the detection limit is relatively poor compared to analytical SIMS systems, which use broader oxygen and cesium ion beams.

3.2. Sputtering

Physical sputtering or ion milling for removal of material is the most universal application of focused ion-beam tools. Because the target atoms are removed by a purely physical process, involving collision cascades, all solid materials can be ion milled, if enough energy is transferred to surface atoms to overcome their binding energy. The number of removed atoms per incident ion, called the sputter yield, depends on material properties such as the binding energy of the surface atoms and the displacement energy for atoms within the lattice. The sputter yield is also a function of the mass and energy of the incident ion. While the sputter yield shows an initial increase with increasing ion energy, the yield decreases again at higher ion energies because with increasing penetration depth, more energy is deposited further away from the surface. Therefore, Ga^+ FIB systems for micromachining operate with energies between 25 and 50 keV. Sputter yield for various materials milled with a focused ion beam are given in Table 1. The removal rate in $\mu m^3 \, sec^{-1} \, nA^{-1}$ (calculated using the yield and tabulated material densities) is also included. For silicon this rate is around $0.24 \, \mu m^3$ per second for 1 nA of beam current. This rate has to be multiplied by the beam current, which ranges from <10 pA to 100 pA for high-

TABLE 1   Focused Ga^+ Ion Beam Sputter Yields and Calculated Removal Rates (For a Beam Current of 1 nA)

Material	Energy (keV)	Yield (atoms/ion)	Rate ($\mu m^3 \, sec^{-1} \, nA^{-1}$)	Ref.
Si	25	1.8–3.9	0.23–0.49	(4,5,8)
	30	1.8–3.1	0.23–0.39	(6,9)
SiO_2	25	0.84	0.24	(8)
	30	0.80–0.85	0.23–0.24	(6,9)
	35	0.84	0.24	(8)
Al	30	1.8	0.19	(9)
W	25	5 ± 0.7	0.49	(8)
Au	25	18–23	1.90–2.43	(8,8)
	30	5.6–15	0.59–1.59	(9)
	40	15.7 ± 1.3	1.7	(1)
SiC	30	2.4	0.32	(7)
	50	2.5	0.33	(7)
GaAs	25	5.3	1.49	(4)
	30	4.9–5.3	1.38–1.49	(3,9)
InP	25	6.7		(4)
	30	7.4–7.6		(3,9)
InAs	30	6.4	2.22	(3)
GaN	30	6.2–7.6	0.88–1.08	(2,7,9)
	50	6.6	0.94	(7)
GaP	30	5.0	1.26	(3)

Reference List
1. Blauner, P.G., Butt, Y., Jae Sang Ro, Thompson, C.V., Melngailis, J., J. Vac. Sci. Technol., B 7, 1816 (1989).
2. Flierl, C., White, I.H., Kuball, M., Heard, P.J., Allen, G.C., Marinelli, C., Rorison, J.M., Penty, R.V., Chen, Y., Wang, S.Y., Mrs Internet Journal of Nitride Semiconductor Research 4, U776–U781 (1999).
3. Menzel, R., Bachmann, T., Wesch, W., Nucl. Instrum. Methods Phys. Res. B 148, 450 (1999).
4. Pellerin, J.G., Griffis, D.P., Russell, P.E., J. Vac. Sci. Technol., B 8, 1945 (1990).
5. Pellerin, J.G., Shedd, G.M., Griffs, D.P., Russell, P.E., J. Vac. Sci. Technol., B 7, 1810 (1989).
6. Santamore, D., Edinger, K., Orloff, J., Melngailis, J., J. Vac. Sci. Technol., B 15, 2346 (1997).
7. Steckl, A.J., Chyr, I., J. Vac. Sci. Technol., B 17, 362 (1999).
8. Xu, X., Della Ratta, A.D., Sosonkina, J., Melngailis, J., J. Vac. Sci. Technol., B 10, 2675 (1992).
9. Young, R.J., Cleaver, J.R.A., Ahmed, H., J. Vac. Sci. Technol., B 11, 234 (1993).

resolution milling (beams size <10 to ~50 nm) up to 10–50 nA for fast material removal at beam sizes of several hundreds of micrometers. The experimental sputter yields given in Table 1 vary to a certain degree. Some of the variation might be caused by differences in the target material density. However, there are several other factors affecting the sputter yield—some of them intrinsic to focused ion-beam milling—which will now be discussed.

3.2.1. Angle of Incidence

The sputter yield is a function of the ion incident angle. As this angle increases from normal incidence, a larger number of collision cascades reach the surface leading to an increased sputter yield. After reaching a maximum the sputter yield will decrease again as the ion beam approaches glancing incidence, because of the increase in reflected ions and the fact that more and more collision cascades terminate at the surface before they are fully developed (15). The optimal angle varies with the material and beam conditions and is generally in the range of 60–80°. Even if the ion beam is macroscopically at normal incidence to the target, local variations of the microscopic incident angle with surface topography will have a significant effect in focused ion-beam milling (16).

Because sloped areas will sputter faster than flat areas, ion milling inherently will reduce the overall surface roughness and smooth out topography in homogenous material. However, because the edges of raised, mesalike structures will mill faster, it becomes very difficult to remove these structures smoothly by focused ion-beam milling. This effect, which results in the trench formed at the edge of the removed mesa, is referred to as "river-bedding" (17).

Given the approximately gaussian current distribution in the beam, one would expect a corresponding sloped sidewall at the edge of a milled trench. However, because of the dynamic dependence of the sputter yield on the evolving slope angle, the sidewall is much steeper. If the milling depth becomes large compared to the beam diameter, the sidewall angle (relative to the surface normal) reaches a constant value of 2–6°, depending on the material (16). If the sample is tilted at this angle before milling, completely vertical standing cross-sections can be achieved.

3.2.2. Channeling

If the ion beam is aligned with a low-index crystallographic orientation, ions travel along the channel resulting in a reduced sputter yield. This can be a problem especially for a polycrystalline material, where differently oriented grains sputter at different rates. This effect leads to a sputter-induced increase

in surface roughness (18). For certain materials, gas-assisted etching can help
to reduce this effect.

3.2.3. Redeposition

It has been found that the physically sputtered particles are ejected from the
surface with a cosine-squared angular distribution with a maximum normal to
the surface (19). Given this distribution, there is a certain probability that
particles, ejected from the bottom of a milled trench, will collide with the
sidewalls and become redeposited, which reduces with the aspect ratio (e.g.,
depth:width ratio) of the trench. Redeposition can be greatly reduced if the
trench is milled with fast repetitive rasters, where each successive scan removes
redeposited material from the previous scan (20,21). However, if the trench
becomes very narrow and deep almost all material is redeposited onto the
sidewalls. Consequently, the aspect ratio (e.g., depth:width) for ion-milled
trenches is limited to approximately 5:1. For certain materials redeposition can
be greatly reduced by gas-assisted etching, which generates volatile particles
(compare Fig. 4).

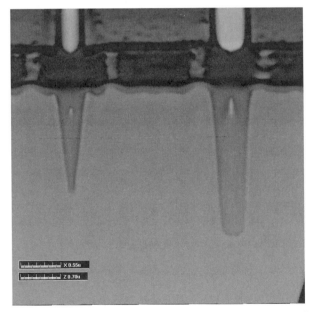

FIGURE 4 Comparison of the aspect ratio achieved by physical sputtering (left side) and gas-
assisted etching (right side). 300 nm wide trenches have been milling in silicon using an ion dose of
$20 nC/\mu m^2$ for sputtering and $2 nC/\mu m^2$ for Cl_2GAE. To preserve the shape of the trenches, they
have been back-filled with FIB-deposited tungsten before cross-sectioning with the FIB.

3.3. Ion-Induced Surface Chemistry

The principle of focused ion-beam induced surface chemistry is to adsorb suitable precursor gas molecules on the substrate and expose them to an ion beam. In the case of ion-beam-induced deposition (IBID), the incident focused ion beam leads to a fragmentation of the adsorbed molecules leaving behind a deposit. Precursor gases usually include organometallic compounds for metal deposition and silanes or siloxanes for insulator deposition. In gas-assisted etching (GAE) applications, a potentially reactive precursor gas is used, which should not react spontaneously with the substrate. The ion beam causes a reaction between the adsorbate molecules and the substrate, forming volatile compounds and thus an enhanced removal of material. The basic components of a system for focused ion beam (FIB)-induced surface chemistry are shown in Fig. 1. The precursor gas exits the gas feed through a nozzle in close proximity of the area where the ion beam is incident, thus creating a local gas ambient with pressures in the range of a few mTorr.

3.3.1. Energy Transfer

It has been demonstrated that the deposition yield in ion-induced deposition is larger than unity and correlates with the nuclear stopping power of the incident ion. From this observation Dubner et al. (22) concluded that ion-beam-induced deposition is a substrate-mediated process—that is, the energy lost by the incident ion excites the substrate locally in the vicinity of the point of impact. The authors further investigated the energy transfer from the ion to the substrate and to the adsorbed molecule. Two possible mechanisms have been discussed: the binary collision model (linear cascade theory) and inter-action with the lattice as a whole (thermal spike model). By correlating experimental broad ion-beam data for different ion masses and ion energies with Monte Carlo simulations, the authors conclude that the energy transfer can be best described by a binary collision model (22,23). Similar experiments for gas-assisted etching have not been reported. However, it can be assumed that the energy-transfer mechanism is dominated by binary collisions as well (24).

3.3.2. Reaction Mechanism

Most of the investigations into ion-beam-induced chemistry have been performed using broad, low-energy (<5 keV) ion beams. Several conclusions about the different mechanisms involved in IBID and GAE can be made from these experiments.

In ion-beam-induced deposition the reaction occurs at the top surface, causing fragmentation of adsorbed precursor molecules. For organometallic precursors, this reaction should lead to an immobilized, nonvolatile metal deposit and desorption of the carbon containing by-products. However, if the by-products do not desorb readily, they can be fragmented further by successive ion impacts and become partially incorporated into the growing film (25).

In gas-assisted etching the formation of volatile products from surface atoms requires that back bonds to the substrate be broken. It has been shown that the ion-induced reactions occur in a surface layer that is partially modified by the etch gas. This intermediate layer can be rather thick (for the Si/F system the fluorosilyl layer is assumed to be 10–20 Å). Interactions of the ion beam with this surface layer are rather complicated. Different reaction paths need to be considered and their relative contribution to the total etch yield will depend on parameters such as ion energy, ion mass, current density, substrate treatment and gas/ion flux ratio (for a review on the subject see Winters and Coburn (24)). Possible mechanisms and reaction paths for ion-beam-induced chemistry include: *chemical sputtering*, where ion bombardment causes a chemical reaction which produces volatile products (26); *chemically enhanced physical sputtering*, where ion-induced reactions create more weakly bound species on the surface. This will lead to an increase in the physical sputter rate (27); *enhanced spontaneous etching*, where the spontaneous etch rate of a material increases during (or after) ion-beam bombardment (28).

3.3.3. Macroscopic Kinetic Model

In a digitally scanned FIB system the ion beam is scanned in discrete steps or "pixels" across the surface and the amount of time the beam is stationary at a given point is called the *dwell time*. Because of the high ion current density in a focused ion-beam system, the instantaneous ion flux J is comparable or higher than the neutral gas flux. Therefore, the surface area under the ion beam is rapidly depleted of adsorbed molecules through ion-induced reactions. After the ion beam moves on, J is zero and the surface will be replenished with adsorbate molecules until the raster is repeated. The time between successive exposures of a given pixel is called the *refresh time*. As a result of this dynamic process, the yield for the ion-assisted etching strongly depends on the experimental conditions such as dwell and refresh time, current density, and gas pressure.

Several authors have proposed macroscopic models for ion-beam-induced chemistry. These models are based on the assumption that the yield (Y) of the reaction depends linearly on the precursor surface coverage (e.g., $Y = s * N(t)/N_0$), where N is the number of adsorbed molecules per surface area, N_0 is the number of available adsorption sites and s represents the maximum reaction yield. This linear approximation implies that intermediate reactions are occurring on a time scale, which is fast compared to the time scale of two successive ion impacts. With this approximation, the reaction yield can be determined using a rate equation for $N(t)$, which includes appropriate terms for precursor adsorption, consumption by ion induced reaction, and spontaneous desorption. This equation is then solved, using boundary conditions given by the FIB processing parameters (29–31).

The general trends in the reaction yield for ion-induced deposition or gas-assisted etching can be summarized as follows. The precursor-depletion effect due to ion-induced reactions can be seen in Fig. 5. Starting with a fully saturated surface ad-layer, the reaction yield decreases with increasing dwell time, until a steady state is reached, where the adsorbate surface coverage only depends on the ratio of the gas flux to the ion beam flux. Therefore, to obtain maximum etch (or deposition) yields the dwell time, or more generally the ion dose per pixel, should be as small as possible. In order to obtain an initially fully saturated surface the refresh time between successive ion exposures has to be long enough. This can be seen in Fig. 6 where after a minimum refresh time, the etch yield reaches a constant maximum value corresponding to a saturated adsorbate layer.

The combination of the two effects leads to some important limitations concerning the maximum ion current and the minimum feature size in focused ion-beam induced chemical processes. First, because the time between succes-

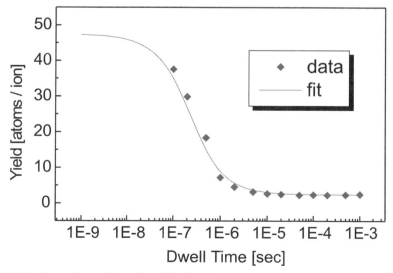

FIGURE 5   Dependence of the etch yield on the dwell time for gas-assisted etching of Si with Cl_2. From Ref. 31, with permission.

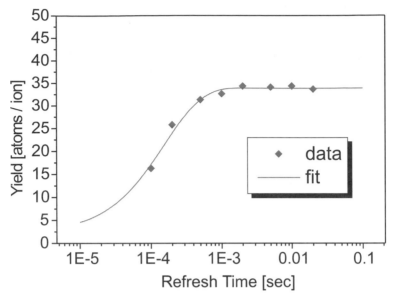

FIGURE 6 Dependence of the etch yield on the refresh time for gas-assisted etching of Si with Cl_2. From Ref 31., with permission.

sive exposures of a given pixel (i.e., refresh time) depends on the size of the feature (i.e., number of pixels multiplied by the dwell time), the reaction yield will drop sharply if the processed area is too small (i.e., $<10\,\mu m^2$). While some FIB systems allow control of the minimum refresh time independently by turning the ion beam "off" (e.g., beam blanking) between successive exposures, the constant blanking might cause a loss in process control and will increase the overall process time.

Second, because of the increasing depletion effect with ion dose, using a higher ion current to achieve faster deposition or etching rates, has to be balanced by increasing the step size (decreasing the number of pixels). A good estimate of the limits imposed by both effects can be obtained by using the concept of an average current density (ion beam current divided by the pattern size). Although the maximum value will depend on the specific process, averaged densities should be in the range of $5–10\,pA/\mu m^2$ and below to obtain optimum process yields.

3.3.4. Gas-Assisted Etching

While physical sputtering or ion milling can be used to remove essentially any kind of solid material, gas-assisted etching (GAE) can be only applied to certain materials. A precursor gas must exist that forms volatile products with the material. Furthermore, to obtain the desired spatial control, the precursor gas must not (or only marginally) spontaneously react with the material. Frequently used precursor gases used in FIB applications include halides (XeF_2, Cl_2, Br_2, I_2) for etching various metals and insulators relevant to semiconductor processing and water vapor for etching of carbon-based materials (32). No etch chemistry exists (or has yet been developed) for certain metals such as Pt, Au, and most metal-oxide compounds used in superconductors or ferroelectrics. With the increasing substitution of aluminum with copper for metal interconnects in high-performance integrated circuits, a suitable etch chemistry for copper has become of great importance and research in this area is currently pursued by both FIB manufacturers and universities (33,34). In the following sections, the characteristics of GAE are summarized. (For a review of the subject see, for example, Young et al. (35).)

3.3.4.1. Enhanced Removal Rate

Some experimental results, reported in the literature for ion-assisted etching using FIB, are listed in Table 2. Shown are the etch-rate enhancement factors (e.g., ratio of gas-assisted milling rate compared to physical sputtering). The large variations reported by different authors are a result of the fact that in contrast to physical sputtering, the yield for the ion-assisted etching strongly depends on the experimental conditions such as dwell and refresh time, current density, and gas pressure (see Section 3.3.3). Because the enhanced

TABLE 2   Gas-Assisted Etching: Enhancement Factors (Removal Rate with Gas vs. Without Gas) for Various Materials and Gas Chemistries

Gas	Si	SiO_2	Al	W	GaAs	InP	Ref.
Cl_2	11 (1)	1 (2)	5–10 (2)	1 (2)	50 (1)	4	(1)
I_2	5–10 (3)	1 (3)	5–15 (3)	—	—	11–13	(4)
XeF_2	7–12 (2)	7–10 (2)	1 (2)	7–10 (2)	—	—	

Reference List
1. Dzioba, S., Cook, J. P. D., Herak, T. V., Livermore, S., Young, M., Rousina, R., Jatar, S., and Shepherd, F. R. *Wafer scale processing of InGaAsP/InP lasers.* J. Vac. Sci. Technol., B 12 (4), 2848 (1994).
2. Casey, J. D. Jr., Doyle, A. F., Lee, R. G., Stewart, D. K., and Zimmermann, H. *Gas-assisted etching with focused ion beam technology.* Microelectron. Eng. 24 (1–4), 43 (1994).
3. Thayer, M. L. IRPS Tutorial (1996).
4. Yamaguchi, A. and Nishikawa, T. *Low-damage specimen preparation technique for transmission electron microscopy using iodine gas-assisted focused ion beam milling.* J. Vac. Sci. Technol., B 13 (3), 962 (1995).

etch rate translated into a lower ion dose needed for micromachining, adverse effects of Ga ion milling such as damage and implantation to underlying structures are also greatly reduced.

3.3.4.2. No Redeposition

Because the products of the ion-induced etching reaction are volatile and therefore not likely to stick to the surface, redeposition will be avoided or greatly reduced. The absence of redeposition permits machining of deep structures with a higher aspect ratio (<1:10 for GAE versus <1:5 for ion milling; see Fig. 4) (21,36–38). Another important advantage is the ability to cut metal lines without creating a conductive path through the redeposited metal.

3.3.4.3. Selectivity

Because GAE is a chemical process, it discriminates between materials (compare Table 2). For example in silicon-based semiconductor processing XeF_2 enhances the etching of Si, SiO_2, and W but does not enhance Al removal, while Cl_2 only increases the etch rate for Si and Al. This selectivity is essential for selectively removing one material in a multilayer structure. The material on top can be rapidly etched with a precursor gas only selective to this material and the removal rate will drop sharply once the lower layer is reached (see Fig. 7). Because of the large difference in the removal rate, "overetching" is not as problematic, thus increasing the process latitude.

3.3.5. Ion-Beam-Induced Deposition

Focused ion-beam-induced deposition (IBID) is mostly used in integrated-circuit (IC) testing and "circuit editing," where flawed prototype designs are repaired by rerouting interconnects through FIB-deposited metal lines and insulating layers. IBID is used also in photolithographic mask repair to add missing adsorber material. Table 3 lists most of the conductor and insulator materials that have been deposited using a focused ion beam (for a review of ion-induced deposition see, for example, Melngailis (39)). Because organo-metallic compounds are used for metal deposition, the FIB deposits contain rather large amounts of carbon and the resistivity of these deposits is generally about one to two orders of magnitude higher than those of pure metal. For most practical applications this is not critical because the connections are rather short and the resistance can be adjusted by increasing the film thickness. For longer connections (a 1 mm long 250 Ω connection takes about 1 hour to deposit) FIB deposition becomes increasingly impractical and other methods

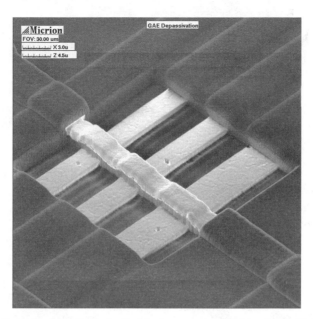

FIGURE 7   Example of the chemical selectivity in gas-assisted etching (GAE). Using XeF_2 GAE, the passivation layer has been removed exposing the underlying aluminum interconnects (Al removal is not enhanced by XeF_2). Picture by Ganesh Sandaram.

such as laser-induced deposition are used (40,41). Interestingly, for reasons that are not completely understood, the minimum width of deposited lines is around 100 nm, even if ion beams with much smaller beam diameter are used for deposition. The aspect ratio of the deposits can be quite high, for example, a 10 μm high array of pillars with a diameter of 200 nm has been deposited using the gold precursor (42).

Most of the commercial FIB systems are equipped with either tungsten-hexacarbonyl [$W(CO)_6$] or (methylcyclopentadienyl) trimethyl-platinum [$(MeCp)Pt(CH_3)_3$]. These compounds are well characterized in terms of deposition yield and resistivity but data about the deposition mechanism is not available. On the other hand, the precursor molecules for Au and Cu are based on a β-Diketonate ligand (acac) commonly found in metal organic chemical vapor deposition (MOCVD) and more mechanistic data is available for this system. The acac-based compounds show some common behaviors, not found in other precursor gases: (1) the microstructure of ion-beam-deposited Au and Cu films consist of small metal islands (20–50 nm) embedded in a carbon matrix (23,43), as opposed to the Pt deposit which shows an amorphous platinum–carbon mixture (44); (2) if the substrate is heated during deposition, the resistivity of the obtained film drops sharply

TABLE 3 Precursor Gases Used in Focused Ion-Beam-Induced Deposition of Conductors and Insulators

Precursor gas Conductors	Yield (atoms/ion)	Composition (M:C:Ga)	Resistivity[a] ($\mu\Omega$cm)	Ref.
$W(CO)_6$	2	75:10:10	150–225	(1)
$Mo(CO)_6$	2–3	67:19:12	200	(2)
$(CH_3)_3NAlH_3$	~5	N/A	900	(3)
$(MeCp)Pt(CH_3)_3$	2.5–35	45:24:28	70–700	(4)
$Au(hfac)(CH_3)_2$	3–8	50:35:15	500	(5)
Cu(hfac) TMVS	10–30	50:50	100	(6)
Insulators		(Si:O:C:Ga)	GΩ cm	
TEOS	~ 2	11:64:11:13	<0.1	(7)
$TEOS/O_2$	2	24:48:0:28	0.08	(8)
$TMCTS/O_2$	17	29:59:0:12	0.2	(8)
$OMCTS/O_2$	11	27:56:0:17	0.6	(8)
$PMCPS/O_2$	11	N/A	6	(9)

[a]The resistivity of FIB deposited oxides depends on the applied electric field. Because the quoted values were measured at different electric fields, they cannot be compared directly.

Reference List
1. Stewart, D. K., Stern, L. A., and Morgan, J. C. *Focused-ion-beam induced deposition of metal for microcircuit modification*. Proc. SPIE **1089**, 18 (1989).
2. Pan, Y., Edinger, K., and Orloff, J. unpublished.
3. Gross, M. E., Harriot, L. R., and Opila, R. L. J. Appl. Phys. **68**, 4820 (1990).
4. Tao, T., JaeSang, Ro, Melngailis, J., Ziling, Xue, and Kaesz, H. D. *Focused ion beam induced deposition of platinum*. J. Vac. Sci. Technol., B 8 (6), 1826 (1990).
5. Blauner, P. G., Jae Sang Ro, Butt, Y., and Melngailis, J. *Focused ion beam fabrication of submicron gold structures*; J. Vac. Sci. Technol., B 7 (4), 609 (1989).
6. Della Ratta, A. D., Melngailis, J., and Thompson, C. V. *Focused-ion beam induced deposition of copper*. J. Vac. Sci. Technol., B **11** (6), 2195 (1993).
7. Young, R. J. and Puretz, J. *Focused ion beam insulator deposition*; J. Vac. Sci. Technol., B 13(6), 2576 (1995).
8. Campbell, A. N., Tanner, D. M., Soden, J. M., Stewart, D. K., Doyle, A., Adams, E., Gibson, M., and Abramo, M. ISTFA '97. Proceedings of the 23rd International Symposium for Testing and Failure Analysis. p. 223 (1997).
9. Edinger, K., Melngailis, J., and Orloff, J. *Study of precursor gases for focused ion beam insulator deposition*; J. Vac. Sci. Technol., B 16(6), 3311 (1998).

(5 $\mu\Omega$ cm at 80 °C for Cu (43), 3–10 $\mu\Omega$ cm at 120 °C for Au (45), compared to resistivities of 200–1,000 $\mu\Omega$ cm when deposited at room temperature). This has been attributed to the increased desorption of the carbon byproducts at elevated temperature. No similar temperature dependence was found for the Pt and W compounds (46).

Insulating material in form of SiO_2 has been deposited using various siloxane precursors and oxygen. Given the fact that Ga ions are used for deposition, the resistivity observed for these compounds is remarkably high. Because of the nonlinear dependence of the resistivity on the applied electric field (e.g., voltage and film thickness) the resistivity values of these compounds cannot be directly compared (47). For thin films (e.g., 100 nm) and low voltages (e.g., 5 V) resistivities are in the range of 10^8 to $10^{12}\,\Omega\,cm$ (47–49).

4. APPLICATIONS

4.1. LOW-ENERGY FIB SYSTEMS

Focused ion-beam systems for micromachining combine precise material removal and addition and allows *in situ* imaging for navigation, alignment, and inspection. In the semiconductor industry these systems are well-established tools for mask repair, failure analysis, and IC prototype rewiring. The following examples of commercial and research applications will illustrate the potential of this technology as well as its limits.

4.1.1. Failure Analysis and Material Science

4.1.1.1. Cross-Sectioning

One important application of FIB is precise cross-sectioning of defect sites to examining subsurface defects. Using the scanning ion microscope mode and circuit layout data, the suspected defect site is positioned under the beam and a rectangular pit, bordering the defect, is milled into the material. The sample is then tilted and the cross-section can be examined with either scanning ion or scanning electron microscopy (in some systems both an FIB and an SEM are mounted onto the specimen chamber). In addition, chemical information can be obtained using, for example, Auger spectroscopy (AES) or secondary ion spectroscopy (SIMS) (50). FIB cross-sectioning not only allows precise location of the defect site, a task that would be very difficult with conventional mechanical polishing techniques, but also offers the possibility to produce a series of cross-sections at the same defect site by successively removing thin slices of material. This ensures that one of the sections will intersect the defect and it also allows the construction of a quasi-three-dimensional image of the site.

4.1.1.2. TEM Sample Preparation

With the ever-shrinking feature size of integrated circuits there is a fast-growing demand to prepare site-specific samples for high-resolution TEM observation. This technique is similar to the FIB cross-sectioning except that two recessions are milled back-to-back, leaving a thin (300–100 nm thick) membrane in between (see Fig. 8). Using the so-called "lift-out" technique, first reported by Overwijk et al., the membrane can be fully or partially released by cutting around the perimeter with the FIB and then it can be transferred onto a TEM specimen holder using a micromanipulator with a sharp needle (51). The total preparation time is around 3–5 hours and represents a huge improvement over conventional TEM preparation techniques. Consequently, FIB-assisted TEM sample preparation is a fast-growing area in IC failure analysis as well as in general material science applications (52).

4.1.1.3. Electrical Testing

Electrical testing of integrated circuits often requires micromachining on a fully functional and packaged chip. In this case focused ion-beam milling is used to access buried signal lines for electron beam or mechanical probing without degrading the electrical performance of the circuit. This task becomes especially difficult in the so-called flip-chip configuration, where the chip is mounted with the structured side facing the circuit package and signal access has to be obtained from the backside. Because removing such large quantities

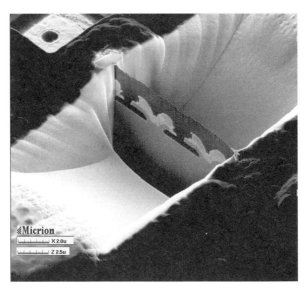

FIGURE 8   Example of a TEM specimen preparation by FIB. Two pits are milled at either side of the area of interest. The remaining wall is trimmed to a thin membrane using a smaller beam current. Picture by Ganesh Sandaram.

of silicon with a focused ion beam is impractical, the entire silicon wafer is first mechanically thinned and then material above the region of interest is locally removed by scanning-laser chemical etching. Finally the last $\sim 10\,\mu m$ of silicon is milled with high precision by FIB (41). The tight geometry and multilevel interconnect wiring (modern ICs contain six and more stacked layers of dielectric and structured metalization) requires delicate machining. Gas-assisted etching (GAE) is frequently used, which allows milling of the high-aspect ratio holes that are necessary to access deep-lying signal lines without destroying adjacent lines. In addition the chemical selectivity of GAE improves end-point detection by increasing the process latitude (for example enhanced removal of dielectric over an aluminum signal line with XeF_2 GAE).

4.1.1.4. Circuit Editing

During the design phase of a new IC, FIB micromachining is used to cut and reroute signal lines of malfunctioning chips. This allows design modifications to be tested without going through a long and expensive maskmaking and production cycle. The requirements for the FIB machining are essentially the same as those just described for IC testing. The rewiring requires the deposition of metal lines and dielectric material to reconnect different parts of the circuit (see Fig. 9). The "repair" can be fairly complex and the capability to deposit dielectric material becomes important for insulating the deposited metal from other exposed metal lines.

4.1.2. Mask Repair

Except for direct-writing strategies, microstructures are usually fabricated by exposing a radiation-sensitive resist through a patterned photolithographic mask. In general, these masks can have two types of defects. If absorber is missing, radiation can pass through and expose the resist (clear defect) or areas that should be transparent can contain unwanted absorber (opaque defect). Both types of defects can be repaired with an FIB by either removing or adding material and given the fact that advanced masks can cost in excess of $50,000, dedicated FIB mask-repair tools have been developed for this purpose. Although the repair process seems to be straightforward, a serious problem in the repair of opaque defects is the reduction in transmission due to ion-beam-induced damage and Ga ion implantation. For standard optical lithography masks, consisting of chrome absorber on a quartz substrate, this "staining effect" can be reduced by using a reactive gas etch for Cr removal (current processes have a 2.5× enhancement in Cr removal rate) or by removing the shallow implanted layer (either locally using XeF_2 assisted FIB etching or globally by wet or dry etching).

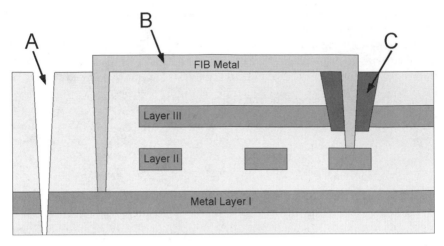

FIGURE 9 Schematic of FIB procedures in circuit editing. A, FIB sputtering or GAE is used to cut metal lines; B, FIB-induced deposition of conductors is used to rewire parts of the circuit, while insulator material is deposited to isolate the newly deposited line from exposed parts of the circuit (C).

Despite these problems, focused ion beams are a well-established tool for repair of conventional and phase-shift masks (in this type of mask the transmitted light is phase shifted in certain areas to reduce diffraction effects). State-of-the-art FIB repair systems allow a placement accuracy of <75 nm (limited by electrical charging of the insulating quartz substrate) and optical transmission after repair is greater than 95% with a phase-shift error of less than 30° (for a review see for example (17,53)). Below a wavelength of 157 nm current lithography will most likely deviate from refractive optics. All of the lithography types currently under consideration—extreme ultraviolet (EUV), X-ray, ion projection, or electron projection (SCALPEL)—will use either reflective multilayer or stencil-thin-membrane-type masks. It can be expected that FIB-based mask repair will play a major role in repair of these masks and research and development of repair techniques is ongoing.

4.1.3. Micro- and Nanofabrication

4.1.3.1. Semiconductor Laser Optics

Focused ion beams have been used to fabricate optical quality laser mirrors and other optical elements in III-V semiconductors. This process produces optically smooth facets, does not degrade laser efficiency, and can be applied in any geometry, whereas conventional cleaving methods usually are restricted to certain crystallographic orientations (54–57).

4.1.3.2. Superconducting and Ferroelectric Materials

High-temperature superconducting (HTS) and ferro- or piezoelectric materials usually consist of transition metal-oxide compounds, which are difficult to pattern with conventional methods such as reactive ion etching (RIE). Using focused ion-beam sputtering, HTS Josephson junctions have been successfully fabricated by patterning either the HTS material directly or the underlying substrate prior to HTS deposition (58–60). Ferroelectric materials are considered as dielectrics for capacitors in nonvolatile random access memory. For investigating the scaling effects at nanometer-size dimensions, functional test capacitors with sizes as small as $0.04\,\mu m^2$ have been built by focused ion-beam sputtering (61).

4.1.3.3. Magnetic Read-Write Heads

One factor that limits the data-storage density on magnetic disks is the width of the write head, which consists of two magnetic poles. By using focused ion-beam trimming, this width can be controlled with high accuracy (62,63). Trimming of magnetic recording heads for hard drives is the first application where a focused ion-beam micromachining is used on a commercial production scale. Fully automated, high-current-density FIB systems with pattern-recognition software have been developed for this process. By first generating a scanning ion image, the software automatically detects and measures the targeted part. Then a milling pattern is generated, which adapts to the individual shape of the inspected part (64). Typical machining time, including stage navigation, is about 5 seconds per head, and typical pole widths are ~100 nm with a placement accuracy better than 25 nm.

4.1.3.4. Proximal Probes and MEMS-Type Objects

Focused ion-beam micromachining has been applied to the fabrication of various types of proximal probes such as high-aspect ratio scanning tunneling microscopy (STM) tips (65), high-resolution scanning force (AFM) tips (66–68) (FIB-machined AFM probes are commercially available) and high-sensitivity cantilevers (69,70), magnetic force microscopy (MFM) sensors (71), and high-resolution apertures for scanning near-field optical microscopy (SNOM) (72,73). In addition, nanometer-sized apertures for various near-field applications have been demonstrated (74–76).

The fabrication of three-dimensional objects has been pursued by Vasile and coworkers (77). In analogy to a conventional lathe they use a rotating sample chuck, perpendicular to the ion beam, to fabricate micron-sized machining tools and microsurgical manipulators. In another approach computer-controlled local variations in sputter time are used to fabricate three-dimensional surface structures used as masters in molding applications. Most of the applications discussed in this paragraph rely on the ability of the focused ion-beam to process nonplanar geometries with submicrometer precision. With the evolution of micro-electromechanical systems (MEMS) this aspect of FIB machining is likely to become more important. MEMS-type devices modified by FIB and reported so far include micromachining of a read-out gap into an accelerometer structure (78) and frequency tuning of micromechanical resonators (79,80).

In some circumstances, an alternative approach to structuring of material by the focused ion beam is to use ion implantation and ion-induced damage to obtain local functionality. In contrast to ion-beam sputtering, the necessary ion dose is generally orders of magnitudes lower, resulting in increased throughput and less implantation and damage to adjacent areas. For electronic and opto-electronic applications, functionality can be achieved through focused ion-beam implantation as outlined in the next chapter. However, ion-induced chemical modifications can also be used to obtain structural functionality. For example, if $\langle 100 \rangle$ oriented silicon is implanted with Ga ions with a minimum dose of $\sim 10^{15}$ ions/cm^2 it becomes insoluble in an anisotropic Si $\langle 100 \rangle$ wet etch. Accordingly, focused ion-beam implantation and subsequent wet etching in KOH has been utilized to fabricate freestanding structures. Cantilevers and bridges with a minimum thickness of 30 nm (corresponding to the penetration depth of 30 keV Ga$^+$ ions) and widths as small as 100 nm have been demonstrated (81,82). In a reversed process, focused ion-beam-induced amorphization, which requires ion doses in the range of $1.5–3 \times 10^{13}$ ions/cm^2 can be used to enhance the etch rate of certain materials. This effect has been used to fabricate FIB-patterned structures in Si (83) GaAs (83–85), InP (86), GaN (87), and SiC (88). Another recent example of the utilization of ion-induced modification of chemical and physical material properties at low-ion doses ($\sim 10^{14}$ ions/cm^2) is the transformation of insulating noble metal-oxide films (PtO$_x$, AuO$_x$) into conducting metal lines by focused ion-beam direct writing (89,90).

4.2. HIGH-ENERGY FIB SYSTEMS

High-energy focused ion-beam systems are used in research for direct, maskless implantation and for ion lithography (91). These systems are usually equipped with an alloy liquid metal ion source and a mass filter and can deliver, for example, the main dopants for Si-based (B, As) and III-IV (Be, Si) semiconductors. However, a large variety of ions can be produced and other applications include, for example, implantation of cobalt ions into silicon to form metal-like $CoSi_2$ layers (92) or implantation of Er and Pr for opto-electronical devices (93).

Besides the fact that electronic devices can be fabricated in a maskless and resistless process, the dose of ions delivered to any given point can be controlled within the limits of the beam diameter and the lateral straggle of the ions (i.e., 50 nm). This allows fast prototyping and device optimization by using different doping levels for identical devices on the same wafer as well as the introduction of a lateral doping gradient. Both configurations are very difficult to achieve with conventional lithography and broad-beam implantation, and would require an extended number of additional masks for each doping level.

In silicon, new or improved devices by FIB implantation have been demonstrated. For example, by implanting a lateral doping gradient, tunable Gunn diodes (94) and fast charge-coupled devices (CCD) (95,96) have been built. A fast 1 GHz flash analog-to-digital converter has been demonstrated by locally adjusting the threshold voltages for different MOS transistors in the circuit (97).

FIB implantation for III-V semiconductors to fabricate novel electrical and opto-electronical devices, such as semiconductor lasers and waveguides, has attracted more interest recently, partly because of the ability to build low-dimensional quantum structures and nanodevices using MBE-grown heterostructures. Electrical functionality in these devices can be obtained by either rendering a conductive layer insulating or semi-insulating through ion-induced damage or Ga implantation (98–100) or by forming a doped, conductive layer through ion implantation (98,101–103). In quantum well structures FIB implantation can also be used to introduce local shifts in the bandgap by ion-induced thermal intermixing (104,105).

In some cases the focused ion-beam system has been combined with a molecular beam epitaxy system into one ultrahigh-vacuum system, which allows direct, in situ fabrication of multilayer devices, such as buried nanostructures. Because no conventional lithographic steps involving resist coating and exposure to air are required, this all-vacuum direct-patterning process results in cleaner interfaces and improved device characteristics. Because the stacked layers in those devices can be very thin (sometimes only a few atomic layers), ion-induced damage of underlying layers can be significant. More recently this problem has been addressed by developing low energy (<200 eV) focused ion beam systems, where the energy of the ions is reduced just above the target by applying a retarding field (106,107). Due to the retarding field, the beam diameter increases and values of 400 nm at 40 pA ion current have been reported (108). Because of low landing energy of the ions, these systems can also be used for direct deposition of metal layers. Direct deposition of metal films from a liquid metal ion source have been demonstrated for Al, Au, Cu, and Nb (108).

5. CONCLUSIONS

Focused ion-beam systems combine precise material modification at the nanometer scale utilizing removal, deposition, and implantation with *in situ* imaging capabilities for navigation, alignment, and inspection. For direct-writing applications, one may characterize the performance of the technology in terms of resolution, writing speed, and material compatibility.

The achievable resolution depends on the selected ion current, which in turn determines the writing speed. At very low beam current (<10 pA) sub-ten nanometer resolution has been demonstrated for imaging and lithography. However, especially in sensitive electronic materials, the practical resolution limit is determined by secondary effects such as ion implantation and ion-induced damage, which for 30 keV Ga affects an area of ~10 nm lateral and ~30 nm in depth. These effects, together with the low writing speed, process stability, and control issues at low currents, might set a more realistic resolution limit on the order of 20–50 nm.

The writing speed for sputter removal and deposition depends on the material and the selected beam current. Typical values are on the order of 0.3 μm³ of material removed or deposited per second for an ion beam current of 1 nA (~100 nm beam size). Maximum ion currents for commercial systems are around 10–50 nA at beam sizes approaching 0.5 μm and larger. At these rates, micromachining of features much larger than 100 μm becomes impractical. To illustrate this point, a 100 μm × 100 μm and 1 μm deep recess milled into silicon with 10 nA beam current will take about 1 hour. Because the beam diameter increases with ion current, other methods such as laser-induced deposition and etching provide higher throughput at comparable resolution and have therefore been used in combination with, and complementary to, focused ion-beam machining for patterning of larger areas.

Although focused ion-beam writing provides unique capabilities and is not restricted to certain materials or a planar geometry, the fact that it is a relatively slow serial process puts a serious constraint on medium-to-large-scale fabrication, which limits its commercial usefulness to high-value products. This is clearly the case for applications in the semiconductor industry, where focused ion beams are very successfully used in circuit analysis and mask repair. In these applications, high throughput is not of critical importance and the benefits, in terms of shorter development cycles and reduced process down time, by far outweigh the operational cost of the FIB system (~$400 per hour at independent commercial FIB services). In terms of commercial scale production, FIB processing is not likely to compete directly with parallel lithographic processes. However, it is useful if the machining is restricted to a limited number of critical areas, which increases the performance of the device and thus adds substantially to the value of the product. Examples for this approach are high-accuracy tuning of MEMS-type resonators and trimming of magnetic write heads for hard disk drives. Commercial interest in the second area has already led to automated, high-current FIB tools, which include automatic pattern recognition. The software considerably reduces overhead time and allows automatic real-time inspection combined with flexible milling-pattern adjustments. While until recently FIB-based production has not been seriously considered, these developments might make the process more feasible and spur further applications.

Research in the area of high-energy focused ion-beam application, particularly in maskless ion implantation, has demonstrated the unique capabilities of this technique, enabling novel devices and improving the performance of existing ones. However, so far, commercial applications for high-energy focused ion beams have not evolved. This is largely due to the fact that in contrast to FIB-assisted IC analysis, IC production is dominated by conventional lithography, which may not produce the same high-performance devices, but allows high throughput at low fabrication cost, as required for the high-volume IC market. In addition, commercial production requires high repeatability and reliability, and FIB process-stability issues, such as stable ion-source operation, need to be addressed. However, given the industry's demand for ever smaller and faster devices and the recent, fast-growing effort in nanotechnology research, one might expect a (limited) number of applications to evolve, where the gain in performance will justify the increased production cost.

REFERENCES

1. Prewett, P. D. and Mair, G. L. Focused Ion Beams from Liquid Metal Ion Sources. New York: John Wiley & Sons, Inc.; 1991.
2. Orloff J. (editor). Charged Particle Optics. Boca Raton, FL: CRC Press; 1997.
3. Orloff, J. High-resolution focused ion beams; Rev. Sci. Instrum. 64, 1105 (1993).

4. Melngailis, J. *Focused ion beam technology and applications*; J. Vac. Sci. Technol., B **5**, 469 (1987).

5. Ziegler, J. F., Biersack, J. P., and Littmark, V. The Stopping and Range of Ions in Solids. New York: Pergamon; 1985.

6. Eriksson, L., Davis, J. A., and Mayer, J. W. *Radiation Effects in Semiconductors*, Vook, F. L., Editor. New York: Plenum Press; 1968 p. 398.

7. Mladenov, G. M. and Emmoth, B. *Polymethyl methacrylate sensitivity variation versus the electronic stopping power at ion lithography exposure*; Appl. Phys. Lett., **38**, 1000 (1981).

8. Ohta, T., Kanayama, T., Tanoue, H., and Komuro, M. *Focused ion beam lithography using Al_2O_3 as a resist for fabrication of X-ray masks*; J. Vac. Sci. Technol., B **7**, 89 (1989).

9. Koshida, N., Ichinose, Y., Ohtaka, K., Komuro, M., and Atoda, N. *Microlithographic behavior of transition metal oxide resists exposed to focused ion beam*; J. Vac. Sci. Technol., B **8**, 1093 (1990).

10. Gierak, J., Vieu, C., Launois, H., Benassayag, G., and Septier, A. *Focused ion beam nano-lithography on AlF_3 at a 10 nm scale*; Appl. Phys. Lett., **70**, 2049 (1997).

11. Lee, H. Y. and Chung, H. B. *Low-energy focused-ion-beam exposure characteristics of an amorphous Se75Ge25 resist*; J. Vac. Sci. Technol., B **15**, 818 (1997).

12. Nikawa, K. *Applications of focused ion beam technique to failure analysis of very large scale*; J. Vac. Sci. Technol., B **9**, 2566 (1991).

13. Barr, D. L. and Brown, W. L. *Contrast formation in focused ion beam images of polycrystalline aluminum*; J. Vac. Sci. Technol., B **13**, 2580 (1995).

14. Dunn, D. N. and Hull, R. *Reconstruction of three-dimensional chemistry and geometry using focused ion beam microscopy*; Appl. Phys. Lett., **75**, 3414 (1999).

15. Xu, X., Della Ratta, A. D., Sosonkina, J., and Melngailis, J. *Focused ion beam induced deposition and ion milling as a function of angle of ion incidence*. J. Vac. Sci. Technol., B **10** (6), 2675 (1992).

16. Ishitani, T. and Yaguchi, T. *Cross-sectional sample preparation by focused ion beam: a review of ion-sample interaction*; Microsc. Res. Tech. **35**, 320 (1996).

17. Morgan, J. C. *Focused ion beam mask repair*; Solid State Technol., **41**, 61 (1998).

18. Kola, R. R., Celler, G. K., and Harriott, L. R. *Roughness effects during focused ion beam repair of X-ray masks with polycrystalline tungsten absorbers*; Mater. Res. Soc. Symp. Proc., **279**, 593 (1993).

19. Muller, K. P. and Petzold, H. C. *Microstructuring of gold on X-ray masks with focused Ga^+ ion beams*; Proc. SPIE, **1263**, 12 (1990).

20. Yamaguchi, H., Shimase, A., Haraichi, S., and Miyauchi, T. *Characteristics of silicon removal by fine focused gallium ion beam*; J. Vac. Sci. Technol., B **3**, 71 (1985).

21. Pellerin, J. G., Griffis, D. P., and Russell, P. E. *Focused ion beam machining of Si, GaAs, and InP*; J. Vac. Sci. Technol., B **8**, 1945 (1990).

22. Dubner, A. D., Wagner, A., Melngailis, J., and Thompson, C. V. *The role of the ion-solid interaction in ion-beam-induced deposition of gold*; J. Appl. Phys., **70**, 665 (1991).

23. Ro, J. S., Thompson, C. V., and Melngailis, J. *Mechanism of ion beam induced deposition of gold*. J. Vac. Sci. Technol., B **12** (1), 73 (1994).

24. Winters, H. F. and Coburn, J. W. *Surface science aspects of etching reactions*; Surf. Sci. Rep. **14**, 161 (1992).

25. Chiang, T. P., Sawin, H. H., and Thompson, C. V. *Surface kinetic study of ion-induced chemical vapor deposition of copper for focused ion beam applications*; J. Vac. Sci. Technol., A **15**, 3104 (1997).

26. Tu, Y. Y., Chuang, T. J., and Winters, H. F. *Chemical sputtering of fluorinated silicon*; Phys. Rev. B **23**, 823 (1981).

27. Mauer, J. L., Logan, J. S., Zielinski, L. B., and Schwartz, G. C. *Mechanisms of silicon etching by CF₄ plasma*; J. Vac. Sci. Technol., **15**, 333 (1978).

28. Coburn, J. W. and Winters, H. F. *The role of energetic ion bombardment in silicon-fluorine chemistry*; Nucl. Instrum. Methods Phys. Res., B **27**, 243 (1987).

29. Takahashi, Y., Madokoro, Y., and Ishitani, T. *Focused ion beam induced deposition in the high current density region*. Jpn. J. Appl. Phys., Part 1, **30** (11B), 3233 (1991).

30. Harriott, L. R. *Digital scan model for focused ion beam induced gas etching*; J. Vac. Sci. Technol., B **11**, 2012 (1993).

31. Edinger, K. and Kraus, T. *Modeling of focused ion beam induced surface chemistry*; J. Vac. Sci. Technol., B **18**, 3190 (2000).

32. Stark, T. J., Shedd, G. M., Vitarelli, J., Griffis, D. P., and Russell, P. E. *H₂O enhanced focused ion beam micromachining*; J. Vac. Sci. Technol., B **13**, 2565 (1995).

33. Phillips, J. R., Griffis, D. P., and Russell, P. E. *Channeling effects during focused-ion-beam micromachining of copper*; J. Vac. Sci. Technol., A **18**, 1061 (2000).

34. Edinger, K. *Gas assisted etching of copper with focused ion beams*; J. Vac. Sci. Technol., B **17**, 3058 (1999).

35. Young, R. J., Cleaver, J. R. A., and Ahmed, H. *Characteristics of gas-assisted focused ion beam etching*; J. Vac. Sci. Technol., B **11**, 234 (1993).

36. Casey, J. D. Jr., Doyle, A. F., Lee, R. G., Stewart, D. K., and Zimmermann, H. *Gas-assisted etching with focused ion beam technology*. Microelectron. Eng., **24** (1-4), 43 (1994).

37. Ochiai, Y., Gamo, K., and Namba, S. *Pressure and irradiation angle dependence of maskless ion beam assisted etching of GaAs and Si*; J. Vac. Sci. Technol., B **3**, 67 (1985).

38. Takado, N., Asakawa, K., Arimoto, H., Morita, T., Sugata, S., Miyauchi, E., and Hashimoto, H. *Chemically-enhanced GaAs maskless etching using a novel focused ion beam etching system with a chlorine molecular and radical beam*. Mater. Res. Soc. Symp. Proc., **75**, 107 (1987).

39. Melngailis, J. *Focused ion beam induced deposition—a review*. Proc. SPIE, **1465**, 36 (1991).

40. Van Doorselaer, K., Van den Reek, M., Van den Bempt, L., Young, R., and Whitney, J. *How to Prepare Golden Devices Using Lesser Materials*; ISTFA '93. Proceedings of the 23rd International Symposium for Testing and Failure Analysis, 1993 p. 405.

41. Livengood, R. H., Winer, P., and Rao, V. R. *Application of advanced micromachining techniques for the characterization and debug of high performance microprocessors*; J. Vac. Sci. Technol., B **17**, 40 (1999).

42. Menzel, R., Bachmann, T., Wesch, W., and Schmidt, C. *Preparation of stopping masks for the P-LIGA technique with focused ion beams*; Nucl. Instrum. Methods Phys. Res., B **139**, 359 (1998).

43. Della Ratta, A. D., Melngailis, J., and Thompson, C. V. *Focused-ion beam induced deposition of copper*; J. Vac. Sci. Technol., B **11** (6), 2195 (1993).

44. Tao, Tao, JaeSang, Ro, Melngailis, J., Ziling, Xue, and Kaesz, H. D. *Focused ion beam induced deposition of platinum*. J. Vac. Sci. Technol., B **8** (6), 1826 (1990).

45. Blauner, P. G., Butt, Y., Jae Sang Ro, Thompson, C. V., and Melngailis, J. *Focused ion beam induced deposition of low-resistivity gold films*. J. Vac. Sci. Technol., B **7** (6), 1816 (1989).

46. Melngailis, J. (personal communication); Edinger, K. (unpublished).

47. Edinger, K., Melngailis, J., and Orloff, J. *Study of precursor gases for focused ion beam insulator deposition*; J. Vac. Sci. Technol., B **16**, 3311 (1998).

48. Young, R. J. and Puretz, J. *Focused ion beam insulator deposition*; J. Vac. Sci. Technol., B **13**, 2576 (1995).

49. Campbell, A. N., Tanner, D. M., Soden, J. M., Stewart, D. K., Doyle, A., Adams, E., Gibson, M., and Abramo, M. *Electrical and chemical characterization of FIB-deposited insulators*; ISTFA '97. Proceedings of the 23rd International Symposium for Testing and Failure Analysis, 1997 p. 223.

50. Verkleij, D. *The use of the focused ion beam in failure analysis*; Microelec. Rel. **38**, 869 (1998).

51. Overwijk, M. H. F., van den Heuvel, F. C., and Bulle-Lieuwma, C. W. T. *Novel scheme for the preparation of transmission electron microscopy specimens with a focused ion beam*; J. Vac. Sci. Technol., B **11**, 2021 (1993).

52. Giannuzzi, L. A., Drown, J. L., Brown, S. R., Irwin, R. B., and Stevie, F. *Applications of the FIB lift-out technique for TEM specimen preparation*; Microsc. Res. Tech., **41**, 285 (1998).

53. Stewart, D. K., Doherty, J. A., Doyle, A. F., and Morgan, J. C. *State of the art in focused ion beam mask repair systems*. Proc. SPIE, **2512**, 398 (1995).

54. Harriott, L.R., Scotti, R.E., Cummings, K. D., and Ambrose, A. F. *Micromachining of Integrated Optical Structures*; APL, **48**, 1704 (1986).

55. Puretz, P., De Freez, R. K., Elliot, R. A., and Orloff, J. *Focused-ion-beam micromachined AlGaAs semiconductor laser mirrors*; Electron. Lett., **22**, 700 (1986).

56. Mack, M. P., Via, G. D., Abare, A. C., Hansen, M., Kozodoy, P., Keller, S., Speck, J. S., Mishra, U. K., Coldren, L. A., and Denbaars, S. P. *Improvement of GaN-based laser diode facets by FIB polishing*; Electron. Lett., **34**, 1315 (1998).

57. Katoh, H., Takeuchi, T., Anbe, C., Mizumoto, R., Yamaguchi, S., Wetzel, C., Amano, H., Akasaki, I., Kaneko, Y., and Yamada, N. *GaN based laser diode with focused ion beam etched mirrors*; Jpn. J. Appl. Phys., Part 2, **37**, L444–L446 (1998).

58. Chen, C.-H., Jin, I., Pai, S. P., Dong, Z. W., Sharma, R. P., Lobb, C. J., Venkatesan, T., Edinger, K., Orloff, J., Melngailis, J., Zhang, Z., and Chu, W. K. *Combined method of focused ion beam milling and ion implantation techniques for the fabrication of high temperature superconductor Josephson junctions*; J. Vac. Sci. Technol., B **16**, 2898 (1998).

59. Morohashi, S., Wen, J. G., Enomoto, Y., and Koshizuka, H. *Fabrication process and interfacial study of high-T-C Josephson junctions fabricated using the focused ion beam technique*; Jpn. J. Appl. Phys., Part 1, **38**, 698 (1999).

60. Kim, S. J., Latyshev, Y. I., and Yamashita, T. *Submicron stacked-junction fabrication from $Bi_2Sr_2CaCu_2O_8$-Deltawhiskers by focused-ion-beam etching*; Appl. Phys. Lett., **74**, 1156 (1999).

61. Stanishevsky, A., Aggarwal, S., Prakash, A. S., Melngailis, J., and Ramesh, R. *Focused ion-beam patterning of nanoscale ferroelectric capacitors*; J. Vac. Sci. Technol., B **16**, 3899 (1998).

62. Khizroev, S. K., Kryder, M. H., Ikeda, Y., Rubin, K., Arnett, P., Best, M., and Thompson, D. A. *Recording heads with track widths suitable for 100 Gbit/in(2)density*; IEEE Trans. Magn., **35**, 2544 (1999).

63. Koshikawa, T., Nagai, A., Yokoyama, Y., Hoshino, T., and Ishizuki, Y. *A new write head trimmed at wafer level by focused ion beam*; IEEE Trans. Magn., **34**, 1471 (1998).

64. Athas, G. J., Noll, K. E., Mello, R., Hill, R., Yansen, D., Wenners, F. F., Nadeau, J. P., Ngo, T., and Siebers, M. *Focused ion beam system for automated MEMS prototyping and processing*; Proc. SPIE, **3223**, 198 (1997).

65. Hopkins, L. C., Griffith, J. E., Harriott, L. R., and Vasile, M. J. *Polycrystalline tungsten and iridium probe tip preparation with a Ga^+ focused ion beam*. J. Vac. Sci. Technol., B **13** (2), 335 (1995).

66. Olbrich, A., Ebersberger, B., Boit, C., Niedermann, P., Hanni, W., Vancea, J., and Hoffmann, H. *High aspect ratio all diamond tips formed by focused ion beam for conducting atomic force microscopy*; J. Vac. Sci. Technol., B **17**, 1570 (1999).

67. Ximen, H. and Russell, P. E. *Microfabrication of AFM tips using focused ion and electron beam techniques*. Ultramicroscopy, **42–44** (pt.B), 1526 (1992).

68. Vasile, M. J., Grigg, D. A., Griffith, J. E., Fitzgerald, E. A., and Russell, P. E. *Scanning probe tips formed by focused ion beams (IC testing application)*; Rev. Sci. Instrum., **62**, 2167 (1991).

69. Fabian, J. H., Scandella, L., Fuhrmann, H., Berger, R., Mezzacasa, T., Musil, C., Gobrecht, J., and Meyer, E. *Finite element calculations and fabrication of cantilever sensors for nanoscale detection*; Ultramicroscopy, **82**, 69 (2000).

70. Kageshima, M., Ogiso, H., Nakano, S., Lantz, M. A., and Tokumoto, H. *Atomic force microscopy cantilevers for sensitive lateral force detection*; Jpn. J. Appl. Phys., Part 1, **38**, 3958 (1999).

71. Folks, L., Best, M. E., Rice, P. M., Terris, B. D., Weller, D., and Chapman, J. N. *Perforated tips for high-resolution in-plane magnetic force microscopy*; Appl. Phys. Lett., **76**, 909 (2000).

72. Pilevar, S., Edinger, K., Atia, W., Smolyaninov, I., and Davis, C. *Focused ion-beam fabrication of fiber probes with well-defined apertures for use in near-field scanning optical microscopy*; Appl. Phys. Lett., **72**, 3133 (1998).

73. Veerman, J. A., Otter, A. M., Kuipers, L., and Van Hulst, N. F. *High definition aperture probes for near-field opticalmicroscopy fabricated by focused ion beam milling*; Appl. Phys. Lett., **72**, 3115 (1998).

74. Shinada, S., Koyama, F., Nishiyama, N., Arai, M., Goto, K., and Iga, K. *Fabrication of micro-aperture surface emitting laser for near field optical data storage*; Jpn. J. Appl. Phys., Part 2, **38**, L1327–L1329 (1999).

75. Yoshikawa, H., Andoh, Y., Yamamoto, M., Fukuzawa, K., Tamamura, T., and Ohkubo, T. *7.5-Mhz data-transfer rate with a planar aperture mounted upon a near-field optical slider*; Opt. Lett., **25**, 67 (2000).

76. Voigt, J., Iline, A., Shi, F., Hudek, P., Rangelow, I., Mariotto, G., Shvets, I., Gunther, P., Loschner, H., and Edinger, K. *Progress on nanostructuring with nanojet*; J. Vac. Sci. Technol., B **18**, 3525 (2000).

77. Vasile, M. J., Nassar, R., and Xie, J. S. *Focused ion beam technology applied to microstructure fabrication*; J. Vac. Sci. Technol., B **16**, 2499 (1998).

78. Daniel, J. H., Moore, D. F., Walker, J. F., and Whitney, J. T. *Focused ion beams in microsystem fabrication*; Microelectron. Eng., **35**, 431 (1997).

79. Prewett, P. D. *Focused ion beams—microfabrication methods and applications.* Vacuum, **44** (3–4), 345 (1993).

80. Syms, R. R. A. and Moore, D. F. *Focused ion beam tuning of in-plane vibrating micro mechanical resonators*; Electron. Lett., **35**, 1277 (1999).

81. Schmidt, B., Bischoff, L., and Teichert, J. *Writing FIB implantation and subsequent anisotropic wet chemical etching for fabrication of 3D structures in silicon*; Sens, Actuators, A, **61**, 369 (1997).

82. Brugger, J., Beljakovic, G., Despont, M., Derooij, N. F., and Vettiger, P. *Silicon micro/nanomechanical device fabrication based on focused ion beam surface modification and KOH etching*; Microelectron. Eng., **35**, 401 (1997).

83. Komuro, M., Hiroshima, H., Tanoue, H., and Kanayama, T. *Maskless etching of a nanometer structure by focused ion beams*; J. Vac. Sci. Technol., B **1**, 985 (1983).

84. Templeton, I. M., Fallahi, M., Charbonneau, S., Champion, H. G., and Allard, L. B. *Focused-ion-beam damage-etch patterning for isolation of quantum structures in AlGaAs/GaAs.* J. Vac. Sci. Technol., B **11** (6), 2416 (1993).

85. Shiokawa, T., Ishibashi, K., Pil Hyon Kim, Aoyagi, Y., Toyada, K., and Namba, S. *Fabrication of periodic structures in GaAs by focused-ion-beam implantation*; Jpn. J. Appl. Phys., Part 1, **29**, 2864 (1990).

86. Konig, H., Reithmaier, J. P., and Forchel, A. *Highly resolved maskless patterning on InP by focused ion beam enhanced wet chemical etching*; Jpn. J. Appl. Phys., Part 1, **38**, 6142 (1999).

87. Schiestel, S., Molnar, B., Carosella, C. A., Stroud, R. M., Knies, D., and Edinger, K. *Patterning of GaN by ion implantation-dependent etching*; Mater. Sci. Eng., B **82**, 111 (2001).

88. Wesch, W., Heft, A., Menzel, R., Bachmann, T., Peiter, G., Hobert, H., Hoche, T., Dannberg, P., and Brauer, A. *Ion beam processing of SiC for optical application*; Nucl. Instrum. Methods Phys. Res. B, **148**, 545 (1999).

89. Machalett, F., Edinger, K., Ye, L., Melngailis, J., Venkatesan, T., Diegel, M., and Steenbeck, K. *Focused ion beam writing of electrical contacts into platinum oxide films*; Appl. Phys. Lett., 76, 3445 (2000).
90. Machalett, F., Edinger, K., Melngailis, J., Diegel, M., Steenbeck, K., and Steinbeiss, E. *Direct patterning of gold oxide thin films by focused ion-beam irradiation*; Appl. Phys. A, 71, 331 (2000).
91. Melngailis, J. *Focused ion beam lithography*; Nucl. Instrum. Methods Phys. Res. B, 80/81, 1271 (1993).
92. Teichert, J., Bischoff, L., and Hausmann, S. *Ion beam synthesis of cobalt disilicide using focused ion beam implantation*; J. Vac. Sci. Technol., B, 16, 2574 (1998).
93. Chao, L. C., Lee, B. K., Chi, C. J., Cheng, J., Chyr, I., and Steckl, A. J. *Rare earth focused ion beam implantation utilizing Er and Pr liquid alloy ion sources*; J. Vac. Sci. Technol., B, 17, 2791 (1999).
94. Lezec, H. J., Ismail, K., Mahoney, L. J., Shepard, M. I., Antoniadis, D. A., and Melngailis, J. *A tunable-frequency Gunn diode fabricated by focused ion-beam implantation*; IEEE Electron Device Lett., 9, 476 (1988).
95. Lattes, A. L., Munroe, S. C., Seaver, M. M., Murguia, J. E., and Melngailis, J. *Improved drift in two-phase, long-channel, shallow buried-channel CCDs with longitudinally nonuniform storage-gate implants*; IEEE Trans. Electron Devices, 39, 1772 (1992).
96. Murguia, J. E., Shepard, M. I., Melngailis, J., Lattes, A. L., and Munroe, S. C. *Increase in silicon charge coupled devices speed with focused ion beam implanted channels*; J. Vac. Sci. Technol., B 9, 2714 (1991).
97. Lee, J. Y. and Kubena, R. L. *Threshold adjustments for complementary metal-oxide-semiconductor optimization using B and As focused ion beams*; Appl. Phys. Lett., 48, 668 (1986).
98. Averback, R. S., Rehn, L. E., Okamoto, P. R., and Cook, R. E. *Effects of cascade energy density on radiation induced segregation and ion beam mixing*; Nucl. Instrum. Methods Phys. Res. A, 182–183, 79 (1915-1981).
99. Hirayama, Y., Saku, T., and Horikoshi, Y. *Electronic transport through very short and narrow channels constricted in GaAs by highly resistive Ga-implanted regions*; Phys. Rev. B, 39, 5535 (1989).
100. Wieck, A. D. and Ploog, K. *In-plane-gated quantum wire transistors fabricated with directly written focused ion beams*; Appl. Phys. Lett., 56, 928 (1990).
101. Sasa, S., Miller, M. S., Li, Y. J., Xu, Z., Ensslin, K., and Petroff, P. M. *Localized two-dimensional electron gas formation by focused Si ion beam implantation into GaAs/AlGaAs heterostructures*; Appl. Phys. Lett., 57, 2259 (1990).
102. Arimoto, H., Kawano, A., Kitada, H., Endoh, A., and Fujii, T. *In situ two-dimensional electron gas fabrication by focused Si ion beam implantation and molecular beam epitaxy overgrowth*; J. Vac. Sci. Technol., B 9, 2675 (1991).
103. Fujisawa, T., Saku, T., Hirayama, Y., and Tarucha, S. *Sub-mu m wide channels with surface potential compensated by focused Si ion beam implantation*; Appl. Phys. Lett. 63, 51 (1993).
104. Oshinowo, J., Dreybrodt, J., Forchel, A., Mestres, N., Calleja, J. M., Gyuro, I., Speier, P., and Zielinski, E. *Photoluminescence study of implantation-induced intermixing of $In_{0.53}Ga_{0.47}As/InP$ single quantum wells by argon ions*; J. Appl. Phys., 74, 1983 (1993).
105. Ishida, K., Takamori, T., Matsui, K., Fukunaga, T., Morita, T., Miyauchi, E., Hashimoto, H., and Nakashima, H. *Fabrication of index-guided AlGaAs MQW lasers by selective disordering using Be focused ion beam implantation*; Jpn. J. Appl. Phys., Part 2, 25, L783 (1986).
106. Yanagisawa, J., Goto, T., Hada, T., Nakai, M., Wakaya, F., Yuba, Y., and Gamo, K. *Carrier profile of the Si-doped layer in GaAs fabricated by a low-energy focused ion beam/molecular beam epitaxy combined system*; J. Vac. Sci. Technol., B 17, 3072 (1999).

107. Sazio, P. J. A., Thompson, J. H., Jones, G. A. C., Linfield, E. H., Ritchie, D. A., Houlton, M., and Smith, G. W. *Use of very low energy in situ focused ion beams for three-dimensional dopant patterning during molecular beam epitaxial growth.* J. Vac. Sci. Technol., B, 14(6), 3933 (1996).
108. Nagamachi, S., Ueda, M., and Ishikawa, J. *Focused ion beam direct deposition and its applications;* J. Vac. Sci. Technol., B 16, 2515 (1998).

CHAPTER 13

Laser Direct-Write Micromachining

ALBERTO PIQUÉ AND DOUGLAS B. CHRISEY

Naval Research Laboratory, Materials Science and Technology Division, Washington, D.C.

C. PAUL CHRISTENSEN

Potomac Photonics Inc., Lanham, Maryland

1. Introduction
2. Trends in Microfabrication
3. Overview of Laser-Matter Interactions
4. Laser Micromachining
5. Summary

1. INTRODUCTION

Lasers are unique energy sources characterized by their spectral purity, spatial and temporal coherence, and high average and peak intensity. Each of these attributes has led to applications that take advantage of these unique qualities. From an industrial applications point of view, the high-power densities achievable with lasers made them ideal tools for material removal and heating very early on. Some of the initial applications involved cutting, drilling, and welding at macroscopic scales (1–3). It was later realized that by taking advantage of the tunability of the laser wavelength and pulse length, it becomes possible to control its interaction with matter very precisely.

The use of lasers for direct-write applications began with the development of laser micromachining systems. Laser micromachining is a subtractive direct-write process that takes advantage of the small feature sizes that can easily be achieved by focusing a laser beam onto a very small area. Despite the fact that laser micromachining is a serial process that results in slower fabrication speeds for large parts, it is ideally suited for small-batch production, prototyping, and customization (4). On small scales, lasers are capable of manufacturing

Direct-Write Technologies for Rapid Prototyping Applications

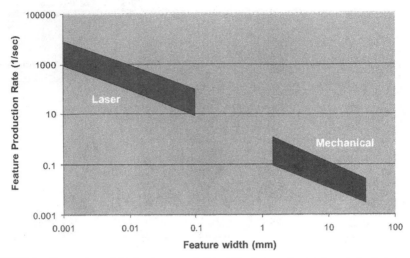

FIGURE 1   Comparison of fabrication rates for blocklike features in stainless steel substrates by laser micromachining and mechanical and drilling processes (5).

throughput rates greater than those achievable by mechanical means such as milling or drilling. Figure 1 shows a plot comparing the time required to remove a blocklike feature with an aspect ratio of 10:1 or less, in a material like stainless steel utilizing lasers and conventional milling and drilling techniques. As feature sizes decrease below 100 μm, the use of lasers results in significantly faster manufacturing rates than mechanical machines operating on millimeter-scale sizes are capable of (5).

This chapter provides an introduction to the use of lasers for direct-write micromachining applications. It begins with a brief discussion on the role of lasers for microfabrication applications. Then it provides an overview of how lasers interact with the surface of solids and a description of the ablation process. This is followed by a discussion of laser micromachining, a subtractive direct-write technique with a proven industrial base of applications that has enjoyed ample success and continues to find new applications in the microelectronic and medical industries. Laser micromachining incorporates many of the aspects that are essential for other laser-based direct-write tools. Examples are laser beam manipulation and control, motion stages, and computer control. To illustrate the level of sophistication currently achieved with laser micromachining systems, examples of various applications have been provided. By the end of this chapter the reader will emerge with a good understanding of the capabilities of laser micromachining and how the success and acceptance of these tools at an industrial level offer similar opportunities for other types of additive laser-based direct-write technologies.

2. TRENDS IN MICROFABRICATION

Today's technologies require the fabrication and processing of structural and functional parts and components at smaller and smaller scales. In order to be able to accommodate this demand new miniaturization and microfabrication tools are required. An example that illustrates this trend is provided by the semiconductor industry and the manufacture of integrated circuits (ICs) and other semiconductor devices. For ICs in particular, more functions and capabilities per unit of volume have been incorporated while their size and their total unit cost has decreased.

Miniaturization plays a role not only in the electronics industry, but also in the communications, medical, aerospace, and military fields, to name a few. In each of these areas there is a growing need to develop components and systems with smaller dimensions. Several technologies, such as photolithography, are currently available for the fabrication of miniature components. Photolithographic techniques are well suited for large production runs of identical parts given the fact that they are parallel and high throughput processes. However, there are many types of materials, geometries, and applications of miniaturized components that are not compatible or cost effective if processed with photolithographic tools. In those cases, other techniques such as direct-write processes might offer unique advantages and capabilities.

The use of lasers for direct-write applications offers many advantages for microfabrication applications. Laser tools are increasingly being used for the fabrication of miniature components, most of them electronic. For example, laser drilling has become the dominant means of producing microvia holes in high-density interconnect circuits. In fiberoptic telecommunications, lasers are used to manufacture filter Bragg gratings and to optimize packaging applications. In the microelectromechanical systems arena, the prototyping of microfluidic devices, tuning of resonant structures, and integration of numerous microsystems is being done with the help of lasers. Finally, in the medical device sector, lasers are used to mill stents and drill gas- and liquid-flow-control orifices for advanced drug-delivery catheters and aspirators. The growing number of companies developing laser micromachining systems for an expanding number of applications is proof that laser micromachining is well established.

3. OVERVIEW OF LASER-MATTER INTERACTIONS

The interaction of a laser beam with the surface of a solid gives rise to a series of complex phenomena that originate in the various mechanisms by which the energy carried by the photons can be absorbed within the solid. Depending on the laser radiation wavelength, pulse length and energy, and its absorption by the solid, physical and/or chemical changes to the surface might occur resulting in photophysical and/or photochemical processes respectively (6).

The absorption of the laser pulse by the solid gives rise to electronic and/or vibrational excitation modes. Depending on the relaxation path taken by these excited modes, various effects can be observed. Most of the excited modes can relax typically through electron–phonon and phonon–phonon scattering. These thermal processes result in heating followed by the atoms or molecules in the solid returning to their ground state. As a result, most if not all of the atoms or molecules in the lattice are involved and as such these are known as nonlocalized processes. On the other hand, if the photoexcited electrons give rise to electronic or vibrational excitation modes that are restricted to single sites within the lattice during timescales exceeding a few lattice vibration periods, localized processes such as radiative recombination, desorption, and ablation might occur. The relative strength between nonlocalized and localized relaxation mechanisms for the photoexcited electrons depends on the electron-lattice coupling characteristics of the solid. For example, ionic ceramics display high electron-lattice coupling strengths, which favor localized excitations, while nonlocalized relaxation processes are more likely in metals due to electron-lattice scattering mechanisms (7). As the intensity of the laser pulse increases, however, the relaxation processes will have a strong localized component independent of the type of material.

Examples of localized relaxation processes include radiative recombination (i.e., luminescence), desorption, and ablation. Radiative recombination is rarely a major channel of energy dissipation due to its long lifetime (>ns). Considering the effects due to single laser pulses, desorption processes involve low sputtering yield of isolated atoms or molecules from the surface with no measurable changes to the solid, while ablation processes involve significant material removal from the surface resulting in obvious changes to the solid.

Following the photoexcitation of the atoms and/or molecules on a surface, a competition between localized and delocalized relaxation processes begins in order to dissipate the absorbed energy. Determining which localized or delocalized path dominates during the relaxation process is quite difficult due to the fact that the excitation, relaxation, and transport phases for each process vary strongly with materials properties and depend on the laser wavelength and laser pulse duration. The exact mechanisms are only known in general terms for a very few cases (7). The various energy dissipation flow-paths, together with the timescales involved, are illustrated in Fig. 2 and further discussed by Miller and Haglund (8).

From a macroscopic point of view, the aforementioned discussion indicates that high-intensity laser pulses can be used for heating, melting, and vaporizing a solid surface. The fraction of the laser pulse that is absorbed by the

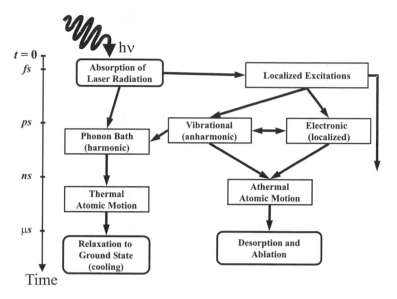

FIGURE 2 Schematic showing the dynamics of laser–solid interactions (7).

surface leads to very fast heating rates which result in localized melting and vaporization of the surface. While the melt front propagates into the solid, the vaporized material leaves the surface, leaving a void behind. If the laser pulse intensity is high enough, the vaporized material will become ionized and give rise to a plasma plume. This dense plasma will strongly absorb the trailing edge of the laser pulse, effectively shielding the surface from the laser beam. These steps take place during the laser ablation process and are shown schematically in Fig. 3.

3.1. ABLATION

From a laser micromachining point of view, the most important laser–solid interaction phenomena is ablation. The conversion of the initial localized electronic or vibrational photoexcitation into kinetic energy of nuclear motion leading to the ejection of species from a surface results in the phenomena of laser-induced desorption and ablation. The former is an extremely gentle process, generally involving individual atomic sites, which has no measurable effect on the solid's surface composition or structure, and plays little role in laser-based direct-write processes. Laser ablation, on the other hand, is a collective phenomenon involving the excitation of a large number of atomic sites, with a clear energy threshold accompanied by the formation of a dense

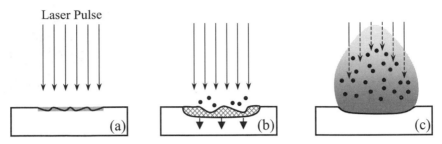

FIGURE 3   Schematic diagram showing the physical processes induced by a laser pulse. (a) Absorption of the laser radiation, followed by (b) melting and vaporization. The melting front propagates into the solid (indicated by the filled arrows). Finally, (c) a plasma is generated which interacts with the remaining laser pulse.

plume of gas and resulting in the removal of material from a surface as a consequence of the laser irradiation. Under the right ablation conditions, it is possible to achieve extremely well-defined boundaries between the ablated regions and those not exposed to the laser pulse, with minimal heating of the neighboring regions. Best results are usually achieved when the incident radiation is in the form of a pulse of relatively short duration and the radiation is strongly absorbed in a thin region near the surface of the workpiece, as shown schematically in Fig. 4. Under these conditions, there is little time for heat to diffuse from the absorption region into the material and produce

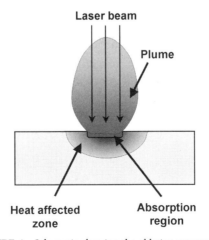

FIGURE 4   Schematic showing the ablation process.

thermal damage and localized melting. Clearly, such characteristics make laser ablation ideal for micromachining applications.

During laser ablation, the photoexcitation energy does not propagate far beyond the surface. The depth of this region can be estimated by noting that the thickness of the laser affected region, t, is of the order of the heat penetration depth, l_h (i.e., thermal diffusion length), or the optical penetration depth, l_α, whichever is larger. That is

$$t \approx \max(l_h, l_\alpha) \tag{1}$$

with $l_h \approx (4D\tau)^{1/2}$ and $l_\alpha = \alpha^{-1}$, where τ is the laser pulse duration, D is the thermal diffusivity and α is the optical absorption coefficient of the material respectively. For laser pulse widths in the nanosecond scale or longer, this is valid only initially, while the density of the photoejected species which give rise to the plume is low, and the laser plume interactions can be neglected. Despite this, Eq. (1) indicates that in general, for materials with higher thermal conductivity—metals and semiconductors—shorter laser pulses will work better, and for wide-bandgap materials, that is, dielectrics and glasses such as SiO_2 with low optical absorption coefficients, shorter wavelengths will perform better. In either case, the shorter the laser pulse and the shorter its wavelength, the better. Because ablation processes depend on many other parameters besides the thermal and optical properties of the material, the preceding discussion should only be used as a starting point. For a more thorough treatment of the subject the reader is directed to various excellent texts on the subject, such as the work by Miller and Haglund (8), and that of Bäurle (9).

The composition of vaporized material in the ablation plume can be very complex and may include molecular fragments of the worksurface, clusters and particles, ions, and products of reactions between surface materials in the plume and the surrounding ambient gas (10). Some of these components may fall back onto the surface, while others diffuse away into the surrounding ambient atmosphere.

After a single laser pulse, the ablated surface ideally has a profile similar to that shown in Fig. 5. Material is removed to a controlled depth, edges are sharply defined by the geometry of the focal spot, and ablated material is fully vaporized and diffuses away. In general, at the worksurface the ablation depth, δ, for a single pulse is a strong function of the laser fluence, F, defined as the laser pulse energy over the laser spot area (J/cm^2). A simple analysis based on the use of the Beer–Lambert law leads to

$$\delta = \frac{1}{\alpha} \ln \left(\frac{F}{F_T} \right) \tag{2}$$

where F_T is the threshold fluence, below which no ablation is observed. F_T is called the ablation threshold. Equation (2) predicts that δ should be linearly

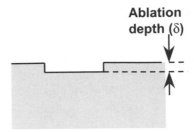

FIGURE 5   Ideal surface profile after a single pulse.

related to log F with a slope equal to α^{-1} (11). This analysis agrees with experimental observations for laser etching of thin films of complex oxide ceramics such as $YBa_2Cu_3O_7$ (12) and $La_{0.75}Ca_{0.25}MnO_x$ (13). However, Eq. (2) does not work well for metals and polymers because it does not take any thermal effects into consideration.

Typically, there is no measurable material removed below the ablation threshold. Above threshold, the depth of ablation first increases logarithmically and then slows to a weak dependence at high fluence levels. For micromachining using ultraviolet (UV) lasers, the fluences employed are above the ablation threshold, typically between 0.1 J/cm^2 and several J/cm^2, depending on the particular material and laser parameters. The corresponding ablation rates are between 0.01 μm/pulse and several μm/pulse. Figure 6 shows that for fluences below ≈0.15 J/cm^2 no substantial ablation takes place in polyimide using wavelengths of 248 nm. Above the threshold, the ablation depth increases linearly in many materials with the number of laser pulses at a given laser fluence (14). For numerous applications, the ablation depth can be controlled with great precision, as in the case of laser micromachining of biological tissues for medical purposes. An example of tissue micromachining is the use of UV laser ablation for sculpturing human corneas to correct nearsightedness (15).

Figure 6 shows the functional dependence of ablation depth on fluence for a common polymer (16). The ablation depth also depends on material parameters such as heat and/or optical penetration depth, enthalpy of vaporization, crystallinity, defect density, etc., and on laser parameters such as wavelength and pulse duration. Ablation rates may also change as a result of effects produced by the laser such as modifications of surface composition or texture. Rates also tend to decrease at the lower portion of deep holes or channels and to increase in some materials at high pulse-repetition rates.

Usually, the laser energy density or fluence is utilized to specify ablation parameters such as the ablation threshold. However, the fluence does not take into consideration the laser pulse length which affects the instantaneous power delivered by the laser pulse. A more useful parameter is the intensity, defined as

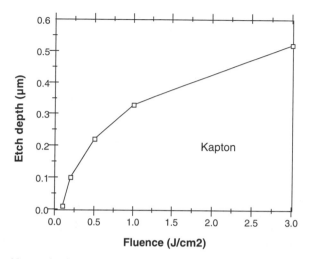

FIGURE 6   Ablation depth as a function of fluence for ablation of polyimide (Kapton) at a wavelength of 248 nm (16).

the peak laser power over the laser spot area (W/cm^2). Nevertheless, fluences are the most commonly reported quantities in the literature on ablation processes. The absolute laser energies required to achieve ablation on small areas on most materials are relatively small as shown on Fig. 7. Notice that as

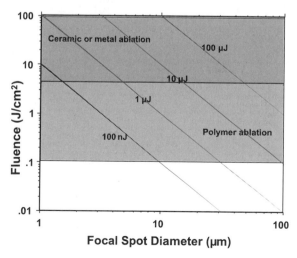

FIGURE 7   Fluence required for ablating different types of materials as a function of spot size. Diagonal lines indicate various laser energies.

long as the feature size is kept under 10 microns in diameter, lasers with energies as low as 10 µJ per pulse can be utilized to ablate polymers, ceramics, and metals.

From an applications point of view, it is obvious that in order to achieve the optimum ablation performance it is important that the proper laser be utilized for a given process. Laser micromachining of a specific material requires that the various laser parameters such as wavelength, pulse energy and duration, beam quality, and also cost per photon be considered against the process constraints that depend on the material properties, process quality, processed area or volume, minimum feature size, and required throughput.

4. LASER MICROMACHINING

Lasers have been employed for materials processing for more than two decades. Applications in micromachining include cutting, welding, heat treatment, etching, lithography, photopolymerization, deposition, etc. From this list it is clear that lasers can be used for subtractive as well as additive processes. In many cases, the laser can often serve as its own process monitor. This multiuse, multirole and *in situ* capability is not offered by any other advanced materials processing techniques such as molecular beam epitaxy (MBE), chemical vapor deposition (CVD), or focused ion-beam etching (FIB) (4).

Laser micromachining capitalizes on the ability of lasers to be focused to small spots on a worksurface, highly localizing the thermal or electronic processes produced by the radiation. For micromachining purposes, the spot size, corresponding to the system's resolution, the depth of focus, and the average laser beam intensity are three of the most important listed laser parameters. For a Gaussian laser beam of circular cross-section, the focused minimum radius, w_o is normally determined by the diffraction limits of the imaging system, so the theoretical resolution at the focal point is given by

$$w_o = \frac{2\lambda f}{\pi d} \qquad (3)$$

where λ is the wavelength of the laser radiation, d is the diameter of the limiting aperture before the focusing lens, and f is the focal length of the lens. Thus the principal way of increasing the resolution is by reducing either the wavelength or the ratio of f/d. Furthermore, the rules of diffraction dictate that a focused laser spot cannot remain so as it propagates in free space. Rather, it will diverge again the same way as it converged. The actual distance around focus where the spot size decreases from $\sqrt{2}w_o$ to w_o and then increases to

$\sqrt{2}w_o$ is known as the depth of focus. The depth of focus (DOF) is given by (17)

$$DOF = \pm \frac{\pi w_o^2}{\lambda} \tag{4}$$

For micromachining applications, the DOF indicates how much variation can be tolerated in the distance between the sample and the focal point. From Eqs. (3) and (4) it follows that there is an inverse relationship between the need for high resolution and a practical depth of field. Material processing with a very short depth of field requires very flat surfaces. If the surface has a nonplanar, nonuniform topology, a servo-loop connected with an interferometric auto-ranging apparatus must be used (18).

Figure 8 shows schematically some relevant aspects that must be considered when utilizing laser micromachining tools for a particular application. These aspects will now be discussed.

Debris. Similar to mechanical machining processes, laser ablation usually generates residual debris. If the debris layer does not adhere strongly to the surface, ultrasonic or solvent cleaning may be adequate for removal. Use of an appropriate ambient can minimize debris formation. For example, Kuper and Brannon (19) have shown that a helium or hydrogen ambient can significantly reduce debris in ablation of polyimide. Other approaches include coating of the

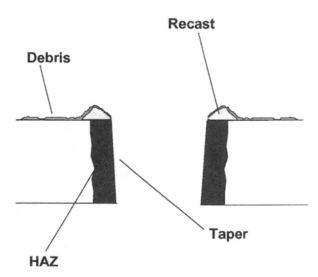

FIGURE 8   Cross-section of a laser micromachined channel showing the various byproducts of the processing.

surface with a protective layer that can be removed, along with overlaying debris, after processing.

Heat-affected zone. Heat produced by laser irradiation diffuses into the substrate raising the temperature of adjacent regions to levels that may produce permanent changes in material properties. The size of this heat-affected zone (HAZ) can be estimated from the characteristic thermal diffusion distance for a time comparable to the laser pulse duration. This characteristic distance is approximately equal to the square root of the product of the thermal diffusivity of the material and the duration of the optical pulse as shown in the preceding section. Figure 9 shows the estimated size of the HAZ in several materials for various laser pulse durations. The figure demonstrates that reduction of laser pulse duration can substantially decrease the size of the HAZ.

Recast. Molten material generated during the ablation process tends to be ejected from the illuminated region by reaction forces associated with the vapor jet leaving the surface. Accumulation of solidified melt at the edge of an ablated hole can produce a burr of recast material that resembles the burr produced by many mechanical machining processes. Recast can be minimized by reducing the volume of the melt pool produced by the optical pulse. Strong optical absorption, high fluence, and short pulse duration all tend to reduce the depth of the melt pool and suppress recast formation.

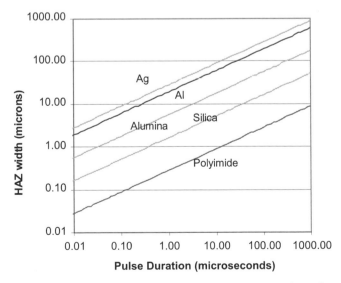

FIGURE 9   Approximate width of the heat-affected zone in various materials as a function of laser pulse duration.

Taper. Optical diffraction and shadowing effects at the top edges of an ablated hole or channel can lead to taper of sidewalls. The angle of sidewall taper has been shown to be a function of the laser fluence at the worksurface, the effective numerical aperture of the focusing optic, and the position of the beam focal point relative to the surface. In general, high fluence and high numerical aperture minimize taper (and can sometimes produce negative taper) when the beam focal point is positioned optimally above the substrate. Of course, in thick materials there must be some depth at which fluence decreases, leading to high taper angles and eventual self-termination of the process. Depth-to-width aspect ratios of 10 for holes and channels are generally achievable in most materials without extensive set-up effort. Much higher aspect ratios have been demonstrated using very high brightness lasers (20) and lasers with ultrashort pulse duration (21).

4.1. LASER SOURCES

For laser micromachining applications, the most commonly used types of laser are: (1) Nd:YAG solid state lasers with nanosecond-to-microsecond pulses emitting at the fundamental (1.06 microns) or in the UV (3rd harmonic, 355 nm); (2) CO_2 gas lasers emitting at 10.6 microns in CW or pulsed mode (milliseconds); and (3) excimer lasers with nanosecond pulses emitted at 248 or 193 nm. All are normally operated in a pulsed mode to achieve best machining quality. The long wavelength associated with CO_2 lasers prevents their focusability to spot diameters smaller than about 20 microns with available optics. As a consequence, CO_2 sources normally are not used in applications requiring very small feature sizes. Excimer lasers are the most cost-effective and reliable sources of UV photons, however the relatively low pulse-repetition rates and limited beam quality of typical excimer sources sometimes makes them difficult to adapt to direct-write processes in industrial applications.

Two new types of laser sources have shown promise for improving the quality and range of applications of laser micromachining. For micromachining materials like Teflon, fused silica, or sapphire, which are relatively transparent in the visible and near ultraviolet, shorter wavelengths than those achieved with excimer lasers are required. In those cases F_2 lasers which emit at 157 nm work very well (22,23). F_2 lasers are gaining popularity in the semiconductor industry as the photolithography resolution limits are pushed to submicron dimensions despite their more complex beam-delivery arrangement due to air absorption. Figure 10 shows the dramatic differences between machining of a thin, fused silica layer on silicon with 193 nm and with 157 nm radiation.

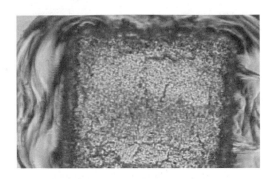

FIGURE 10   Laser ablation of a thin fused silica layer deposited on silicon using 193 nm and 157 nm radiation (23).

Pulsed lasers with femtosecond pulse duration are finding many applications in micromachining given their advantages for laser materials processing (23,24). Many of these lasers use Ti:sapphire as the lasing medium and operate at wavelengths of 775 nm. Laser pulses of even modest energy produced by these ultrashort pulse sources are characterized by extremely high peak power and large electric fields. They are capable of producing high-order multiphoton absorption and avalanche breakdown in nearly all materials. Compared to longer pulse sources, they are much less dependent on linear optical absorption effects for energy deposition and can machine most substrates. Because the laser pulses are shorter than many fundamental energy-transfer processes, there is little opportunity for generation of heat during the pulse, and the majority of the ablation products leave the substrate surface as a plasma plume. Generation of adverse secondary effects is minimized, and there is little opportunity for thermal damage to surrounding regions, almost no recast or burr, and very little debris (21). (See Fig. 11.)

Femtosecond laser photons continue to be relatively expensive, but femtosecond laser systems are transitioning from being expensive laboratory tools to

FIGURE 11 Laser-ablated hole in stainless steel produced by a femosecond Ti : Sapphire laser. (Photo courtesy Laser Zentrum Hannover.)

becoming commercially available systems. Broader availability will increase their use in micromachining applications in the future.

4.2. Matching the Laser to the Application

The availability of the various laser sources just mentioned allows selection of laser parameters for process optimization. In general it is desirable to select a wavelength that is strongly absorbed by the substrate and compatible with machined feature sizes, a pulse duration appropriate for the desired process quality, and a pulse energy that is appropriate for the desired focal spot size and fluence. Processing speed is usually directly proportional to the laser pulse-repetition rate or average power. Photon cost is also an important consideration for industrial processes; frequently, the laser that produces the best quality is not the laser that processes at the lowest cost.

4.3. Laser Microfabrication Tools

Figure 12 shows a photograph of a typical laser micromachining system together with a schematic showing the basic elements of such a system. We have already mentioned the various lasers commonly used in these systems. We will now describe some of the other components.

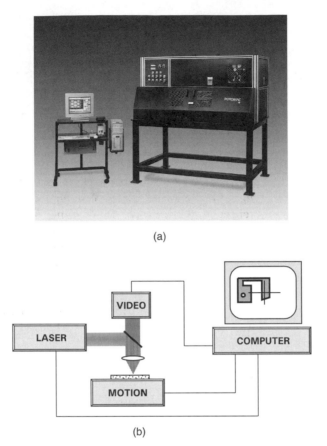

FIGURE 12   (a) Photograph of a commercially available laser micromachining system. (b) Schematic showing the main components of the system. (Courtesy of Potomac Photonics.)

4.3.1. Motion Control

Direct-write micromachining utilizes relative motion between the focused beam and the workpiece to remove material in shapes or patterns and locate these at the desired position on the part. Laser micromachining systems are normally used to produce features in the 2-to-200-micron range, so that relative motion must have high resolution and accuracy as well as good repeatability. Resolution and repeatability needs to be substantially better than the feature size to produce smooth edges and accurate shapes, so that submicron precision is often required. Relative motion can be achieved by moving the part or by moving the beam, and in some hybrid systems, both can be moved independently.

4.3.1.1. Moving Part

Relative motion can be generated by mounting the part on precision translation stages and transporting it under the focused beam. This approach capitalizes on the availability of highly precise linear stages that, in extreme cases, can be moved in increments of tens of nanometers. Precision stages can be driven by leadscrews and rotary dc motors or by linear motors. Performance depends strongly on the accuracy by which the associated motion encoders can track and resolve changes in position. Stages with the highest accuracy and resolution often use encoders that incorporate an optical interferometer, although the performance of these is sometimes rivaled by glass scale encoders. Stages driven by leadscrews may use rotary encoders on the leadscrew for position feedback, although these are susceptible to thermal expansion of the leadscrew and backlash in the drive train.

To produce motion in more than one axis, linear stages must be stacked and the combined weight of upper stages, the part itself, and any necessary part fixturing can result in a fairly large mass and system inertia. Moving part systems therefore tend to be used where emphasis is on precision of motion rather than on speed or acceleration.

4.3.1.2. Moving Beam

The position of the focal spot on the substrate also can be controlled by using galvanometer mirrors to change the direction of the beam entering the final focus lens. Because the moving mass in this case is quite small, the focal spot can be scanned over the substrate with very high speed and acceleration. However, the numerical aperture of the focusing lens imposes a limit on the size of the addressable area on the worksurface as well as on the size of the focal spot. As a consequence, small focal spots can only be produced in conjunction with small scan areas.

4.3.1.3. Hybrid Systems

The high acceleration of the laser beam relative to the worksurface associated with the galvanometer beam delivery and the large range of motion available from linear translation stages is sometimes combined in high-speed industrial systems. Such systems are designed to fabricate hundreds of features per second over large planar surfaces. These systems are capable of moving the beam to cut small individual features, such as hole patterns, while the entire substrate is moved more slowly under the focused beam.

4.3.2. Imaging of the Worksurface

Video microscopy has become an essential element of laser micromachining, allowing location of fiducials, alignment of features, monitoring of the machining process, and inspection of completed operations. Figure 13 shows schematics of the several approaches to video imaging found in laser micromachining stations. The most popular are viewing through the laser focus lens, viewing of an area adjacent to the laser focus with a parallel separate viewing

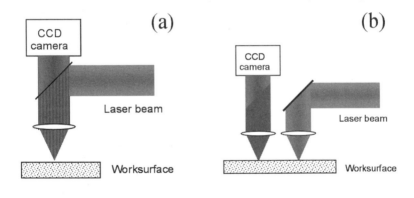

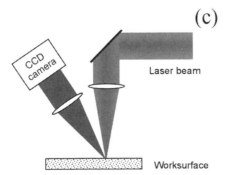

FIGURE 13   Three methods of video imaging of the worksurface: (a) Through-the-lens viewing, (b) parallel viewing, and (c) side viewing.

system, and viewing of the laser processing area from the side with a separate viewing system.

Through-the-lens viewing allows direct observation of the laser process and simplifies registration of the process zone relative to substrate fiducials, compared to other direct-write techniques such as ink-jet, micropen, thermal spray, laser guided direct write, etc. However, focusing lenses that are capable of imaging at deep UV or far IR wavelengths—as well as the visible wavelengths used by the imaging camera—can be complex and expensive. As a consequence, imaging systems used in conjunction with excimer, F_2, and CO_2 lasers often use parallel or side-viewing systems. Parallel viewing does not allow real-time observation of the laser process and requires the ability to accurately translate the substrate from the processing area to the viewing area. However, it is relatively inexpensive, provides high image quality, and allows location of fiducials. Viewing of the process area from the side provides a simple method of observing the laser process, but seldom allows the full image field to be in focus. Precise location of fiducial features is also complicated by the interplay of horizontal and vertical substrate motion in the camera image.

4.3.3. Computer and Software

Most laser micromachining systems use some form of computer control to coordinate laser firing with substrate and beam motion, and to provide a user interface. Because most processes occur on scales too small to be directly observed, the control computer and its software strongly influence the look and feel of the machine for the operator. Digital images produced by video systems can be directly fed to image processing and machine vision programs to allow automatic registration of fiducials and measurement of critical dimensions. Dimensional data thereby available in near real time can be used to improve yield and the quality of the micromachining operation and reduce the need for costly handling of miniature parts in later quality-control inspections (25).

Available micromachining software packages accept engineering data in the form of CAD or Gerber files from a network, and then generate the motion and laser control commands necessary for part fabrication. The combination of input data manipulation, video image processing, and automated machine operations results in a tool that allows a relatively unskilled operator to produce miniature, high-quality parts. With computer-controlled machine loading of raw materials and unloading of finished parts, these flexible micromanufacturing tools are ready for operation in a fully automated, lights-out environment.

TABLE 1 Typical Single-Pulse Ablation Depths in
Strongly Absorbing Materials for KrF Excimer Lasers
Operating at 248 nm

Material	Single-pulse ablation depth (microns)
Polymers	0.3 to 0.7
Ceramics and glasses	0.1 to 0.2
Diamond	0.05 to 0.1
Metals	0.1 to 0.2

4.4. MACHINING SPEED AND THROUGHPUT

Estimates of machining speed or part throughput are frequently required for
cost analysis and feasibility evaluations. Reasonably good estimates can be
made from single-pulse ablation data for the laser tool and material of interest.
We have seen here that most materials exhibit a relationship between ablation
depth and fluence that is qualitatively similar to that of Fig. 6. Material is
removed most efficiently when the laser fluence is near the "knee" of the
ablation curve, and the single-pulse ablation depth in that fluence range is
easily determined. This single-pulse ablation can be used to characterize the
interaction of the laser with the material. Table 1 shows typical single-pulse
ablation depths for excimer laser ablation of strongly absorbing materials.

Using the ablation depth, δ, the focal spot diameter, d_s, and the pulse-
repetition rate of the laser, R, the time required to carry out any micromachining process can be estimated. For example the cutting speed, S, for material
thickness, W is approximately

$$S = \frac{d_s R \delta}{W} \tag{5}$$

and the time, T, required to drill a hole through the material is approximately

$$T = \frac{W}{R\delta} \tag{6}$$

4.5. APPLICATIONS

4.5.1. 3-D Micromachining

Submicron single-pulse ablation depths in strongly absorbing materials engen-
ders the possibility of three-dimensional (3-D) machining with good surface

quality and depth control. Best results are achieved with highly absorbing substrates such as those that do not exhibit a molten phase. UV laser machining of diamond and polyimide, for example, can produce 3-D structures of good quality (26).

An effective approach to 3-D machining is shown in Fig. 14. The laser is operated at a constant pulse-repetition rate and turned on and off at appropriate moments as the focal spot is scanned over the substrate. Material is removed layer-by-layer, revealing the desired 3-D shape. Figure 15(a), a four-level hologram fabricated by KrF laser ablation of a polyimide film, is an example of the depth control available with highly absorbing materials. In this case the depth resolution among the pixels is approximately 250 nm.

When the spatial energy distribution in the focal spot is reasonably uniform and pulse overlap is optimized, very smooth surfaces can be produced. Figure 15(b) shows an example of optically smooth surfaces on a 3-D structure fabricated in diamond by scanned laser ablation.

Direct-write processes are serial in nature, and consequently are often more costly than are competing processes that fabricate large numbers of parts simultaneously. However, an attractive approach to low-cost, high-volume fabrication of microstructures uses direct-write micromachining to fabricate micromolds for high-volume replication of laser-produced microstructures. Figure 16 shows an approach to replication of a laser-machined structure using a metallic inverse.

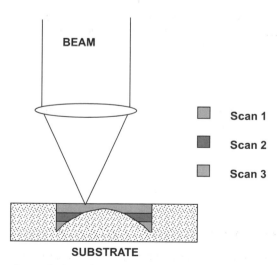

FIGURE 14   3-D machining by layered removal of materials using scanned laser ablation.

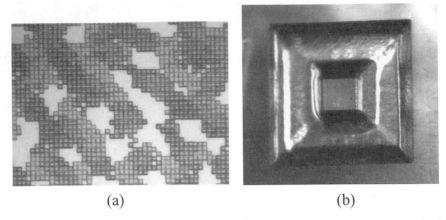

(a)   (b)

FIGURE 15 Examples of 3-D structures fabricated by scanned laser ablation. (a) Four-level hologram with 10-micron pixels formed by KrF laser ablation of polyimide. (b) 3-D pyramid structure in diamond with optically smooth surfaces.

4.5.2. Micromachining of Layered Materials

Several important industrial applications of laser micromachining involve selective machining of layered materials. These applications capitalize on the fairly large differences in ablation threshold fluence among many polymers, metals, and ceramics. When a material overlayer has a significantly lower threshold for ablation than that of its underlying layers, the laser fluence on the worksurface can be adjusted to selectively remove the overlayer without damage to the material below, as shown schematically in Fig. 17(a). Structures that have been shown to be well suited to selective overlayer machining include most polymers on metals, polymer composites, indium tin oxide on glass, and silicon dioxide on aluminum. Applications have included drilling of microvias in flexible interconnect circuits and integrated circuit packages (27,28), stripping of small wires (see Fig. 17(b)), and patterning of flat panel displays.

4.5.3. Micromarking

In general, marking is one of the largest applications of lasers, and, as miniaturization becomes pervasive, laser micromarking of small components is desirable for identification and tracing of miniature medical, electronic, and optical devices. Figure 18 shows small laser-generated marks that are used to identify glass display components and microfluidic devices. In these applica-

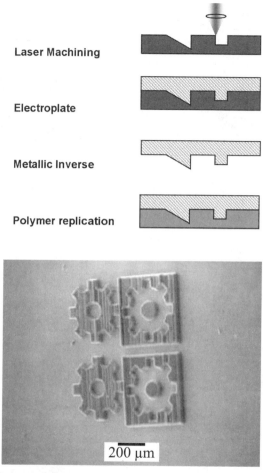

Laser Machining

Electroplate

Metallic Inverse

Polymer replication

200 µm

FIGURE 16 Replication of direct-write laser-machined structures by micromolding. In this example, a micromold of negative and positive topologies of a gearlike structure was fabricated in polyimide using an ultraviolet laser. A metallic inverse mold was formed by electroless deposition of nickel. The nickel mold was then used to replicate the shapes using a UV-cured polymer. The replicated components, shown on the bottom, faithfully reproduce even micron-scale imperfections of the original mold.

tions the process is optimized to gently produce a shallow mark that is unlikely to be a future source of microcracks. Similar laser techniques are used to mark semiconductor wafers and diced chips, allowing them to be tracked through the fabrication and assembly process. Lasers also are used to produce very small marks invisible to the naked eye on high-value substrates, like diamonds

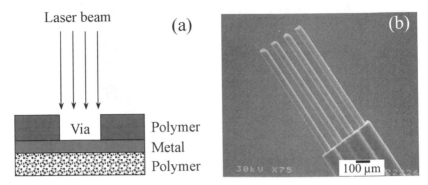

FIGURE 17   (a) Schematic laser via drilling in flex circuit materials. (b) Laser stripping of insulation from a cable with 50-micron dia. wires.

and other gemstones (see Fig. 19). In this case, laser micromarking is a means of identifying a component without detracting from its appearance or otherwise reducing its value.

4.5.4. Drilling of Precision Holes and Orifices

Holes and orifices with diameters smaller than about 100 microns are difficult to drill mechanically in many materials. Laser drilling has become the method of choice in many applications. Laser drilling of orifices in ink-jet printer heads

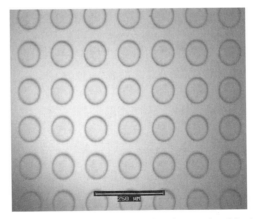

FIGURE 18   100-micron dots on flat-panel display glass produced by 193-nm ablation.

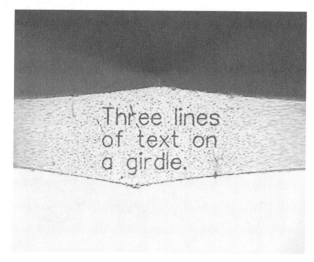

FIGURE 19 Identifying text laser inscribed on the girdle of a diamond gemstone. Height of the characters is 30 microns.

has found broad application, and precision holes in medical and pharmaceutical devices such as drug-delivery catheters, pills, inhalers, and oxygen regulators are best fabricated by laser micromachining (5). Figure 20 shows a micrograph of a polyurethane catheter with laser-drilled holes of various shapes and sizes.

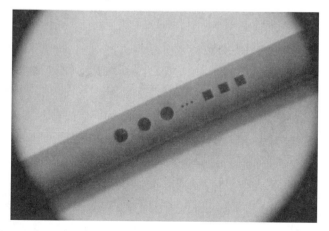

FIGURE 20 Holes with widths from 15 to 100 microns drilled in a polyurethane catheter.

FIGURE 21   Laser cutting of a stent pattern in a 1.5-mm stainless steel tube.

4.5.5. Stents and Miniature Tube Structures

Because laser processing is not limited to planar geometries, it is often used to cut and shape miniature tube components. For example, each year lasers are used to fabricate millions of vascular stents, which are used to prevent blockage of human arteries. These cylindrical metal mesh structures are fabricated by lasers cutting complex patterns in the walls of stainless steel or in NiTi tubes with diameters of approximately 1 mm, as shown in Fig. 21. Many other miniature tube components used in minimally invasive surgery and implantable medical devices are well suited to laser processing.

4.5.6. Sheet Part Fabrication

At small-scale sizes, excision of complex shapes from metal or polymer sheets becomes increasingly difficult. Features smaller than approximately 1 mm are difficult to punch and are not easily produced by other mechanical processes. Laser direct-write processing becomes increasingly cost effective as scale sizes decrease, and lasers are widely used to excise complex miniature shapes from larger sheets. Figure 22 provides some examples of parts cut from sheet materials using laser direct-write techniques.

4.5.7. Trimming and Repair

Trimming of thick- and thin-film resistors has long been an important application of lasers. As the size of electronic and mechanical components

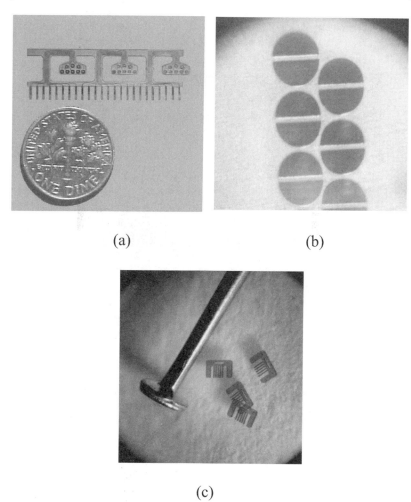

(a)   (b)

(c)

FIGURE 22   Laser-excised sheet components. (a) Miniature flex interconnect circuit excised by an ultraviolet laser (courtesy Stanford University). (b) D-shaped holes laser cut from a 50-micron stainless steel sheet using an Nd : YLF laser. Thickness of the bars between the holes is 50 microns. (c) Comb structures cut from a polyimide sheet using an ultraviolet laser.

shrinks, laser trimming is finding a variety of new applications that reduce cost and increase yield (29). Trimming of miniature surface-mount capacitors allows their specification to tight tolerances. Similarly, lasers can be used to trim and tune quartz resonators and surface acoustic wave devices (30). Micromechanical resonators frequently must be trimmed to compensate for process variations by selective material removal on miniature struts and beams.

Lasers also have long been used in semiconductor memory repair to blow links connecting redundant elements (31). A similar approach is used for repair of flat panel displays. All of these trimming operations require very precise material removal in areas that may be as small as a few microns.

4.5.8. Fabrication of Miniature Assemblies

Lasers promise to play an important role in the fabrication of microsystems and miniature assemblies. Direct-write systems have been used to fabricate microfluidics systems by direct machining of substrates (32) and by patterning sheets which can be laminated together to produce complex, 3-D microstructures (33). Batch-assembly techniques based on laser ablation of bonding materials have been demonstrated for handling and orienting collections of miniature parts (34). Laser drilling of microvias (27) has become a very important application and is essential in the fabrication of the electronic interconnect structures found in miniature and high-performance electronic products. Lasers are used in the automated assembly of photonic components, where they have been shown to allow bonding of small components with submicron precision (35). Because they are capable of processing such a wide range of material with a precision substantially beyond that associated with mechanical systems, laser tools play important roles in building the assemblies needed to cluster semiconductor or MEMS components into useful assemblies and viable products.

5. SUMMARY

Laser tools can be used for micromachining a wide array of polymers, metals, ceramics, and glasses. They operate in ambient air and are applicable to both planar and nonplanar substrates. These features make the laser a valuable tool in micromachining operations that require feature sizes smaller than those produced mechanically. Several types of lasers are used in micromachining tools and matching of the laser parameters, such as wavelength and pulse duration, to the optical and thermal characteristics of the workpiece is essential to process quality and economy.

Direct-write laser micromachining requires integration of the laser source, beam delivery optics, digital imaging systems, and precision motion components, and this is generally accomplished with a control computer and appropriate software. Effective integration of these components results in tools capable of machining 3-D structures, layered materials, miniature tubes, and sheet parts, as well as of marking and trimming operations. The pervasive evolution toward miniaturization in many industries assures a growing demand for these valuable microfabrication tools.

REFERENCES

1. J. F. Ready, *Industrial Applications of Lasers*, Academic Press, 1978.
2. W. W. Duley, *Laser Processing and Analysis of Materials*, Plenum, 1983.
3. L. Migliore, (ed.), *Laser Materials Processing*, Marcel Dekker, Inc., 1996.
4. M. Madou, *Fundamentals of Microfabrication*, CRC Press, 1997.
5. C. P. Christensen, *Laser tools for micromachining: reducing part size, time, and cost*, Medical Device and Diagnostic Industry, Sept. 1998.
6. J. Mazumder and A. Kar, *Theory and Application of Laser Chemical Vapor Deposition*, Plenum, 1995.
7. R. F. Haglund, Mechanisms of Laser-Induced Desorption and Ablation, in *Laser Ablation and Desorption*, J. C. Miller and R. F. Haglund, (eds.), p. 15, Academic Press, 1998.
8. J. C. Miller and R. F. Haglund, (eds.), *Laser Ablation and Desorption*, Academic Press, 1998.
9. D. Bäurle, *Laser Processing and Chemistry*, 3rd Edition, Springer, 2000.
10. J. C. Miller and D. B. Geohegan, (eds.), *Laser Ablation: Mechanisms and Applications*, Am. Inst. of Physics, Woodbury, NY, 1994.
11. W. W. Duley, *UV Lasers: Effects and Applications in Materials Science*, Cambridge University Press, 1996.
12. A. Inam, X.D. Wu, T. Venkatesan, S.B. Ogale, C.C. Chang and D. Dijkkamp, Appl. Phys. Lett., Vol. 51, p. 1112 (1987).
13. A. M. Dhote, R. Sheekala, S.I. Patil, S.B. Ogale, T. Venkatesan, C.M. Williams, Appl. Phys. Lett., Vol. 67, p. 3644 (1995).
14. G. P. Behrmann and M.T. Duignan, *Excimer laser micromachining for rapid fabrication of diffractive optical elements*, Appl. Opt., Vol. 36, No. 20, p. 4666 (1997).
15. G. H. Pettit, M. N. Ediger, and R. P. Weiblinger, *Excimer laser ablation of the cornea*, Opt. Eng. Vol. 34, p. 661 (1995).
16. R. Srinivasan and B. Braren, Ablative Photodecomposition of Polymers by UV Laser Radiation, In *Lasers in Polymer Science and Technology*, CRC Press, Boca Raton, FL, 1990.
17. I. W. Boyd, *Laser Processing of Thin Films and Microstructures*, Springer-Verlag, 1987.
18. H. Helvajian, Laser Materials Processing: A Multifunctional In-Situ Processing Tool for Microinstrument Development, in *Microengineering for Space Systems*, H. Helvajian, (ed.), AIAA and the Aerospace Press, Reston, VA, Monograph 97–02, 1997, p. 67.
19. S. Kuper and J. Brannon, *Ambient gas effects on debris formed during KrF laser ablation of polyimide*, Appl. Phys. Lett. 60, 30 (1992), p. 1633.
20. L. J. Marabella, *Diode-pumped lasers and applications*, Proc. SPIE-Int. Soc. Opt. Eng., Vol. 2214, p. 178 (1994).
21. H. K. Tonshoff, et. al., *Microdrilling of metals with ultrashort laser pulses*, J. Laser App. 12, p. 23 (2000).
22. P. R. Herman, et. al., *Processing applications with the 157-nm fluorine excimer laser*, Proc. SPIE, Int. Soc. Opt. Eng., Vol. 2992, p. 86–95 (1997).
23. C. P. Christensen, *Adaptable, efficient lasers expand their role in microfabrication*, Photonics Spectra, p. 106, (2000).
24. C. Momma, et. al., *Short-pulse laser ablation of solid targets*, Opt. Commun. 129, 134 (1996).
25. E. Teoman, *Automatic microinspection improves quality*, Optoelectron. World, (1999).

26. C. P. Christensen, *Waveguide excimer laser fabrication of 3D microstructures*, Proc. SPIE, Int. Soc. Eng., Vol. 2045, p. 141 (1994).
27. C. Dunsky, H. Matsumoto, and B. Larson, *High Quality Microvia Formation with Imaged UV YAG Lasers* Proc. IPC Printed Ckt. Expo., San Diego, CA, April 4, 1000, p. S15-5-1.
28. P. R. Herman, et. al., *Advanced-laser processing of photonic and microelectronic components at Photonics Research Ontario*, Proc. SPIE-Int. Soc. Opt. Eng. Vol. 3618, p. 240 (1999).
29. W. Bloomstein, *Marketwatch: laser trimming meets IC demands*, Laser Focus World, Dec. 1999.
30. S. M. Metev and V. P. Veiko, *Laser-Assisted Micro-Technology*, 2nd Ed., Springer, 1998.
31. Y. Sun, et. al. *Optimization of memory redundancy laser link processing*, Proc. SPIE-Int. Soc. Opt. Eng. Vo. 2636, p. 152 (1995).
32. M. S. Talary, et. al., *Microfabrication of biofactory-on-a-chip devices using laser ablation technology*, Proc. SPIE-Int. Soc. Opt. Eng. Vol. 3680, p. 572 (1999).
33. P. Martin, et. al., *Laser-micromachined and laminated microfluidic components for miniaturized thermal, chemical, and biological systems*, Proc. SPIE-Int. Soc. Opt. Eng., Vol. 3680, p. 826 (1999).
34. Holmes A. S, *Laser fabrication and assembly processes for MEMS* presented at SPIE Photonics West LASE 2001, San Jose, CA, 2001, p. 19.
35. K. Mobarhan and R. Heyler, *To manufacture more, automate the production of optoelectronic devices*, Vol. 34, no. 6, p. 156 (2000).

CHAPTER 14

3D Microengineering via Laser Direct-Write Processing Approaches

HENRY HELVAJIAN

Center for Microtechnology, The Aerospace Corporation, Los Angeles, California

1. INTRODUCTION

Material processing via a direct-write (DW) action is conceptually appealing because it offers a direct means for variegating control in the processes and actions to be performed. These attributes become valuable in materials processing when a high degree of customization is to be exercised—as during rapid prototyping operations. Direct-write processing is dictated by the maskless patterning and serial processing action of the interacting tool. The processing tool can be any device that alters the material properties of the underlying matter, be it embedded in the volume or merely on the surface layer. The required force for the action to be performed nominally dictates the type of tool to be used with the paramount desire being to control the delivery of the energy. The overall usefulness of any such tool is typified by its processing speed, precision, accuracy, and its potential use as a multiple processing implement. The laser, as a unique material processing tool, can fulfill all these criteria. It can simplify the design of a direct-write tool because it can be used to alter a material's inherent property or its topology on the surface or in the volume. It can accomplish this by the subtle alteration of the material phase or chemical bonding or by the more brute force surgical schemes that either remove or deposit matter. It can also serve as a diagnostics probe of the operation. Laser direct-write (LDW) approaches have the additional feature of *in situ* modification of the material during exposure to the processing environment. LDW methods are also nonintrusive offering line-of-sight view of the processing zone. By far the most important aspect of the LDW method is the ability to deposit energy with site specificity and induce material processing under nonequilibrium conditions.

Planar two-dimensional (2D) material processing has been exemplified, for many years, by the processing techniques used in microelectronics fabrication. These batch-process-compatible approaches continue to offer many advantages in materials processing, most notably, process uniformity and repeatability over a large number of cofabricated components which make possible the manufacturing of low-cost devices. Parallel or batch processing is a major advantage in manufacturing that cannot easily be overlooked when evaluating the merits of the DW or LDW approaches. Both DW and LDW are serial fabrication approaches with likely throughput limits on piece-part production. To somewhat mitigate this handicap, one can employ, at least for the LDW approach, very high-repetition lasers (e.g., MHz) (1) and multiple-beam-processing techniques. But unless attention is given to the efficiency of the specific photon-processing step, these efforts will not necessarily result in a cost-effective fabrication process. A processing domain where DW or especially LDW have clear advantage is in the area of *true* three-dimensional (3D) fabrication, rapid prototyping, and end-of-process custom modification. Custom or end-of-process modifications can most easily be done via a DW approach and best by an LDW approach where the laser's action-at-a-distance feature is helpful. True 3D fabrication is the ability to make piece parts via a free-form style. Implementing these complex shapes via a 2D planar process approach would require masking steps too numerous to be viable. As for the case of undertaking rapid prototyping operations, the up-front cost of mask design and fabrication hinders quick turn-around testing operations. Although these application examples show clear advantages for using direct-write approaches, material processing in the micro- and nanometer-scale domain can benefit from a "merged" processing approach. The merged-process approach, as applied to an LDW process, would incorporate batch fabrication techniques that are solely designed to aid or assist the laser direct-write process. By such a scheme, processing throughput could be increased without loss of advantages as in customization and rapid prototyping proffered by the LDW approach.

Microengineering is a discipline dealing with the design, materials synthesis, micromachining, assembly, integration, and packaging of miniature 2D and 3D sensors, microelectronics, and microelectromechanical systems (MEMS) (2). The physical structures and components have nominal dimensions, from the nanometer to the micrometer scale, and up to the millimeter scale. A recent key development in microengineering has been the ability to extend basic microelectronics processing techniques to enable fabrication of suspended microstructures, in silicon material, where the shapes are prismatic extensions of 2D patterns (i.e., 2.5D). These microstructures with cofabricated drive electronics are the MEMS components that perform the necessary energy transduction in inertial, chemical, and biological microsystem devices (3). The worldwide market for these devices in 1998 totaled $16 billion and is predicted to approach $38 billion by the year 2002 (4). Microengineering has received worldwide attention because it promises to enable development of "intelligent" microinstruments with wide-ranging applications in medicine, transportation, communications, and housing. This promise can certainly be met if complimentary microfabrication techniques can be developed for materials other than semiconductors—commonly used materials such as polymers, glasses (5), and ceramics. Laser material processing technology can further advance microengineering development by virtue of being a processing tool for varied materials. Laser microfabrication techniques and processing "windows" are being developed for alumina and silicon carbide (6), for polymers (7), and glasses (8). Furthermore, the recent exposé on the need to develop "soft" processing techniques for advanced inorganic materials (e.g., nanocrystalline $BaTiO_3$, $SrTiO_3$, $LiNbO_3$, $LiCoO_3$, CdS, ZnS, FeS) has highlighted the need for nondestructive (material additive) material processing approaches (9).

Another aspect of microfabrication technology that can strongly propel microengineering technology is the ability to fabricate microstructures in true 3D shapes (10). There are several applications where true 3D microstructures are preferred. In certain fluid flow applications (e.g., gas flow through a nozzle) true 3D shaping of the channel/orifice can reduce losses and lead to better flow control. In the development of very high frequency (i.e., mm wave) communication systems, true 3D shaping (e.g., for a miniaturized antenna horn) can lead to better-defined electromagnetic wave far-field patterns. In a complementary application, there is better acoustic suppression if microhole shapes can be tailored in 3D. Similarly, for applications where suspended microstructures are used as microscaffolding for holding other microsensors or trapping specific microbiological materials, a true 3D scaffold would offer certain device advantages (e.g., a pyramid shape is preferred over a rectangular prismatic shape for emitter tips and precision sensing tips). Furthermore, the ability to fabricate true 3D cavities could enable multilayer packaging schemes with complex via hole patterning. Finally, there are instances in microdevice assembly and packaging where an extended 3D shape patterned on top of a 2D structure can enhance pick-and-place operations or permit physical confinement of an unsupported microstructure during packaging operations. In general the ability to fabricate true 3D shapes in micro scale is useful and, when applied to certain critical components, can lower a device noise floor or make it operate more efficiently. In this regard, LDW material processing techniques constitute an enabling technology.

In general, the range of operability (i.e., the processing "window") of a particular laser material process is often defined by the physics/chemistry of the laser material interaction using the materials *initial* state properties. As for example in a thermally initiated process like a laser-induced phase change (e.g., laser-induced melting for welding operations), the thermal properties of the irradiated material at ambient temperature might be used to set the power. In the case of material processing via photochemical or multiphoton (MP) excitation schemes (e.g., in some laser chemical vapor deposition (CVD) and for MP polymerization approaches), the laser wavelength and pulsewidth are chosen based on specific photochemical characteristics of the material or gas. Although this approach is reasonable at the outset, in most cases it does not lend to efficient use of laser photon energy and compounds processing problems. *Material properties change during and after irradiation* (11). For example, thermophysical parameters are a function of temperature. In addition, the irradiated material may change phase resulting in poorer light coupling or the increasing density of photo induced defects can lead to added absorption. Also susceptible to change are processes driven by photophysical mechanisms; they can saturate, and in nonlinear excitation schemes unwanted energy-sink channels can open. As a result, the coupling efficiency between the laser and material can be reduced either during the laser pulse or for subsequent laser shots (12). For LDW fabrication of 3D structures and especially where sequential layering operations are necessary, these dynamic changes in the material, if not recognized, can be a significant impediment to achieving process control and developing a workable process "window."

In the past, laser systems could be treated as essentially monochromatic light sources with little hope of gaining control on the quality of the laser light. However, with technological enhancements such microprocessors, sensors, feedback control, and a better choice of materials for the laser, current laser systems are substantially more reliable than their predecessors and offer an enticing means for dynamically "tailoring" the laser beam temporal and spectroscopic characteristics. We believe that the applicability of laser processing techniques in general and especially for LDW will rest on the feasibility of "tailoring" the energy delivery to match that of the changing material. In this regard, implementing a chirp on the laser (i.e., temporal variation of the carrier frequency) to coherently drive molecular oscillators to population inversion represents the ultimate in control (13). If this can be accomplished with sufficient control and under processing conditions, it would essentially lead to processing with minimal wasted energy and enable "fabrication" within a nonequilibrium environment. These approaches might be necessary in the development of nanoengineered systems and nanoelectromechanical systems (NEMS) where the desire is to "fabricate" or assemble structures on the nanometer scale (14) utilizing molecular, cluster, and protein units as building blocks.

This chapter provides an overview of laser direct-write material processing approaches for the microfabrication of true 3D structures. The overview is meant to provide the engineering principles for developing a 3D LDW microfabrication tool for laser microengineering applications. The review lacks extensive discussion regarding the fundamental photophysical processes that drive laser-material-interaction phenomenon. These topics have been presented in numerous excellent reviews (15). This review is by no means meant to be a comprehensive document, it is designed to present a commonly held view of what is necessary and desired for implementing laser direct-write approaches for 3D microfabrication and what may be required if LDW is to be applicable in future nanotechnology development. Many of the pertinent equations and aspects of the fundamental laser photophysical processes can be found in the chapters in the edited work by D. J. Ehrlich and J. Y. Tsao (16), the review by C. I. H. Ashby (17), and in numerous other works on laser material interaction phenomenon (18).

2. THE LASER DIRECT-WRITE 3D PROCESSING TOOL

A LDW material processing tool comprises a pulsed or a continuous wave (CW) laser, a computer-controlled shuttering or triggering scheme, a laser-beam shaping and delivery system (may include laser-beam characterization) and a means for "writing" the desired pattern in the material. Pattern generation is typically accomplished by means of a computer-controlled XYZ stepper, XY galvanometers with auto-focusing capabilities, or a computerized robot-arm. There is the option to either move the sample or the laser beam, and the choice often depends on various engineering criteria such as a particular processing speed requirement, the scale size of the piece part to be fabricated, the desired resolution, and the complexity of the 3D structures to be fabricated. Regardless of the motion-control scheme used, the motion pattern is synchronized to the laser energy delivery rate. In the more advanced LDW tools, the patterns are first drawn using a standard computer-assisted drawing (CAD) program and then translated into motion-control language (e.g., G-code). In the case where the fabrication is to be in true three dimensions, the 3D pattern can be drawn either as stacked layers of 2D patterns (also called "masks") or as a solid 3D drawing much like the output generated from solid-modeling software programs. Regardless of the drawing approach used, the information from the drawing program (i.e., CAD program) must first be translated into a serial program sequence that describes the precise "tool" path in 3D coordinates. Advanced LDW tools take the output from the "tool" path program and direct it to a motion-control driver that can perform coordinated motion on at least three axes. The actual LDW tool may actually have additional axes (>3 axes) to perform a full 3D motion (e.g., including tip-tilt) sequence and may actually use inverse-kinematics features ("virtual axis") to get superposition moves. If the LDW tool requires the precise delivery of "feedstock" or a reagent material (e.g., LDW via photo-induced molecular self-assembly, laser-induced forward transfer, MAPLE DW) additional control axes will also be necessary that must also be coordinated with the primary motion motors. Regardless of the number of axes employed, the ultimate resolution of any coordinated multiaxis motion system will be determined by the accuracy of the displacement measurement system. The dynamic measurement of the rigid body motion and the derived kinematics are nontrivial for a multiaxis manipulation system—as for example in a six-degree-of-freedom (DOF) displacement (three translational and three rotational) system that can provide complete information for both motion and vibration. In general, optical sensing methods are superior displacement measurements because they are noncontact and provide resolution at the wavelength of light. Furthermore, they are immune to electromagnetic interference and can be made compact. Recent experiments show that a laser source, a grating (operating on three orders −1, 0, +1), and three 2D position sensors with optical elements can be configured to provide a 6-DOF displacement measuring system (19). The experimental results, which depend on the types of sensors and the multiaxis drive motors used, show that the measurement system has a maximum error of ±10 μm for translation and ±0.012° for rotation (repeatability is also on the same order). With better use of sensors, the accuracy can be further increased. To appreciate the difficulty of the experiment, one must realize that the quoted error value is for a 6-DOF displacement and includes the effect of compounding errors on all axes.

Fashioning 3D micro- or nanostructures by LDW techniques can be accomplished by a number of material modification approaches. The laser beam DW "tool" can be used to:

- remove material (e.g., via laser ablation (20) or etching (21)),
- add material (e.g., via laser CVD (22) or laser induced forward transfer (LIFT) (23)),

- directly induce a phase change in the material as it "writes" a pattern (e.g., laser crystallization) (24),
- expose a material that enables a patterned phase change that can be utilized for material processing in a postprocessing step (e.g., photoresist technology, volumetric patterning of photostructurable glass) (25),
- sinter a powder to form a solid (e.g., laser sintering of powders),
- cross-link and polymerize a photosensitive polymer to harden into the desired shape (e.g., 3D laser stereolithography patterning) (26), and
- photo-initiate complex molecular (e.g., protein) self-assembly processes (27) for nanometer scale 3D engineering.

In all of these approaches, the crucial objective is controlling the laser beam properties (e.g., focal volume, energy density, coherence) within the relevant part of the laser-material interaction zone. In laser processing the desire is to initiate and maintain a laser material-interaction process without incurring additional complications from *competing* photophysical process events.

3. LASER MATERIAL INTERACTION PHYSICS

In the past twenty-five years, laser material processing has evolved from its use in cutting and welding/forging applications to the current applications in nanometer scale processing, controlled atomic layer deposition, and 3D patterning. This transformation from using lasers for more than just directed heating sources is a result of better understanding of the gas- and liquid-phase photochemistry, more insight into the physics of the laser solid interaction process, and the desire to apply this knowledge for altering materials in a controllable fashion. The following sections give a brief overview of the basic processes that are necessary to understand laser processing.

3.1. BEAM PROPAGATION, ENERGY DELIVERY ON TARGET, AND COHERENCE

Under most laser-processing conditions the criteria for optimum laser-material interactions are described by a handful of equations. These equations describe the propagation of the laser beam energy through the beam delivery optics, the photophysical interaction with the surface (i.e., absorption, surface chemical interactions), and the subsequent surface modification as a result of electronic and thermal excitation. The equations are simplified for a Gaussian laser beam propagating in a diffraction-limited optical system, though for most material processing a top-hat or flat-top homogenized beam is used (28). A Gaussian beam can be described by the radius function $\omega(z)$ and the wavefront curvature function $R(z)$ along the propagation direction z. The functions $\omega(z)$ and $R(z)$ are given by the Eqs. (1)–(3) (29),

$$\omega^2(z) = \omega_o^2 \left[1 + \left(\frac{2z}{b} \right)^2 \right] \tag{1}$$

$$R(z) = z[1 + (b/2z)^2] \tag{2}$$

$$b = 2\pi\omega_o^2/\lambda \tag{3}$$

where λ is the wavelength, ω_o is the beam radius at the waist, and b is the confocal parameter (i.e., distance within which the diameter of a focused beam remains nearly constant: $-b/2 < z < +b/2$). The Gaussian beam contracts to a minimum diameter $2\omega_o$ at the beam waist where the phase front is a plane wave.

Consider a collimated and strongly focusing Gaussian beam with the criteria that 99% of the energy is to be transmitted by the focusing lens of focal length f, through an aperture D, and 86% of the energy is to be contained within a diameter $d_o = 2\omega_o$. The spot size d_o for this beam is given by Eq. (4) (30). Equation (4) can be recast using the more familiar $f^\#$ parameter (where $f^\#$ is called the f number and defined as f/D). Lenses with $f^\# > 2$ are relatively inexpensive while lenses with $f^\# < 1$ are commonly multielement designs and very expensive. Equation (4) states that for a given fixed-focal-length lens to arrive at a smaller d_o means using shorter wavelengths and larger Gaussian beam diameters at the lens (i.e., aperture).

$$d_o \approx 2f\lambda/D \rightarrow 2f^\#\lambda \tag{4}$$

The depth of focus for this Gaussian beam is given by the confocal parameter, b, given in Eq. (3) but can also be approximated as $\approx 2\pi(f^\#)^2\lambda$. With a 248-nm-wavelength laser beam and a focusing lens with $f^\# = 2$ (i.e., $f/2$) optics, the result gives a minimum spot diameter of ~ 1 μm and a depth of focus of ~ 6 μm for processing in air. If the medium surrounding the processing surface is of the material other than air, the minimum spot size is further reduced by the index n. In the given example, 86% of the incident energy is focused onto the minimum diameter spot, which leaves 14% of the energy distributed over a diameter ~ 2.3 μm. The 14% energy "spill-over" may be damaging for some 3D microfabrication applications—especially when operating well above the energy threshold. To guarantee greater than 98% energy confinement within the design spot size, Eq. (4) should be multiplied by 2.3. There are practical reasons why very small $f^\#$ may not be desirable for some microfabrication applications. With decreasing $f^\#$ the minimum spot size declines by the same factor as opposed to the depth of focus, which declines by the larger factor,

$\pi(f^{\#})^2$. Material processing with a very short depth of focus requires a very flat surface or a servo-loop connected to an interferometric autoranging device to maintain the focus as the sample is moved.

The spot size and depth of focus are also called the resolution and the depth of field. The resolution, R, of a diffraction-limited imaging system is given by Eq. (5) (31), where C is a constant (derived from the first Fraunhofer diffraction minimum for a circular aperture) and is related to the coherence of the illumination source. The value of C goes from 0.61 (incoherent illumination) to 0.77 (coherent illumination). The $N.A.$ is the numerical aperture ($\equiv \eta \sin\theta_{max}$; η is the index of refraction and θ_{max} is 1/2 the maximum acceptance cone angle), which specifies the "light-gathering power" of an optical system. A low $N.A.$ has a value of ~ 0.25, while a high $N.A.$ is ~ 0.5 (e.g., Schwarzschild lens (32)). The depth of field for such an imaging system is given by Eq. (6).

$$R = \frac{C\lambda}{N.A.} \tag{5}$$

$$Z = \frac{\lambda}{N.A.^2} \tag{6}$$

Again, the $N.A.$ emphasizes the inverse relationship between the need for high resolution and a practical depth of field. Equations (4)–(6) show that the shorter the wavelength the better the resolution and the smaller the minimum spot size achievable, which argues for using UV over IR light. However, most materials show increasing dispersion in the index n $(dn/d\lambda)$ at the shorter wavelengths and especially in the deep UV. For transmissive optics, this results in chromatic aberration in the focusing/imaging, which must be taken into account for UV laser sources with large spectral bandwidths. Current UV excimer lasers without injection-locking wavelength stabilization have bandwidths that make less than half-micron processing difficult. The defocusing error df as a function of the source bandwidth is given in Eq. (7) (33).

$$df = \left(\frac{dn}{d\lambda}\right) \Delta\lambda \frac{f}{1-n} \tag{7}$$

Table 1 gives the measured dispersion for fused silica for the excimer laser wavelengths and the maximum allowable source spectral linewidth, $\Delta\lambda_{max}$, for achieving a defocusing error of less than 1 µm given an $f = 1$ cm lens (34). Current excimer laser linewidths are typically 0.8 nm for wavelength at 248 nm. Injection-locking schemes can be used to narrow the emission linewidths $\Delta\lambda$, to < 0.003 nm and thereby also reduce the defocusing errors. In contrast to the excimer laser the current femtosecond (fs) lasers, which are being used for 3D microfabrication of silica, have natural bandwidths ranging

TABLE 1   Dispersion Values for Fused Silica, $dn/d\lambda$, and Maximum Allowable Spectral Source Linewidth for Achieving a Defocusing Error of Less Than 1 μm, $\Delta\lambda_{max}$. Given an $f = 1$ cm Lens.

	193 nm	248 nm	308 nm	351 nm
$dn/d\lambda$ ($\times 10^{-4}$ nm^{-1})	8.9	6	3.2	1.8
$\Delta\lambda_{max}$ ($\times 10^{-2}$ nm)	6	8	16	29

from 30–50 nm. For these lasers and the given optical example, there would be a minimum defocusing error of ~ 600 microns (for $\Delta\lambda \sim 50$ nm and ~ 248 nm irradiation). A defocusing error of this magnitude, without some form of compensation, would be intolerable for many LDW 3D microfabrication applications.

Fundamentally, a reduction in the laser linewidth means an increase in the coherence properties of the laser. In essence, it means maintaining the distinct time and phase relationship of the emitted photon wavetrains. In general the coherence properties of a laser are delineated as arising from either the spatial or temporal domain. Spatial coherence refers to the time and phase relationship of wavetrains originating at spatially separated points (i.e., light from extended optical sources). For lasers, the phenomenon is related to the number of laser transverse cavity modes which can extract energy from the gain curve. Reducing the number of operating cavity modes increases the spatial coherence. Temporal coherence is the consequence of the finite duration, $\Delta\tau$, of the light emission from the atoms that translates to a finite frequency distribution spread, Δv (i.e., finite bandwidth as a result these truncated sinusoidal wavetrains). For pulsed lasers, the phenomenon is related to the pulsewidth of the laser (i.e., the duration of the population inversion), the atomic energy decay modes, and the average number of times the laser energy is allowed to recirculate prior to exiting (i.e., cavity round-trip times). Of the two coherence properties the spatial coherence influences the imaging and focusability of the laser beam, while the temporal coherence would be of importance in inducing novel photochemical or photophysical processes on surfaces or in bulk media. In practicality, the spatial coherence is the more important and most precision processing lasers operate only in the fundamental transverse cavity mode TEM_{oo} (i.e. lowest order self-reproducing hermite-gaussian cavity mode). Maintaining the spatial coherence of a laser also requires that the paraxial optical beam delivery system be specially designed to match the confocal parameter of the laser with that of the cascading transfer optics.

In general most pulsed lasers are best described as partially coherent sources while CW lasers can be designed as nearly perfect coherent sources. The temporal coherence property of pulsed lasers is typically given by the coherence length, l_c. The l_c defines a distance, usually measured from the exit window of the laser, where the laser electromagnetic field amplitude and phase front changes in a consistently predictable way. In effect the laser is considered a temporally coherent source for interactions at a distance less than l_c. Using the time-bandwidth product theorem or Heisenberg uncertainty principle (which states that $\Delta v \Delta \tau \sim 1$, where $\Delta \tau$ is the laser pulse width and Δv is the spectral bandwidth), the corresponding coherence length l_c is given by Eqs. (8) and (9). Where c is the speed of light and λ, v are respectively the wavelength and frequency.

$$l_c = c\Delta \tau \approx \frac{c}{\Delta v} \qquad (8)$$

given that $c = \lambda v$ and $\Delta v = (c/\lambda^2)\Delta \lambda$

$$l_c \approx \lambda \left(\frac{v}{\Delta v}\right) \approx \frac{\lambda^2}{\Delta \lambda} \qquad (9)$$

Current state-of-the-art solid state lasers with injection-seeded oscillators have typical spectral linewidths of 50–90 MHz. For example, a Nd-Yag laser operating at a laser wavelength of 1064 nm with a spectral linewidth of 90 MHz gives a coherence length of 3.3 meters. Novel nonthermal processing of materials would be possible within this coherence length. Because most of these lasers have a pulse width of 10 nsec—which corresponds to an optical length of ~3.3 meters, laser photons from the leading edge of the pulse would be coherent in relation to laser photons at the trailing edge of the pulse. Examples of nonthermal processing include the coherent "driving" of energy into excitonic particles and of surface electrons. These techniques would be useful in "3D" fabrication in the submicron to the nanometer scale. By implementing coherent excitations it could be possible to induce "coherent" self-assembly of atoms/molecules on a surface. A recent calculation shows that by introducing coherence between molecular energy levels in conjunction with two frequency-related electromagnetic fields, it becomes possible to initiate a molecular deposition of arbitrary pattern (35). Coherent excitation schemes would also be important in chirped lasers that are designed to activate and maintain an excitation. Recent computer simulations (36) and experiments (37) show that application of chirped ultrashort pulses can generate large-amplitude excitations in ground-state molecular systems.

3.2. PROCESSING SPEED AND PROCESS WINDOW

There are cases in laser-material processing where the laser repetition rate or the stepper/scanner speed is not the limiting factor. In such cases, the processing throughput is determined by the fundamental process speed,

V_{sp}(μm/sec). The V_{sp} depends on many factors but is intrinsically dependent on the fundamental photophysical interaction (e.g., electronic, thermal, plasma), the character of the surface under irradiation, and the properties of the incident laser light. The photophysical interaction is a function of the laser fluence (J/cm^2) or the intensity (W/cm^2). In general, at very low fluences the photophysical process is primarily induced by electronic excitations, at intermediate fluences thermal processes dominate, and for high laser fluences the processing is dictated by the above-surface laser-initiated plasma. Similarly, the morphology of the surface changes with increasing laser fluence. In general, at low fluences there is surface and near-surface defect formation, at intermediate fluences there is surface melting and rapid recrystallization leading to amorphization, and with high fluences there is plasma sputtering, spallation, and shock-induced damage. For all laser fluences, knowledge of prior irradiation dose is critical to predicting additional changes. Therefore, for controlled processing, the photophysical interaction must be maintained within the domain of interest. Table 2 displays pertinent photophysical processes and the factors that impact laser processing. The table illustrates the many factors that characterize a process window and ultimately the processing speed, V_{sp}. A process window is characterized by measurement of the phenomenological

TABLE 2 Examples of Photophysical Processes Used in "Direct-Write" Laser Processing and Some Pertinent Factors

Photophysical processes	Critical factors
Chemical (deposition/etch)	Reaction initiator (gas-phase absorption or substrate absorption)
	Optical absorption coefficient
	Heterogeneous reaction rate at gas–solid interface
	Diffusion of reactants/products (mass transport)
	Substrate thermal conductivity
	Nucleation rate
	Point and line defect densities
Ablation	Laser fluence and intensity
	Irradiation dose
	Surface morphology
	Bulk defect density
	Thermal conductivity (thermal diffusion length)
Laser-induced desorption	Wavelength
	Optical absorption coefficient
	Adsorbate binding energy
	Fluence and intensity
	Point defect density
	Metals (excited plasmon density)

processing rate, Γ (e.g., $\mu m/sec$ of material etched, ablated, deposited). A simple model can be derived to relate V_{sp} to Γ. Assume the laser spot size diameter on the workpiece is $D(\mu m)$ and the required processing thickness is ℓ (μm) (i.e., material etched, ablated, or deposited). Then ℓ/Γ gives the time to process one spot size. The stepper/scanner must then move a distance D per unit time of ℓ/Γ. The stepper/scanner speed or material processing speed is then given by Eq. (10)

$$V_{sp} = (D/\ell)\Gamma \qquad (10)$$

The projected processing speed, whether the laser is used for etching, annealing, or forming novel material like "luminescent silicon," (38) depends on the laser parameters and the experimental conditions employed. The major items are as follows:

- Type of laser (CW or pulsed)
- Laser wavelength
- Processing technique or chemistry employed (e.g., chlorine-etch or ablation)
- Laser polarization vector with respect to the scan direction (i.e., E ‖: parallel or E ⊥ perpendicular)

For example, a CW laser (Ar^+ ion) operating at 514 nm with 1.5 W output power, with a chlorine etch chemistry and polarization set to E ‖ can cut trenches in silicon with a process speed, V_{sp} of several mm/sec (39). For the opposite polarization, the etching rate is a factor of 2 slower and results in nonuniform etching of the side walls. In using pulsed-laser ablation (excimer laser, 248 nm, $1.3\,J/cm^2$, 5 Hz repetition rate) to cut silicon, a V_{sp} of 0.6 mm/sec has been achieved with hole depths of 150 μm (40). In comparing the two techniques, the laser-assisted chlorine-etch technique produces a smoother wall (41), but the ablation technique has a wider dynamic range of trenching depth.

The Γ for numerous materials and laser processes (e.g., semiconductors, insulators, and metals) can be found in the literature (42). Because the process conditions influence Γ, one expects different Γ for different laser irradiation conditions (i.e., pulsewidth, wavelength, etc.). For example, for process windows where Γ is limited by the diffusion of reactants or products into and out of the processing zone, using smaller spot sizes can lead to an increase in Γ by factors approaching 10 (4). This is primarily a result of diffusion geometry. As the spot size is *reduced*, there is a transition in the reactant and product diffusion from a 3D expansion process to a 1D process. A consequence of the reduced dimensionality is that the reaction fluxes in the active zone increase leading to an effective larger Γ. For laser-assisted chemical processing, the transition from 1D to 3D expansion appears for spot sizes near 80 μm (43).

Diffusion-limited processing can be deleterious in the fabrication of high aspect ratio structures (hole-depth/hole-width $\gg$ 1). Under these circumstances the limitation is mass transport to and from the surface and the derived simple processing speed model as shown in Eq. (10) is not valid; it must be modified to include parameters for diffusion. Equation (11) (44) relates V_{sp} to Γ for processing in the diffusion-limited regime. L is a constant related to the diffusion properties of the product. Equation (11) reduces to Eq. (10) for small ℓ/L ratios.

$$V_{sp} = D\Gamma/L[(\exp(\ell/L) - 1)]^{-1} \qquad (11)$$

Processing speed, V_{sp}, is also influenced by light scattering out of the active zone or waveguiding into/out of an active zone. For example, if the material processed is by a CW or long pulse laser and the products include cluster particles, then the intensity at the processing interface could be reduced as a result of the light shadowing or scattering from the ejecta. The Mie scattering theory applied to large opaque particles (with diameters $\approx \lambda$) show that the loss of laser energy is proportional to twice the particle diameter. Another effect in the laser fabrication of high aspect-ratio structures like deep (> 50 μm) via-holes is that of waveguiding (45) by means of total internal reflection. Waveguiding can lead to side wall nonuniformity, because in-plane or polarized light incident into a hole will not be reflected but absorbed at the wall (46).

3.3. Optical Absorption, Thermophysics, and Laser-Induced Plasmas

The fundamental mechanism in the laser material interaction event is governed by the materials optical absorption properties. Leaving out the specifics, these absorptions can have bandwidths with near molecular origin ($< cm^{-1}$) as for certain adsorbates, or bandwidths more akin to those found in solids ($\gg cm^{-1}$) as for periodic lattice crystals, or bandwidths that resemble near continuum absorption as for a free-electron gas metal or a plasma. In the first two cases spectroscopic measurements can define the absorption character properties, while in the third case specific models and the material dielectric properties can be used to get general absorption behavior. For example, the optical properties of the metals are derived from the dielectric constant. At laser wavelengths where a metal behaves like an ideal free-electron metal, its dielectric properties can be approximated by the Drude model (47). The dielectric constant ε is temperature dependent and can be related to the angle-dependent reflectivity, R. Equations (12) and (13) describe the change in reflectivity for the TE (transverse electric—electric vector perpendicular to plane of incidence = s-polarized) and the TM (transverse magnetic—electric vector parallel to plane of incidence = p-polarized) polarized waves as a function of incident angle θ. The angle θ as defined is relative to the surface normal. At normal incidence ($\theta = 0$) the TM and TE electric field reflectivity merge into one equation which, for completeness, is given in Eq. (14). The n and k are the real and imaginary parts of the complex index of refraction. They are related to the dielectric constant by the equation $\varepsilon^{1/2} = n + ik$. The extinction coefficient, k, is related to the optical absorption coefficient $\alpha(\mathrm{cm}^{-1})$ as given by Eq. (15). Equation (15) also defines the optical absorption coefficient and the resulting attenuation of the laser intensity (I) over a distance (z).

$$R_{\mathrm{TM}} = \frac{[\cos(\theta) - \cos(\theta)/\varepsilon^{1/2}]}{[\cos(\theta) + \cos(\theta)/\varepsilon^{1/2}]} \tag{12}$$

$$R_{\mathrm{TE}} = \frac{[\cos(\theta) - \varepsilon^{1/2}\cos(\theta)]}{[\cos(\theta) + \varepsilon^{1/2}\cos(\theta)]} \tag{13}$$

$$R_{\mathrm{TM/TE}} = \frac{(n-1)^2 + (k)^2}{(n+1)^2 + (k)^2} \tag{14}$$

$$\alpha = 4\pi k/\lambda \tag{15}$$

where $dI(z)/dz = -\alpha I$.

Regardless of the initial absorption process, the absorbed energy quickly spreads via numerous decay channels to result in bulk heating. However there are cases where the selective deposition of energy in a specific absorption feature induces a specific action without the consequences of heat. These nonthermal phenomena are generally observed in femtosecond laser-pulse experiments (48) or in low-fluence laser material interaction experiments (49). In both cases the material is "processed" on the atomic scale with processing yields so low as to have little use for practical applications. However, with the ever-increasing laser repetition rates (kHz → MHz) low-yield, species-selective processes can be viable for certain applications where atomic level of control is necessary. Even so, for most laser-material processing the consequences of bulk sample heating by the laser must be considered because it can influence both the processing resolution and the fabrication throughput. The time scale and the nature of the heat-flow characterizes the type of processing. For long processing times (i.e., CW lasers) the temperature distribution is in steady state and the heat-flow problem is only tractable via a 3D solution. One consequence of this for LDW fabrication is that the effect of high temperatures on adjacent features will need to be taken into account. In contrast, for short processing times (i.e., short-pulse lasers), the heat-flow is primarily a one-dimensional problem where the temperature gradient is into the bulk and normal to the surface. Under these conditions, the effect on the surrounding area can be ignored. Because lasers can induce a wide range of heating rates up to as high as 10^{15} K/sec (e.g., femtosecond pulse excitation), the processing thermophysics is governed by the thermal properties of the irradiated material, including the thermal conductivity (κ; W/cm-K), the heat capacity (c_p; J/cm^3-K) and the temperature-dependent optical properties. Equation (16) defines the thermal diffusion length (cm) where τ is the processing duration (i.e., the laser-pulse duration). The ratio (κ/c_p) is called the thermal diffusivity, D_T (cm^2/sec), and can be used to calculate the time for reaching a steady-state temperature within a processing zone of size **D**. This time to reach steady-state is given by the equation $T \approx D^2/(4D_T)$.

$$\chi = (4\kappa\tau/c_p)^{1/2} \tag{16}$$

Using the metals as an example, the optical absorption depth ($1/\alpha$) for most metals, in the visible and the near IR regions, is only a few hundred angstroms. On the other hand, for nanosecond pulsed lasers the thermal diffusion length χ is on the order of 1 μm. So for most *pulsed* laser processing of metals, the optical absorption depth is much shorter than the thermal diffusion length ($1/\alpha < \chi$). Under these circumstances the temperature at the surface can be calculated if the laser pulse shape is known (50). However, for the purpose of LDW material processing on the micrometer scale, the key issue is whether the thermal diffusion length, χ, is greater or less than the feature size, **D**, to be processed and likewise, if the processing time, D/Γ, for the feature is greater or less than $T \approx D^2/(4D_T)$. If the *processing time is greater* than T, then the solution requires a 3D analysis in which the laser intensity radial distribution must be identified. The 3D analysis is complicated, but for a circular aperture and a CW irradiation zone, the affected area is a hemisphere of diameter $(\pi D)^{1/2}$. On the contrary, if the *processing time is less* than T, then the heat diffusion is a one-dimensional problem and the maximum temperature rise at the surface, ΔT_{max}, can be approximated by Eq. (17) (51).

$$\Delta T_{max} = [F(1 - R_{sol})]/(c_p\chi) \tag{17}$$

where F is the laser fluence (J/cm^2) and R_{sol} is the surface optical reflectance. The numerator on the right-hand side of the equation describes the laser fluence absorbed. In most cases the thermochemistry is governed by the temperature fall time, which is given by $\Delta T_{fall} \sim \chi^2/(4D_T)$ (52). For weak absorbing materials where $1/\alpha > \chi$, then $1/\alpha$ replaces χ in both Eq. (17) and in the equation for ΔT_{fall}. In the particular case of weak absorbers (i.e., wide bandgap insulators or semiconductors) or thin films, the laser material interaction may create defects (53) in the material or infuse stress or strain in both the irradiated and surrounding area (54). To analyze the stress and strain distribution, the material thermoelastic equations must be solved to quantify the effect of the laser heating (55). Understanding of the residual stress is important to the laser annealing technique (56) or for any laser direct-write processing technique. The annealing irradiation dose and the scan speed have an effect on the residual stress distribution. A highly stressed material commonly quenches by atom dislocation and microcracking, while a low-stressed material shows a shift in the phonon spectrum. In both cases the material is ridden with defects that can be used to advantage to induce particle emission from the surface via a nonthermal laser excitation scheme (57,58).

The description of the laser-material interaction as given by the optical absorption and subsequent thermal processes is valid as long as the photo-ejected species density is small. With increasing laser fluence, more material evaporates and the likelihood for photoionization and thermionic emission (59) increases. Equation (18) gives the Richardson-Smith equation, which estimates as a function of temperature the thermionic ion emission current.

$$J_+ = A_p T^2 \exp(-(I_p + \varphi_o - U_{ce})/kT) \qquad (18)$$

The A_p is a constant, T is the local temperature, and I_p, φ_o, and U_{ce} are, respectively, the ionization potential (eV/atom), the electron workfunction (eV), and the cohesive energy (eV/atom). With further increase in the laser fluence, both the photoionized and the photoemitted electrons absorb energy from the laser beam via the inverse Bremsstrahlung process. The process is described as a three-body interaction with nearby ions whereby the electron is raised to a higher electronic kinetic energy state. The higher kinetic energy electron ionizes additional atoms via electron impact excitation. The resulting effect is an avalanche of ionization with less light actually delivered to the target and more into the protoplasma. The absorption coefficient for the Bremsstrahlung process can be calculated and is given in Eq. (19) in cgs units (60).

$$\begin{aligned} K_v &= (4/3)(2\pi/3kT)^{1/2}(n_e n_i Z^2 e^6/hcm^{3/2}v^3)[1 - \exp(-hv/kT)] \\ &= 3.69 \times 10^8 (Z^3 n_i^2/T^{1/2}v^3)[1 - \exp(-hv/kT)] \end{aligned} \qquad (19)$$

where n_i and n_e are, respectively, the ion and electron densities in a plasma of average charge Z and temperature T. The c, e, m, h, and k are, respectively, the velocity of light, the electronic charge, the electron mass, Planck's constant, Boltzmann's constant, and v is the frequency of light which is related to the wavelength, λ, by the equation $c/(N\lambda)$ (where N is the plasma optical index). The term $1/K_v$ defines the light absorption pathlength (cm) into the plasma, while the term $[1 - \exp(-hv/kT)]$ accounts for losses by stimulated emission. For specific conditions Eq. (19) can be approximated. For $hv \gg kT$ (e.g., UV wavelength laser) the $K_v \sim (T^{1/2}v^3)^{-1}$, while for $hv \ll kT$ (e.g., high-temperature plasma), the absorption coefficient is approximated by $K_v \sim (T^{3/2}v^2)^{-1}$. All other parameters being equal, in both extreme cases the shorter wavelength laser is the more preferable as it results in a smaller K_v. If the laser fluence is such that an above-surface plasma does form, then only optical frequencies higher than the plasma frequency, $v_p = 8.9 \times 10^3 n_e^{1/2}$, can penetrate the plasma. Conversely, given a laser with frequency v, the laser can penetrate the plasma for electron densities $n_e < (v/8.9 \times 10^3)^2$.

The plasma temperature T, which appears in Eq. (19), is difficult to measure for a plasma not in local thermodynamic equilibrium. However, where thermodynamic equilibrium can be assumed (e.g., for long pulsewidth lasers or laser-induced plasma densities), the temperature can be determined by spectroscopic measurement of the emission intensities and the coupled Saha equations (61). Regardless of the plasma temperature, a laser-induced plasma absorbs power from the laser. The absorbed power is radiated primarily via the Bremsstrahlung or lost via the plasma thermal conductivity. For microprocessing the result can be a reduction in resolution. On the other hand, for certain types of macroscopic processing, the plasma can be "tailored" to delivering maximum energy transfer to the surface (62).

The radiated power (W/cm^3) is given in Eq. (20) and the thermal conductivity (W cm^{-1} K^{-1}) is given in Eq. (21) (63). For completeness, Eq. (22) shows the time for equilibrating the electron and ion temperatures. The term $(\ln \Lambda)$ is a function of plasma parameters (64) and is of the order of 10, and A is the ion atomic weight in amu.

$$P = 1.42 \times 10^{-34} Z^3 n_i^2 T^{1/2} \tag{20}$$

$$\kappa = (1.95 \times 10^{-11} T^{5/2})/(Z \ln \Lambda) \tag{21}$$

$$\tau_{eq} = 252 \, A \, T^{3/2}/(n_e Z^2 \ln \Lambda) \tag{22}$$

Assume a laser irradiates an aluminum surface with spot diameter of 1 μm. Assume also that the fluence is such that a plasma temperature of 2×10^4 K (~ 1.7 eV) is established with $n_i = n_e \sim 10^{17}$ cm^{-3} (considered a weak plasma: 0.001% of solid state density, or $\sim 1\%$ vapor density at 1 atm). For a 10 nsec pulse laser the plasma thickness is $\sim 2.7 \times 10^{-3}$ cm ($\sim 2.7 \times 10^{-7}$ cm for a 1 psec laser) which results in a volume of $\sim 2 \times 10^{-11}$ cm^3 and a total radiated power ~ 1 μW. The thermal conductivity of this plasma becomes $\sim 10^{-2}$ W cm^{-1} K^{-1}, which is similar to that of an insulator (e.g., TiO$_2$). The time for establishing temperature equilibrium is ($\tau_{eq} \sim 0.1$ nsec. Now consider increasing the ion and electron densities by a factor of 100. The power radiated from the small volume increases by 10^4 (10 mW) while the time for reaching equilibrium decreases by 100 (1 psec). This reradiated power if not controlled could be detrimental in 3D microfabrication—where it could affect nearby nanostructures.

3.4. MULTIPHOTON ABSORPTION AND PROCESSING

The general equation for the transmitted intensity of light I(W/cm^2), at wavelength λ, propagating in the z direction within a media possessing multiple photon absorption processes may be written in the form shown in Eq. (23). The α (cm^{-1}) are the absorption coefficients for the linear (α_1) and the two photon (α_2) processes. For small α_2 Eq. (23) reduces to the standard Beer's Law for absorption. For the linear absorption α_1 can be related to the extinction coefficient, k, and the laser wavelength λ as given in Eq. (15). For two-photon absorption and for an isotropic material (a practical advantage when processing true 3D microstructures), α_2 can be related to the complex part of the third-order nonlinear susceptibility, $\chi^{(3)}$ and c is the speed of light, ε_σ is the permitivity of free space and n_o is the linear refractive index. Spectroscopic studies usually express the two-photon absorption in terms of the cross-section, σ_2 (units of $cm^4s\ photon^{-1}\ molecule^{-1}$) rather than α_2. The two quantities are related as, $\sigma_2 = (hc/\lambda N)\ \alpha_2$ where N is the number density of molecules. A recent review of two-photon absorption by Kershaw (65) includes an assembled list of measured cross-sections for organic materials. The two-photon absorption (or higher photon processes) processes have unique applications in true 3D microfabrication—for example, where microstructures are to be embedded in the bulk or a higher precision volumetric processing is desired.

$$d(I)/dz = -\alpha_1 I - \alpha_2 I^2 \ldots \tag{23}$$

$$\alpha_1 = \frac{4\pi k}{\lambda} \tag{24}$$

$$\alpha_2 = \frac{3\pi}{\varepsilon_o c \lambda n_o^2} \text{Im}[\chi^{(3)}] \tag{25}$$

4. TOPICS RELEVANT TO 3D LASER MICROENGINEERING

In reviewing the preceding equations with the specific aim toward 3D material processing, we arrive at the following conclusions. The Gaussian profile, as described by Eqs. (1) and (2) has a mathematical representation that makes possible the characterization of certain engineering parameters of a laser-material processing tool (i.e., spot size, confocal parameter). Notwithstanding, many laser processing applications utilize a super-Gaussian (66) or top-hat beam profile. These applications are mostly in material processing, where one would prefer a "cutting" tool that has a well-defined edge and a uniform fluence across the irradiated surface. The downside of a top-hat beam profile is that this profile exists for a short distance around a specific "working" focal length. Outside this zone, the beam profile and quality can be severely altered and any interaction with a material can lead to nonuniform processing. A mathematical representation of this phenomenon can be obtained by looking at the Fourier domain. Recall, that the far-field or the Fraunhofer diffraction pattern at the focal point of a lens is the Fourier transform of the aperture function (where the aperture function can be related to the source strength/unit area over the object or input plane) (67). In other words to generate a circular top-hat field distribution at focus (e.g., the rectangle function $\Pi(q/2a)$ where q is the spatial frequency and a is constant), one needs to "generate" a Bessel function of order 1 (i.e., $aJ_1(2\pi ar)/r$, where r is radial distance and a is constant) at the object plane. Bessel functions oscillate but are not periodic (except in the limit $r \rightarrow$ infinity). In contrast, the two-dimensional Fourier transform (i.e., Hankel transform) of a Gaussian is one that enables easy characterization of the laser beam in the object plane (i.e., via the knife edge test), instilling confidence that the far-field spatial intensity distribution will be equally characterized.

From the perspective of conducting precision 3D processing Eqs. (4)–(6) suggest the use of shorter wavelengths. In general the shorter the wavelength the larger the optical absorption (Eq. (15)), but the smaller the laser beam spot size (Eq. (4)) and depth of field (Eq. (3)$\approx 2\pi(f^{\#}2\lambda)$), consequently a smaller activated volume of material. However, when focusing energy into a small volume, there is always the likelihood of generating a local plasma as a result of a fluence-threshold reducing process such as a material impurity or defect (e.g., a microcavity). The "efficiency" of forming this microplasma is a strong function of the wavelength. As the wavelength is reduced the Bremsstrahlung absorption coefficient (Eq. (19)) increases as a cubic function of λ. Therefore, when processing in the UV wavelengths, more care must be exercised to mitigate microplasma initiation. Also, with decreasing laser wavelength, materials show increasing dispersion in the index, n, and this results in an error in the focal distance (Eq. (7)) that must be compensated. Finally, if the material "processing" approach requires a large temporal coherence, then for a given laser bandwidth, the coherence length decreases as the square of the laser wavelength (λ^2) (Eq. (9)).

4.1. OPTICAL ABSORPTION, EXCITATION, AND ENERGY DECAY

There are various photophysical processes that can drive laser material-interaction phenomenon but the genesis is fundamentally an electronic photoexcitation process with the primary step being a single- or a multiple-photon absorption event. The absorption can be from free conduction band electrons (e.g., in a metal), valence band electrons (e.g., in semiconductors), molecular orbital electrons (e.g., in gases and liquids), and electrons trapped at a defect-site. Even though the initial excitation may be electronic, the representative driving process that leads to nuclear motion and atomic/lattice re-arrangement or reaction (i.e., etching) may in fact be thermal. In most laser processing the representative process is identified with the final energy conversion step, where the absorbed laser energy is channeled into nuclear motion of the atoms. In solid materials, the electronic excitation often decays by exciting lattice vibrations (i.e., phonons) and thereby generating localized heating. Material modification/removal can then proceed via a thermally initiated process. In some materials, such as in insulators, the initial excitation relaxes to form an excited metastable state comprising of a self-trapped exciton (i.e., trapped electron-hole pair) or a self-trapped hole/electron that can absorb additional photons or directly induce an atom displacement and/or a desorption event. In this case, material modification/removal is more akin to a nonthermal process. In laser-irradiated material, the dynamics of the energy-decay processes play an important role in defining the driving photophysical process. The energy decay mechanisms critically depend on the specific interactions between the "hot" electron and the lattice, the efficiency of this coupling as a function of excitation energy and the number of available energy decay pathways (e.g., electron–phonon scattering, emission, exciton formation, lattice re-arrangement). For pure (i.e., no impurities), uniform-chemical-composition crystalline materials, the parameters that characterize a photophysical process sequence are typically the applied photon flux density and the resulting density of electronic excitation. However, most materials used in laser processing are not well-ordered materials. They could have large variations in chemical composition. Furthermore, they may contain impurities, voids, and other inhomogeneities (e.g., be comprised of multiple phases). Under these circumstances the optical absorption properties are delineated by scattering phenomenon. The Mie theory (68) could be used to calculate the light propagation (i.e., reflection, transmission, and absorption) if the average separation of the scattering centers is greater than the wavelength of light and multiple scattering events are not included. If the scattering density is high then a Monte Carlo simulation (69) can be used to give insight into particular key parameters (i.e., reflectance, absorption, and transmittance).

4.2. LASER EXPOSURE TIME HISTORY AND RESIDUAL STRESS

The laser parameters that are considered relevant for most 3D laser micro-engineering applications are similar to a list that might be assembled for any laser-processing application. Table 3 shows this typical list. However, in 3D microprocessing the laser-material interaction is designed to shape a micro-

TABLE 3   Laser Parameters for Microengineering Applications

Parameters	Impact
Wavelength	Radiation coupling efficiency to surface; spatial resolution
Average power	Influences intensity, background heating
Pulse energy	Influences peak intensity, photophysical process
Pulse duration	Influences peak intensity, thermal diffusion length
Focused spot size	Influences intensity, processing area, spatial resolution
Intensity	Influences type and extent of surface processing
Fluence	Influences type and extent of processing for pulse applications
Pulse repetition rate	Influences processing rate, direct current (DC) heating
Polarization	Influences coupling of energy to surface; process anisotropy (70)

scopic volume element of matter. Small volumes have small heat capacities and physically small pathways to heat sink—especially if the microstructures are suspended. As a result, the laser exposure time history becomes another parameter to monitor. For example, in fashioning a 3D microstructure by an LDW process like ablation or etching (i.e., processing via a material-subtraction approach), it is important to keep an account of the heat-flow history of the fashioned structure that remains. The intent is to minimize residual stress in the microstructure. Laser ablation and thermally activated etching techniques applied to an LDW process can locally induce temperature spikes in excess of $> 10^{10}$ K/sec. These temperature jumps, if they are nonuniform spatially, can leave high residual stress in the surrounding region. Current mitigation strategies involve thermophysical modeling of the patterned microstructures and the use of cofabricated test microstructures. Thermal modeling is usually nontrivial when the microstructures have complex shapes and are arrayed with close near neighbors. Unlike for most laser-processing applications, where a one-dimensional heat-flow analysis is found to be sufficient, for small 3D microsystems a complex 3D heat-flow analysis is necessary. These are typically carried out by thermal-analysis software where the microstructure pattern is represented in a solid model and, to save computer processing time, only the relevant portions are adequately analyzed. The difficult part is estimating the amount of heat removed (i.e., by the ablated products or volatile etch products) during a subtractive material processing operation. As microstructure size is reduced, estimating the heat removal by material exeunt becomes more relevant.

4.3. THERMAL GRADIENTS AND 3D GROWTH INSTABILITIES

Fabrication processes that are additive like laser CVD, volumetric exposure-patterning, or two-photon polymerization, have the following problems that are pertinent to 3D microfabrication. The thermal gradient as a result of LDW beam must be evaluated in terms of its effect on previously deposited/exposed/polymerized layers. For example, can the prior deposited layers sustain the thermal shock (i.e., gradient) imposed by processing of the top layer or is the thermal gradient sufficient to permit thermophysical adhesion to the many prior processed layers. The answers are more tractable for thin films ($\ll 1$ μm) than for thick films ($\gg 1$ μm) (71).

In multiple-layer deposition, it is known that even during homoepitaxial growth, there is a coarsening of the surface topology with increasing film thickness (72). This process nominally starts in the formation of mounds and has been observed in semiconductors (73), metals, and metal alloys (74). Surface coarsening reflects an increase in the number of defects as a result of a growth instability. For homoepitaxial growth, the physics of mound formation is currently understood in terms of an adatom-diffusion bias that leads to an "uphill current" of mass-flow toward ascending step-edges and increased probability for nucleation (75). It could be argued that by increasing the temperature (i.e., for LDW approach applying more laser power) this coarsening phenomenon could be mitigated, but both experimental (76) and Monte Carlo calculations (77) show that the coarsening exponent, n (where mound feature size is given by r_c and scales as h^n with h being the average film thickness), is larger with increasing temperature at least for some systems. Surface coarsening, during the growth phase, appears to be determined by the strength of the step barrier and the diffusion rate around corners and not necessarily the surface symmetry. The control of these growth instabilities could lead to fabrication of atomically flat or nanostructured surfaces. A recent calculation shows that the application of an AC field, parallel to the surface during the growth phase leads to surface smoothing ($\sim 60\%$ reduction in maximum mound height was estimated) (78). The presence of the periodic external field (e.g., $E \sim 10^8$ V/m and $v_o \sim 10^{13}$ s^{-1} on for 1 μs) leads to a "downhill" mass-flow current that partially compensates for the mound generation process (i.e., diffusion bias and "uphill current"). This technique would be useful in LDW material processing, if it were not for the large electric field required. If the compensation process could be conducted with the electric field reduced by a factor of 100, then thick films may be grown atomically flat via the LDW process.

4.4. UNIFORM VOLUMETRIC EXPOSURE

In LDW processing where the desired material transformation is photolytically initiated (e.g., volumetric exposure-patterning, two-photon polymerization) the optical absorption depth takes on special relevance. For high-resolution volumetric patterning (i.e., for 3D processing), the precise volume element where the laser "activates" the material must be characterized. In general, this critical volume element is a function of the laser intensity, fluence, and exposure (i.e., number of laser shots or irradiance duration) while its shape is dictated by the spatial/temporal properties of the laser beam. Using beam-shaping optics and temporal smoothing techniques (e.g., use of a Raman scattering cell), the spatial qualities of the laser beam can be tailored at the focus. Given this refinement, it is then possible to microfabricate 3D structures by a sequence of digital volume element bits (i.e., voxels or volume pixels). Insofar that each voxel can be uniformly microfabricated, this 3D patterning approach is conceptually appealing and naturally lends to integration with solid modeling software. The resulting 3D LDW processing tool would enable solid modeling designs to be first simulated and the results to be directly converted into fabricated pieces. The key to uniform voxel fabrication is understanding and characterizing the following optical processes.

- Identifying the optical absorption process that "activates" the material.
- Characterizing the product state of the activated material and its dependence on additional laser irradiation.
- Characterizing the effect of the transmitted light on near-neighbor voxels and unprocessed media.
- Characterizing the light-induced changes in the material optical index (e.g., thermal on nonlinear intensity effects).
- Characterizing the light scatter as a result of spatial nonuniformity in the material.
- Characterizing the generation and propagation of acoustic waves from thermoelastic expansion (i.e., nonradiative decay under pulsed-laser excitation results in localized heating and an acoustic wave).

5. 3D MICROFABRICATION BY 2D DIRECT-WRITE PATTERNING APPROACHES

Three-dimensional laser direct-write fabrication technology made its introduction by showcasing material processing via laser ablation (e.g., in complex marking applications). As a processing tool, laser ablation works in applications where high resolution and aspect ratio are not stringent requirements and the effects of thermally initiated stress (i.e., residual stress) can be overlooked. In microengineering applications residual stress, if not minimized, can impair the desired shape of a microstructure and reduce its reliability. Residual stress comes about because mechanical constraints (e.g., rigid end supports) limit expansion/contraction of the microfabricated component upon heating/cool-cooling. The magnitude of the stress, σ, can be derived to first order for a simple isotropic solid rod that is heated/cooled uniformly. Equation (26) shows this relationship for a temperature change from T_i to T_f and for a material with an elastic modulus, E, and linear coefficient of thermal expansion, α_l (79).

$$\sigma = E\alpha_l(T_i - T_f) = E\alpha_l \Delta T \qquad (26)$$

Upon heating, which is usually the case in laser processing, the ΔT is negative and the resulting stress is compressive ($\sigma < 0$) signifying the rod is constrained from expanding. Conversely, with cooling ΔT is positive and the stress is tensile ($\sigma > 0$). Laser ablative processes can increase the surface temperature of a constrained microstructure during fabrication by hundreds to thousands of degrees. Equation (26) can be recast in parameters useful for laser processing by incorporating Eqs. (16) and (17) (i.e., thermal diffusion length and maximum temperature rise for a 1D thermal analysis). This result is shown in Eq. (27) where F is the laser fluence (J/cm^2), R_{sol} is the surface optical reflectance, κ is the thermal conductivity (W/cm-K), where τ is the processing duration (i.e., the laser-pulse duration), and c_p is the heat capacity ($J/cm^3 = K$).

$$\sigma = E\alpha_l \left(T_i - \left(\frac{F(1 - R_{sol})}{\sqrt{4\kappa\tau c_p}} \right) \right) \qquad (27)$$

The thermophysical representation of the stress equation suggests that reducing residual stress means using materials that have large heat capacity, conductivity, and surface optical reflectance. It also suggests using lasers with longer temporal pulse widths.

A more complex thermophysical problem arises if the 3D microstructure is either a thermal insulator, large in comparison to the thermal diffusion length (see Eq. (16)), or is subjected to rapid heating/cooling cycles. Stress can then arise from the cycling temperature gradients with the likelihood that the 3D microstructure will mechanically fail as a result of thermal shock. The capacity of a material to withstand this type of failure is termed its thermal shock resistance (TSR). Equation (28) defines the requirements for large TSR. A material must have high fracture strength, σ_f, high thermal conductivity, κ, low elastic moduli, and low coefficient of thermal expansion (80).

$$\text{TSR} \cong \frac{\sigma_f \kappa}{E\alpha_l} \tag{28}$$

Much of the thermal-induced stress can be reduced by temperature annealing (i.e., either via laser direct-write approaches for local annealing or via bulk heating). As stress relief, temperature annealing is a viable option for large- or mesoscale structures but is not as easy to implement in the case of microscale 3D components and, especially, suspended structure components. The difficulty is calculating the thermal diffusivity (κ/c_p) of the 3D microcomponent and the microbeam supporting structures. Thermal simulation programs can give some insight, but microstructure thermophysical properties depend on many variables such as the the grain size (if noncrystalline), the shape, and the existence of cracks/defects.

Even though many laser 3D patterning processes are thermally driven, there are some applications mostly in the submicron and in the nanometer scale where nonthermal processes would be of benefit. One advantage to processing in the nonthermal regime is that thermal stress effects can be partially mitigated. In this laser-fluence regime materials can be processed either by direct photolytic bond dissociation (81), by electronic excitation of surface or bulk states that leads to material removal (82), or by a multiphoton absorption process as that found in femtosecond laser ablation (83). There is experimental evidence for nonthermal ablation/desorption. In the case of UV laser ablation of certain polymers there is evidence to suggest that a nonthermal photolysis process is active (84). Similary, in the VUV laser ablation of GaN and quartz, there is evidence of nonthermal excitation from trapped charge defect states that lead to efficient ablation (85). Finally, there is experimental evidence that nonthermal ablation/desorption can lead to atomic layer-by-layer removal (86). One practical outcome of a nonthermal ablation process is the increased resolution during 3D fabrication. This aspect already has been identified by the much sharper edge acuity in the patterns generated by femtosecond laser ablation, where nonthermal excitation is more likely, and in ablation processes in which the excitation is from a laser-induced defect or an impurity state.

A major application area in 3D laser ablation micromachining is the need to process glass and hard materials like ceramics (e.g., Si_3N_4, Al_2O_3, and ZrO_2). These materials and other wide bandgap materials are not easily processed by traditional approaches and it appears that laser ablation techniques may be one viable approach. Recent experiments have shown that with a diode-pumped solid state Q-switched Nd-Yag laser (1 kHz pulse-repitition rate, > 100 J/cm^2), operating in the third harmonic (i.e., 355 nm), Si_3N_4, Al_2O_3, and ZrO_2 can be micromachined with ablation rates of 0.02, 0.2, and 0.01–0.015 mm^3/min respectively (87). The interesting aspect is that a surface roughness of

$R_a < 1$ μm could be achieved in the process. GaN and SiC, both difficult materials to micromachine, have been processed with high acuity and free of cracks using a combination VUV-UV wavelength laser (estimated etch rate ~ 35 nm/pulse) (88). Fused silica has been similarly processed with VUV (157 nm) laser light (89). These results suggest that the use of laser wavelengths above the material bandgap does lead to more efficient ablation. However, laser wavelengths below the bandgap can still be used to induce efficient ablation if electrons from valence bands can be excited to defect states below the conduction band. The material then ablates by subsequent excited state absorption. As a result, excimer lasers and Nd-Yag lasers operating in the UV harmonics are now proving to be highly successful for microfabricating industrial components with feature sizes in the range of 0.05–1, 000 μm (90).

Ultrafast (UF) femtosecond lasers also play a large role in the 3D microfabrication of transparent and wide bandgap materials. In comparison to nanosecond pulse and VUV laser ablation, the mechanism of laser ablation with UF lasers is different. UF lasers can drive nonlinear processes that can lead to enhanced energy absorption. This is observed in a lowering of the ablation threshold for both single-shot and repetitive-shot mass-removal operations. A reduction in ablation threshold had been previously observed in data taken by nanosecond and picosecond pulses. Recent experimental results for sapphire (Al_2O_3) show this reduction in the fluence threshold continues when comparing data taken with picosecond and subpicosecond pulses. For a 3.7 ps laser pulse, the ablation threshold is nearly $4 J/cm^2$ for single-shot ablation and ~ $3.5 J/cm^2$ for 30 laser shots. The comparative data for 0.2 ps laser pulse is $3.5 J/cm^2$ and ~ $2 J/cm^2$. One application area for UF lasers is clear: fabrication of 3D microstructures that are intricate and intertwined. Using a two-photon polymerization technique, photonic crystals have recently been developed, using a layer-by-layer process and a focused UF laser (91). In another novel application, a UF laser has been used to fabricate optical waveguides (i.e., patterned refractive index changes) in various glasses (92) with the intent to develop a 4D (i.e., 3 euclidean axis plus use of spectral hole burning in the dopant Sm^{2+}) ultrahigh density ($10 TB/cm^3$) optical memory (93).

Laser ablation techniques will continue to be used in laser microengineering of 3D components, whether they be by direct material removal similar to a milling machine or via block material transfer as typically done in the LIFT technique. Either process leaves the resulting material under stress, so it must be annealed in a postprocessing step.

5.1. LASER CVD AND ETCHING

It is the opinion of this author that laser direct-write material processing came to be widespread as a result of the development and commercialization of two high-resolution laser-material processing tools. Namely, laser chemical vapor deposition (LCVD) and laser chemical-assisted etching (LCAE). These tools have a complementary function in processing—for LCVD a specific material of interest (e.g., metals, semiconductors, oxides) is deposited, while in LCAE the material is selectively removed or etched. Both tools represent a collection of techniques, chemistries, and processes that employ the laser as an activator of a precursor molecule that is either photolyzed, pyrolyzed, or merely vibrationally/rotationally excited. The latter approach when applied to a surface is essentially local heating of the surface to increase the reaction rate. With this approach, exceedingly complex microstructures have been fabricated either by mass-removal or by mass-addition (94). Both techniques can potentially process materials with diffraction limited resolution and, by using interferometric-lithography schemes, can process materials in the tens-of-nanometer range (95).

The development period for both of these tools spans nearly twenty years and includes a large body of fundamental research in laser gas-phase spectroscopy and photochemistry, molecular precursor synthesis, photo- and thermophysics of light absorption, reactive material transport and kinetics, and epitaxy. A significant compendium of this work has been assembled in the edited work of D. J. Ehrlich and J. Y. Tsao (96). Tables 4–7 present abbreviated lists of some of the processes possible via LCVD and LCAE.

For metal deposition, the precursor molecules are typically carbonyls, alkyls, halides, and oxyhalides, while for semiconductors they are either hydrides, alkyls, or alkyl halides. For etching, the precursors are halogens, hydroxides, and alkyl halides. The precursor molecules are designed to be "carriers" of the species to be deposited or to form radicals upon dissociation and initiate subsequent chemistry.

The UV absorption spectrum of most of the pertinent precursor molecules is identified by a threshold absorption in the UV with increasing absorption for shorter wavelengths. Some band structure may be evident but it is superimposed on a continuum. Pulsed UV laser irradiation within this continuum leads to efficient dissociation but not necessarily to desired product states. In general, the UV photochemistry involves an electronic excitation that leads to photolysis with a requirement for additional photons to either sever more ligands or to induce additional radical chemistry. The metal carbonyls, on the other hand, show distinct bands in the UV and the strong absorption corresponds to a metal-to-ligand charge transfer. An electron is promoted from a molecular orbital localized on the metal to an antibonding π^* molecular orbital localized on the carbonyl ligand. The outcome is the severing of a ligand. There is at least one experimental result that complete fragmentation of the carbonyl, leaving a bare metal atom, is possible through a sequential multiphoton absorption process (97). In general and for practical

TABLE 4 Semiconductor Etching Rates via Laser CAE (from Ref (98))

Semiconductor	Ambient	Laser	Intensity/energy	Etch rate	Reference
Si	Cl_2, HCl	Ar^+	$\geq 5\,MW/cm^2$	7 µm/sec	Ehrlich (1981) (99)
Si	KOH	Ar^+	$\sim 10^7\,W/cm^2$	15 µm/sec	von Gutfeld (1982) (100)
Si	NaOH	Nd : YAG	NA	4 µm/min	Bunkin (1985) (101)
Si	NaOH	CO_2	NA	2 µm/min	Bunkin (1985) (102)
α-Si	KOH	Ruby	$0.5\,J/cm^2$	500 Å/pulse	Krimmel (1985) (103)
Si	Cl_2 (0.1 torr)	N_2	$0.12\,J/cm^2$	$\sim 1\,Å/pulse$	Sesselmann (1985) (104)
Ge	Br_2	Ar^+	$0.1\,KW/cm^2$	36 µm/sec	Sullivan (1968) (105)
Ge	Br_2 (10^{-4} torr)	DYE	$0.1–0.3\,J/cm^2$	$0.2\,Å/pulse$	Davis (1984) (106)
Ge	Br_2 (2.5 torr)	Ar^+	40 W	860 Å/sec	Baklanov (1974) (107)
GaAs	H_2SO_4	Ar^+	$0.2\,MW/cm^2$	30 µm/min	Osgood (1982) (108)
GaAs	HNO_3	Ar^+	$60\,MW/cm^2$	2 µm/sec	Tisone (1983) (109)
GaAs	HF	Kr^+	$60\,MW/cm^2$	0.7 µm/min	Tsukada (1984) (110)
GaAs	HNO_3	Ar^+ (257 nm)	$1\,W/cm^2$	12 µm/min	Podlesnik (1986) (111)
GaAs	Cl	Ar^+	$1.4\,KW/cm^2$	5 Å/sec	Ashby (1984) (112)
GaAs	CH_3Br (750 torr)	Ar^+ (257 nm)	$1\,KW/cm^2$	60 Å/sec	Ehrlich (1980) (113)
GaAs	HBr	ArF (193 nm)	$32\,mJ/cm^2$	8.2 µm/min	Brewer (1985) (114)
GaAs	CCl_4	Ar^+	$0.1\,MW/cm^2$	6 µm/sec	Takai (1985) (115)
GaAs	Cl_2	Ar^+ (488 nm)	$2\,MW/cm^2$	33 µm/sec	Tucker (1984) (116)
InP	CH_3Br (750 torr)	Ar^+ (257 nm)	$100\,W/cm^2$	9.4 Å/sec	Ehrlich (1980) (117)
InP	H_3PO_4	Ar^+	$1.6\,MW/cm^2$	140 µm/sec	Bjorkholm (1983) (118)
GaP	KOH	Ar^+ (351 nm)	$3.5\,KW/cm^2$	600 Å/sec	Johnson (1984) (119)

TABLE 5 Insulator Etching via Laser CAE (from Ref (120))

Insulator	Ambient	Laser	Intensity/energy	Etch rate	Reference
SiO_2	CF_3Br	CO_2 (1 Hz)	0.4 J/pulse	17 Å/min	Steinfeld (1980) (121)
SiO_2	$SF_6,H_2 : H_2O$	CO_2	3.5 J/cm^2	100 Å/min	Ambartsumyan (1982) (122)
SiO_2	$NF_3 : H_2$	ArF	7.7 mJ/cm^2	0.12 nm/sec	Yokoyama (1985) (123)
SiO_2	SiH_4 (Plasma)	KrF	0.3 J/cm^2	40 Å/min	Gee (1984) (124)
Al_2O_3/TiC	KOH	Ar^+	1 MW/cm^2	200 µm/sec	von Gutfeld (1982) (125)
Diamond	Cl_2, O_2, NO_2	ArF	30 J/cm^2	1,400 Å/pulse	Rothschild (1986) (126)
Polyimide	KOH	Ar^+	0.03 MW/cm^2	0.3 µm/sec	Moskowitz (1984) (127)
Polyimide	Air	KrF	0.3 J/cm^2	0.15 µm/pulse	Brannon (1985) (128)
Polyimide	Air	CO_2	1.7 J/cm^2	0.8 µm/pulse	Brannon (1986) (129)
Polyethylene terephthalate	Air	ArF	0.37 J/cm^2	0.12 µm/pulse	Srinivasan (1982) (130)
Polyethylene terephthalate	Air	XeCl	1.1 J/cm^2	1 µm/pulse	Andrew (1983) (131)
Polymethyl-methacrylate	Air	ArF	0.25 J/cm^2	0.3 µm/pulse	Srinivasan (1983) (132)

TABLE 6 Metal Etching via Laser CAE (from Ref (133))

Metal	Ambient	Laser	Intensity/energy	Etch rate	Reference
Ag	Cl_2 (0.1 torr)	N_2 (337 nm)	$0.12\,J/cm^2$	$500\,\text{Å}/min$	Sesselmann (1985) (134)
Ag	Air	XeCl	$0.05\,J/cm^2$	$0.10\,\mu m/pulse$	Andrew (1983) (135)
Al	Cl_2 (0.1 torr)	N_2 (337 nm)	$0.12\,J/cm^2$	$330\,\text{Å}/sec$	Sesselmann (1985) (136)
Al	Air	XeCl	$0.2\,J/cm^2$	$0.12\,\mu m/pulse$	Andrew (1983) (137)
Al	Cl_2 (0.1 torr)	XeCl	$1.0\,J/cm^2$	$1.4\,\mu m/sec$	Koren (1986) (138)
Au	Air	XeCl	$0.03\,J/cm^2$	$0.05\,\mu m/pulse$	Andrew (1983) (139)
Cr	Air	XeCl	$0.24\,J/cm^2$	$0.08\,\mu m/pulse$	Andrew (1983) (140)
Cu	Air	XeCl	$0.08\,J/cm^2$	$0.10\,\mu m/pulse$	Andrew (1983) (141)
Fe	Cl_2 (0.1 torr)	N_2 (337 nm)	$0.12\,J/cm^2$	$850\,\text{Å}/min$	Sesselmann (1985) (142)
Mo	NF_3	ArF	$0.057\,J/cm^2$	$0.22\,\text{Å}/pulse$	Loper (1985) (143)
Mo	Cl_2 (0.1 torr)	N_2 (337 nm)	$0.12\,J/cm^2$	$16\,\text{Å}/sec$	Sesselmann (1985) (144)
Mo	Air	Ar^+	$> 10^8\,W/cm^2$	$20\,\mu m/sec$	Koren (1986) (145)
Ni	Air	XeCl	$0.24\,J/cm^2$	$0.12\,\mu m/pulse$	Andrew (1983) (146)
Ti	NF_3	ArF	$0.115\,J/cm^2$	$0.29\,\text{Å}/pulse$	Loper (1985) (147)
W	Cl_2 (0.1 torr)	N_2 (337 nm)	$0.12\,J/cm^2$	$240\,\text{Å}/min$	Sesselmann (1985) (148)

TABLE 7 Examples of Localized Photochemical Deposition via Laser CVD (from Ref (149))

Deposit	Precursor	Laser	Comments/References
Ti*	$TiCl_4$	Ar^+ (SH)	Tsao et al. (1983) (150)
Ti/Al*	$TiCl_4$, $Al(CH_3)_3$	Ar^+ (SH)	Ehrlich and Tsao (1983) (151); Tsao and Ehrlich (1984) (152)
Cr	$Cr(CO)_6$	Ar^+ (SH)	Deposits contaminated by C, O. Jackson and Tyndall (1988) (153), Ehrlich et al. (1981) (154), Gluck et al. (1987) (155)
Mo	$Mo(CO)_6$	Ar^+ (SH)	Deposits contaminated by C, O. Jackson and Tyndall (1988) (153), Gluck et al. (1987) (155)
Mo	$Mo(CO)_6$	Ar^{++}, 351–364 nm	Laser thermal and photochemical. Gilgen et al. (1987) (156)
W	$W(CO)_6$	Ar^+ (SH)	Deposits contaminated by C, O. Jackson and Tyndall (1988) (153), Ehrlich et al. (1981) (154), Gluck et al. (1987) (155), Jackson and Tyndall (1987) (157), Chiu et al. (1985) (158)
W	$W(CO)_6$	Ar^{++}, 351–364 nm	Laser thermal and photochemical. Gilgen et al. (1987) (156)
Mn	$Mn_2(CO)_{10}$	Kr^{++}, 337–356 nm	Kitai and Il Wolga (1983) (159)
Fe*	$Fe(CO)_5$	Ar^+ (SH)	Ehrlich et al. (1981) (154)
CrO_x	CrO_2Cl_2	Ar^+, 488/514 nm	Laser thermal and photochemical. Arnone et al. (1986) (160)
Cu	$Cu(hfac)_2$	Ar^+ (SH)	Films contaminated by carbon. Wilson and Houle (1985) (161), Houle et al. (1986) (162), Jones et al. (1985) (163)
Pt	$Pt(hfac)_2$	Ar^{++}, 351–364 nm	Laser thermal and photochemical. Gilgen et al. (1987) (156), Braichotte and van den Bergh (1985) (164)
Zn*	$Zn(CH_3)_2$	Ar^+ (SH)	Osgood and Ehrlich (1982) (165), Brueck and Ehrlich (1982) (166), Ehrlich et al. (1982) (167), Ehrlich et al. (1981) (168)
Zn*	$Zn(C_2H_5)_2$	ArF*, KrF*	Focused pulse laser. Aylett and Haigh (1984) (169)
Zn*	$Zn(C_2H_5)_2$	Ar^+ (SH)	Krchnavek et al. (1987) (170)
Cd	$Cd(CH_3)_2$	Ar^+ (SH)	Osgood and Ehrlich (1982) (165), Brueck and Ehrlich (1982) (166), Ehrlich et al. (1982) (167), Ehrlich et al. (1981) (168), Deutsch et al. (1979) (171), Wood et al. (1983) (172), Ehrlich et al. (1980) (173), Ehrlich and Osgood (1981) (174)
Al*	$Al(CH_3)_3$	Ar^+ (SH)	Deutsch et al. (1979) (171), Ehrlich et al. (1980) (173), Ehrlich and Osgood (1981) (174), Rytz-Froidevaux et al. (1983) (175), Osgood and Gilgen (1985) (176)

(continued)

TABLE 7   (*continued*)

Deposit	Precursor	Laser	Comments/References
Al*	Al(i-C$_4$H$_9$)$_3$	Ar$^+$ (SH)	Mingxin et al. (1984) (177)
Ga*	Ga(CH$_3$)$_3$	Ar$^+$ (SH)	Rytz-Froidevaux et al. (1983) (175)
In	In(CH$_3$)$_3$	Ar$^+$ (SH)	Osgood and Gilgen (1985) (176)
Sn	Sn(CH$_3$)$_4$	Ar$^+$ (SH)	Tsao and Ehrlich (1984) (178), Braichotte et al. (1985) (179)
Sn*	Sn(C$_2$H$_5$)$_4$	ArF*, KrF*	Focused pulsed laser. Aylett and Haigh (1984) (169)
Pb*	Pb(C$_2$H$_5$)$_4$	Ar$^+$ (SH)	Braichotte et al. (1985) (179), Tsao et al. (1985) (180)

*No indication of deposit purity.

interest, LCVD/LCAE processing requires UV wavelengths for photolysis, while pyrolysis and ensuing chemistry is done using visible or IR wavelengths.

The LCVD and LCAE techniques essentially involve material processing on a layer-by-layer level. Fabrication of structures greater than a few microns is accomplished by repeated scan operations under computerized motion control. In comparing the effectiveness of LCVD or LCAE in applications for 3D microfabrication, the LCAE technique is found to be more accommodating when microfabricating very thick (> 10 μm) structures or structures with large volume (> 1 mm^3). It is fundamentally more difficult to *deposit* thick films layer-by-layer and maintain effective mechanical and electrical properties. The application areas where LCVD and LCAE techniques have been used reflect this disparity. For example, while LCAE has been applied to 3D etching of microstructures and volumetric cutting operations (181), LCVD has been applied in the fabrication of complex material, multilayer thin-films (e.g., fabrication of soft X-ray tungsten-silicon mirrors with a laterally varying film thickness (182)). However, as nanoengineering technology develops and nanometer-scale structures or "scaffolding" is required, direct-write LCVD should become a viable technique because it enables site-specific deposition with near atomic control. This is not to say that the LCAE technique will not be as useful in nanoengineering applications; on the contrary, the LCAE technique becomes important in applications where a material surface needs texturing on the nanometer scale. Experiments show that by controlling the pressure of an etchant gas mixture, the laser fluence, and the number of pulses, it is possible to controllably remove material at the angstrom/pulse rate via a "dry-etching" ablation process. Using this approach and with Cl$_2$/He etch-gas mixtures (total pressures < 100 mTorr), grating structures (with minimum feature sizes 0.3–0.5 μm) have been fabricated on InP (183). Similar "digital" etching control has been shown with Cl_2 reacting on GaAs (184). Nanometer-scale surface texturing, with site-specific control, will be a growth area in advanced material processing. Both LCAE and LCVD could be used in these applications, and with both techniques the laser fluence/power will have to be low to minimize "damaging" the nanometer structures. For these applications to become viable for nanometer-scale fabrication, the effect of the generated surface electromagnetic waves (185) on the processing chemistry/physics will have to be investigated. It is already known that these surface excitations can be used to alter surface topology (186).

6. DIRECT-WRITE VOLUMETRIC (3D) PATTERNING

Volumetric patterning by laser photopolymerization of resins has blossomed from an interesting curiosity (ca 1980s) to a growth industry for rapid 3D fabrication of models (187). The technique applies a UV laser beam (e.g., He-Cd) to cross-link and polymerize (i.e., harden) a liquid polymer. Complicated 3D structures are formed by sequential stacking-patterned layers of hardened polymer. Typical applications use stereolithographic approaches to enhance the precision of the hardening polymer volume unit (voxel). What follows are some technical issues that arise in applying these techniques to the fabrication of 3D structures in the microscale.

- The viscosity of the liquid increases for small volumes leading to deformation/destruction of the hardening polymer.
- The fabricated microstructure must be removed from the base plate and the force for removal is not negligible.
- Nonuniformity of monomer density in the liquid polymer leads to process variability.

These issues are not insurmountable but do require additional care and preparation. By controlling these parameters, true 3D polymer fluidic components have been fabricated (e.g., a meander bent pipe $100 \times 100 \times 1,000$ µm, 3D connected pipes with 30 µm inner diameter) (188). In addition a complete integrated fluidic system has been produced that, by design, is mass producible (189).

Recent experimental evidence shows that 3D microfabrication by volumetric patterning is also possible by two-photon photopolymerization of resins. The fabrication of a microgear assembly with an external diameter of 7 µm, thickness of 2.3 µm, and resolution of ~ 0.5 µm only required a laser threshold energy $\sim 0.2-0.5$ µJ (190). The method uses a tightly focused pulsed laser where the laser wavelength energy is below that required to excite the initiator in the photopolymer through a conventional one-photon absorption process. Two photons are required to excite and cross-link the polymer. Because of the quadratic dependence of the two-photon absorption process, the excitation volume, and hence the polymer solidification volume, is further reduced. This permits spatial resolution that is smaller than the limits set by optical diffraction (i.e., submicrometer range). There are two approaches to implementing this technique. One is to use ultra-short-pulse lasers with high peak powers to "overcome" the small two-photon absorptivities. The second approach is to develop specific compounds with large two-photon absorption cross-sections (191). There has been some successful research in this area. As for example, the investigation of the (bis-(diphenyl-amino) substituted polyenes, the (bis-(diphenylamino) stilbene repeat unit dendrimers and the molecular "engineering" of push-pull dipolar molecules for enhanced two-photon absorption (192). Perhaps the most intriguing application of volumetric patterning is the potential use of holographic confocal microscopy to pattern or impregnate a 3D image directly into a media.

6.1. A MERGED-PROCESS APPROACH EXAMPLE

It is possible to take the advantages of LDW processing and merge them with those of batch processing if the direct-write segment is utilized only for volumetric patterning of the material and not for material deposition or removal. Such a merged-process approach increases the net processing speed and retains the main advantages of each approach:

- direct-write processing features maskless processing and true 3D processing.
- batch processing features processing ease, cost-effectiveness because of parallel processing, and wafer-scale uniformity.

A merged-process LDW microfabrication technique has been developed at The Aerospace Corporation for the 3D microfabrication of glass/ceramic materials (193). The technique relies on use materials that include photolytically active ingredients that upon exposure retain the image (194). Patterns of any complex shape can be "written" via LDW processing and then chemically batch processed to remove the exposed portion. The approach can be likened to polymer chemical resist processing technology, which is in routine use in the microelectronics industry, except in this case the material is a glass/ceramic and structures of several millimeters in height can be fabricated. The lesson learned from this material-processing development is that a merged-process approach propels the use of LDW into application areas not normally feasible.

The crux of developing a merged-process for other materials is "engineering" materials that can be photoactivated to be chemically reactive.

There is a class of photostructurable glass/ceramic materials that incorporate photoactivators and can therefore be volume patterned. These materials have the unique property that the photactivated sites can be made to undergo devitrification (i.e., crystallization) at low temperatures. Two potential outcomes are then possible: (1) the devitrified phase is soluble and can be removed (i.e., etched away) with exposure to hydrofluoric acid (HF); (2) the devitrified material can be converted from a glass state into a full ceramic state (195) by undergoing a second programmed bake-step. There are over 5,000 varieties of these glass/ceramics that go by various generic names such as photositalls and photocerams. Donald Stookey of Corning Corporation is credited with doing the initial (ca 1948) (196) research on photositalls. Currently, there are two manufacturers that make photositalls suitable for microfabrication applications: *Foturan* (Schott Corporation) and *PRG-3 Photoceram* (HOYA Corporation). To our knowledge, microstructure fabrication in these photositalls is currently done by UV lamp lithography through a mask followed by HF chemical etching (197).

The Aerospace Corporation's microengineering process differs from the UV lamp exposure process in that a focused, high-repetition-rate, pulsed laser is used to volumetrically pattern the glass. The pattern design and implementation is controlled by standard CADCAM software. To make the laser process viable, specific measurements were first needed to identify the process "window." For example, at selected UV laser wavelengths, the required photon dose for exposure was measured along with the effect that dose rate has on the chemical etch efficiency (i.e., between exposed and unexposed areas) (198). Also measured was the minimum photon dose required to trigger high-efficiency chemical etching (199). In the latter investigation a nonlinear dependence on the laser fluence was identified in the measurement. Using the measured data we have fabricated numerous true 3D microstructures in glass/ceramic material. Microstructures have been fabricated that are designed to operate either

- as solitary units,
- in large arrays, or
- as an interconnected pattern of assemblies that can function as complex microdevices (e.g., fluidic).

Furthermore, microstructure components have been fabricated that are

- large (i.e., cm scale) and to an extent free-standing (i.e., but supported by ribs/springs),
- small (i.e., 50-micron scale) that have aspect ratios $\gg 10$,

- undercut and can be made to move, or
- embedded (e.g., patterned embedded channels within a 3-mm-thick wafer).

Embedded channels have a particularly interesting use, as for example, in connecting reservoirs capability to save an expensive sealing step in packaging. Using this we have microengineered several prototype microthruster systems that are designed to maintain attitude control of a 1-kg class nanosatellite (200). The microthruster impulse and thrust efficiency have been experimentally measured (201) and one particular design flew in space on NASA Shuttle mission STS-93 in July 1999 (202).

6.2. THE FABRICATION PROCESS

Photositalls function via a three-step process: illumination, ceramization, and preferential isotropic etching (203). For Foturan, the photosensitive character arises from trapped Ce^{3+} (admixture CeO_2) and Ag^+ (admixture Ag_2O) ions that are stabilized by Sb_2O_3 in a lithium aluminosilicate mixture host (204). Using the conventional linear absorption model, Ce^{3+} can be photoionized to form Ce^{4+} and a free electron at photon energies near 3.97 eV (318 nm). The free electron neutralizes a nearby Ag^+ ion (i.e., $Ag^+ + e^- \rightarrow Ag^0$) leaving a latent image of the absorption event. In the ceramization step, migration and local clustering of the Ag^0 nuclei lead to formation of lithium silicate crystals. In a 5% solution of HF these crystals etch 20–40 times faster than the unexposed amorphous material. An aspect that is critical to the surface finish of the final microstructure or the degree of ceramization is the growth rate of the crystals and the maximum bake temperature. Both growth rate and phase of the lithium silicates can be controlled during the bake-step. A low temperature ($\sim 600\,^\circ C$) bake results in multiphase crystals that dissolve in HF acid, while a high-temperature bake ($> 700\,^\circ C$) forms a true ceramic phase that is resistant to HF. Figure 1 shows data from X-ray diffraction analysis of two processed samples: an exposed but not baked sample (top: glassy-amorphous state) and a sample that has undergone a programmed bake sequence for ceramization (bottom).

The Aerospace Corporation microfabrication process utilizes the wavelength dependence of the UV absorption to control the volume of material that is exposed. By changing the laser wavelength from 248 nm (OD = 3.0) to 355 nm (OD = 0.1) it is possible to vary the penetration depth of the laser light from less than 100 microns to over several millimeters. In addition, to insure against thermal runaway damage within the focal volume region and also permit better precision in the depth of exposure, certain laser parameters are

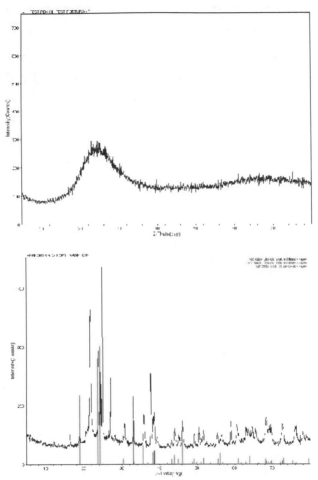

FIGURE 1   X-ray 2Θ scans of amorphous (exposed but not baked) sample (top), and crystalline (exposed and baked) (bottom). The identified bands in the bottom scan are for lithium silicate.

controlled. The incident laser fluence and the applied dose (i.e., number of laser shots) are controlled during exposure and finally, great care is taken to control the shape of the laser beam, the focal volume, and the depth of focus (i.e., confocal parameter). Figure 2 shows that for a constant laser fluence the exposure depth depends on the number of shots applied. This data suggests that by controlling the laser shot number (i.e., for constant fluence) the material can be "cut" to a predetermined depth.

The aforementioned controls are integrated with an XYZ microstepper that is accurate to 5 microns over 100 mm XY translation and an automated system that can transfer 1 of 4 incident laser wavelengths to the sample surface. Three

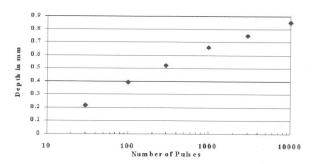

FIGURE 2   Etch depth versus number of laser pulses (266 nm) for constant per-pulse laser fluence and constant etch protocol.

individually selectable microscope objectives can be used to focus the laser beam onto the sample (5X, 10X, and 20X). Two high-repetition lasers "feed" the exposure tool. A 2 kHz excimer laser (Potomac Photonics SGX1000) and a 1 kHz diode seeded Nd-Yag laser (Continuum HPO-1000). Automated fast shutters flag on/off the various laser beams and power meters are used for average power readings. Local dose control can be set through software commands to the stepper motor-speed parameter or via burst-commands to the laser. For a multicolor exposure process, the various wavelength patterns are drawn separately, as layers, using AutoCAD (AutoDesk Co.) software. The DXF format output of the AutoCAD software is converted to machine-control language using a translator program. All or any selected number of wave-length-specific layers can be run automatically. Figure 3 shows two free-

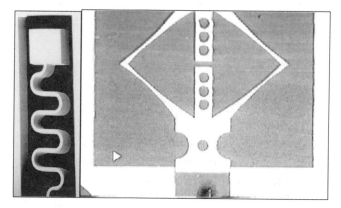

FIGURE 3   SEM of a free-standing meander spring (200 μm deep, 80 μm wide, and 4 mm long) with proof mass (left) and an optical microscope photograph of a bowspring loading a piston, 10 mm long with bow tapering to 80 μm wide (right).

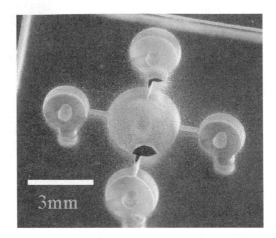

FIGURE 4 Prototype fluidic mixing chambers. Wafer is 1 mm thick.

FIGURE 5 Prototype fluidic device patterned and etched and then converted into ceramic state (top), and partially converted to ceramic state (center square) (bottom).

standing structures that can be used for prototyping resonant microstructures and springs, while in Fig. 4 a prototype design for a 4-way fluidic mixing and pass-through device is shown. The etched pattern utilized 4 DXF-layers. In Fig. 5 we show a similar device that has been transformed into the full ceramic state (top) and a coupon that has been partially ceramicized (bottom). In the bottom figure, the center orange portion is a full ceramic (i.e., modulus of rupture $150\,N/mm^2$) while the outer, "white" portion is a partial ceramic (i.e., modulus of rupture $90\,N/mm^2$). Feature resolution of better than 20 microns can be maintained during the glass-to-ceramic transformation process and by implementing a direct-write approach, selective areas of a coupon can be fully ceramicized.

6.3. NONLINEAR FLUENCE DEPENDENCE AND FABRICATION

Early experimental results from our laboratory suggested that a critical dose of UV light is necessary to *form a connected etchable network of lithium silicate crystals*. The goal was to utilize this property of the material as a means for precisely controlling energy deposition for not only 3D fabrication but for embedded structure fabrication. This hypothesis could be cast in the form shown in Eq. (29).

$$D_c \propto F^m(r, z)N \qquad (29)$$

where D_c defines the critical dose for *forming a connected etchable network of lithium silicate crystals*, F is the per laser pulse fluence (J/cm^2) with radial dependence, r, and depth dependence, z, and power dependence, m. N is the number of laser shots.

To test this hypothesis an experiment was devised that used a pulsed diode seeded Nd-Yag laser system operating on the 3rd and 4th harmonics to irradiate 1 mm thick Foturan samples with a known spot size. The analysis consisted of precisely measuring the etch depth (for 266 nm irradiation) and hole diameter (for 355 nm irradiation) as a function of the number of laser pulses with a proven Gaussian profile. Both hole depth and diameter define a boundary region where the material is barely exposed but has not developed a connected network of crystals to promote etching. This region is easily identifiable in cross-section scanning electron microscope (SEM) pictures in which the images show isolated crystals. For a Gaussian spatial distribution, it is possible to derive the total irradiated laser fluence (and dose) at the boundary region presuming there is measurement of the irradiated laser-pulse energy.

In our experiment, the spot size on the sample was measured using a knife edge on a translation stage and in front of a power meter. For an incident Gaussian-beam profile, spot size ω_o, and power P_o, the transmitted power past a knife edge is given by the complementary error function (e.g., *erfc*) as shown in Eq. (30).

$$P = \frac{P_o}{2} \, erfc\left(\frac{2z}{\omega_o\sqrt{2}}\right) \qquad (30)$$

Consequently, by measuring the laser power and the number of laser shots administered per sample, it is possible to derive the laser fluence and the irradiated total dose at the boundary regions. Figure 6 plots this derived fluence as a function of laser shot number on a log–log scale. The fit is for $m = 2$. The result indicates a squared dependence on the laser fluence. We cannot yet experimentally prove whether the nonlinear fluence dependence results from a photoabsorption process (i.e., via a true sequential two-photon process or a process that involves a long-lived intermediate state and some form of energy pooling) or is merely a consequence of the required density of etchable sites. Further experiments are underway in our laboratory to elucidate the underlying nonlinear mechanism in the photolytic process. Regardless, these results are surprising and they enable the laser to microfabricate embedded structures. By regulating the dose to near the measured critical value and by appropriately shaping the laser focal volume, we have micro-fabricated embedded stacked channels. Figure 7 shows two rectangular stacked channels that were exposed from the same side. Note that there is no exposure

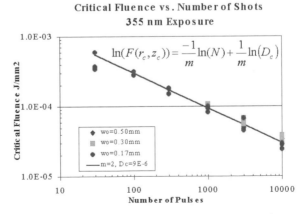

FIGURE 6   Nonlinear exposure process. Data for 3 spot sizes (0.5, 0.3, and 0.17 mm) are plotted. The results for 266 nm are similar but not shown (205).

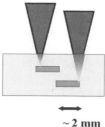

~2 mm

FIGURE 7   Two channels exposed from the top surface without inducing any exposure in between. This postbaked Foturan sample clearly shows which regions have been exposed.

above and below each channel and also none in between. The material is only exposed in the volume region where the administered laser dose is above a critical value. Finally, Fig. 8 shows two SEMs that show a series of fully processed embedded tunnels that connect microfluidic reservoirs. The merged process was used to fabricate both the tunnels and the reservoirs utilizing both the linear and nonlinear processing aspects. The direct-write exposure patterns were run sequentially. The processing technique removes a package-sealing step in microfluidics applications and enables fabrication of undercut structures that can serve as "scaffolding" for novel MEMS devices made of glass/ceramic material.

This capability for embedding microchannels or microstructures has numerous applications. Specifically, it reduces a potential wafer-bonding step in a microfluidic application and enables the selective undercutting of supported

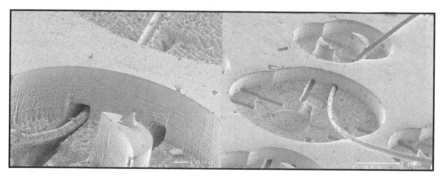

FIGURE 8   Two scanning electron micrographs show a series of embedded microtunnels that connect fluidic reservoirs. A human hair is threaded through the tunnels and serves to illustrate the tunnel size (~100 microns wide by design). The left image is a close-up of the tunnel while the right image shows an application for a multichamber microfluidic mixing system. Open reservoirs are mm in diameter and ~300 microns deep.

structures. An embedded exposure does not necessarily lead to an embedded cavity. The length and aspect ratio of any embedded microcavity structure will strongly depend on getting the etchant into the exposed region and removing the by-products. This aspect could be simulated by calculations of mass transport through a liquid.

6.4. MICROTHRUSTER APPLICATIONS

Microthrusters (1–10 mN force) are used for propulsion and attitude control in small space satellites. Smaller microthrusters in the 10–1,000 µN class can also be used for dynamic suppression/damping of vibrations in extended space structures. Microthruster propellants can be a high-pressure gas, a liquid, a sublimating solid, or a chemically reactive solid. Regardless of the type of propellant, the efficient use of the limited supply is always a paramount issue to space system designers, because spacecraft lifetime is usually dictated by the availability of the onboard propellant. This concern becomes more acute for pico- and nanosatellites (≤ 1 kg class) because of the overall limited size and volume. Regardless of the type, microthrusters typically have mesoscopic (cm-to-mm-scale) dimensions (e.g., fuel tank volume), microscopic (micron-scale) segments (e.g., fuel lines), and even nanoscopic (nanometer-scale) structures (e.g., surface coatings). The large dynamic range in dimension over which a material must be processed places a constraint on the accuracy of the fabrication tools when developing an *integrated* microthruster. For example, an integrated ion propulsion system would require the fabrication of a fuel tank (cm-scale) that is "cofabricated" with an array of field emitter tips (nanometer scale). Laser processing offers one approach to bridging this wide dynamic range of precision processing. By implementing such high-fluence techniques as laser ablation with medium-fluence techniques like direct-write volumetric patterning and low-fluence techniques that lead to surface texturing, it is possible to use a laser to cofabricate cm-scale structures with nanometer-scale structures.

Using the aforementioned LDW technique, we have fabricated several prototype microthrusters. First, a bidirectional cold gas microthruster (1 mN class) was developed in which one key element of the unit was the precise microfabrication of the hourglass-shaped exit nozzle (206). The hourglass shape is easy to profile with an appropriate f number ($f^{\#}$) objective lens and the setting of the laser dose to initiate exposure within both the focusing and diverging focal volume. By using this approach, the hourglass shape can be fabricated from one side of the glass/ceramic wafer and uses only one pulse from a 10 ns Nd-Yag laser operating in the third harmonic. No trepanning operations are required.

Another type of microthruster has been developed that is essentially an array of individually addressable microsolid thrusters (207). It is a stack of three wafers comprising two silicon wafers sandwiching a glass ceramic wafer. The bottom wafer contains a patterned array of resistors that act as igniters, the middle wafer contains an array of fueled microchambers that have been fabricated using the laser direct-write volumetric patterning technique, and the top layer is a laser patterned array of SiN membranes that are designed to contain the rising combusting gas pressure to a predetermined value. The advantage of the thruster design is its ease of fabrication and the ability to simultaneously make a large number of one-shot microthrusters on wafer-scale dimensions. Figure 9 shows a picture of an assembled 15 "one-shot" micro-thruster array on a U.S. penny. The per-pixel thrust provided is measured to be 80 mN on average, and roughly 60 W of power is released during the roughly 0.4 msec combustion event (208). The glass/ceramic fuel layer plays an important role in this microthruster design. The firing of one microchamber must not induce fratricide to an adjoining chamber. With a brittle material such as single crystal silicon, there is the potential for cleaving and inducing near-neighbor ignition as a result of the microexplosion shockwave. However, by using a glass/ceramic material for the fuel container, where the material strength can be controlled to fit the application, the device is more robust with a reduced potential for near-neighbor fratricide. Measurements done in our laboratory show that the modulus of rupture (MOR) of the vitreous glass state

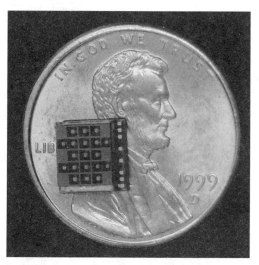

FIGURE 9 A (3 × 5) array of digitally addressable microthrusters (the 4 side holes are used for alignment during assembly). Thruster coupon flew on U.S./NASA Shuttle mission STS-93 and was tested for mechanical survivability using an inert propellant.

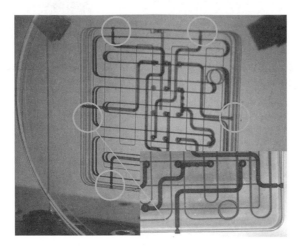

FIGURE 10   Complete liquid/gas fluid distribution system for the COSA. The photograph was taken prior to etching. All dark portions etch releasing the square coupon. Circled regions locate the microthruster nozzles. Inset shows a blowup of two of the micronozzle regions.

is 60 N/mm^2 and increases to 98 N/mm^2 after partial ceramization. The laser fabricated digital thruster was flown in a passive experiment aboard the NASA shuttle (STS-93) and in an active experiment (with firing of thrusters) aboard a suborbital rocket (i.e., Scorpius) (209). It is also scheduled to fly on a follow-on Picosat (250 gm) mission.

Figure 10 shows a recent advancement of the application of LDW volumetric patterning to the fabrication of an integrated miniature satellite (100 gm class). The figure shows a six-wafer stack where each individual wafer has undergone laser volumetric patterning. The actual assembled unit comprises 7 wafers (50 × 50 × 8 mm); inclusion of the very dense plenum layer would obscure all the other layers. The mass-producible system represents the complete propulsion system for a Co-Orbiting Satellite Assistant (COSA) and includes the fuel tank, micro nozzles, fuel filter, atomizer, sensor nodes, and fuel heater. The total unit is fabricated out of glass ceramic material and implements both 2D and 3D volumetric patterning techniques using 3 UV laser wavelengths (248 nm, 266 nm, and 355 nm).

7. TAILORING THE MATERIAL TO ADVANTAGE

In LDW microfabrication the technologically interesting aspects of the process are not only the laser excitation process but also the particular energy-decay mechanisms and the state of the material just following the laser pulse. The energy-decay mechanisms that are of most interest are those that lead to lattice re-arrangement (i.e., defect formation), trapped excitation (e.g., long-lived excitons), and electronic defects in the bulk material. More specifically, the processes of interest are those that do not necessarily lead to mass removal but leave the material in a "sensitized" state. A material that is "sensitized" commonly has a lower laser threshold for processing (210) as a result of the laser induced defects. The material is also more photon efficient for subsequent laser-processing operations. This feature also called incubation (211) has been investigated but never used to full advantage. For example, in the recent study of the decomposition of the material poly-(methyl methacrylate) by pulsed UV laser ablation, it was shown that incubation proceeds via photoinduced formation of defect centers (i.e., backbone cleavage of polymer leading to the formation of $-CH_2-C(CH_3)COOCH_3$ and $C-(CH_3)COOCH_3-CH_2$ terminated chains), while the subthreshold fluence ablative pulse is a thermally driven phenomena (212). Controlled incubation or lowering of the fluence threshold for a process can be used to overcome the serial and therefore limited-speed processing nature of the LDW technique, especially in process operations where the laser is used to *physically remove matter*. By separating the patterning and mass-removal steps it is possible to increase processing speed and resolution. For example, an LDW processing tool can be used to impregnate a 3D image in a material by merely sensitizing. This sensitized portion of the material is then removed via a batch process (e.g., a chemical processing step). Another approach would be that presensitized material is patterned via a very low-power LDW tool where the 3D shape is directly formed. The wafer is then batch processed. In both of these examples the unique capabilities of the LDW approach (i.e., maskless processing, true 3D processing) are used to advantage without the recognized limitation of speed. This concept is at core of The Aerospace Corporation merged-process for 3D material processing of glass/ceramic materials. Although photocerams are a unique class of materials, some thought should be given to "engineering" materials to enable laser direct-write patterning in the most photon-efficient manner. An example is the recent work to increase the ablation rate in micromachining of transparent materials. Transparent materials such as fused silica, quartz, and calcium fluoride are transparent in the UV excimer laser wavelengths. By applying an organic solution containing pyrene to the surface, these materials could be etched with 248 nm laser irradiation (~ 1 J/cm^2) with etch rates on the order of tens of nanometer/pulse (213). A more direct effort to change the material property to suit a laser process is the recent work to increase absorption at specific laser wavelengths by chemically altering polymers (214) (e.g., commercial polycarbonate and polymers that have photolabile $N = N - X$ groups in the backbone chain, where N is nitrogen and X is traizeno-, diazo-sulfido, pentazadieno-, and daizo-phosphonate groups).

Maximum ablation rates of 2 μm/laser shot have been realized with fluence thresholds below 150 mJ/cm^2. Furthermore, ablation results from these altered polymers show clean dry-etching contours with no surface contamination products.

The future applicability of LDW processing will strongly depend on developing an "engineered" substrate for laser processing. The "engineered" substrate material must be such that, upon wavelength specific laser excitation, the material is altered in a fashion that can be further processed either chemically, or via a much-reduced fluence laser ablation/desorption process. The latter processing approach would use very large area ($\gg$ cm) low-fluence lasers to selectively remove material, somewhat mimicking a standard batch process but employing an environmentally friendly dry mass-removal processing step.

8. SUMMARY AND CONCLUSIONS

A broad-brush review has been presented on the feasibility of using laser direct-write approaches for the fabrication of true 3D microstructures. The overview presents the pertinent optical and thermophysical equations that can be used as a guide for developing a particular laser microfabrication process. The equations are applicable for material processing via mass-subtraction (e.g., the laser ablation and laser chemical-assisted etching processes) or processing via mass-addition (e.g., laser polymerization and laser chemical vapor deposition processes). A volumetric patterning technique was also presented for the processing of glass/ceramic materials. This particular processing approach merges aspects of LDW and batch-processing. Finally, where possible we have identified issues that are particular to LDW micropatterning and microfabrication and have discussed the need for developing "engineered" substrate and materials that are "tailored" solely for LDW applications (215). That withstanding, this review misses on several fronts. First, there is a lack of specific details for each discussed laser material process. Second, this review also does not discuss several novel laser "processing" techniques that show some applicability to nanotechnology development. Not discussed are laser volumetric patterning and the epitaxial growth techniques implemented by laser trapping and control of atomic beams (216), the techniques that use lasers for inducing molecular self-assembly or preparing precise dimension nanoclusters for potential deposition. Finally, not mentioned are new techniques emerging from biophysics and protein chemistry that use light sources to induce biophysical processes in protein/bacteria for self-assembly. These are clearly experimental techniques but appear to have wide-ranging applications in nanotechnology.

The large market growth in the development of MEMS and microsystems and the desire for customization, if not fabrication-on-demand, has opened new opportunities for laser direct-write material processing. Microengineering will be a growth discipline as miniaturization and systems integration become more the norm. The ability to fashion microsystems and nanosystems in three dimensions will be necessary to realize the goal of completely integrated microsystems. As laser-light-source wavelengths continue to get shorter (e.g., X-rays) there are a host of industries and applications that require 100 nm and sub-100 nm lithography/processing. A short list might include semiconductors, molecular electronics, optical communication, microphotonics, magnetic information storage (e.g., MRAM), quantum effect electronics, biological research, and MEMS. The interactions of optical processes at the nanoscale is called nanophotonics. A year 2000 review covers this growing field (217).

With the advent of polymer electronics and the growing trend for direct-write deposition of electronic components, it will not be long before direct-write electronics technology is available. Direct-write photonics may require a longer development period but are still possible. Regardless, lasers in general will play a role in direct-write material processing (218). The key is developing a substrate material whereby laser direct-write processing permits the co-fabrication of micromechanical, microfluidic, electronic, and photonic components. The resulting device would be a true direct-write processed microsystem.

ACKNOWLEDGMENTS

The author thanks the contributions and many discussions with D. Taylor, P. Fuqua, B. Hansen, M. Abraham, S. Janson, and J. Barrie of The Aerospace Corporation and the many colleagues in both the laser-material processing and microengineering communities. Acknowledgment is also given for support by The Aerospace Corporation Corporate Research Initiative Program and The Air Force Office of Scientific Research, Dr. Howard Schlossberg, Program Manager.

REFERENCES

1. M. J. Kelley, H. F. Dylla, G. R. Neil, L. J. Brillson, D. P. Henkel, and H. Helvajian, "UV FEL light source for industrial processing" SPIE Vol. 2703, Lasers as Tools for Manufacturing of Durable Goods and Microelectronics, L. R. Migliore, C. Roychoudhuri, R. D. Schaeffer, J. Mazumder and J. J. Dubowski, Eds. (1996) 15.
2. J. Brannon, J. Greer, and H. Helvajian "Laser Processing for Microengineering Applications" in Microengineering Aerospace Systems, H. Helvajian, Ed. (AIAA and Aerospace Press, Reston, Va, 1999) pg. 145.
3. E. Thielicke and E. Obermeier, "Microactuators and their technologies", Mechatronics, 10, (2000) 431.

4. NEXUS (The Network of Excellence in Multifunctional Microsystems) Task Force Report: Market Analysis for Microsystems 1996–2002, NEXUS Office c/o Fraunhofer-IsiT, Trachen-bergring 11A, D-12249 Berlin.

5. A. J. Ikushima, T. Fujiwara, and K. Saito, "Silica glass: A material for photonics" J. Appl. Phys. **88**, No. 3 (2000) 1201.

6. D. Sciti, C. Melandri, and A. Bellosi, "Excimer laser-induced microstructural changes of alumina and silicon carbide" J. Mat. Sci. **35** (2000) 3799.

7. S. Maruo and S. Kawata, "Two-photon-absorbed near-infrared photopolymerization for three dimensional microfabrication" J. Microelectromechanical Systems, 7, No. 4 (1998) 411.

8. K. Shihoyama, A. Furukawa, Ph. Bado, and A. A. Said, "Micromachining with ultrafast lasers" in 1st Int. Symp. on Laser Precision Microfabrication, I. Miyamoto, K. Sugioka and T. W. Sigmon, Eds. SPIE Vol. 4088 (2000) 140; W. W. Hansen, S. W. Jansen, and H. Helvajian, "Direct-write UV laser microfabrication of 3D structures in lithium-alumosilicate glass" SPIE Proc. Vol. 2691, "Laser Applications in Microelectronic and Optoelectronic Manufacturing" II, J. J. Dubowski, Ed. (1997), 104.

9. Soft Processing for Advanced Inorganic Materials, MRS Bulletin Vol. 25, No. 9, Sept. 2000, and articles therein.

10. Proceedings; Manufacturing three-dimensional components and devices at the meso and micro scales: A workshop, cosponsored by the US National Institute of Standards and Technology and the National Science Foundation, May 18–19 Gaithersburg, MD, (1999).

11. M. von Allmen, and A. Blatter, "Laser-beam Interactions with Materials: Physical Principles and Applications" Springer-Verlag series in Material Science (Springer-Verlag, NY, 1995) pg. 78.

12. S. Preuss, H. -C. Langowski, T. Damm, M Stuke, Appl. Phys. A54, (1992) 360; E. Matthias and Z. L. Wu, " Non-destructive reading of laser-induced single shot incubation in dielectric coatings" Appl. Phys. A56 (1993) 95.

13. J. Cao, C. J. Bardeen, and K. R. Wilson, "Molecular π pulses: Population inversion with positively chirped short pulses", J. Chem. Phys. **113**, No. 3 (2000) 1898; S. Ruhman and R. Kosloff, J. Opt. Soc. Am. B, 7, (1990) 1748; G. Cerullo, C. J. Bardeen, Q. Wang, C. V. Shank, Chem. Phys. Lett. **262**, (1996) 362.

14. Interagency Working Group on Nanoscience, Engineering and Technology. Workshop Report: Nanotechnology Research Directions—Vision for Nanotechnology R&D in the Next Decade, (1999) National Science and Technology Council/Committee on Technology, White House, (1999).

15. R. F. Haglund, Jr., and N. Itoh, "Electronic Processes in Laser Ablation of Semiconductors and Insulators", in Springer Series in Materials Science, Vol. 28 *Laser Ablation* J. C. Miller, Ed. (Springer-Verlag, Berlin, 1994) pg. 11; Laser-Beam Interactions with Materials: Physical Principles and Applications, M. von Allmen, and A. Blatter (Springer-Verlag, Berlin, 1987).

16. Laser Microfabrication: Thin Film Processes and Lithography, D. J. Ehrlich and J. Y. Tsao Eds. (Academic Press, NY, 1989).

17. C. I. H. Ashby, "Laser driven etching" in *Thin Film Processes II*, (Academic Press, NY) (1991) pg. 783.

18. *Pulsed Laser Deposition of Thin Films*, D. B. Chrisey and G. K. Hubler, Eds. (John Wiley & Sons, NY, 1994); *Laser Ablation: Principles and Applications*, J. C. Miller, Ed. Springer Series in Materials Science Vol. 28 (Springer-Verlag, Berlin, 1994); *Laser-Beam Interactions with Materials: Physical Principles and Applications* M. von Allmen, and A. Blatter (Springer-Verlag, Berlin, 1987); *Interfaces Under Laser Irradiation*, L. D. Laude, D. Bäurle, M .Wautelet, Eds. NATO Advanced Science Institute Series (Martinus, Jijhoff, Boston, 1987).

19. J-A Kim, K-C Kim, E. W. Bae, S. Kim, Y. K. Kwak, "Six-degree-of-freedom displacement measurement system using a diffraction grating" Rev. Sci. Instr. 71, No. 8 (2000) 3214.

20. K. Zimmer, D. Hirsch, and F. Bigl, "Excimer laser machining for the fabrication of analogous microstructures" Appl. Surf. Sci. Vol. 96–98 (1996) 425.
21. D. J. Ehrlich, R. M. Osgood, Jr., and T. F. Deutsch "Laser chemical technique for rapid direct writing of surface relief in silicon" Appl. Phys. Lett. 38, (1981) 1018.
22. R. L. Jackson, T. H. Baum, T. T. Kodas, D. J. Ehrlich, G. W. Tyndall, and P. B. Comita "Laser deposition: Energetics and chemical kinetics" in Laser Microfabrication: Thin Film Processes and Lithography, D. J. Ehrlich and J. Y. Tsao, Eds. Academic Press (1989) 385.
23. S. M. Metev, and V. P. Veiko, "Laser-assisted microtechnology" In Springer Series in Material Science Vol. 19, (Springer-Verlag, NY, 1994) pg. 78.
24. J.-H. Jeon, M.-C. Lee, K.-C. Park and M.-K. Han, "New excimer laser recrystallization of poly-Si for effective grain growth and grain boundary arrangement", Jpn. J. Appl. Phys. 39 (2000) 2042; C. Palma and C. Sapia, "Laser pattern-write cyrstallization of amorphous SiC alloys" J. Elect. Mat. 29, No. 5 (2000) 607.
25. P. Fuqua, S. W. Janson, W. W. Hansen, and H. Helvajian, "Fabrication of true 3D microstructures in glass/ceramic materials by pulsed UV laser volumetric exposure techniques" in SPIE Proc. Laser Applications in Microelectronic and Optoelectronic Manufacturing IV, J. J. Dubowski, H. Helvajian, E. W. Kreutz, K. Sugioka, Eds. Vol. 3618 (2000) 213.
26. K. Yamguchi, T. Nakamoto, and P. Abraha, "Consideration on the accuracy of fabricating microstructures using UV laser induced polymerization" 5th Int. Symp. on Micromachine and Human Science Proceedings, IEEE Vol. 94th 0707–0 (1994) 171; E. E. Said-Galiev, L. N. Nikitin, A. L. Rusanov, N. M. Belomoina, and E. S. Krongauz, "Laser-induced thermochemical synthesis and crosslinking of polymers in the solid phase" High Energy Chemistry, Vol. 28, No. 4, (1994) 332.
27. G. Steinberg-Yfrach, J.-L. Riguad, E. N. Durantini, A. L. Moore, D. Gust, and T. A. Moore, "Light driven production of ATP catalyzed by F_0F_1-ATP synthase in an artificial photosynthetic membrane" Nature Vol. 392, (1998) 479.
28. A. G. Cullis, H. C. Webber, and P. Bailey, "A device for laser beam diffusion and homogenization" J. Phys. E. Sci. Instrum. 12, (1979) 688.
29. H. Kogelnik and T. Li, "Laser beam resonators" Proc. IEEE 54 (1966) 1312.
30. A. E. Siegman, Lasers (University Science Books, Mill Valley CA, 1986) pg. 676.
31. M. Born and E. Wolf, Principals of Optics, 5th Ed. (Pergamon Press, NY, 1975) pg. 419.
32. D. J. Ehrlich, J. Y. Tsao, and C. O. Bozler, "Submicron patterning, by projected excimer laser-based beam induced chemistry" J. Vac. Sci. Technol. B3, (1994) 2555.
33. Y. S. Liu "Sources, optics, and laser microfabrication systems for direct writing and projection lithography" in Laser Microfabrication: Thin Film Processes and Lithography, D. J. Ehrlich and J. Y. Tsao, Eds. (Academic Press, NY, 1989) pg. 3.
34. Ibid.
35. B. K. Dey, M. Shapiro, and P. Brummer, "Coherently controlled nanoscale molecular deposition" Phys. Rev. Lett. 85, No. 15, (2000) 3125.
36. S. Ruhman and R. Kosloff "Application of chirped ultrashort pulses for generating large-amplitude ground-state vibrational coherence: a computer simulation" J. Opt. Soc. Am B7, No. 8, (1990) 1748.
37. G. Cerullo, C. J. Bardeen, Q. Wang, and C. V. Shank, "High power femtosecond chirped pulse excitation of molecules in solution" Chem. Phys. Lett. (1996) 362.
38. K. M. A. El-Kader, J. Oswald, J. Kocka, and V. Chab, "Formation of luminescent silicon by laser annealing of a-Si:H" Appl. Phys. Lett. 64, (1994) 2555.
39. G. V. Treyz, R. Beach, and R. N. Osgood, Jr., "Rapid direct writing of high aspect-ratio trenches in silicon" Appl. Phys. Lett. 50, (1987) 475.
40. G. B. Shinn, F. Steigerwald, H. Stiegler, R. Sauerbrey, F. K. Tittle, and W. L. Wilson, Jr., "Excimer laser photoablation of silicon" J. Vac. Sci. Technol. B4 (1986) 1273.

41. M. Ishii, T. Meguro, T. Sugano, K. Gamo, and Y. Aoyagi, "Digital etching by using a laser beam: On the control of digital etching products" Appl. Surf. Sci. **79/80** (1994) 104.

42. *Laser Microfabrication—Thin Film Processes and Lithography*, J. Y. Tsao, D. J. Ehrlich, Eds. (Academic Press, NY, 1989); R. Haglund, Jr., and R. Kelly, "Electronic processes in sputtering by laser beams," in *Fundamental Processes in Sputtering of Atoms and Molecules*, P. Sigmund, Ed. (Munksgaard, Copenhagen, 1993).

43. M. Eyett and D. Bauerle, "Influence of the beam spot size on ablation rates in pulsed-laser processing" Appl. Phys. Lett. **51**, (1987) 2054.

44. C. I. H. Ashby, "Laser driven etching" in *Thin Film Processes II* (Academic Press, NY, 1991) pg. 783.

45. D. V. Podlesnik, H. H. Gilgen, and R. M. Osgood, Jr., "Waveguiding effects in laser-induced aqueous etching of semiconductors" Appl. Phys. Lett. **48**, (1986) 496.

46. R. J. Wallace, M. Bass, and S. M. Copley, "Curvature of laser machined grooves in Si_3N_4" J. Appl. Phys. **59**, (1986) 3555.

47. C. Kittel, *Introduction of Solid State Physics*, 6th Ed. (John Wiley & Sons, NY, 1986).

48. D. Ashkenasi, A. Rosenfeld, H. Varel, M. Wahmer, and E. E. B. Campbell, "Laser processing of sapphire with picsond and sub-picosecond pulses", Appl. Surf. Sci., Vol. 120 (1997) pg. 65.

49. L. Wiedeman and H. Helvajian, "Laser photodecomposition of sintered YBCO: Ejected species population distributions and initial kinetic energies for the laser ablation wavelengths 351, 248, and 193" J. Appl. Phys. **70**, (1991) 4513; H. Helvajian and R. Welle, "Threshold level laser photoablation of crystalline silver: Ejected ion translational energy distributions" J. Chem. Phys. **91**, (1989) 2616; H. Helvajian, "Surface excitation mediated physics in low-fluence laser material processing" SPIE Proc. **2403** (1995) pg. 1; R. H. Ritchie, J. R. Manson, and P. M. Echenique, "Surface plasmon-ion interaction in laser ablation of ions from a surface" Phys. Rev. B., Vol. 49 (1994) 2963; D. P. Taylor, W. C. Simpson, K. Knutsen, M. A. Henderson, and T. M Orlando, "Photon stimulated desorption of cations from yttria-stabilized cubic ZrO_2 (100)" in *Laser Ablation*, R. E. Russo, D. B. Geohegan, R. F. Haglund, Jr., K Murakami, Eds. (Elsevier, NY, 1997) pg. 101.

50. J. F. Ready, *Effects of High-Power Laser Irradiation*, (Academic Press, NY, 1971).

51. C. I. H. Ashby and J. Y. Tsao, "Photophysics and thermophysics of absorption and energy transport in solids" in *Laser Microfabrication—Thin Film Processes in Lithography*, J. Y. Tsao and D. J. Ehrlich, Eds. (Academic Press, NY, 1989), pg. 272.

52. Ibid.

53. J. T. Dickinson, S. C. Langford, J. J. Shin, and D. L. Doering, "Positive ion emission from excimer laser excited MgO surfaces" Phys. Rev. Lett. **73**, (1994) 2630.

54. F. Wood, and D. H. Lowndes, "Laser processing of wide band gap semiconductors and insulators" *Cryst. Latt. Def. Amorph. Mat.* **12**, (1986) 475; A. H. Guenther and J. K. Mciver, "The role of thermal conductivity in the pulsed laser damage sensitivity of optical thin films" *Thin Solid Films*, **163**, (1988) 203.

55. L. P. Welsh, J. A. Tuchman, and I. P. Herman, "The importance of thermal stresses and strains induced in laser processing with focused Gaussian beams" J. Appl. Phys. **64**, (1988) 6274.

56. A. L. Dawar, S. Roy, T. Nath, S. Tyagi, and P. C. Mathur, "Effect of laser annealing on electrical and optical properties of n-Mercury Cadmium Telluride" J. Appl. Phys. **69**, (1991) 3849.

57. M. Raff, M. Schutze, C. Trappe, R. Hannot, and H. Kurz, "Laser-stimulated nonthermal particle emission from InP and GaAs surfaces" Phys. Rev. B. **50** (1994) 11031.

58. H. Helvajian and R. Welle, "Threshold level laser photoablation of crystalline silver: Ejected ion translational energy distributions" J. Chem. Phys. **91**, (1989) 2616.

59. L. P. Smith, "The emission of positive ions from Tungsten and Molybdenum" Phys. Rev. **35**, (1930) 381.

60. J. F. Ready, *Effects of High-Power Laser Irradiation*, (Academic Press, NY, 1971).

61. Ibid.

62. A. N. Pirri, R. G. Root, and P. K. S. Wu "Plasma energy transfer to metal surfaces irradiated by pulsed laser ablation" J. AIAA **16**, (1978) 1296.

63. J. F. Ready, *Effects of High-Power Laser Irradiation*, (Academic Press, NY, 1971).

64. L. Spitzer, *Physics of Fully Ionized Gases* (Wiley-Interscience, NY, 1956).

65. S. Kershaw, "Two-Photon Absorption" in *Two-Photon Absorption, Characterization Techniques and Tabulations for Non Linear Organic Optical Materials*, M. G. Kuzyk and C. W. Dirk, Eds. (Dekker, New York, 1998) pg. 515.

66. A. E. Siegman, *Lasers* (University Science Books, Mill Valley, CA, 1986), pg. 738.

67. R. N. Bracewell, *The Fourier Transform and its Applications*, 2nd (McGraw-Hill, NY, 1978).

68. *Principles of Optics*, M. Born and E. Wolf, 5th Edition (Pergamon Press, NY, 1975) pg. 633.

69. P. A. Atanasov, S. E. Imamova, H. Hügel and T. Abeln "Optical parameters of silicon carbide and silicon nitride ceramics in 0.2–1.3 μm spectral range" J. Appl. Phys. Vol. 88, No. 8, (2000) 4671.

70. Y. Lu et al., "Wet-chemical etching of Mn-Zn ferrite by focused Ar^+-laser irradiation in H_3PO_4," Appl. Phys. **A47**, 319 (1988).

71. *Epitaxial Growth—Principles and Applications* Mat. Res. Soc. Symp. Proc. **570**, A-L.Barabasi, M. Krishnamurthy, F. Liu, T. Pearsall, Eds. (1999) and papers therein.

72. J. G. Amar, "Mechanisms of mound coarsening in unstable epitaxial growth" Mat. Res. Soc. Symp. Proc. **570**, (1999) pg. 11.

73. M. D. Johnson, C. Orme, A. W. Hunt, D. Graff, J. Sudijono, L. M. Sander, and B. G. Orr, Phys. Rev. Lett. **72** (1994) pg. 116; J. E. Van Nostrand, S. J. Chey, M.-A. Hasan, D. G. Cahill, and J. E. Greene, Phys Rev. Lett. **75**, (1995) pg. 1127.

74. H.-J. Ernst, F. Fabre, R. Folkerts, and J. Lapujoulade, Phys. Rev. Lett. **72** (1994) pg. 112; J. A. Stroscio, D. T. Pierce, M. Stiles, A. Zangwill, and L. M. Sander, Phys. Rev. Lett. **75**, (1995) pg. 4246; L. C. Jorritsma, M. Bijnagte, G. Rosenfeld, and B. Poelsema, Phys. Rev. Lett. **78** (1997) pg. 911.

75. G. Ehrlich and F. Hudda, "Atomic view of surface self-diffusion: Tungsten on tungsten," J. Chem. Phys. **44** (1966) pg. 1039; R. L. Schwoebel, "Step motion on crystal surfaces II," J. Appl. Phys. **40**, (1969) 614.

76. F. Tsui, J. Wellman, C. Uher, and R. Clarke, Phys. Rev. Lett. **76** (1996) pg. 3164.

77. J. G. Amar and F. Family, Phys. Rev. **54** (1996) 14742.

78. C.-S. Lee, I. Derényi, and A.-L. Barabási "Smoothing surfaces by an AC field: An application of the ratchet effect" Mat. Res. Sox. Symp. Proc. **570**, (1999) pg. 39.

79. W. D. Callister, Jr., *Material Science and Engineering: An Introduction* (John Wiley & Sons, NY, 1985) pg. 499.

80. Ibid, pg. 500.

81. R. Srinivasan and B. Braren "Ultraviolet laser ablation of organic polymers" Chem. Rev. **89** (1989) 1303.

82. J. Kanasaki, I. K. Yu, Y. Nakai and N. Itoh, "Laser induced electronic emissions of Si atoms from Si(100) Surfaces" Jpn. J. Appl. Phys. Vol. 32. (1993) L859; M. Raff, M. Sachutze, C. Trappe, R. Hannot, and H. Kurz, "Laser stimulated nonthermal particle emission from InP and GaAs surfaces" Phys. Rev. B. Vol. 50. (1994) 11031; J. T. Dickinson, S. C. Langford, and J. J. Shin, "Positive ion emission from excimer laser excited MgO surfaces" Phys. Rev. Lett. Vol. 73, (1994) 2630; H. Helvajian, L. Wiedeman and H.-S. Kim "Photophysical processes in low-fluence UV laser-material interaction and the relevance to atomic layer processing" Adv. Mat. for Opt. Elect. Vol. 2, (1993) 31.

83. D. Ashkenasi, A. Rosenfeld, H. Varel, M. Wahmer, and E. E. B. Campbell, Appl. Surf. Sci. **120**, (1997) 65; S. Preuss, M. Spath, Y. Zhang, and M. Stuke, Appl. Phys. Lett. **62**, (1993) 3049.

84. J. H. Brannon, "Excimer laser ablation of polyimide: A 14 year IBM perspective" SPIE Vo. 2991, Laser Applications in Microelectronic and Optoelectronic Manufacturing II (1997)146.

85. K. Sugioka, S. Wada, H. Tashiro, K. Toyoda, T. Sakai, H. Takai, H. Moriwaki, and A. Nakamura, "Direct patterning of quartz substrates by laser ablation using VUV antistokes Raman pulses" Mat. Res. Soc. Symp. Proc. Vol. 285 (1993) 225.

86. J. Reif, H. Fallgren, H. B. Nielsen, and E. Matthias, "Layer-dependent laser sputtering of BaF$_2$ (111)" Appl. Phys. Lett. 49, No. 15, (1986) 930.

87. D. Hellrung, Li.-Y. Yek, F. Depiereux, A. Gillner, and R. Poprawe, "High-accuracy micromachining of ceramics by frequency-tripled Nd: Yag lasers" SPIE, Proc. *Laser Applications in Microelectronic and Optoelectronic Manufacturing IV* Vol. 3618, J. Dubowski, H. Helvajian, K. Sugioka, and E. Kreutz, Eds. (1999) 348.

88. K. Sugioka and K. Midorikawa, "Novel technology for laser precision microfabrication of hard materials" *SPIE Proc. 1st Int. Symp. on Laser Precision Microfabrication*, I. Miyamoto, K. Sugioka, T. W. Sigmon, Eds. Vol. 4088 (2000) 110.

89. P. R. Herman, K. P. Shen, P. Corkum, A. Naumov, S. Ng, and J. Zhang, "Advanced laser microfabrication of photonic components" *SPIE Proc. 1st Int. Symp. on Laser Precision Microfabrication*, I. Miyamoto, K. Sugioka, T. W. Sigmon, Eds. Vol. 4088 (2000) 345.

90. M. Gower, "Excimer laser microfabrication and micromachining" *SPIE Proc. 1st Int. Symp. on Laser Precision Microfabrication*, I. Miyamoto, K. Sugioka, T. W. Sigmon, Eds. Vol. 4088 (2000) 124.

91. H. Misawa, S. Juodkazis, H.-B. Sun, S. Matsuo and J. Nishii, "Formation of photonic crystals by femtosecond laser microfabrication" *SPIE Proc. 1st Int. Symp. on Laser Precision Microfabrication* I. Miyamoto, K. Sugioka, T. W. Sigmon, Eds. Vol. 4088 (2000) 28.

92. K. Hirao, "Internal modification of glass materials with a femtosecond laser" *SPIE Proc. 1st Int. Symp. on Laser Precision Microfabrication*, I. Miyamoto, K. Sugioka, T. W. Sigmon, Eds. Vol. 4088 (2000) 33.

93. Ibid.

94. B. Shen, R. Izquierdo, and M. Meunier "Laser fabrication of three-dimensional microstructures, cavities and columns" SPIE Vol. 2045, *Laser-Assisted Fabrication of Thin Films and Microstructures*, I. W. Boyd, Ed. (1993) 91.

95. S. H. Zaidi and S. R. J. Brueck, "Interferometric lithography for nanoscale fabrication" SPIE Proc. Vol. 3618, *Laser Applications in Microelectronics and Optoelectronics Manufacturing IV*, J. J. Dubowski, H. Helvajian, E. W. Kreutz and K. Sugioka Eds. (1999) 2.

96. *Laser Microfabrication: Thin Film Processes and Lithography*, D. J. Ehrlich and J. Y. Tsao (Academic Press, New York, NY) 1989.

97. G. W. Tyndall and R. L. Jackson, "UV multiple photon dissociation of Cr(CO)$_6$ to Cr* and CO: Evidence for direct and sequential dissociation processes" J. Am. Chem. Soc. 109 (1987), 582.

98. J. J. Ritsko "Laser Etching" in *Laser Microfabrication—Thin Film Processes and Lithography*, J. Y. Tsao and D. J. Ehrlich, Eds. 1989 (Academic Press, NY) pg. 334.

99. D. J. Ehrlich, R. M. Osgood, Jr., and T. F. Deutsch, "Laser chemical technique for rapid direct writing of surface relief in silicon" Appl. Phys. Lett. 38, (1981) 1018–1020.

100. R. J. von Gutfeld and R. T. Hodgson, "Laser Enhanced Etching in KOH" Appl. Phys. Lett. 40, (1982) 352–354.

101. F. V. Bunkin, B. S. Lukyanchuk, G. A. Shafiev, E. K. Kozlova, A. I. Portniagin, A. A. Yeryomenko, P. Mogyorosi, and J. G. Kiss, "Si etching affected by IR laser irradiation" Appl. Phys. A37, (1985) 117–119.

102. F. V. Bunkin, B. S. Lukyanchuk, G. A. Shafiev, E. K. Kozlova, A. I. Portniagin, A. A. Yeryomenko, P. Mogyorosi, and J. G. Kiss, "Si Etching Affected by IR Laser Irradiation" Appl. Phys. A37, (1985) 117–119.

103. E. F. Krimmel, A. G. K. Lutsch, R. Swanepoel, and J. Brink, "Contribution to Time Resolved Enhanced Chemical Etching and Simultaneous Annealing of Ion Implantation Amorphized Silicon under Intense Laser Irradiation" Appl. Phys. A38, (1985) 109–115.
104. W. Sesselmann and T. J. Chuang, "Chlorine surface interaction and laser induced shift etching reactions" J. Vac. Sci. Tech. B3, (1985) 1507–1512.
105. M. V. Sullivan and G. A. Kolb, "Direct photoetching of evaporated germanium and its use in mask fabrication" Electrochem. Technol. 6, (1986) 430.
106. G. P. Davis, C. A. Moore, and R. Gottscho "Ti dynamics of laser simulated etching of germanium by bromine" J. Appl. Phys. 56, (1984) 1808–1811.
107. M. R. Baklanov, J. M. Beterov, S. M. Repinskii, A. V. Rzhanov, V. P. Chebotaev, and N. I. Yurshina, "Initiation of a surface chemical reaction between single-crystal germanium and bromine gas by using a powerful argon laser" Sov. Phy. Dokl. 19, (1974) 312–314.
108. R. M. Osgood, A. Sanchez-Rubio, D. J. Ehrlich, and V. Daneu, "Localized laser etching of compound semiconductors in aqueous solution" Appl. Phys. Lett. 40, (1982) 391–393.
109. G. C. Tisone and A. W. Johnson, "Laser controlled etching of chromium doped and n doped [100] Ga⁻As" Mat. Res. Soc. Symp. Proc. 17, (1983) 73.
110. N. Tsukada, S. Sugata, H. Saitoh, K. Yamanaka, and Y. Mita, "Grating formation of gallium arsenide by one step laser photochemical etching" in Second European Conference on Integrated Optics, Florence, Italy, IEE, (1984).
111. D. Podlesnik, H. H. Gilgen, and R. M. Osgood, "Waveguiding effects in laser induced aqueous etching of semiconductors" Appl. Phys. Lett. 48, (1986) 496–598.
112. C. I. H. Ashby, "Photochemical dry etching of GaAs" Appl. Phys. Lett. 45, (1984) 892–894.
113. D. J. Ehrlich, R. M. Osgood, Jr., and T. F. Deutsch, "Ti laser induced microscopic etching of GaAs and InP" Appl. Phys. Lett. 36, (1980) 698–700.
114. P. Brewer, D. McClure, and R. M. Osgood, "Dry, laser assisted rapid HBR etching of GaAs" Appl. Phys. Lett. 47, (1985) 310–312.
115. M. Takai, H. Nakai, J. Tsuchimoto, K. Gamo, and S. Namba, "Local temperature rise during laser induced etching of gallium arsenide in SiCl₄ atmosphere" Jpn. J. Appl. Phys. 24, (1985) L705–L708.
116. A. W. Tucker and M. Birnbaum, "Laser chemical etching method for drilling vias in GaAs" SPIE 385, (1983) 131–140.
117. D. J. Ehrlich, R. M. Osgood, Jr., and T. F. Deutsch, "Ti laser induced microscopic etching of GaAs and InP" Appl. Phys. Lett. 36, (1980) 698–700.
118. J. E. Bjorkholm and A. A. Ballman, "Localized wet-chemical etching of INP induced by laser heating" Appl. Phys. Lett. 43, (1983) 574–576.
119. A. W. Johnson and G. C. Tisone, "Laser photochemical etching of GaP in KOH aqueous Solutions" Mat. Res. Soc. Symp. Proc 29, (1984) 145–150.
120. J. J. Ritsko "Laser Etching" in Laser Microfabrication—Thin Film Processes and Lithography, J. Y. Tsao and D. J. Ehrlich, Eds. (Academic Press, NY 1989) pg. 363.
121. J. I. Steinfeld, T. G. Anderson, C. Reiser, D. R. Denison, L. D. Hartsough, and J. R. Hollahan, "Surface etching by laser generated free radicals" J. Electrochem. Soc. 127, (1980) 514–515.
122. R. V. Ambartsumyan, Y. A. Gorokhov, A. L. Gritsenko, and V. N. Lokhman "Selectivity of the etching of the basic microelectronics materials through many photon dissociation of SF₆ molecules in an intense CO₂ laser beam" Sov. Tech. Phys. Lett. (USA) 8, (1982) 276–277.
123. S. Yokoyama, Y. Yamakage, and M. Hirose, "Laser induced photochemical etching of SiO₂ studied by X-ray photoelectron spectroscopy" Appl. Phys. Lett. 47, (1985) 389–391.
124. J. M. Gee and P. J. Hargis, "Laser induced etching of insulators using a DC glow discharge in silane" SPIE 459, (1984) 132–137.

125. R. J. von Gutfeld and R. T. Hodgson, "Laser enhanced etching in KOH" Appl. Phys. Lett. **40**, (1982) 352–354.
126. M. Rothschild, C. Arnone, and D. J. Ehrlich, "Excimer laser etching of diamond and hard carbon films by direct writing and optical projection" J. Vac. Sci. Tech. **B4**, (1986) 310–314.
127. P. A. Moskowitz, D. R. Vigliotti, and R. J. von Gutfeld, "Laser micromachining of polyimide materials" In *Polyimides* (Mital, K. Ed.)" Plenum, New York, Vol. **1**, (1984) 365–376.
128. J. H. Brannon, J. R. Lankard, A. I. Baise, F. Burns, and J. Kaufman, "Excimer laser etching of polyimide" J. Appl. Phys. **58**, (1985) 2036–2043.
129. J. H. Brannon and J. R. Lankard, "Pulsed CO_2 laser etching of polyimide" Appl. Phys. Lett. **48**, (1986) 1226–1228.
130. R. Srinivasan and V. Mayne-Banton, "Self-developing photoetching of poly (ethylene terephthalate) films by far ultraviolet excimer laser radiation" Appl. Phys. Lett. **41**, (1982) 576–578.
131. J. E. Andrew, P. E. Dyer, D. Forster, and P. H. Key, "Direct etching of polymeric materials using a XeCl laser" Appl. Phys. Lett. **43**, (1983) 717–719.
132. R. Srinivasan, "Kinetics of ablative photodecomposition of organic polymers in the far ultraviolet (193 nm)" J. Vac. Sci. Tech. **B1**, (1983) 923–926.
133. J. J. Ritsko "Laser etching" in *Laser Microfabrication—Thin Film Processes and Lithography*, J. Y. Tsao and D. J. Ehrlich, Eds. (Academic Press, NY 1989) pg. 372.
134. W. Sesselmann and T. J. Chuang, "Chlorine surface interaction and laser induced shift etching reactions" J. Va. Sci. Tech. **B3**, (1985) 1507–1512.
135. J. E. Andrew, P. E. Dyer, R. D. Greenough, and P. H. Key, "Metal film removal and patterning using a XeCl laser" Appl. Phys. Lett. **43**, (1983) 1076–1078.
136. W. Sesselmann and T. J. Chuang, "Chlorine surface interaction and laser induced shift etching reactions" J. Va. Sci. Tech. **B3**, (1985) 1507–1512.
137. J. E. Andrew, P. E. Dyer, R. D. Greenough, and P. H. Key, "Metal film removal and patterning using a XeCl laser" Appl. Phys. Lett. **43**, (1983) 1076–1078.
138. G. Koren, F. Ho, and J. J. Ritsko, "XeCl laser controlled chemical etching of aluminum in chlorine gas" Appl. Phys. **A40**, (1986) 13–23.
139. J. E. Andrew, P. E. Dyer, R. D. Greenough, and P. H. Key, "Metal film removal and patterning using a XeCl laser" Appl. Phys. Lett. **43**, (1983) 1076–1078.
140. J. E. Andrew, P. E. Dyer, R. D. Greenough, and P. H. Key, "Metal film removal and patterning using a XeCl laser" Appl. Phys. Lett. **43**, (1983) 1076–1078.
141. J. E. Andrew, P. E. Dyer, R. D. Greenough, and P. H. Key, "Metal film removal and patterning using a XeCl laser" Appl. Phys. Lett. **43**, (1983) 1076–1078.
142. W. Sesselmann and T. J. Chuang, "Chlorine surface interaction and laser induced shift etching reactions" J. Va. Sci. Tech. **B3**, (1985) 1507–1512.
143. G. L. Loper and M. D. Tabat, "UV laser generated fluorine atom etching of polycrystalline Si, Mo, and Ti" Appl. Phys. Lett. **46**, (1985) 654–656.
144. W. Sesselmann and T. J. Chuang, "Chlorine surface interaction and laser induced shift etching reactions" J. Va. Sci. Tech. **B3**, (1985) 1507–1512.
145. G. Koren, F. Ho, and J. J. Ritsko, "XeCl laser controlled chemical etching of aluminum in chlorine gas" Appl. Phys. **A40**, (1986) 13–23.
146. J. E. Andrew, P. E. Dyer, R. D. Greenough, and P. H. Key, "Metal film removal and patterning using a XeCl laser" Appl. Phys. Lett. **43**, (1983) 1076–1078.
147. G. L. Loper and M. D. Tabat, "UV laser generated fluorine atom etching of polycrystalline Si, Mo, and Ti" Appl. Phys. Lett. **46**, (1985) 654–656.
148. W. Sesselmann and T. J. Chuang, "Chlorine surface interaction and laser induced shift etching reactions" J. Va. Sci. Tech. **B3**, (1985) 1507–1512.

149. R. L. Jackson, T. H. Baum, T.T. Kodas, D. J. Ehrlich, G.W. Tyndall, and P. B. Comita "Laser deposition: energetics and chemical kinetics" in *Laser Microfabrication—Thin Film Processes and Lithography*, J. Y. Tsao and D. J. Ehrlich, Eds. (Academic Press, NY 1989) pg. 414.

150. J. Y. Tsao, R. A. Becker, D. J. Ehrlich, and F. J. Leonberger, "Photodeposition of Ti and application to direct writing of Ti:LINBO₃ waveguides" Appl. Phys. Lett. 42, (1983) 559–561.

151. D. J. Ehrlich and J. Y. Tsao, "UV laser photodeposition of patterned catalyst films from adsorbate mixtures" Appl. Phys. Lett. 46, (1985) 198–200.

152. J. Y. Tsao and D. J. Ehrlich, "UV-laser photodeposition from surface-adsorbed mixtures of trimethylaluminum and titanium tetrachloride" J. Chem. Phys. 81, (1984) 4620–4625.

153. R. L. Jackson and G. W. Tyndall, "Kinetics and mechanism of laser-induced photochemical deposition from the group 6 hexacarbonyls" J. Appl. Phys. 64, (1988) 2092–2102.

154. D. J. Ehrlich, R. M. Osgood, Jr., and T. F. Deutsch, "Direct writing of refractory metal thin film structures by laser photodeposition" J. Electrochem. Soc. 128, (1981) 2039–2041.

155. N. S. Gluck, G. J. Wolga, C. E. Bartosch, W. Ho, and Z. Ying, "Mechanisms of carbon and oxygen incorporation into thin metal films grown by laser photolysis of carbonyls" J. Appl. Phys. 61, (1987) 998–1005.

156. H. H. Gilgen, T. Cacouris, P. S. Shaw, R. R. Krchnavek, and R. M. Osgood, "Direct writing of metal conductors with near-UV light" Appl. Phys. B42, (1987) 55–66.

157. R. L. Jackson and G. W. Tyndall, "Quartz crystal microbalance measurements of absolute laser photodeposition rates: Application to 257-nm deposition from W(CO)₆" J. Appl. Phys. 62, (1987) 315–317.

158. M. S. Chiu, Y. G. Tseng, and Y. K. Ku, "Pressure-dependent rate saturation in photodeposition from W(CO)₆" Opt. Lett. 10, (1985) 113–115.

159. A. Kitai and G. J. Wolga, "Selective Mn doping of thin film ZnS:MN electroluminescent devices by laser photochemical vapor deposition" In *Laser Diagnostics and Photochemical Processing for Semiconductor Devices*, Osgood, R. M., Brueck, S. R. J., and Schlossberg, H. R., Eds. Mater. Res. Soc. Symp. Proc. 17, (1983) 141–147.

160. C. Arnone, M. Rothschild, J. G. Black, and D. J. Ehrlich, "Visible-laser photodeposition of chromium oxide films and single crystals" Appl. Phys. Lett. 49, (1986) 1018–1020.

161. R. J. Wilson and F. A. Houle, "Composition, structure, and electric field variations in photodeposition" Phys. Rev. Lett. 55, (1985) 2184–2187.

162. F. A. Houle, R. J. Wilson, and T. H. Baum, "Surface processes leading to carbon contamination of photochemically deposited copper films" J. Vac. Sci. Technol. A4, (1986) 2452–2458.

163. C. R. Jones, F. A. Houle, C. A. Kovac, and T. H. Baum, "Photochemical generation and deposition of copper from a gas phase precursor" Appl. Phys. Lett. 46, (1985) 97–99.

164. D. Braichotte and H. van den Bergh, "Schottky diodes and ohmic contacts formed by thermally assisted photolytic laser chemical vapor deposition" Proc. Int. Conf. Lasers '85, (1985) 688–696.

165. R. M. Osgood, Jr. and D. J. Ehrlich, "Optically induced microstructures in laser-photodeposited metal films" Opt. Lett. 7, (1982) 385–387.

166. S. R. J. Brueck and D. J. Ehrlich, "Stimulated surface-plasma-wave scattering and growth of a periodic structure in laser-photodeposited metal films" Phys. Rev. Lett. 48, (1982) 1678–1681.

167. D. J. Ehrlich, R. M. Osgood, Jr., and T. F. Deutsch, "Photodeposition of metal films with ultraviolet laser light" J. Vac. Sci. Technol. 21, (1982) 23–32.

168. D. J. Ehrlich, R. M. Osgood, Jr., and T. F. Deutsch, "Spatially delineated growth of metal films via photochemical pronucleation" Appl. Phys. Lett. 38, (1981) 946–948.

169. J. R. Aylett and J. Haigh, "Laser photolytic deposition on indium phosphide" in *Laser Processing and Diagnostics* Beauerle, D., Ed. Springer Ser. Chem. Phys. 39, (1984) 263–268.

170. R. R. Krchnavek, H. H. Gilgen, J. C. Chen, P. S. Shaw, T. J. Licata, and R. M. Osgood, Jr., "Photodeposition rates of metal from metal alkyls" J. Vac. Sci. Technol. B5, (1987) 20–26.

171. T. F. Deutsch, D. J. Ehrlich, and R. M. Osgood, Jr., "Laser photodeposition of metal films with microscopic features" Appl. Phys. Lett. 35, (1979) 175–177.

172. T. H. Wood, J. C. White, and B. A. Thacker, "Ultraviolet photodecomposition for metal deposition: gas versus surface phase processes" Appl. Phys. Lett. 42, (1983) 408–410.

173. D. J. Ehrlich, R. M. Osgood, Jr., and T. F. Deutsch, "Laser microphotochemistry for use in solid-state electronics" IEEE J. Quant. Electron. QE-16, (1980) 1233–1243.

174. D. J. Ehrlich and R. M. Osgood, Jr., "UV photolysis of van der Waals molecular films" Chem. Phys. Lett. 79, (1981) 381–388.

175. Y. Rytz-Froidevaux, R. P. Salath, and H. H. Gilgen, "Laser-initiated Ga-deposition of quartz substrates" in *Laser Diagnostics and Photochemical Processing for Semiconductor Devices* Osgood, R. M., Brueck, S. R. J., and Schlossberg, H. R., Eds.) Mater. Res. Soc. Symp. Proc. 17, (1983) 29–34.

176. R. M. Osgood and H. H. Gilgen, "Laser direct writing of materials" Am. Rev. Mater. Sci. 15, (1985) 549–576.

177. Q. Mingxin, R. Monot, and H. van den Bergh, "Structure and formation of metal film by deposition of laser chemistry" Scientia Sinica (Ser. A) 27, (1984) 531–539.

178. J. Y. Tsao and D. J. Ehrlich, "Patterned photonucleation of chemical vapor deposition of Al by UV-laser photodeposition" Appl. Phys. Lett. 45, (1984) 617–619.

179. D. Braichotte, K. Ernst, R. Monot, J. -M. Philippoz, M. Qiu, and H. van den Bergh, "Laser induced surface reactions in microelectronic technology" Helv. Phys. Acta 58, (1985) 879–882.

180. J. Y. Tsao, H. J. Zeiger, and D. J. Ehrlich, "Measurement of surface diffusion by laser-beam-localized surface photochemistry" Surf. Sci. 160, (1985) 419–442.

181. T. M. Bloomstein and D. J. Ehrlich, "Laser-chemical three-dimensional writing for micro-electromechanics and application to standard-cell microfluidics" Appl. Phys. Lett. 61 (1992) 708; R. Livengood, P. Winer, J. A. Giacobbe, J. Stinson, and J. D. Finnegan, Conference Proceedings of the 25th ISTFA 99, "Advanced micro-surgery techniques and material parasitics for debug of flip-chip microprocessor" Santa Clara, CA.

182. K. Mutch, Y. Yamada, S. Takeyama, T. Miyata, "Excimer laser beam scanning chemical vapor deposition of ultra-thin multilayer films for soft x-ray mirrors" SPIE Proc. Vol. 2403, *Laser Induced Thin Film Processing*, J. J. Dubowski, Ed. (1995) 270.

183. M Prasad, J. J. Dubowski, and H. E. Ruda, "Laser assisted etching for micropatterning of InP" SPIE Proc. Vol 2403, *Laser-Induced Thin Film Processing*, J. J. Dubowski, Ed. (1995) 414.

184. M. Ishii, T. Meguro, T. Sugano, K. Gamo, and Y. Aoyagi, " Surface reaction control in digital etching of GaAs by using a tunable UV laser system: Reaction control mechanism in layer-by-layer etching" Appl. Surf. Sci, 86 (1995) 554.

185. S. R. J. Brueck and D. J. Ehrlich, "Stimulated surface plasma wave scattering and growth of a periodic structure in laser photodeposited films", Phys. Rev. Lett. 48, No. 24, (1982) 1678; H. Helvajian, "Surface excitation mediated physics in low-fluence laser material interactions and applications to thin-film material processing" SPIE Proc. Vol. 2403, *Laser-Induced Thin Film Processing* J. J. Dubowski, Ed. (1995) 2.

186. V. Ya. Panchenko, V. N. Seminogov, and A. I. Khudobenko, "Utilization of surface electromagnetic wave excitation for increase of submicron diffraction grating depth on n-InP fabricated by holographic we etching" SPIE Proc. Vol. 2403, *Laser-Induced Thin Film Processing*, J. J. Dubowski, Ed. (1995) 394.

187. 3D systems invented the stereolithography process in 1986 (*http://www.3dsystems.com*); K. Ikuta and K. Hirowatari, "Real three dimensional micro fabrication using stereo lithography and metal molding" IEEE Int. Workshop on Micro Electro Mechanical Systems (MEMS93) (1993) 42.

188. Ibid.

189. K. Ikuta, T. Ogata, M. Tsubio, and S. Kojima, "Development of a mass producible micro stereo lithography" Proc. IEEE Int. Workshop on Micro Electro Mechanical Systems (MEMS96) (1996) 301.

190. H.-B. Sun, T. Kawakami, Y, Xu, J.-Y. Ye, S. Matuso, H. Misawa, M. Miwa, and R. Kaneko, "Real three-dimensional microstructures fabricated by photopolymerization of resins through two-photon absorption", Optics Lett. **25**, No. 15 (2000) 1110.

191. M. Albota, D. Beljonne, J.-L. Bradas, J. E. Ehrlich, J.-Y. Fu, A. A. Heikal, S. E. Hess, T. Kogel, M. D. Levin, S. R. Marder, D. McCord-Maughon, J. W. Perry, H. Rackel, M. Rumi, G. Subramaniam, W. W. Webb, X.-L.Wu, and C. Xu, "Design of organic molecules with large two-photon absorption cross sections" Science **281**, (1998) 1653.

192. M Barzoukas and M. Blanchard-Desce, "Molecular engineering of push-pull dipolar and quadrupolar molecules for two-photon absorption: A multivalence-bond states approach" J. Chem. Phys. **113**, No. 113, (2000) 3951.

193. W. W. Hansen, S. W. Janson, and H. Helvajian, "Direct-write UV laser microfabrication of 3D structures in lithium alumosilicate glass" Proc. of SPIE Vol. 2991, pg. 104, 1997.

194. P. Fuqua, S. W. Janson, W. W. Hansen, and H. Helvajian "Fabrication of true 3D micro-structures in glass/ceramic materials by pulsed UV laser volumetric exposure techniques" Proc. SPIE Vol. 3618, pg. 213, 1999.

195. A. Berezhnoi, *Glass-Ceramics and Photo-Sitalls* (Plenum Press, NY, 1970).

196. S. D. Stookey Pyroceram, Codes 9606, 9608, 1st Report, Corning Glass Works, Corning, NY, May 1957.

197. D. Hulsenberg, R. Bruntsch, K. Schmidt, and F. Reinhold, "Micromechanische Bearbeitung von Fotoempfindlichem Glas" Silikattechnik, Vol. 41, (1990) 364; *http://www.mikroglas.com*.

198. W. W. Hansen, S. W. Janson, and H. Helvajian, "Direct-write UV laser microfabrication of 3D structures in lithium alumosilicate glass" Proc. of SPIE Vol. 2991, 1997 pg. 104.

199. P. D. Fuqua, D. P. Taylor, H. Helvajian, W. W. Hansen, and M. H. Abraham, "A UV direct-write approach for formation of embedded structures in photostructurable glass-ceramics" in *Materials Development for Direct-Write Technologies*, D. B. Chrisey, D. R. Gamota, H. Helvajian, and D.P. Taylor, Eds. (Mater. Res. Soc. Proc. **624** (2000), in press.

200. S. W. Janson and H. Helvajian, "Batch-fabricated microthrusters for kilo-gram class spacecraft" 1998 IEEE Aerospace Conference, Paper 093c Mar 21–28, Snowmass Village, CO.

201. D. H. Lewis Jr., S. W. Janson, R. B. Cohen, and E. K. Antonsson, "Digital propulsion" Sens. Actuators, A Vol. 80 (2000) pg. 143.

202. D. G. Sutton, R. S. Smith, P. Chafee, S. Janson, L. Kumar, N. Marquez, J. Osborn, B. Weiller, and L. Weideman "MEMS Space Flight Testbed" AIAA paper 99-4604 Proceedings of AIAA Space Technology Conference Sept. 1999.

203. A. Berzhnoi, *Glass-Ceramics and Photositalls* (Plenum Press, NY, 1970).

204. D. Hulsenberg, R. Bruntsch, K. Schmidt, and F. Reinhold, "Micromechanische Bearbeitung von Fotoempfindlichem Glas" Silikattechnik, Vol. 41, (1990) 364; T.R. Dietrich, W. Ehrfeld, M. Lacher, M. Krämer, B. Speit, "Fabrication technologies for microsystems utilizing photo-etchable glass" Microelec. Eng. (1996) 497–504.

205. P. D. Fuqua, D. P. Taylor, H. Helvajian, W. W. Hansen, and M. H. Abraham, "A UV direct-write approach for formation of embedded structures in photostructurable glass-ceramics" in

Materials Development for Direct-Write Technologies, D. B. Chrisey, D. R. Gamota, H. Helvajian, and D. P. Taylor, Eds. Mater. Res. Soc. Proc. **624** (2000), in press.

206. H. Helvajian and S. W. Janson, "Batch-fabricated microthrusters: Initial results" AIAA-96-2988 (1996); S. W. Janson and H. Helvajian, "Batch-fabricated microthrusters for kilo-gram class spacecraft" Proceedings of the GOMAC '98 Conference *Micro-Systems and Their Applications*, March 16–19, 1998, Arlington, VA.

207. D. Lewis, S. Janson, R. Cohen, and E. Antonsson, "Digital micropropulsion" Sens. Actuators, A; Physical, 2000, 80(2), pp. 143–154; S.W. Janson, H. Helvajian, and K. Breuer, "MEMS, Microengineering and Aerospace Systems" AIAA paper 99-3802, 30th AIAA Fluid Dynamics Conference, Norfolk, VA, June 1999.

208. D. Lewis, S. Janson, R. Cohen, and E. Antonsson, "Digital micropropulsion" Proceedings of the IEEE MEMS '99, Jan. 1999.

209. J. V. Berry, R. E. Conger, and J. R. Wertz, "The sprite mini-lift vehicle: Performance, cost and schedule projections for the first of the Scorpius low-cost launch vehicles" 13th Annual AIAA/USU Conference on Small Satellites, Logan, Utah, August 1999; *http://www. design. caltech.edu/micropropulsion/index.html*.

210. J. T. Dickinson, "The role of defects in the laser ablation of wide bandgap materials" Nucl. Inst. Meth. Phys. Res. B, Vol. 91, (1994) 634.

211. S. Kuper and M. Stuke, "UV-excimer-laser ablation of polymethyl methacrylate at 246 nm: Characterization of incubation sites with Fourier ransform IR- and UV spectroscopy," Appl. Phys. A: Solids Surf. **49** (1989) 211.

212. G. B. Blanchet, P. Cotts, and C. R. Fincher, Jr. "Incubation: Subthreshold ablation of poly-(methyl-methacrylate) and the nature of the decomposition pathways" J. Appl. Phys. **88**, No. 5, (2000) 2975.

213. J. Wang, H. Niino, and A. Yabe, "Micromachining of transparent materials by laser ablation of organic solution" SPIE Proc. 1st Int. Symp. on Laser Precision Microfabrication, I. Miyamoto, K. Sugioka, T. W. Sigmon, Eds. Vol. 4088 (2000) 64.

214. T. Lippert, "Laser micromachining of chemically altered polymers" SPIE Vol. 3274 *Laser Applications in Microelectronic and Optoelectronic Manufacturing II*, J. J. Dubowski and P. E. Dyer, Eds. (1998) 204.

215. *Materials Development for Direct-Write Technologies*, D.B. Chrisey, D.R. Gamota, H. Helvajian, and D.P. Taylor, Eds. Mater. Res. Soc. Proc. **624** (2000) in press.

216. G. Timp, "Is atomically precise lithography necessary for nanoelectronics?" in *Physics of Nanostructures* J. H. Davies and A. R. Long, Eds. Proc. Of 38th Scottish Univ. Summer School in Physics, St. Andrews, July 1991, A NATO Advanced Study Institute, (Instit. of Physics Publishing, Philadelphia, PA, 1992) pg. 101.

217. Y. Shen, C. S. Friend, Y. Jiang, D. Jakubczyk, J. Swiatkiewicz, and P. N. Prasad, "Nanophotonics: Interactions, materials and applications" J. Phys. Chem. B, Vol. 104 (2000) 7577.

218. D. Chrisey, "Matrix assisted pulsed laser evaporation direct-write (MAPLE_DW)" in *Materials Development for Direct-Write Technologies*, D. B. Chrisey, D. R. Gamota, H. Helvajian, and D. P. Taylor, Eds. Mater. Res. Soc. Proc. **624** (2000), in press; H. Helvajian "Laser material processing: A multifunctional in-situ processing tool for microinstrument development" in *Microengineering for Space Systems* The Aerospace Press Monograph 97–02 (1997) pg. 67.

CHAPTER 15

Flow- and Laser-Guided Direct Write of Electronic and Biological Components

MICHAEL J. RENN, GREG MARQUEZ, BRUCE H. KING, MARCELINO ESSIEN, AND W. DOYLE MILLER

Optomec, Inc., Albuquerque, New Mexico

1. Motivation
2. Fundamentals
3. Material Results
4. Electronic Components
5. Future Work
6. Conclusion

1. MOTIVATION

There is a constant requirement in the electronics industry to reduce the overall size of electronic systems. Over the past 40 years, the electronics and telecommunications industries have repeatedly demonstrated an unprecedented pace of innovation. Each significant advance has been predicated on a step-wise increase in computing power that itself is founded upon the ongoing miniaturization of the underlying electronic systems.

In many instances, it has been the semiconductor industry that has led this charge. As Moore's Law predicted, on-chip functions and the resulting speed have been doubling roughly every 18 months. Unfortunately, much of the supporting electronics infrastructure, from chip packaging to printed circuit boards, has had difficulty keeping pace. As a result, the industry as a whole now faces a critical juncture where the limitation to progress is not simply the chip itself.

Until now, improving integrated circuit (IC) packaging technologies has long been viewed as secondary to the advantages and savings that are realized by shrinking the wafers themselves. However, the Moore's Law effect, without comparable gains in packaging technology, has created a situation where packaging has become a relatively expensive portion of chip production costs.

For example, it is estimated that semiconductor manufacturers have been able to shrink the functional costs of chips more than a thousandfold over the years. However, chip packaging costs have only decreased from roughly two cents per "pin" (or I/O contact point) down to a half cent per pin. The ITRS roadmap projects this trend will continue, and by 2005, I/O count will increase fourfold, while the cost per pin will decline by only about 25%.

Packaging is but one area where direct-write technologies will have impact. Another is the replacement of discrete surface mount devices on printed circuit boards by direct-write components. It is estimated that 70% of the area of a printed wire board is due to discrete passive components. In addition, by direct writing interconnects between the passive devices, the need for solder joints will be eliminated and the entire package will become significantly more robust.

Similarly the ability to deposit living material, cells, bacteria, and proteins with micron accuracy promises to revolutionize tissue engineering, and biochip and biosensor construction. Volumes of a droplet (1 fL) dispensed by laser direct-write order of magnitude smaller than the mechanical microspotting and microjet ($\sim$1 nL) dispensing technologies. By depositing liquid droplets containing biomolecules such as proteins or nucleic acids, high-density arrays can be generated. Particle fluxes as high as 10,000 Hz with simultaneous placement below five micrometers have been achieved. This compares to 1 Hz for microspotting. A 10,000 address microarray using microspotting requires about 3 cm^2, while laser-guided direct writing would require less than 1 mm^2 with a 10 μm spot size. An example of such a pattern generated using glycerol is shown in Fig. 1 below.

The long-term preservation of tissue-specific functions is important if engineered tissue is to successfully compensate for organ failure. A number of studies have demonstrated the importance of three-dimensional structures on the behavior of cells in culture. For example, hepatocytes cultured as a monolayer lose many of their liver-specific functions within a few days. However, when these same cells are overlaid with a collagen gel to mimic the three-dimensional structure of the liver, they retain many of their liver-specific functions for weeks in culture. Therefore, the ability to spatially organize cells into well-defined three-dimensional arrays that closely mimic native tissue architecture can facilitate the fabrication of engineered tissue. Laser-guided direct writing potentially has this capability.

Regardless of the application, laser-guided direct writing has many advantages over existing methods for surface patterning. In contrast to optical trapping (1), laser-guided direct-writing allows particles to be captured

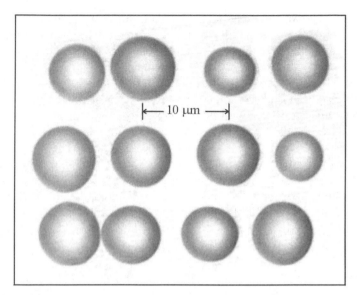

FIGURE 1 Optical micrograph of glycerin droplets deposited by the LGDW approach. Each droplet consists of 5–7 glycerin particles deposited to an accuracy of 2.4 ± 1.0 μm.

continuously from the surrounding fluid and directed onto the substrate. In comparison to photolithography, the process adds material to the surface (as opposed to etching material) and does not require harsh or corrosive chemicals. In contrast to robotic microspotting, deposition, inkjetting, and screen-printing, particles are strongly localized within the laser beam and the deposition accuracy can be below one micrometer. Most importantly, nearly any material in either liquid or aerosol suspension can be captured and deposited as long as convection and gravity are weaker than the guidance forces (typically in the nanonewton range). Potentially, many types of materials can be co-deposited on a single substrate, which will allow simultaneous deposition of both electronic and biological particles.

This chapter summarizes recent results including micron-scale deposition techniques, material developments, laser treatments of various metals and dielectrics, and component performance.

2. FUNDAMENTALS

Flow-guided and laser-guided direct-write (FGDW and LGDW respectively) technologies (2,3) are laser-based processes for dispensing and processing liquid and colloidal materials on virtually any substrate. Shown schematically in Fig. 2, the first step to both deposition processes is to generate an aerosol from a starting liquid precursor or colloid suspension. For pure liquids, liquid solutions, and liquids that contain small colloids (<0.3 micron diameter) the preferred generation method is ultrasonic atomization. A variety of ultrasonic atomization devices are commercially available, most notably the ultrasonic atomizers utilized for room humidification. At a drive frequeny of 1.6 MHz the mean particle size is 2 microns and volume is approximately 1 fL. Optomec aerosol generation units are optimized for extreme stability (<3% variation in output) particle generation under 24/7 operating conditions. It must be noted that only smaller solid particles (<0.3 microns) can be efficiently atomized with this approach. Larger colloids and living biological cells (1–20 microns) are atomized by pneumatic nebulization. A common example of these devices is the hand-held nebulizers used for respiratory therapy. In both generation

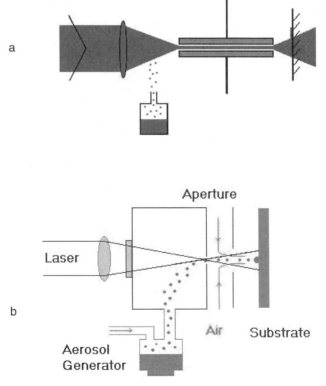

FIGURE 2   (a) Schematic of a laser-guided direct-write (LGDW) system. (b) Flow-guided direct write (FGDW) system.

methods, the output of the atomizer is a dense mist of droplets that contain the material's interest. The particles are fed into either the flow-guided or laser-guided deposition devices and focused by the optical and hydrodynamic forces to a narrow beam.

The laser-guided deposition device utilizes laser-induced optical forces to focus particles into a narrow beam. The particle focusing operates on the principle of momentum transfer from the laser beam to the particle. A one-watt laser focused onto a particle can exert up to 3 nN of force along the beam direction. Two forces act on the particles. Light which is back-scattered from the particle causes the particle to accelerate in the direction of the beam. This force is commonly referred to as radiation pressure (or optical wind). A substantial portion of the laser light penetrates through the particles and is scattered at small angles in the forward direction. The small-angle scattering gives rise to forces perpendicular to the laser beam. For most materials, these forces are such that the particle is pulled to the center of the beam where the intensity is maximal. The combination of axial and radial forces causes particles to be drawn into the center of the beam and pushed in the direction of beam travel.

As depicted in Fig. 2a, the laser beam is coupled into a hollow optical fiber to allow particle guidance over millimeter to centimeter distances. While most optical fibers have a solid core, hollow-core optical fibers (developed for infrared laser delivery) permit transmission of both light and particles through the central hollow region. In the hollow fiber case, the laser light reflects from the walls at a grazing angle of incidence. The reflectivity is not total, as in the solid fiber case, but the light can be guided over several cm in a 20-micron diameter core without substantial loss. The guided-laser beam, in turn, induces the optical forces that guide the particles. The deposition rate is adjustable from 1 particle per second up to 10,000 per second.

The hollow fiber system offers several other advantages besides long-distance light guidance. First, natural convective fluid motion is often large enough to overwhelm optical forces, making free space particle guidance difficult. Hollow fibers alleviate this problem because the fiber interior provides a quiescent environment shielded from the external surroundings. In fact, the fiber exterior can be exposed to air currents (or even to a vacuum) and the particles within are not disturbed. Second, the intensity profile inside the fiber is well defined with the intensity being maximal at the radial center and zero at the fiber wall. The intensity gradient draws particles toward the radial center of the fiber and keeps them from adhering to the fiber walls. Third, the fiber also allows the source and deposition regions to be isolated from each other, assuring that the direct-write patterns are not contaminated by nonguided particles. Fibers from several source chambers can be coupled to the same deposition chamber for co-deposition of multiple materials.

For throughput higher than 10,000 per second, an aerosol assist method is used to increase particle velocity (from 1 cm/s to 10 m/s). This method, called Flow-Guided Direct Write (FGDW), is drawn schematically in Fig. 2b. In FGDW, particles are fed into a sealed chamber by a carrier gas and directed through a millimeter size orifice. The particles emerging from the orifice are combined with a second air stream, such that the particle stream forms the core of the combined stream and the second air stream forms a cylindrical sheath around the aerosol. The combined streams are then forced through a second, submillimeter-sized orifice. The emerging particle stream is focused to significantly smaller diameter than the physical orifice size (approximately 5–10 times smaller). The small beam diameter is maintained without significant divergence over several millimeters from the orifice. The mass throughput is controlled by the aerosol carrier gas-flow rate, while the beam diameter is controlled primarily through the sheath air-flow rate. By adjusting these two parameters, the beam diameter and mass throughput is controlled.

The particle materials are deposited on a variety of substrate materials including alumina, glass, polyimide, barium titanate, plastics, and metals. The physical linewidth depends on several parameters and can approach 1 micron under optimal conditions. One parameter is the particle size. Particles larger than around 0.1 micron follow a straight-line trajectory from the outlet orifice and physically impact the substrate. In this case the deposition width is approximately that of the particle-beam diameter and the profile mirrors the radial distribution of particles within the beam. However, particles smaller than 0.1 micron may have insufficient momentum to impact the substrate in the FGDW process. These particles tend to follow the radial shear flow and are either deposited as satellites outside the deposition zone or are swept away from the substrate. The second parameter affecting lineshape is the viscosity of the droplets. Completely dried droplets, such as solid barium titanate particles, will not flow tangentially to the substrate after physical impaction. Low-viscosity liquids, such as water, will impact the substrate and then flow tangentially depending on substrate wetting conditions and the total amount of material deposited. A key factor to obtaining crisp edge definition is to dry the salt- and particle-laden droplets before deposition, that is, on the fly. In this case the material is deposited as a viscous paste that flows very little (see Table 1).

Once the materials are physically deposited they are then treated either chemically, thermally, or with the laser to form the final desired product. The electronic materials, in particular, require a postdeposition thermal treatment in order to form fully dense, high-quality electronic material. It is desirable that the thermal treatment be performed either at low temperature ($< 200\,^\circ$C) or in a very short time in order to not damage the temperature-sensitive substrate materials. Several low-temperature, metal-organic precursors have recently

TABLE 1   Physical Characteristics of Laser- and Flow-Guided Direct Writing[a]

	Accuracy	Volumetric throughput	Compatible substrates	Compatible materials
LGDW	2 μm	10^{-4} mm³/s	Nonabsorbing glass, plastics, and ceramics	Nonabsorbing droplets, solid particulates, and biological materials
FGDW	25 μm	0.25 mm³/s	Unlimited	Atomizable fluids, colloids, and biological materials

[a] LGDW is limited to primarily nonabsorbing materials and substrates because of laser-induced convective heating.

been developed by Superior MicroPowders, Paralec, Inc., and others. These precursors, amenable to sub-200 °C treatments on various substrates, result in metal deposits when decomposed. To date the low-temperature chemistries developed for metals have not been extended to other classes of materials, such as dielectrics, ferrites, and resistors. For these materials, a sintering step is accomplished by scanning a focused laser beam across the deposited material. In contrast to traditional electronic material processing involving high temperatures ($\sim$1,000 °C) and long time scales (minutes to hours), laser treatments in this process occur at low temperatures ($\sim$200 °C) and short time scales ($<$10 ms). As a result, the deposits are successfully treated on temperature-sensitive substrates such as FR-4 and polyimide.

In the case of biological materials, the laser is not used to treat the material, only to deposit. A major concern is that the laser does not degrade cell viability during transport. Initial studies with the deposition of embryonic chick spinal cord cells found that individual cells (diameter $=$ 9 μm) could be guided by a 450 mW near-infrared laser beam and deposited in arbitrarily defined arrays onto a glass target surface (4,5). The cells that were exposed to the light remained viable and grew normal-appearing neurites. A second concern is that the cells remain viable during the aerosol generation processes. In this regard, ultrasonic atomization apparently causes the cell membranes to rupture and only cell debris is deposited. Pneumatic atomization, on the other hand, leads to very high viability. Shown in Fig. 3 is a sequence of optical micrographs of pneumatically atomized 3T3 mouse fibroblast cells deposited on a treated culture flask (nuclon Δ treatment). The cells have been cultured for various time increments over the course of 72 hours. The cells deposit as round spheres and after approximately 20 minutes they begin to flatten out and adhere to the substrate. Nonadhered cells are considered nonviable. Approximately 87% of these cells are found to be viable compared to 97% of cells in a nonaerosolized control culture. This indicates that the vast majority of cells are

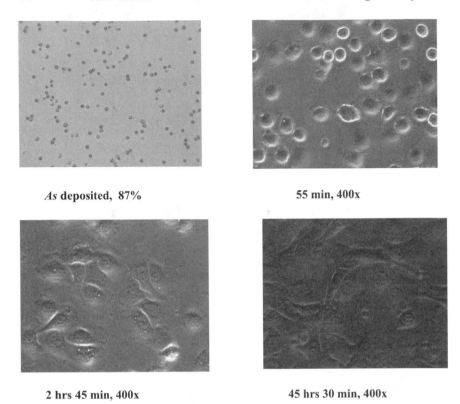

As deposited, 87%   55 min, 400x

2 hrs 45 min, 400x   45 hrs 30 min, 400x

FIGURE 3   Micrographs showing the growth properties of FGDW-deposited 3T3 mouse fibroblast cells. *As deposited* the cells are spherical in shape. During culturing the cells flatten and adhere to the substrate. Nonviable, detached cells account for less than 13% of the deposited cells. After two days a confluent culture is reached.

viable after atomization. Over longer times the cells flatten and extend fibrous outgrowths (possibly collagen). The cells multiply several times before reaching a confluent culture at 72 hours. As mentioned, laser direct writing potentially allows the three-dimensional patterning of cells using multiple cell types with cell placement at arbitrarily selected positions. Efforts toward this goal are continuing.

3. MATERIAL RESULTS

Most of the metal precursors are simple organic and inorganic metal salts, such as silver nitrate, dissolved in a suitable solvent. Provided the precursors have

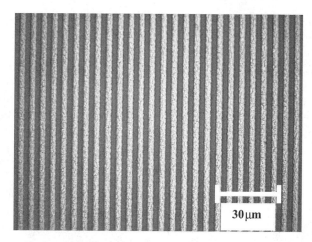

FIGURE 4   Optical micrograph of Pt lines written on alumina. The lines are 2 microns wide and are spaced by 5 microns.

the necessary rheological properties, they are easily atomized and deposited as viscous droplets. As mentioned, it is desirable to deposit the materials as a thick paste. If the material viscocity is too low, the line definition will be poor. If the particles are too dry, the deposits will be loosely connected and extremely porous. Both extremes lead to degraded electronic performance. Precursors of several metals including Pt, Au, Cu, Ag, and Pd have been successfully developed. The discrete lines of platinum on an alumina substrate in Fig. 4 show the excellent uniformity and line density that can be achieved. These lines are 2 microns wide and separated by 5 microns. X-ray-diffraction studies have shown that all these materials, when treated either thermally or with the laser, are converted to metal. The resistivities are generally less than $10\times$ the resistivity of bulk material, but often are closer to $2\times$ bulk. The electrical resistivity measurements on various substrates are summarized in Table 2.

TABLE 2   Table of Resistivity Values for Well-Adhered Conductive Traces on Various High- and Low-Temperature Substrates[a]

	Alumina	Kapton	FR4	Teflon	Glass	Epoxy	Silicon
Ag		3.0	3.0	2.4	3.0		
Au		4.5					
Cu		3.4			3.1		
Pt	20	32	53		31	32	21

[a] The resistivities are in units of 10^{-8} Ohm-m.

Adhesion to the substrate varies but generally is very good. In particular, Pt has been shown to deposit and adhere well to Kapton, glass, $BaTiO_3$, FR-4, and Si.

Compatible precursors may be mixed to produce alloys. For example, Pt–Rh alloys have been deposited and shown to exhibit thermocouple behavior, as discussed in the devices section. The short-time-scale laser treatments are particularly advantageous for processing air-sensitive materials such as copper and palladium. During short times, little oxygen from the ambient environment diffuses to the heated metals to cause oxidation. We routinely treat copper precursor under ambient conditions to form dense metallic copper. XRD reveals no apparent oxidation in the bulk, although a thin surface oxide is usually evident. Similarly, laser treatments are effective in treating silver-palladium precursor to form oxide-free alloys. The oxides readily form by oven treatment but not by laser treatment.

While various metal precursor systems have been developed and shown to yield dense metals, analogous precursors for dielectric, ferrite, and resistor systems have not yet been developed. In these cases the preferred approach is to deposit colloidal suspensions of ceramic powder along with a low-temperature binder material. The powders are typically high-quality single-phase particles produced by various traditional processing methods. The binder material is typically a low-melting-temperature low-loss glass. A particle size of <300 nm works best for ultrasonic atomization, while sizes of 0.3 to 5 microns are best for pneumatic atomization. In some cases, such as with dielectrics and ferrites, the final material performance is related to the particle size of the starting materials. Pneumatic atomization is then the preferred approach. The binder and powder materials are intended to have vastly different melting temperatures. For example, $BaTiO_3$ melts at approximately 1,600 °C, whereas glass-softening temperatures can be as low as 400 °C. The deposits are treated by heating with a laser to melt the glass. The local temperature is sufficient to melt the glass but is low enough to not affect the high-temperature material. The treatment is short enough in duration that the underlying substrate is not damaged. The resulting deposits are well adhered to underlying substrates and electrodes, and pass qualitative scrape and tape tests.

An example of a laser-treated $BaTiO_3$/glass deposit is shown in Fig. 5. The deposit consists of a bimodal distribution of $BaTiO_3$ (1 micron and 0.1 micron) as well as low-temperature glass particles. As evidenced by the SEM micrograph, the laser has melted the glass and fused the barium titanate. The dielectric layer in this case is 2 microns and the measured dielectric constant is 160. The loss tangent at 500 kHz is 0.9%. The layer thickness for this deposit is significantly below that obtainable by screen printing, approaching the thickness obtainable via thin-film-processing routes. The capability to deposit thin layers gives rise to high capacitance per unit area, in this case, $C/A = 7.2 \times 10^{-4} F/m^2$. However, thin-film approaches typically involve high-

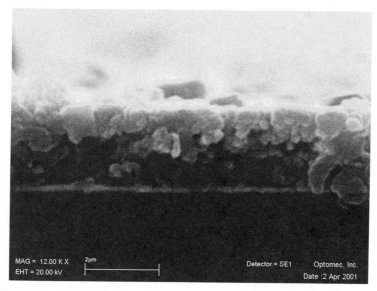

FIGURE 5   SEM micrograph of BT/glass. The 2-micron-thick layer has a capacitance per unit area
$C/A = 7 \times 10^{-4}$ F/m^2 and a loss of 0.8% @ 500 kHz.

temperature processing, limiting their use to substrates that can withstand the
processing temperature. The direct-write approach has been applied to
substrates such as Kapton and solder mask.

Similar to the dielectrics, traditional metal-oxide resistor material, RuO_2,
has been co-deposited with glass and densified. Dense ruthenate resistors on
glass, Kapton, FR4, and other substrates have been demonstrated. The
resistivity of the RuO_2-glass system spans several decades and is centered at
10 Ohm-m, 7 orders more resistive than conductor metals.

Resistors with resistivity intermediate to the conductors and ruthenates are
fabricated from metal precursors and high-temperature glass. In this case the
glass does not act as an aid to densification. Instead, the metal precursor
decomposes around the glass which provides insulating inclusions in the
metal. Resistivities are shown to span from that of good conductor
($\sim 3 \times 10^{-8}$ Ohm-m) with no glass additive to 4 orders greater with 30
volume % glass. Figure 6 shows an optical micrograph of a 5 kOhm silver/glass
resistor. The resistor trace is 60 microns wide as has been terminated with
contact pads of Ag–Pd alloy. The inset shows the excellent connectivity
between the resistor trace and contact pad. Glass loadings greater than 30%
per volume of metal generally lead to the formation of brittle resistor traces

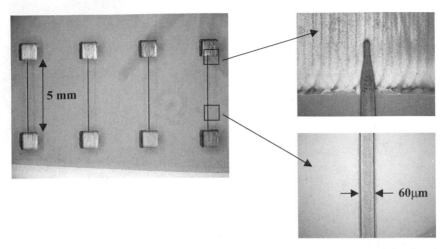

FIGURE 6   Optical micrographs of silver/glass resistor traces. The traces are terminated with pads of silver/palladium alloy (18% palladium by volume). The resistance is 6.0 kOhm, the length is 5 mm, and width is 60 μm.

that are not well adhered. The preferred system for higher resistivity is the aforementioned Ruthenate-glass system.

4. ELECTRONIC COMPONENTS

With the materials presented, several electronic components have been produced. These have primarily been made using either Ag, Cu, or Pt precursors deposited on either glass, Kapton, or FR4, and include interconnects, capacitors, thermocouples, and antennae. Because the process is capable of writing 10 μm wide conductor traces with ~2× bulk resistivity, it is ideally suited to produce interconnects. An example of interconnects for bond pad redistribution is shown in Fig. 7.

The conductor traces in this case are 35 μm wide with a 70 μm pitch. Another device, which has been demonstrated, is an antenna, constructed of copper on Kapton. Shown in Fig. 8, the antenna is a 1.5-turn Archimedean spiral that starts with a 150 μm wide strip and tapers to a 10 μm line. This antenna operates with a center frequency of 1.35 GHz and an efficiency of 99.8%. Compared to a traditional patch antenna, the stripline antenna is approximately 1,000 times smaller in volume.

Figure 9 shows a direct-write pattern of silver electrodes on a barium neodymium titanate block for an RF filter application. The frequency sweeps for three direct-write filters on both receive and transmit bands are shown

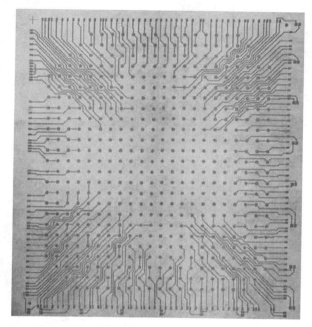

FIGURE 7   Cu interconnects on Kapton for bondpad redistribution (35 µm wide lines and 70 µm pitch).

below the micrograph. Compared to a reference filter, fabricated by screen printing, the direct-write filters perform very closely to the component that was processed with traditional processes.

Pt and a Pt–10%Rh alloy have been used to demonstrate direct-write thermocouples. The output voltage of one of these thermocouples is shown in Fig. 10 along with the output voltage of a standard Pt/Pt–Rh thermocouple. Overall, the output voltage follows the standard curve very well. The discrepancy between the measured and expected curves is attributed to the test apparatus; a reference junction was not used, accounting for the offset, and the temperature was not measured directly at the Pt/Pt–Rh junction.

5. FUTURE WORK

Although much of the work with the Optomec Direct Write process has focused on electronic devices, micron-scale structures have also been demonstrated. Figure 11 shows a series of posts that have been built using the process. The posts are silver and built, on glass, by alternatively depositing

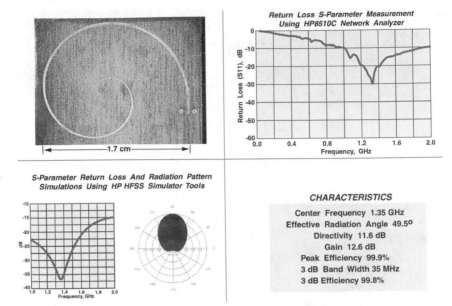

FIGURE 8   Spiral Archimedean antenna fabricated by depositing 3-micron-thick copper onto 3 mil Kapton. The copper strip tapers from 150 microns at the beginning to 10 microns at the center. The 2 via holes were drilled with the laser and filled with Cu. The copper ground plane was written by sequential rastering. The measured operating frequency (upper right) is in excellent agreement with the simulated performance (lower left). The summary characteristics are shown in the lower right box.

silver precursor and then decomposing the precursor with the laser. These posts are 80 μm in diameter and have an aspect ratio of 25 : 1.

Commercialization of laser direct writing is the primary goal of the project. The commercial alpha tool is shown in Fig. 12 along with various attributes. The system will be available for rapid prototyping and rapid manufacturing applications for electronic and biological applications.

6. CONCLUSION

A new, laser-based direct-write process has been developed for direct writing electronic components and mechanical structures with extreme accuracies and in a maskless process. This is truly "art to part." This process is capable of depositing a wide range of metals, ceramics, oxides, and biological materials

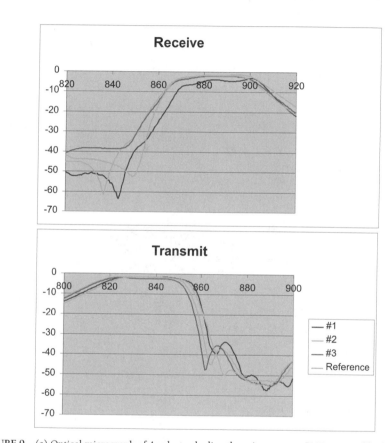

FIGURE 9 (a) Optical micrograph of Ag electrode directly written onto a BNT ceramic block for a wireless RF filter application. (b) and (c) show the RF response function of three direct-write filter components compared to a reference component. The reference was fabricated by current screen printing and trimming processes. As can be seen the direct-write components compare favorably and direct write has the additional advantage of being a maskless, single-step process.

490

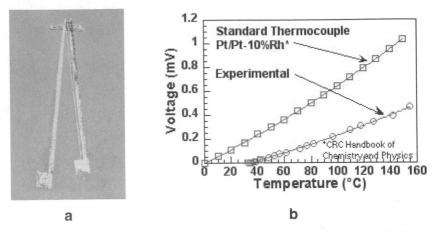

a   b

FIGURE 10 (a) Pt/Pt–Rh thermocouple and measured voltage response. The Pt/Rh alloy was made by mixing compatible precursors and writing the lines by FGDW. (b) Measured emf when the junction is heated with a resistor wire. The CRC handbook values are shown for comparison. The measured emf increases with temperature, similar to a normal thermocouple. The lack of a 0 °C reference junction leads to the offset between the measured and handbook values. The difference in slopes is attributed to heating of the probe pads during the measurement.

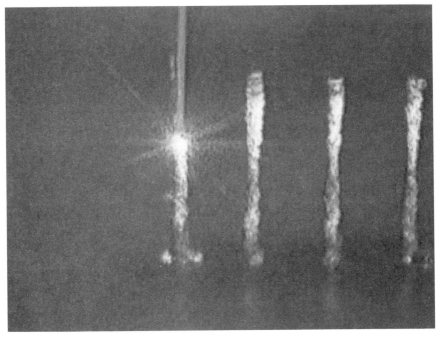

FIGURE 11   Silver posts built by FGDW (80 μm diameter and 25 : 1 aspect ratio). Silver precursor is deposited and simultaneously treated by the laser to form the posts.

Machine Specifications

- Inline Conveyor System
- 12" × 12" (305 mm × 305 mm) Deposition Area, Auto-Vision Camera and Contact Surface Sensor
- Z-Axis Head for Multiple Deposition
- Composite Frame Construction
- X, Y, Z Overhead Gantry
- 3 Axis Brushless Servo Motors with Closed-Loop Encoders
- Small Footprint—36.5"W × 47"D × 59.5"H
- Weight—1,700 lb (771 kg)
- Utilities—220 VAC, 30 A, Compressed air

FIGURE 12   The overall system provides a work area of 12″ × 12″. This size work area is ideal for board manufacturing, boats/carriers and individual components. The overall footprint of the system is 36.5″ × 47″ × 59.5″ and will ultimately incorporate multiple material capability for electronic and biological materials (courtesy Optomec, Inc., Albuquerque, New Mexico).

with micron-scale feature size. High material throughput has been demonstrated, which makes the process attractive for both rapid prototyping and rapid manufacturing of components and circuits.

Laser-guided direct writing is an emerging technology for high-throughput deposition of micrometer-, and submicrometer-sized particles. This is a simple system that can be set up at low cost and will deposit virtually any material with micrometer-scale accuracy. Multiple applications are anticipated in tissue engineering, hybrid electronic/biological systems, biochip array fabrication, and basic scientific research.

Ultimately, perhaps the most intriguing potential for direct-write technologies is to enable the combination of many of today's disparate electronics manufacturing processes into an integrated production system. This system would coincidentally manufacture the printed circuit board and necessary components, and as well perform the packaging and assembly functions, eliminating many costly steps and serving as the gateway to fully three-dimensional electronic systems and hybrid electronic/biological systems.

ACKNOWLEDGMENT

This work is supported by the Defense Advanced Research Projects Agency under contract #N00014-99-C-0258.

REFERENCES

1. Ashkin, A. (1970) *Phys. Rev. Lett.* 24, 156–159.
2. Renn, M. J. & Pastel, R. (1998) *J. Vac. Sci. Technol.* B16, 3859–3863.
3. Renn, M. J., Pastel, R. & Lewandowski, H. (1999) *Phys. Rev. Lett.* 82, 1574–1577.
4. Odde, D. J. & Renn, M. J. (1998) *Ann. Biomed. Eng.* 26, S-141.
5. Odde, D. J. & Renn, M. J. (2000) *Biotechnol. Bioeng.* 67.

CHAPTER 16

Laser-Induced Forward Transfer: An Approach to Single-Step Microfabrication

IOANNA ZERGIOTI,* G. KOUNDOURAKIS,* N. A. VAINOS,*
AND C. FOTAKIS†*

*Foundation for Research and Technology-Hellas (FORTH), Heraklion, Crete, Greece,
†Department of Physics, University of Crete, Greece

---

1. An Overview of the Laser-Induced Forward
   Transfer Process
2. Deposition of Single Elements
3. Deposition of Oxide Compounds
4. Transfer Mechanisms
5. Applications of LIFT
6. Summary and Conclusions

1. AN OVERVIEW OF THE LASER-INDUCED FORWARD TRANSFER PROCESS

Laser-induced forward transfer (LIFT) denotes the selective forward ablation and deposition of materials using lasers (1). This technique usually utilizes pulsed lasers to remove a thin film of target material from a transparent supporting plate and deposit it onto a receiver substrate. The thin-film target, which has been previously prepared on a transparent substrate, is placed in close proximity to the receiver. The effectiveness of LIFT relies on defining the critical parameters of the laser-solid interaction. Its optimization depends on the specific optical, thermophysical, and mechanical properties of the materials involved. Various metals and oxides have been used in LIFT applications, together with a variety of laser sources, from the near infrared to the

Direct-Write Technologies for Rapid Prototyping Applications

ultraviolet. In most cases, transfer of material is achieved using single laser pulses, although it has also been demonstrated by the use of continuous wave (cw) lasers.

The general principle of the LIFT method is outlined in Fig. 1. The target material is deposited on a quartz wafer or other laser-transparent substrate. The distance between the target and the receiver substrate can be varied from near-contact to several micrometers. Vacuum, low-pressure gas or ambient air conditions have been used in transfer experiments. Laser light is focused at the thin-film/receiver interface, vaporizing a fraction of the target material. The fraction vaporized depends on the laser wavelength, laser intensity, and the optical extinction coefficient of the target. During this process, the high vapor pressure formed expels the remaining material at a high speed, prior to complete vaporization. This picture seems to apply well in the case of most metal targets, and agrees with a "ballistic" model of laser ablation (3). In this latter approach, melting and vaporization are initiated at the metal target/substrate interface. The melt-front propagates from the interface to the free surface of the film. Within a short time scale, mechanical forces exceed the mechanical strength of the film and a violent lift-off process takes place. A mixture of solid and melt is ejected at high speed and impacts on the receiver substrate. This picture provides a qualitative insight of the process dynamics applicable at least for metallic "targets" at relatively low laser energy density.

This chapter reviews research efforts made in this field, placing emphasis on cases of potential technological interest. An overview of this work is outlined in Table 1. Following an historical account of LIFT, selected examples of current developments having technological interest will be presented. The potential of the technique will be discussed together with its advantages and inherent impediments.

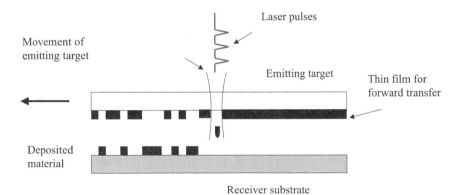

FIGURE 1   Scheme of the Laser-Induced Forward Transfer method.

TABLE 1 Selected Works on Laser-Induced Forward Transfer Material Deposition

Transferred material	Substrate	Feature size (μm)	Laser type (λ)	Reference	Ref. no.
Cu	Silicon	50	ArF (193 nm)	Bohandy (1986)	[1]
Cu	Silica	50	ArF (193 nm)	Adrian (1987)	[3]
Cu, Ag	Fused silica	15	2ω Nd:YAG (532 nm)	Bohandy (1988)	[2]
Al	Silicon	200	Nd:YAG (1.06 μm)	Schultze (1991)	[10]
Au	Quartz, glass		Nd:YAG (532 nm), KrFl (248 nm)	Baseman (1990)	[27]
Ti, Cr	Glass	10	Ruby (694 nm)	Kantor (1992)	[6]
W, Cr	Glass	10	Ar$^+$ (515 nm), Nd:YAG (1.06 μm)	Toth (1993)	[4]
YBaCuO	MgO, silicon	100	ArF (193 nm), Nd:YAG (1.06 μm, 5 nsec), 2ω Nd:YAG (530 nm, 100 nsec)	Fogarassy (1989)	[18]
BiSrCaCuO					
Au, Al	Silicon, quartz	7–10	ArF (193 nm)	Latsch (1994)	[8]
Pd	Quartz, ceramics, polymers	150	ArF, KrF, KrCl, XeCl, XeF (193 nm, 248 nm, 222 nm, 308 nm, 351 nm)	Esrom (1995)	[9]
Diamond	Silicon	10	Cu laser (510 nm, 20 ns, 10 kHz), KrF laser (248 nm, 15 ns)	Pimenov (1995)	[17]
C, Cr	Glass	20	Nd:YLF (1,047 nm)	Tam (1996)	[7]
Cr, Pt, Mo, In, In$_2$O$_3$	Glass, silicon	1	KrF (248 nm, 500 fs)	Zergioti (1998)	[12]
Pt, Cr, In$_2$O$_3$	Glass	3	KrF (248 nm, 500 fs)	Papakonstantinou (1999)	[21]
Ag, Au, NiCr, BaTiO$_3$, SrTiO$_3$, Y$_3$FeO$_{12}$	Glass, alumina, silicon, printed circuit boards	25	KrF (248 nm, 30 ns)	Piqué (1999)	[29]
Au/Sn	Silicon	30	Ti:sapphire (775 nm, 0.1–8 ps, 1 kHz)	Bahnisch (2000)	[11]

LIFT may compete with established schemes in terms of operational simplicity and novelty for materials growth. Its potential for producing specialized elements and complex structures and materials has been established in several cases. These examples confirm its potential for relatively simple and low-cost solutions in direct-write applications, in areas where no other available methods have been demonstrated. On the other hand, LIFT is applicable only for a limited range of materials that possess the proper thermophysical properties.

2. DEPOSITION OF SINGLE ELEMENTS

The LIFT process was first shown by Bohandy et al. (1,2) to be capable of producing direct writing of 50 μm wide Cu and Ag lines by using single pulses of a nanosecond ArF excimer laser (193 nm) under high vacuum conditions (10^{-6} mbar). The analysis of Bohandy et al. (3) has shown that, for typical LIFT conditions, the metal at the target/substrate interface, exposed to the laser irradiation, reaches the boiling point before the film melts through to the surface. This observation is consistent with a process involving vapor-driven propulsion of metal from the film onto the receiver.

A systematic study of laser-induced transfer of W, Ti, and Cr patterns using lasers of different pulse widths, at atmospheric pressure, has been performed by Toth et al. (4,5). W films of 100 nm thickness on glass were used as the target material. The receiver distance from the substrate was 1 μm (measured interferometrically). A Nd:YAG laser with pulse width of 100 μsec to 1,000 μs, and peak power 20 mW to 230 mW, was used as the laser source. In Fig. 2, a scanning electron micrograph (SEM) of 100 nm thick adherent tungsten patterns is presented. These patterns were deposited onto a glass substrate using 500 μs wide triangular pulses at a peak power of 130 mW. Relatively large patterns covering an area of several square millimeters were deposited using Cr and Ti thin-films "targets" by Kantor et al. (6). Ti and Cr films of 200 nm thickness were irradiated through the glass support by means of a 1.8 × 1.8 mm mask using single ruby laser pulses (694 nm, 20 ns, 1.1 J/cm^2). The roles of the target interface and film-to-substrate distance were investigated. The efficiency of materials transfer was assigned with reference to the optical transmittance of both deposited and ablated areas using an He–Ne laser. In the case of poorly adherent films the transfer yield was found to be independent of the film-to-substrate distance for up to 60 μm.

C and Cr deposition has been performed at atmospheric pressure by Tam et al. (7) using an Nd:YLF diode-pumped Q-switched laser (15 μJ, 15 ns). The distance between the receiver and the target film ranged from 25 to 75 μm. The target film was 50 nm and the resulting deposition feature width was 25 μm.

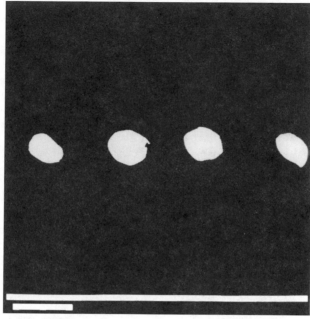

100 μm

FIGURE 2 Backscattered electron microscope images of transferred thin tungsten patterns. The scaling bar indicates 100 μm. (Reprint from *Applied Surface Science* 86, Z. Kantor, Z. Toth, T. Szorenyl, 196–201, Copyright (1995), with permission from Elsevier Science.)

Au and Al deposition by ArF excimer LIFT on silicon and quartz substrates have been published by Latsch et al. (8). These authors achieved feature size resolution of 7–10 μm. Combining LIFT with conventional electroless metal plating Cu line deposition was performed by Esrom et al. (9). They used excimer lasers at various wavelengths (193, 222, 248, 308, and 351 nm) and fluences, and demonstrated direct writing of 300 μm wide Cu lines by electroless plating of the transferred Pd lines on AlN. Al dots of 200 μm width and 0.8 μm height using Nd:YAG and Nd:Glass laser have been published by Schultze et al. (10).

Experimental results for Au/Sn metallic pattern deposition, by a single laser shot of a high-repetition rate laser system, have been published by Bahnisch et al. (11). They have used ultrashort pulses from a Ti:sapphire laser (775 nm, 1 kHz) of 0.1 ps to 8 ps pulse width and up to 0.5 mJ pulse energy.

A wide range of metals including Cr, In, Mo, and Pt has been deposited by Zergioti et al. (12–15). By utilizing a microetching system developed by Vainos et al. (16), a very strong image reduction was achieved enabling for the first time the realization of submicrometer features of Cr. A 500 fs KrF excimer laser (248 nm) has been used to take advantage of the short absorption length and the consequent limited thermal diffusion length, which lowers the ablation threshold and enables high-definition operation. Experiments were carried out in a low-vacuum (0.1 Torr) environment and have resulted in reproducible and adherent structures of submicron size. The receiver materials used were glass and silicon wafers and the process allowed binary amplitude and multilevel optical diffractive structures to be produced. Cr, In, Mo, and Pt thin films of 40, 80, and 200 nm thickness, prepared by sputtering and e-beam evaporation, were used as target materials.

A schematic diagram of the excimer laser microdeposition set-up is shown in Fig. 3. The laser beam was focused using high power (30×) image projection (16) onto the target surface. The fluence of the laser was varied between 50 and 550 mJ/cm^2 on the target. The distance between the target and the receiver surfaces was variable from near contact to 300 μm with an accuracy better than 5 μm. The optical absorption in the target film affects the threshold laser fluence, above which a single pulse leads to film removal and transfer onto the receiver surface. The absorption depth of Cr at 248 nm is 220 Å, implying that the transmission of the excimer laser light through even the thinnest 400 Å target film used is negligible. Various feature spots, lines,

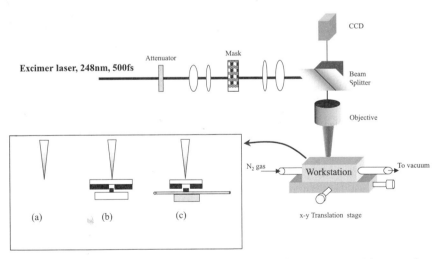

FIGURE 3 Experimental layout for direct excimer laser microfabrication. Inset: (a) microetching, (b) microprinting, (c) microprinting on optical fibers. (Reprint from Applied Optics 38, Mailis, S., Zergioti, I., Koundourakis, G., Ikiades A., Patentalaki, Papakonstantinou, P., Vainos, N.A., Fotakis, C., 11, 2301–2308, Copyright (1999), with permission from Optical Society of America.)

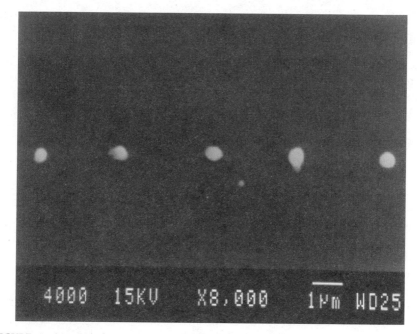

FIGURE 4 Scanning electron microscopy picture of isolated Cr dots deposited on glass by femtosecond laser microdeposition. The target material was 400 Å Cr. The UV illuminated area was 4 μm × 4 μm and the energy density was 100 mJ/cm². (Reprint from Applied Physics A, **66**, Zergioti, I., Mailis, S., Vainos, N. A., Papakonstantinou, P., Kalpouzos, C., Grigoropoulos, C. P., Fotakis, C., 579–582, Copyright (1998) with permission from Springer-Verlag GmbH & Co. KG.)

and patterns have been deposited. Figure 4 shows a SEM micrograph of submicron Cr dots achieved using a 400 Å Cr thick target film and a laser fluence on target of 100 mJ/cm². The microprinting pattern is actually smaller than the illuminated spot, because only the central portion of the focused laser spot is above the ablation threshold.

2.1. LIFT OF DIAMONDS

Selective deposition of diamondlike films by laser-induced forward transfer of ultrafine diamond particles was achieved by Pimenov et al. (17). Two laser sources were used, a copper vapor laser (510 nm, 20 ns, 10 kHz) and a KrF excimer laser (248 nm, 15 ns). Ultrafine diamond powder mixed with photoresist was coated on a transparent quartz material. The laser pulses irradiated the quartz substrate, placed in close proximity to the Silicon receiver substrate.

After the transfer of the diamond-containing photoresist onto the Si substrate, the coated receiver was placed in a CVD reactor. The deposition of the polycrystalline diamond film on the Si substrate was performed using an arc discharge in an atmosphere of CH_4/H_2. The average deposition rate was 10 μm/h and spatial resolution of the method was about 10 μm. This case represents another example of combined operations mediating compatibility with existing processing methods.

3. DEPOSITION OF OXIDE COMPOUNDS

Beyond conventional LIFT of pure metals, direct laser-based microdeposition of compound materials is of great interest. The possibility of direct writing of miniature structures of complex compounds raises exciting technological prospects. In addition, the possibility of building 3-dimensional structures, not achievable by other means, has encouraged further investigations.

Deposition of YBaCuO and BiSrCaCuO precoated thin films has been performed by Fogarassy et al. (18–20) using LIFT. Thin films of 50 to 800 nm thickness of YBaCuO and BiSrCaCuO amorphous compounds were initially deposited at low temperature (< 400 °C) by pulsed laser deposition or dc sputtering on suprasil quartz. The films were then irradiated through their transparent support using laser pulses at wavelengths ranging from the UV to IR regions. An ArF excimer laser (193 nm, 20 ns) at fluences between 0.05 and 0.5 J/cm^2, a pulsed Nd:YAG (1,064 nm, 5 ns) at fluences between 0.1 to 1 J/cm^2, and a high repetition 5 kHz Q-switched frequency doubled Nd:YAG (530 nm, 100 ns) at fluences 0.3 to 2.5 J/cm^2 were used. The precoated films were transferred by a single laser shot onto silicon and MgO single-crystal substrates under vacuum (10^{-4} Torr) or in air. The YBaCuO and BiSrCaCuO films were successfully converted into the superconducting phase with an onset critical temperature of about 90 K and a zero resistance at 80 K, by subsequent thermal annealing in an oxygen atmosphere at 850–900 °C. Patterns of 100 μm width were obtained by mask projection or by direct writing. Laser melting of the precoated layer at the film-receiver interfaces is observed by SEM analysis. Figure 5a,b depicts the surface morphology of the transferred BiSrCaCuO film transferred through a mask, which is characteristic of melting and resolidification phenomena. A small amount of sputtered material is also observed to protrude perpendicular to the 100 μm wide deposited line, which is a clear indication of the presence of a molten phase during the transfer process of the oxide. The mechanisms for transferring YBaCuO and BiSrCaCuO thin films by the LIFT technique were also discussed using the outcome of calculations based on the solution of heat-flow equations.

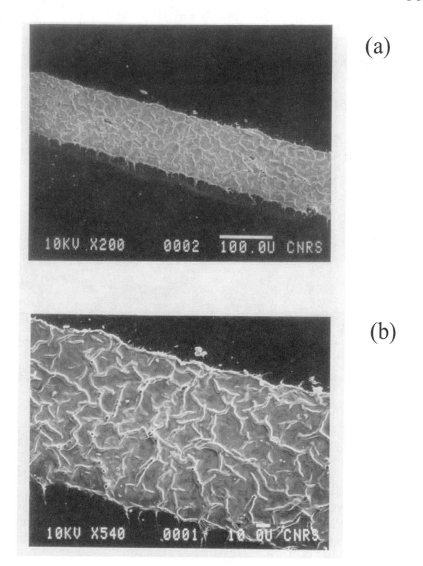

FIGURE 5 (a) Scanning electron micrograph of a BiSrCaCuO line deposited by LIFT technique onto silicon substrate (mask patterning), (b) Surface detail. (Reprint from Materials & Manufacturing Processes 7, Fogarassy, E., Fuchs, C., Unamuno, S. de, Perriere, J., Kerherve, F., (1), 31–51, Copyright (1992), by courtesy of Marcel Dekker Inc.)

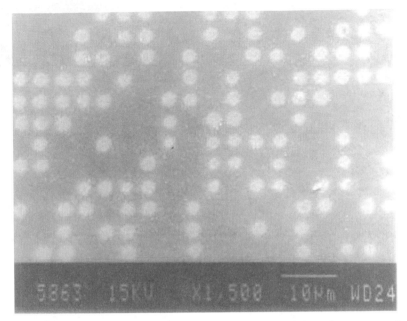

FIGURE 6 Scanning electron microscopy of a computer generated holographic pattern produced on quartz by InO_x microdeposition. (Reprint from Applied Surface Science **86**, Papakonstantinou, P., Vainos, N. A., Fotakis, C., 151, 159–170, Copyright (1999), with permission from Elsevier Science.)

These calculations suggest that, for these two compounds, film removal from the transparent support is achieved above the surface-melting threshold.

The growth of InO_x patterns on glass in a forward transfer mode have been reported by Papakonstantinou et al. (12,21). InO_x exhibits interesting electrical and optical properties, including the possibility for holographic recording in its nonstoichiometric form (InO_x) (22). Thin films of InO_x of 50 nm to 450 nm in thickness were prepared by reactive pulsed laser deposition (23) on quartz and were subsequently used as target materials. Figure 6 shows a holographic pattern having a pixel size of 4 μm × 4 μm produced on glass using a 200 nm thick In_2O_3 target at an energy density of 150 mJ/cm². The capabilities of the process for microprinting in a step-and-repeat operation were investigated. The crystallinity of the deposited features was studied by X-ray diffraction (XRD). The diffraction patterns of the target films grown on quartz are shown in Fig. 7a. These patterns include discrete diffraction peaks superimposed on a continuous background, which is consistent with the presence of crystalline

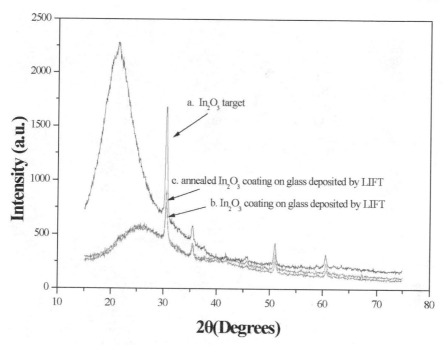

FIGURE 7 XRD patterns for (**a**) the target In_2O_3 film on quartz, (**b**) microdeposited In_2O_3 coating on glass by LIFT, and (**c**) microdeposited In_2O_3 coating on glass by LIFT after annealing at 350 °C. (Reprint from Applied Physics A, **66**, Zergioti, I., Mailis, S., Vainos, N. A., Papakonstantinou, P., Kalpouzos, C., Grigoropoulos, C. P., Fotakis, C., 579–582, Copyright (1998) with permission from Springer-Verlag GmbH & Co. KG.)

grains embedded in an amorphous matrix. All the diffraction peaks could be assigned to the cubic structure of In_2O_3. The absence of metallic In peaks is verified. The forward ablation threshold fluence when using 150–200 nm thick target films was 45 mJ/cm^2. The X-ray pattern of the transferred material on corning glass receiver is also shown in Fig. 7b. It should be mentioned that direct comparison of the peak intensities between indium oxide/quartz and indium oxide/glass systems cannot be made on the basis of the spectra shown in Fig. 7a,b because the two films did not cover the same area. All the diffraction peaks could be assigned to the In_2O_3 phase. Annealing in flowing oxygen at 350 °C improved the transparency and the crystalline quality as evidenced from the enhanced reflection intensities also shown in Fig. 7c. Such operation, however, diminishes the optical activation properties of InO_x.

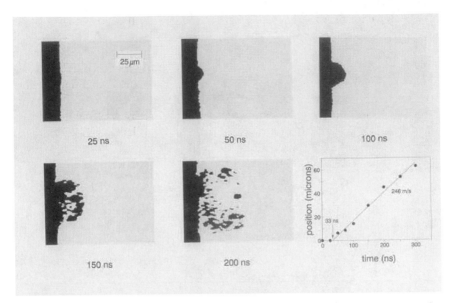

FIGURE 8 Shadowgraphs of black laser ablation transfer film ablation by a 250 nsec pulse. During the image sequence, the coating is seen to fly off the film surface into air (lighter region at right). The graph shows the position of the ablation front versus time. The best linear fit gives a velocity of 246 m/sec. Extrapolation of the fit to zero distance shows that the coating first leaves the surface at $t_d = 33$ nsec. (Reprint from *Journal of Imaging Science and Technology*, **36**, Sandy Lee, I.-Y., Tolbert, W.A., Dlott, D.D., Doxtader, M.M., Foley, D.M., Arnold, D.R., Ellis, E.W., 2, 180–7, Copyright (1992), with permission of IS&T: The Society of Imaging Science and Technology sole copyright owners of The Journal of Imaging Science and Technology.)

4. TRANSFER MECHANISMS

4.1. IMAGING DIAGNOSTICS

The dynamics of the LIFT technique using ultrafast microscopy imaging have been studied by Dlott et al. (24,25). Ablation was caused by a pulsed Nd : YAG laser (1.064 μm, 250 ns) and the laser ablation transfer coating was observed using a 25 ps probe pulse of an Nd : YAG-pumped dye laser, which propagated nearly perpendicular to the 250 ns ablation pulse. An image sequence of the film transfer is shown in Fig. 8. In the image sequence, the coating is seen to fly off the film surface into the air. The graph depicts the position of the ablation front versus time. The best linear fit gives a velocity of 246 m/s (0.75 Mach). The use of ps optical pulses results in a reduction of the laser fluence threshold by one order of magnitude, as compared to that of 100 ns long pulses.

The behavior of Au atoms and emissive species in LIFT processes has been studied by Okata et al. (26), by applying two-dimensional laser-induced fluorescence and imaging of thermal radiation. In order to observe the micron-sized LIFT process, a long working distance microscope was used. Thin film of Au deposited onto a quartz plate, placed in a vacuum chamber, was ablated by backside illuminating with a dye laser (440 nm). The thermal radiation from the ejected emissive species was observed by an image-intensified CCD camera. The ablation energy was changed from 20 µJ to 100 mJ. At 100 µJ laser pulse energy, the fastest components had a velocity of 2 km/s while at 13 µJ atoms were not detected.

4.2. TEMPERATURE PROFILE ANALYSIS

The temperature profile of the metal during the laser pulse have been calculated by Kantor et al. (6). Considering that the lateral dimensions of the processed areas are much larger than the thermal diffusion length in both the substrate and the metal film temperature, changes are described by the one-dimensional heat-conduction equation

$$\frac{\partial T}{\partial t} = \frac{1}{\rho(T)c(T)} \frac{\partial}{\partial z}\left[\kappa(T)\frac{\partial T}{\partial z}\right] + \frac{I(z, t)\alpha(T)}{\rho(T)c(T)} \tag{1}$$

where $T, t, z, k, \rho, c, \alpha$ denote respectively, the temperature, time, depth from the upper substrate of the thin film, the thermal conductivities, densities, specific heat, and optical absorption coefficients of the metal film and the glass support.

The spatial and temporal dependence of the intensity is given as $I(z, t) = I_0(t)(1 - R)\exp(-az)$, where $I_0(t)$ is the intensity of the incident laser beam and R the reflectivity of the sample measured at 694 nm. The intensity profile of the $I_0(t)$ was approximated by a Gaussian shape. The spatial and temporal temperature profiles have been calculated using the method of finite differences taking into account the temperature dependence of the thermophysical data and incorporating melting and vaporization. Ti and Cr films (200 nm thickness) were irradiated through the glass support by single pulses of a ruby laser (694 nm, 20 ns) using a mask of 1.8 mm × 1.8 mm. Calculated maximum temperatures in titanium and chromium films are shown in Figs. 9 and 10 as a function of the processing laser fluence, respectively. The calculated curves in Figs. 10 and 11 refer to both supported and free-standing films.

506

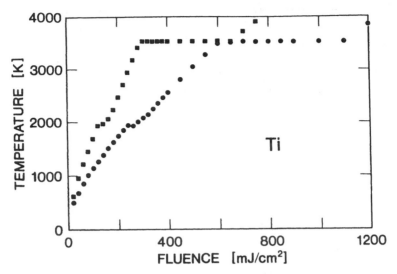

FIGURE 9 Calculated maximum temperature vs. incident fluence for 200 nm thick Ti target supported (●) and free-standing (■) titanium films. (Reprint from Applied Physics A, **54**, Kantor, Z., Toth, Z., Szorenyl, T., 170–175, Copyright (1992) with permission from Springer-Verlag GmbH & Co. KG.)

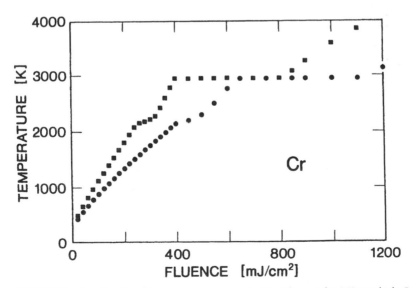

FIGURE 10 Calculated maximum temperature vs. incident fluence for 173 nm thick Cr target supported (●) and free-standing (■) titanium films. (Reprint from Applied Physics A, **54**, Kantor, Z., Toth, Z., Szorenyl, T., 170–175, Copyright (1992) with permission from Springer-Verlag GmbH & Co. KG.)

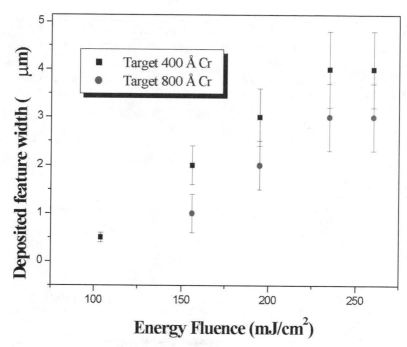

FIGURE 11 Plot of the deposition feature width of Cr dots as a function of the laser energy fluence using different thickness target materials (400 Å and 800 Å Cr). (Reprint from Applied Physics A, 66, Zergioti, I., Mailis, S., Vainos, N. A., Papakonstantinou, P., Kalpouzos, C., Grigoropoulos, C.P., Fotakis, C., 579–582, Copyright (1998) with permission from Springer-Verlag GmbH & Co. KG.)

4.3. DEPOSITION STUDIES

The minimum laser fluences at the wavelengths and pulse widths of KrF (248 nm, 25 ns), and 2ω Nd : YAG (532 nm, 15 ns) lasers required to lift off thin Au films from optical quartz, as a function of the film thickness, were measured by Baseman et al. (27). They concluded that the films lift off when the Au quartz interface reaches the normal boiling point of Au. In other experiments the laser fluence dependence of the width of Cr dots obtained by LIFT has been studied by Zergioti et al. This dependence is shown in Fig. 11 for two different Cr target thickness, of 400 Å and 800 Å. The forward ablation threshold has been defined as the single-pulse energy density value at which complete thin-film material ejection in the direction of propagation of the laser beam is performed. This threshold is influenced by the thickness of the target film. For Cr targets thicker than 2,000 Å the LIFT technique by a single excimer laser pulse even at the highest energy density (500 mJ/cm^2) employed was not applicable. The forward ablation threshold values of the 800 Å and 400 Å Cr target films were respectively 150 mJ/cm^2 ($\pm 20\%$) and 100 mJ/cm^2 ($\pm 20\%$), while the best quality of the deposited dots in terms of uniformity was obtained for the thinner Cr target films, by minimizing the thermal effects and also the extent of the damaged (melted) area.

The spread of the ablated material has also been studied by varying the distance between the target and receiver surfaces from near contact to 500 μm. Figure 12 shows the deposited feature spread size of Cr lines as a function of the distance for a 4 × 4 μm^2 illuminated area of the target Cr film at 156 and 260 mJ/cm^2 energy fluence.

It is worth noting here that for comparison purposes, a series of identical experiments has been carried out using the same optical system with a conventional KrF excimer laser emitting pulses of 20 ns duration. The results

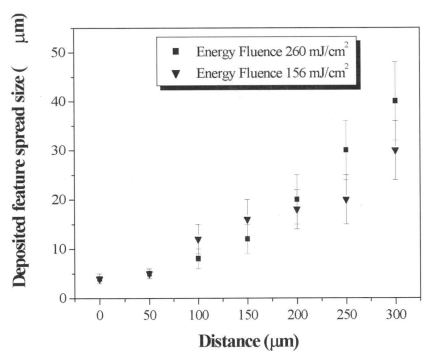

FIGURE 12   Plot of the deposited feature spread size as a function of the distance between the target film and the receiver surfaces. The energy densities were 156 and 260 mJ/cm^2. (Reprint from Applied Physics A, **66**, Zergioti, I., Mailis, S., Vainos, N.A., Papakonstantinou, P., Kalpouzos, C., Grigoropoulos, C.P., Fotakis, C., 579–582, Copyright (1998) with permission from Springer-Verlag GmbH & Co. KG.)

obtained with the nanosecond excimer laser did not yield deposition features of high quality, in comparison to those obtained by means of a subpicosecond KrF laser. In the latter case, the deposited features were superior in terms of their morphology and resolution. The remelting of the transferred material by the trailing part of the laser pulse is often encountered in experiments with ns or longer laser pulses. Recent studies (28) have shown that ultrafast laser pulses have precise breakdown thresholds and minimum thermal diffusion, thus offering advantages in precision micromachining and microdeposition applications.

5. APPLICATIONS OF LIFT

The advantages of LIFT point to numerous applications beyond the existing microfabrication. The ability to fabricate micron and submicron patterns in a single-step operation is recently of great interest. Current trends lead to the demonstration of complex structures for microelectronics and optoelectronics.

Electronics and sensor materials using a novel direct-write technique (MAPLE DW) that combines the laser-induced forward transfer (LIFT) method with the matrix-assisted pulsed laser evaporation (MAPLE) have been demonstrated by Piqué et al. (29). The MAPLE DW technique utilizes all of the advantages associated with LIFT and MAPLE to produce laser-driven direct-write processes capable of transferring materials such as metals, ceramics, and polymers onto polymeric, metallic, and ceramic substrates at room temperature. The overall writing resolution for this technique is currently of the order of 10 μm. A variety of devices have been fabricated, including parallel plate and interdigitated capacitors, flat inductors, conducting lines, resistors, and chemoresistive gas sensors. Figure 13a depicts Au conducting lines (of 30 μm width and 10 μm thickness) deposited by LIFT on a RO4003 circuit board. The lines were generated by overlapping the 25 μm diameter laser spots (248 nm excimer) each with a fluence of 550 mJ/cm^2 and 100 laser passes. The 5 coplanar resistors shown in Fig. 13b were made by LIFT of nichrome ribbons over the RO4003 substrate. Gold lines with a resistivity of 75 μOhm cm were obtained. Furthermore, parallel plate capacitors and inductors with gold electrodes and BaTiO layers with a measured capacitance of the devices ranging from 2 to 40 pF, and effective dielectric constants from 22 to 40 with tan δ between 0.11 and 0.17 have been realized.

Patterns of high-Tc-superconducting YBaCuO and BiSrCaCuO films formed by the LIFT technique have been realized by Fogarassy et al. (19). The experimental details of this work are described above. The LIFT films were electrically insulated and in order to obtain superconducting phases, oxygen annealing treatments were carried out. In addition, subsequent thermal

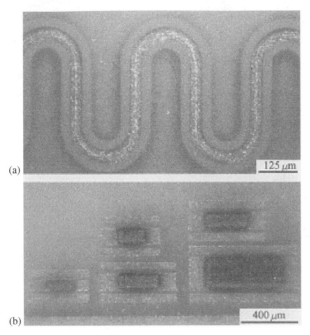

FIGURE 13 (a) Lines deposited by LIFT on RO4003 circuit board. The Au line is approximately 30 μm wide after a final laser trimming step, performed along both sides of the line. (b) Optical micrograph michrome coplanar resistors made by LIFT on RO4003. (Reprint from Applied Physics A, **69**, Piqué, A., Chrisey D.B., Auyeung, R.C.Y., Fitz-Gerald, J., Wu, H.D., McGill, R.A., Lakeou, S., Wu, P.K., Nguyen, V., Duignan, M., S279–284, Copyright (1999) with permission from Springer-Verlag GmbH & Co. KG.)

annealing in an oxygen atmosphere whithin the 850–900 °C temperature range was found by resistivity measurements to be an important process for obtaining superconducting transitions. As shown in Fig. 14 the normalized resistance of the films decreases with temperature and around 100 K a transition is observed, the zero resistivity being reached in the 77 to 85 K range. A correlation between the surface morphology and resistivity was observed because the high density of the needles created on the surface leads to incomplete transitions.

Complicated diffractive optics such as binary amplitude computer-generated holograms have been fabricated by Zergioti et al. (12). The microdeposition was performed either by serial writing (pixel-by-pixel) of the diffractive pattern or by directly projecting a master hologram mask on the target film. This work demonstrated that single and multilevel holographic structure fabrication using LIFT of metal and oxides patterns is possible. Figure 15

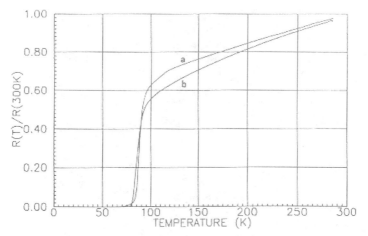

FIGURE 14 Resistivity as a function of temperature in BiSrCaCuO films deposited by LIFT with a pulsed ArF (**a**) and Nd:YAG (**b**) laser, respectively. (Reprint from Materials & Manufacturing Processes 7, Fogarassy, E., Fuchs, C., Unamuno, S. de, Perriere, J., Kerherve, F., (1), 31–51 Copyright (1992), by courtesy of Marcel Dekker Inc.)

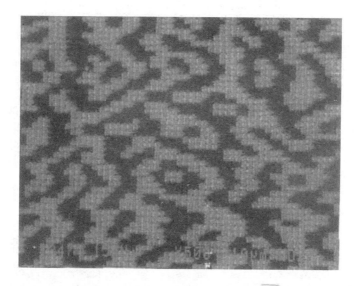

10 μm

FIGURE 15 Scanning electron microscopy of a computer-generated holographic pattern produced by Cr microdeposition on Silicon (100). The target material was 400 Å Cr. The pattern consists of 64 × 64 pixels with pixel size 3 μm × 3 μm. (Reprint from Applied Physics A, **66**, Zergioti, I., Mailis, S., Vainos, N.A., Papakonstantinou, P., Kalpouzos, C., Grigoropoulos, C.P., Fotakis, C., 579–582, Copyright (1998) with permission from Springer-Verlag GmbH & Co. KG.)

FIGURE 16   Reconstruction of the femtosecond excimer laser microdeposited computer-generated hologram using a He–Ne laser. (Reprint from Applied Surface Science **126–127**, Zergioti, I.,
Mailis, S., Vainos, N.A., Ikiades A., Grigoropoulos, C.P., Fotakis, C., 82–86, Copyright (1999), with
permission from Elsevier Science.)

exhibits a hologram pattern produced by the deposition from a 400 Å thick Cr
target on silicon at an energy density of 339 mJ/cm^2 having a pixel size of
3 μm × 3 μm. The reconstructed results are shown in Fig. 16. Under illumination with red laser light (633 nm), the pattern "FORTH" is visible. Figure 17
depicts a multilevel computer-generated holographic structure on glass,
comprising 3 layers of Cr selectively deposited on appropriate areas on a
glass plate. The energy fluence was 372 mJ/cm^2.

Microscopic structures of chromium were deposited on the outer surface of
etched fibers using 500 fs KrF laser pulses by Mailis et al. (15). The laser
fluence was lower than 100 mJ/cm^2. The structures were deposited in a point-
to-point manner by accurate alignment of the fiber axis along the motion axis.
Figure 18 shows close-up views of a chirped gratinglike structure comprising
submicrometer features as deposited on the flat surface of an Andrew D-type
fiber having an outer diameter of 125 μm. The size of the spots deposited
in this case was 2 μm. The structure is used in the ongoing investigation
of guided wave coupling effects. It is clear that for single-mode fibers more
complex asymmetric etching procedures must be applied to approach the well-

100 µm

FIGURE 17   Scanning electron microscopy picture of a computer-generated multilevel structure of Cr on Glass. The pattern pixel size was 5 µm × 5 µm. (Reprint from Applied Physics A, **66**, Zergioti, I., Mailis, S., Vainos, N.A., Papakonstantinou, P., Kalpouzos, C., Grigoropoulos, C.P., Fotakis, C., 579–582, Copyright (1998) with permission from Springer-Verlag GmbH & Co. KG.)

confined field in the fiber core. These results, however, demonstrate the unique potential of the LIFT method for fabricating microstructures onto nonplanar and high-curvature surfaces. Such applications would be proven significant in the further development of optical-fiber sensors and optical telecommunications devices, including alternative Bragg fiber gratings, rocking filters, fiber polarises, and evanescent-wave fiber sensors.

Several drawbacks of the LIFT method should be noted. Although the principles of the technique and its practical application appear to be simple, the mechanisms involved are not fully understood. The precise control of the distance target-receiver is also a crucial requirement. Finally, the quality of the deposited structures depends critically on the thermophysical properties of the materials used. Further studies will determine the operational requirements toward a wider application of the method.

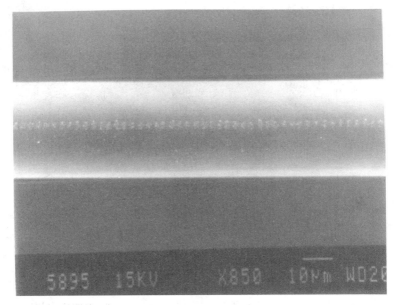

FIGURE 18 Close-up views of grating like structure developed on D-type optical fiber revealing 2 μm features periodically deposited on a 35 μm diameter optical fiber. (Reprint from Applied Optics **38**, Mailis, S., Zergioti, I., Koundourakis, G., Ikiades A., Patentalaki, Papakonstantinou, P., Vainos, N.A., Fotakis, C., 11, 2301–2308 Copyright (1999), with permission from Optical Society of America.)

6. SUMMARY AND CONCLUSIONS

The laser-induced forward transfer (LIFT) is a new technique utilizing pulsed lasers to remove thin-film material from a transparent support and deposit it onto a suitable substrate. LIFT is a method for direct serial writing of well-defined lines and isolated dots, with submicron resolution. The microdeposition results of metals, superconducting oxides, and indium oxide features currently indicate the potential of the scheme for growing of miniature amorphous and crystalline structures. The LIFT process exhibits several advantages over conventional methods, primarily simplicity and novelty. Even though several impediments have been observed, the potential for high accuracy, submicron patterns, and complex structure fabrication exists and must be explored further.

ACKNOWLEDGMENTS

Part of the processes described here have been carried out through projects within the "Ultraviolet Laser Facility" (ULF) which operates at FORTH, with support from the EU Directorate Research Program "Access to Large Installation" at the Foundation for Research and Technology-Hellas. Support through the EU project HOLAUTHENTIC is also acknowledged. Last, the authors are grateful to D. Dlott, E. Fogarassy, A. Piqué, Z. Toth, and their publishers for kindly providing permission for use of published material.

REFERENCES

1. Bohandy, J., Kim, B.F., Adrian, F. J. (1986). Metal deposition from a supported metal film using an excimer laser. *J. Appl. Phys.* **60**, (4), 1538–9.
2. Adrian, F. J., Bohandy, J., Kim, B.F., Jette, A. N., Thompson, P. (1987). A study of the mechanism of metal deposition by the laser-induced forward transfer process. *J. Vac. Sci. Technol.* **B5** (5) 1491–1494.
3. Bohandy, J., Kim, B.F., Adrian, F. J., Jette, A. N. (1988). Metal deposition at 532 nm using a laser transfer technique. *J. Appl. Phys.* **63** (4), 1158–62.
4. Toth, Z., Szorenyi, T., Toth, A.L. (1993). Ar^+ laser-induced forward transfer (LIFT): A novel method for micrometer-size surface patterning. *Appl. Surf. Sci.* **69**, 317–320.
5. Kantor, Z., Toth, Z., Szorenyi, T., (1995). Metal pattern deposition by laser-induced forward transfer, *Appl. Surf. Sci.* **86**, 196–201.
6. Kantor, Z., Toth, Z., Szorenyi, T., (1992). Laser induced forward transfer: The effect of support-film interface and film-to-substrate distance on transfer. *Appl. Phys. A* **54**, 170–175.
7. Poon, C.C., Tam, A.C. (1996). Laser-induced forward transfer of carbon and chromium films in gases of one atmosphere pressure. *CLEO Digest* 377–8.
8. Latsch S., Hiraoka H., Nieveen W., Bargon J. (1994). Interface study on laser-induced material transfer from polymer and quartz surfaces. *Appl. Surf. Sci.* **81** 183–194.
9. Esrom, H., Zhang, J.-Y., Kogelschatz, U., Pedraza, A.J. (1995). New approach of a laser-induced forward transfer for deposition of patterned thin metal films. *Appl. Surf. Sci.* **86**, 202–207.
10. Schultze, V., Wagner, M. (1991). Laser-induced forward transfer of aluminium. *Appl. Surf. Sci.* **52**, 303–309.
11. Bahnisch. R., Gross, W., Menschig, A. (2000). Single-shot, high repetition rate pattern transfer. *Microelectron. Eng.* **50**, 541–546.
12. Zergioti, I., Mailis, S., Vainos, N.A., Papakonstantinou, P., Kalpouzos, C., Grigoropoulos, C.P., Fotakis, C. (1998). Microdeposition of metal oxide structures using ultrashort laser pulses. *Appl. Phys. A*, **66**, 579–582.
13. Zergioti, I., Mailis, S., Vainos, N.A., Fotakis, C., Chen S., Grigoropoulos, C.P. (1998). Micro-deposition of metals by femtosecond excimer laser. *Appl. Surf. Sci.* **127–129**, 601–605.
14. Zergioti, I., Mailis, S., Vainos, N.A., Ikiades A., Grigoropoulos, C.P., Fotakis, C. (1999). Microprinting and microetching of diffractive structures using ultrashort laser pulses. *Appl. Surf. Sci.* **138–139**, 82–86.
15. Mailis, S., Zergioti, I., Koundourakis, G., Ikiades A., Patentalaki, Papakonstantinou, P., Vainos, N.A., Fotakis, C. (1999). Etching and printing of diffractive optical microstructures by a femtosecond excimer laser. *Appl. Opt.* **38**, 11, 2301–2308.

16. Vainos, N.A., Mailis, S., Pissadakis, S., Boutsikaris, L., Parmiter, P.J.M., Dainty, P., Hall, T.J. (1996). Excimer laser use for microetching computer-generated holographic structures. *Appl. Opt.* **35**, 32, 6304–6319.

17. Pimenov, S.M., Shafeev, G.A., Smolin, A.A., Konov, V.I., Vodolaga, B.K. (1995). Laser-induced forward transfer of ultra-fine diamond particles for selective deposition of diamond films. *Appl. Surf. Sci.* **86**, 208–212.

18. Fogarassy, E., Fuchs, C., Kerherve, F., Hauchecorne, G., Perriere, J. (1989). Laser-induced forward transfer: A new approach for the deposition of high Tc superconducting thin films. *J. Mater. Res.* **4**, 5, 1082–6.

19. Fogarassy, E., Fuchs, C., Unamuno, S. de, Perriere, J., Kerherve, F. (1992). High Tc superconducting thin film deposition by laser induced forward transfer. *Mater. Manufac. Process.* **7**(1), 31–51.

20. Fogarassy, E., Fuchs, C., Kerherve, F., Hauchecorne, S., Perriere, J. (1989). Laser-induced forward transfer of high-Tc YBaCuO and BiSrCaCuO superconducting thin films. *J. Appl. Phys.* **66** (1) 457–9.

21. Papakonstantinou, P., Vainos, N.A., Fotakis, C. (1999). Microfabrication by UV femtosecond laser ablation of Pt, Cr and indium oxide thin films. *Appl. Surf. Sci.* **151**, 159–170.

22. Mailis, S., Boutsikaris, L. and Vainos, N.A., Xirouhaki, C., Vasiliou, G., Garawal, N., Kiriakidis, G., Fritzsche, H. (1996). Holographic recording in indium-oxide (In_2O_3) and indium-tin-oxide (In_2O_3Sn) thin films. *Appl. Phys. Lett.* **69**, 2459.

23. Grivas, C., Gill, D.S., Mailis, S., Boutsikaris, L., Vainos, N.A. (1998). Indium oxide thin-film holographic recorders grown by excimer laser reactive sputtering. *Appl. Phys. A*, **66**, 2, 201–4.

24. Sandy Lee, I.-Yin, Tolbert, W.A., Dlott, D.D., Doxtader, M.M., Foley, D.M., Arnold, Dana R., Ellis, E.W. (1992). Dynamics of laser ablation transfer imaging investigated by ultrafast microscopy. *J. Imag. Sci. Technol.* **36** (2) 180–7.

25. Tolbert, W.A., Sandy Lee, I.-Y., Doxtader, M.M., Ellis, E.W., Dlott, D.D. (1993). High-speed color imaging by laser ablation transfer with a dynamic release layer: Fundamental mechanisms. *J. Imag. Sci. Technol.* **37** (4) 411–421.

26. Yoshiki, N., Tashuo, O., Mitsuo, M. (2000). Imaging of the behavior of atoms and emissive species in laser-induced forward transfer process. *CLEO Digest.* 493–494.

27. Baseman, R.J., Froberg, N.M., Andreshak, J.C., Schlesinger, Z. (1990) Minimum fluence for laser blow-off of thin gold films at 248 and 532 nm. *Appl. Phys. Lett.* **56** (15), 1412–1414.

28. Liu, X., Du, D., Mourou, G. (1997). Laser ablation and micromachining with ultrashort laser pulses. *IEEE J. Quantum Electron.* **33**, 10, 1706–1716.

29. Piqué, A., Chrisey D.B., Auyeung, R.C.Y., Fitz-Gerald, J., Wu, H.D., McGill, R.A., Lakeou, S., Wu, P.K., Nguyen, V., Duignan, M. (1999). A novel laser transfer process for direct writing of electronic and sensor material. *Appl. Phys. A* **69** S279–284.

Matrix Assisted Pulsed Laser Evaporation-Direct Write (MAPLE-DW): A New Method to Rapidly Prototype Organic and Inorganic Materials

JAMES M. FITZ-GERALD

University of Virginia, Department of Materials Science and Engineering, Charlottesville, Virginia

PHILIP D. RACK

The University of Tennessee, Department of Materials Science and Engineering, Knoxville, Tennessee

BRADLEY RINGEISEN, DANIEL YOUNG, ROHIT MODI, RAY AUYEUNG

Naval Research Laboratory, Materials Science and Technology Division, Washington, D.C.

HUEY-DAW WU

SFA, Inc., Largo, Maryland

1. INTRODUCTION

In this chapter we present Matrix Assisted Pulsed Laser Evaporation-Direct Write (MAPLE-DW), as a novel laser-based direct-write process that can generate mesoscopic (ranging from 10 μm to 5 mm in feature size) patterns of almost any material, ranging from metals and ceramics, to biological materials and polymers. In general any material that takes the form of a rheological fluid, polymer-based composite or fine powder can be directly written into mesoscopic patterns with MAPLE-DW. This fact presents a wide range of flexibility in terms of starting raw materials and is a primary advantage over similar direct-write methods.

The methodology of the MAPLE-DW technique, along with the current understanding of the writing mechanism will be introduced followed by examples. Discussion of applications will include the direct writing of metals and ceramics, as well as composite resistor materials and phosphor patterns for display applications. Recent results on the direct writing of active proteins and two types of living cells will be presented, underscoring the utility of MAPLE-DW in the nascent field of engineered, biological-based mesoscale technology.

From a historical perspective, MAPLE-DW was initially developed to meet the current need for rapid prototyping of electronic circuitry at low processing temperatures. The exponential growth of the electronics industry has led to increasing demand for smaller, faster, and lower-cost devices. An acceleration of product cycles, an increased number of limited-production products, and the desire for just-in-time manufacturing, all have placed strain on traditional development methods. This has caused the prototyping stage to be a bottleneck and small-lot manufacturing to be prohibitively expensive. While simulations are widely used to partially mitigate this bottleneck, physical prototyping of electronics will always be necessary.

Mesoscopic electronic components are conventionally manufactured with thick-film technology, primarily screen printing and tape-casting techniques. These techniques can be efficient and low cost in many high-volume applications, but still suffer from a number of limitations. For example, they require customized tooling to produce a part, making them expensive and unsuitable for rapid prototyping, and high processing temperatures ($\sim 850\,^\circ$C) preclude the use of polymer substrates. MAPLE-DW has proven capable of overcoming many of these limitations. Successful direct writing patterns of conductor, dielectric, and resistor materials onto both high- and low-temperature substrates has positioned MAPLE-DW as a strong candidate for next-generation electronics rapid prototyping.

The success with electronic materials has generated the current interest in applying MAPLE-DW to other materials systems where mesoscopic patterns are useful; the display and biological applications mentioned here are only a few such possibilities.

2. BACKGROUND

A *mesoscopic direct-write technique* has come to mean a fabrication technology that can produce complex, mesoscopic-scale (10 mm to 10 μm) devices from a specified design in a relatively short period of time and with minimum human participation, customized tooling, or intermediate processing steps. As such, direct-write technologies differ from conventional manufacturing methods. Ideally, a direct-write method must be flexible and effective in the deposition of a wide variety of materials, on virtually any surface, and under ambient conditions. It is also clear that to interact with today's computer-dominated industry, the technique must be easily integrated with CAD/CAM systems. Direct-write methods are often similar to printing technologies, but the goal of direct writing is to produce patterns that possess much more functionality and can have 3-dimensional features. It is important to note that these methods are not meant to replace photolithography and other submicron manufacturing methods. Instead, they are intended for rapid prototyping (and perhaps some mass manufacturing) in the mesoscopic regime. Various direct-write technologies have been developed in recent years including inkjet, plasma spray, focused ion beam, and several liquid microdispensing approaches, which have been described elsewhere in this book.

The aforementioned direct-write techniques all depend on specialized, high-quality *starting materials* (see Chapters 4 and 6). Often, the starting material is not the same as the final *functional material*; patterns of the starting material are produced by the direct-write technique, and then converted into functional material by *ex situ* treatment. With the exception of a few techniques, inorganic starting materials take the form of fluids, inks, pastes, and composites with specially tailored chemical and rheological properties (viscosities, densities, surface tension, chemical reaction route, etc.). They are typically based on an organic vehicle or solvent and can include metalorganic precursors, powders, nanopowders, flakes, binders, solvents, dispersants, and surfactants. These are often converted to a functional material by thermal or photothermal processing. The starting materials for biological MAPLE-DW applications are often complex, aqueous-based solutions that can contain a combination of active biomolecules, living cells, buffers, gels, nutrients, inorganic powders, and other chemicals. These can be transferred in the form of a liquid, room-temperature solid, or frozen solid.

2.1. MATRIX ASSISTED PULSED LASER EVAPORATION (MAPLE)

Shortly after the discovery of lasers, researchers representing all disciplines of science began investigating the interaction of laser radiation with different types of materials. Laser-material interactions can result in a wide range of effects that depend on the properties of the laser, the material, and the ambient environment. These effects can range from simple photothermal heating and photolytic chemical reactions to ablation and plasma formation, to many other possible effects. In the ablation regime of laser-material interaction, one of the best known applications is Pulsed Laser Deposition (PLD) of inorganic thin films. In PLD, a short, intense laser pulse ablates material from a sacrificial target and the resulting vapor is condensed onto a substrate, forming a thin film in a controlled manner. A hybrid process has been developed at the Naval Research Laboratory, dedicated to the soft deposition of polymer and organic materials through a process termed Matrix Assisted Pulsed Laser Evaporation (MAPLE) (1–4). In this process, the laser pulse has a lower fluence ($\sim 0.2\,J/cm^2$) than in conventional PLD (1.5–$5\,J/cm^2$ for metals and ceramics). Unlike conventional PLD, the target is a dilute matrix, made up of a solvent and the organic molecules to be deposited as solute. The matrix is typically frozen at low temperatures. If tuned correctly, the laser pulse will be preferentially absorbed by the solvent molecules. The laser-produced temperature rise is large compared to the melting point of the solvent, but low compared to the decomposition temperature of the solute. When the MAPLE process is optimized, the collective collisions of the evaporating solvent with the organic molecule act to gently desorb the organic molecule intact, that is, with only minimal chemical decomposition. The organic molecule preferentially condenses onto a substrate, while the evaporating solvent has a near-zero sticking coefficient. The solvent is rapidly pumped away or it can be trapped for re-use.

2.2. LASER-INDUCED FORWARD TRANSFER (LIFT)

The LIFT process (covered in Chapter 16) is considered a progenitor of the MAPLE direct-write technique. LIFT employs laser radiation to vaporize and transfer material from a thin-film starting material of metal or oxide, on an optically transparent support, onto a substrate placed in close proximity (5–11). Because this involves a short, intense laser pulse absorbed by an inorganic solid-state target, the laser-material interactions in LIFT are very similar to those of PLD, but have different results. The results of the localized heating and vaporization can either be simple vapor deposition onto the substrate or the physical transfer of a portion of the thin film, as a solid. Patterning is achieved by moving the laser beam and/or the substrate, or by pattern projection. The first two are methods of direct writing patterns. There are several requirements for LIFT to produce useful patterns: the laser fluence should just exceed the threshold fluence for removing the thin film from the transparent support, the target thin film should not be too thick—less than a few 1,000 Å, the target film should be in close contact to the substrate, and the absorption of the target film should be high. Operating outside this regime results in problems with morphology, spatial resolution, and adherence of the transferred patterns. Repetitive transfer of material can control the film thickness deposited onto the substrate. LIFT is a conceptually simple direct-write technique, but has been used to demonstrate sophisticated submicron resolution direct writing. Despite the advantages of LIFT, the stating materials are limited to metals and simple oxides whose useful properties are relatively insensitive to stoichiometric and crystallographic state.

3. MATRIX ASSISTED PULSED LASER EVAPORATION-DIRECT WRITE

The MAPLE-DW technique utilizes the technical approach of LIFT, in that a focused laser pulse is used to transfer material from a transparent carrier onto an acceptor substrate. MAPLE-DW takes advantage of the basic mechanism of MAPLE by using a target material that typically contains a sacrificial organic vehicle that protects the material of interest from laser-induced degradation. The result is a laser-driven direct-write process that is capable of transferring various materials such as metals, ceramics, and polymers onto polymeric, metallic, and ceramic substrates at room temperature and at atmospheric pressure, with a resolution on the order of $10\,\mu m$ (12–14). Because MAPLE-DW is a laser-based direct-writing method, the highly focused laser beam can also be utilized for *in situ* micromachining, surface annealing, drilling, and trimming applications simply by removing the ribbon from the laser path. Thus, MAPLE-DW is both an *additive* as well as *subtractive* process. It can also be adapted to operate with multiple lasers of different wavelengths, whereby the wavelength from one laser has been optimized for the transfer and micromachining operations, that is, a UV laser, while additional lasers are used for modifying the surface as well as annealing or laser sintering of either the substrate or any of the already deposited layers (IR or visible).

3.1. EXPERIMENTAL CONFIGURATION

A schematic of the MAPLE-DW apparatus is shown in Fig. 1. A laser transparent substrate such as a quartz disc or polyester film is coated on one side with a thick film of the material to be transferred, typically 1–20 microns thick depending on the material and application. As was discussed, the film is

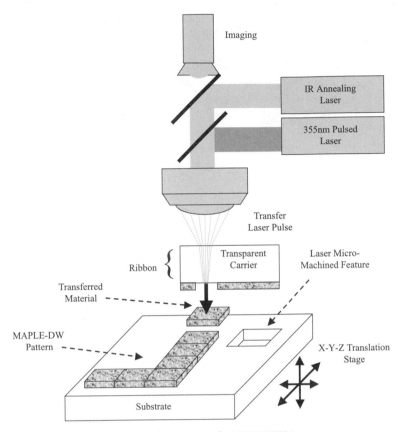

FIGURE 1   Schematic diagram of a MAPLE-DW instrument.

typically a fluid, ink, paste, or composite material (henceforth referred to collectively as *inks*) that consist of a dispersing fluid, polymer, or organic vehicle in combination with powders, soluble chemicals, biological materials, and/or rheological modifiers. The carrier plus thick film of ink is known hereby as a *ribbon* and is placed in close proximity (5 to 100 μm) to the acceptor substrate. As with LIFT, the laser is focused through the transparent substrate onto the ink layer, see Fig. 2. When a laser pulse strikes the ink, a small portion is vaporized at the ink/carrier interface, and the vapor expansion transfers the remainder onto the acceptor substrate. During a MAPLE-DW transfer the majority of the ink is not vaporized. This allows complex suspended powder materials and dissolved chemicals in the organic vehicle to be transferred without substantially modifying their predeposition properties. Furthermore, there is no appreciable heating of the substrate onto which the material is transferred.

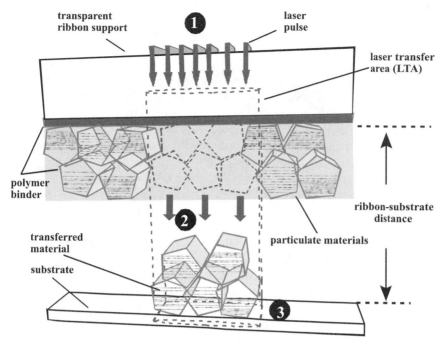

FIGURE 2   Diagram of MAPLE-DW ribbon, ink layer, and substrate during transfer.

The MAPLE-DW process is easily automated by mounting both the acceptor substrate and the ribbon onto stages that move independently of each other by computer-controlled stepper motors. By appropriate control of the positions of both the ribbon and the substrate, with respect to the laser focus, complex patterns can be fabricated. By changing the type of ribbon used (such as a CD carousel), multicomponent structures can easily be produced.

3.2. MATERIAL AND INK ISSUES

Each application of MAPLE-DW requires different final materials properties and different ink design. For example, the temperature limitations in a specific application will strongly affect materials choices and chemistry selection. Another example is that many electronic material properties improve with increasing density. Inks of these materials must be designed to minimize porosity, and this requirement drives much of the powder selection and metalorganic precursor chemistry in these applications. However, battery and sensor applications may greatly benefit from directly written material that is highly porous, yet adherent to the underlying substrate. Inks for these devices might contain the active material in powder form, a crosslinking polymer that rigidly connects the particles together, and an excess of organic solvent that will evaporate away and leave high porosity. A third example is selecting the liquidity or solidity of the ink at room temperature. In circumstances where surface-tension forces are heavily relied on to convert multiple transfers into a single homogeneous layer, the flow properties of a low-viscosity liquid are necessary, and the ink layer is likely to take the form of a rheological fluid. When this is not necessary, a solid, polymer-based composite material may produce a more physically stable ribbon. As a final example, a biological ink must contain the proper buffers or nutrients to maintain the active biomolecules or living cells of interest.

In consideration of starting materials, there is one additional complexity. The laser-based nature of MAPLE-DW introduces a host of possible laser-material interactions that could affect the properties of the final written material—ink properties could be modified depending on the laser energy. Fortunately, practical experience has shown that if an ink has a relatively high optical absorption at the transfer laser wavelength of interest, the ink properties are not substantially modified during the transfer process. This is explained by the fact that the majority of the laser energy is absorbed near the ink-carrier interface and does not interact with the majority of the volume. Therefore, the properties of the resulting functional material depend on the starting material composition and any subsequent thermal, photothermal, or chemical processing, and are essentially independent of the transfer process. This makes the design of an ink for MAPLE-DW very similar to other printing or manufacturing methods (inkjet printing, screen printing, etc.), with the one added caveat that optical absorption must be high for the ink and low for the carrier. Fortunately, many inks already exhibit sufficient optical absorption for MAPLE-DW, because most concentrated powder dispersions are either strongly absorbing or strongly scattering. This means that many inks currently in use by the electronics industry can be used with MAPLE-DW. For example, conductor, dielectric, and resistive inks for screen printing are compatible for MAPLE-DW with little or no modification. The existing body of ink design knowledge can largely be applied directly to the MAPLE-DW process for use in areas such as rheological control, in methods of obtaining and stabilizing homogeneous dispersion, and in obtaining desired functional properties. Needless to say, this has accelerated the development of MAPLE-DW and will facilitate industrial acceptance of the technique.

3.3. RIBBON AND MANUFACTURABILITY ISSUES

Due to the high energy density of the focused laser pulse, a ribbon substrate for MAPLE-DW must possess > 90% transparency at the operating wavelength in order to mitigate energy loss, heating, and optical damage of the ribbon during writing. This effectively limits ribbon material choice to quartz, glasses, and polymers (other transparent materials being too expensive). Final ribbon selection for a specific application must be determined by the laser wavelength and manufacturing considerations. In practice, manufacturing considerations and instrument design will determine the ribbon material and geometry. MAPLE-DW instrument designs can be broken down into two categories: those that use a flat, rigid ribbon and those that use a flexible ribbon, known as the reel-to-reel approach.

The rigid ribbon is made to either raster or rotate on a spindle, independent of substrate movement. The moving ribbon continuously provides fresh starting material at the laser-material interaction zone. A rigid surface facilitates application of the starting material thick film. The primary drawback to this approach is the limited amount of ribbon surface area available for direct writing, necessitating frequent ribbon changes. This type of instrument design is used in research environments because it is easily implemented, and used in biological applications because cells may be easily cultured directly on a rigid ribbon.

The reel-to-reel design provides a large surface area of material for direct writing, which can be made available in a compact volume. Thin, flexible laser-transparent polymer tapes of great length are available commercially, and are considered the only practical ribbon choice for reel-to-reel MAPLE-DW. The primary disadvantage of reel-to-reel applications is formation and stability of the starting material layer. If the coated ribbon is to be made ahead of time, the starting material must be sufficiently stable to be stored on a reel, preventing the use of rheological fluids. Application of starting material just before transfer removes this limitation, but requires more complex instrumentation and expense.

Table 1 summarizes some of the currently available ribbon-manufacture techniques along with relevant parameters.

3.4. UNDERSTANDING THE MAPLE-DW PROCESS

The transfer kinetics of the MAPLE-DW process is a complicated function of the pulsed laser conditions and the ribbon's optical, thermal, and rheological properties. While conceptually similar to the LIFT process, the kinetics and outcome of the MAPLE-DW process can be significantly different. Mechanistically, MAPLE-DW can be broken down into three events: (1) absorption of the laser radiation by the matrix or particles; (2) vaporization and bubble formation of the matrix material by direct absorption or heat transfer from the particles; and (3) expansion of the bubbles at the ribbon/support interface, resulting in plume formation.

TABLE 1 Summary of Ribbon Fabrication Processes Used in MAPLE-DW

Process property	Screen printing	Spin coating powders	Spin coating sols	Inkjet printing	Evaporation	Electro-phoretic	Electro-less	Jet-vapor coating
Thickness resolution	3–50 μm	1–2 μm dependent on particle size	<1.0 μm	Dependent on particle size	Å level	5–10 μm	Å	>2 nm
Thickness variation	Good	Poor particle size distribution, mass	***	***	Controlled by deposition area	2–3 μm	***	2 nm
Coating time	180 sec	30 sec	30 sec	360 sec	3,600 sec	300 sec	8–10 hrs	300 sec
Post-anneal	Yes, polymer removal	Yes, helpful (200–300 °C)	Yes T > 500 °C	Yes, polymer removal	No	No	Specific to system	No
Reaction process	None	No	Yes	None	None	Specific	Yes	No
Ribbon agglomeration	Controlled by slurry dispersion	Controlled by dispersion	None	Controlled by dispersion	None	Controlled by dispersion	Controlled by dispersion	None
Polymers used	Variable	Variable: PBMA, PP, PBMA, PEG	Variable	Variable: PBMA, PP, PBMA, PEG	None	None	None	None
Polymer problems	Laser absorption	Laser absorption, agglomeration	LIFT	Laser absorption, agglomeration	None	None	None	None
Ribbon adhesion	Post HT—good	Good	Good	Good	Specific to material	Poor	Good	High
Materials	Powder forms	Powder and slurry forms	Sol must be available	Powder, colloidal forms	Metals, simple oxides	Conducting substrate	Colloids	Wide classes
Stoichiometry	Powder dependent	Powder dependent	Dependent on sol	Powder, colloid dependent	Poor	Powder dependent	Metal salts, solution dependent	Specific materials

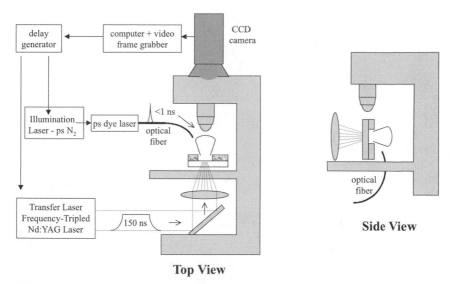

FIGURE 3 Schematic set-up for time-resolved microscopy with top-view and side-view geometries.

In order to better understand dynamics of the laser-induced transfer of inks in ambient conditions, ultrafast time-resolved optical microscopy of the MAPLE-DW process was performed by Young et al. (15). The transfer material in question was an ink of barium zirconium titanate powder (BZT) dispersed in α-terpineol, and a quartz plate was used as a carrier. The ultrafast microscope setup is shown in Fig. 3. A CCD camera and computer/frame-grabber combination triggered a variable-delay generator, which fired a 150 ns, 355 nm frequency-tripled Nd:YAG ablation laser and a 600 ps, 500 nm N_2/Dye strobe laser. The transfer pulse was focused to a 45 μm diameter spot, through the support and onto the ink layer, with a nominal fluence (averaged over the roughly Gaussian beam profile) of 0.15 J/cm². The microscope objective was focused onto the interface spot, which was illuminated by the strobe laser pulse to produce a single image frame captured by the CCD and computer (16,17). One image at each time delay was selected to generate a reconstructed, stop-action movie of the ablation dynamics. To ensure that each image was truly representative, at least three images were examined at each delay setting. As indicated in Fig. 3, the microscope was arranged in two different configurations to view either the BZT/matrix coating from the top or the BZT/matrix plume from the side.

Figures 4 and 5 show the time series of top-view and side-view images of the MAPLE-DW process. Time zero (± 100 ns) refers to a point half-way up the rising edge of the nominally 150 ns ablation pulse. Several distinct stages in

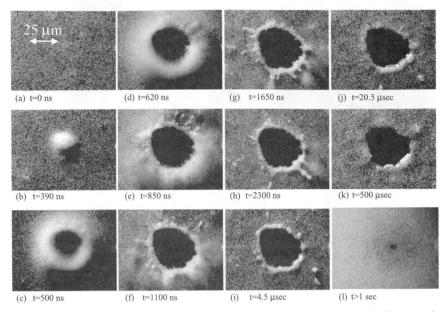

FIGURE 4   Time-gated imaging from a top-view perspective is shown (a–l) for a single BZT/terpineol transfer ($\lambda = 355\,nm$, pulse width = 150 ns, spot size = 25 µm).

the transfer process can be seen by considering Figs. 4 and 5. The first stage that is observed is the initiation of the expansion front which occurs after $\sim 350\,ns$ (from Figs. 4b and 5b). The expansion front propagates and the first evidence of material removal from the ribbon support occurs at $\sim 500\,ns$ (Figs. 4c and 5c). As time progresses up to $\sim 850\,ns$, the ablated spot increases in diameter (Figs. 4b–e) due to the roughly Gaussian laser pulse that is more intense at the center. During the time of maximum spot diameter (about 1–100 µs), there is a pile-up of BZT/α-terpineol material at the edge of the ablated spot. Eventually ($t > 1\,sec$) the fluid BZT/α-terpineol gradually fills in the desorbed area (compare Fig. 4k with 4l). In Fig. 5, the position of the expansion front is plotted versus time. A linear least-squares fit of the data gives a front velocity of $\sim 200\,m/s$.

In comparison of MAPLE-DW with previous LIFT transfer studies of carbon black (18), dye-doped polymers (19), and metal thin films (20), the most notable difference is the delayed response of the MAPLE-DW process. As stated, the first indications of material removal are apparent at $\sim 500\,ns$, and the desorbed region diameter continues to increase until $\sim 1\,\mu s$. Both times are *long after the end* of the UV ablation pulse. These observations are in stark contrast to LIFT-processed events where the first signs of material removal typically appear at about $\sim 50\,ns$ (depending on pulse fluence), and the diameter maximum is observed right at the end of the laser pulse.

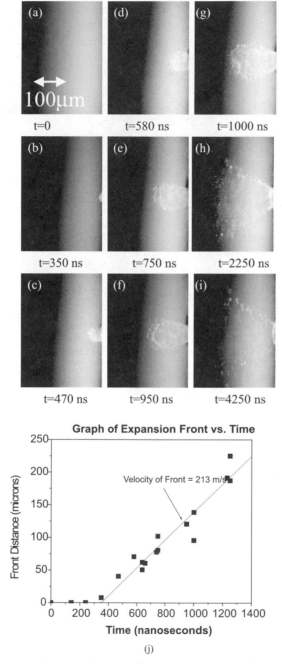

Graph of Expansion Front vs. Time

Velocity of Front = 213 m/s

(j)

FIGURE 5 Time-gated imaging from a side-view perspective is shown (a–i) for a single BZT/terpineol transfer ($\lambda = 355$ nm, pulse width $= 150$ ns, spot size $= 25\,\mu$m). (j) is a plot of the expansion distance vs. time showing a front velocity of ~ 0.2 km/s.

The delayed response of the BZT/α-terpineol composite is attributed to the particle nature of the absorbing medium and the high viscosity of the fluid matrix. The α-terpineol is transparent to the 355 nm laser radiation, therefore the incident 355 nm light is absorbed by the highly concentrated BZT particles. Most of the laser pulse is absorbed by the BZT particles near the ribbon/support interface (21). The heat is subsequently transferred from the BZT particles to the α-terpineol matrix and bubbles form and propagate from the BZT/α-terpineol interface. The formation and expansion of these bubbles is a function of the viscosity of the matrix, which in this case is relatively slow due to the high viscosity of the α-terpineol. The formation and expansion of many bubbles at the support-composite interface is the mechanism of plume generation. As can be seen by this particular example, the transfer mechanism is a function of the optical, thermal, and rheological properties of the ribbon composite material.

4. MAPLE-DW OF INORGANIC MATERIALS

In this section, we discuss MAPLE-DW of inorganic materials: metals, ceramics, resistors, and phosphors. Mesoscopic patterns of these materials are used primarily in the electronics and display industries. In many applications, it is desirable to direct write inorganic materials with a maximum processing temperature of 400 °C. This allows the fabrication of passive electronic components and RF devices with the performance of conventional thick-film materials, but on low-temperature flexible substrates such as plastics, paper, fabrics, etc. Materials such as silver, gold, palladium, and copper conductors, polymer thick film, ruthenium oxide-based resistors, and metal titanate-based dielectrics all have applications in the direct writing of electronics on low-temperature substrates.

Many of the desired electronic properties of inorganic direct-written materials improve with increasing density and crystallinity. Note, however, that the fabrication of fully dense, crystalline materials is usually impossible at these low temperatures. A common strategy to maximize density for conductors and dielectrics is to use a starting material that contains a mixture of polydispersed powder and metalorganic precursor. The polydispersity and shape of the powder particles is optimized such that density is maximized as the particles form a random close-packed network. Upon thermal treatment, the metalorganic precursors decompose and partially fill in the interstitial space between the powder particles. This process also chemically "welds" the powder particles together, improving hardness and adhesion. There is a wide range of chemistries that can be used to accomplish this. Similar effects can be induced by vapor, liquid, and/or gas co-reactant processes. Care must be taken to avoid chemistries that are incompatible with other fabrication-line processing steps, or which cause carbon and hydroxyl-group incorporation that will cause high losses at microwave frequencies. To further improve materials properties, especially of oxide ceramics, laser surface sintering is often used to enhance particle–particle bonding and densification without exposing thermally sensitive substrates to high temperatures. In most cases, individual direct-write techniques make trade-offs between starting material properties that are amenable with the transfer process, optimal functional material properties, and writing properties such as resolution or speed. However, it is possible that well-designed starting materials, for a given direct-write process, can produce a functional material that performs nearly as well as that derived through more conventional manufacturing methods.

4.1. METALS AND CERAMICS

The successful direct writing of metals and ceramics is critical if MAPLE-DW is to be utilized as an electronics rapid-prototyping technology. Conductive metallic traces make up a large portion of the surface area on circuit boards. Dense, high-quality dielectrics are necessary if capacitors are to be directly written. (Note that resistors are necessary as well, but these composite materials are more complicated and will be covered separately in the next section.)

The MAPLE-DW system used to write metallic and dielectric patterns (12,13), is shown in Fig. 6a. The third harmonic of a pulsed Nd:YAG laser (355 nm, 15 ns pulse width) was used for all transfer experiments. By changing the aperture size, beam spots ranging from 20 to 80 μm were generated, and a typical beam spot is shown in Fig. 6b. The laser fluence ranged from 1.6 to 2.2 J/cm^2, and was estimated by averaging the total energy of the incident beam over the irradiated area. A vacuum chuck and computer-controlled stage (x-y-z) was used to control the relative travel and positioning of the substrate relative to the ribbon. The substrate to ribbon gap was set at 50 and 75 μm for the metallic and dielectric transfers, respectively. Various substrates were used for the transfer experiments including silicon, glass, alumina, and polyimide.

Fused silica quartz disks were used as ribbon carriers. Proprietary silver and dielectric inks were applied using the roller-bar deposition technique, with a controlled thickness ranging from 1.5–10 μm. The Ag ink used in this study was composed of an organic vehicle, Ag particles (400 nm to 2 μm diameter), an Ag metalorganic precursor, and a dispersant. The dielectric ink was water based, and composed of barium titanate powder (BTO, 125 to 175 nm diameter), a titanium dioxide precursor, and dispersant.

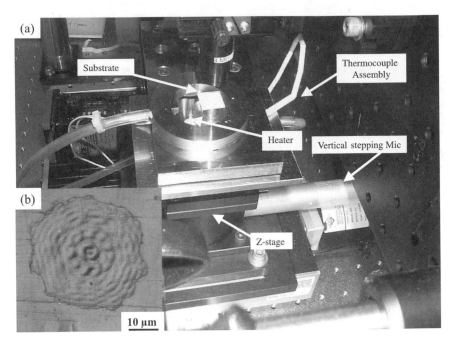

FIGURE 6   MAPLE-DW system incorporating computer controlled X-Y-Z stages for ribbon and substrate manipulation with *in situ* heating (a), and typical appearance of a laser spot for size reference on Kapton (b).

Silver lines were fabricated using a 50 μm square profile beam and 3 μm thick ribbons, and thermally processed to evaporate the organic vehicle and decompose the precursor to metallic silver. Figure 7 shows a 3-dimensional profile of the surface of the Ag ribbon after thermal processing, showing the transferred sections. Figure 8 illustrates a 40 μm wide line that is 6 μm thick, written with 2 passes with MAPLE-DW on a polyimide (Kapton) substrate. Figure 8a shows that the line morphology is uniform and the aspect ratio is within 20% as shown in Fig. 8c. Figure 8b,d shows that the MAPLE-DW process can produce lines with high density and uniform thickness. In addition to density and conductivity, the adhesion of the written devices is important as well. Figure 9 shows the results of a scotch-tape testing experiment performed on Ag lines written on a glass substrate. Figure 9a,b shows scanning electron microscope (SEM) micrographs prior to tape application, whereas Fig. 9c,d shows SEM micrographs after the tape has been removed. It is clear that the Ag lines survived the tape test. On closer inspection, it can be observed that partial polymer glue from the tape is still left on the line from the pull off, showing

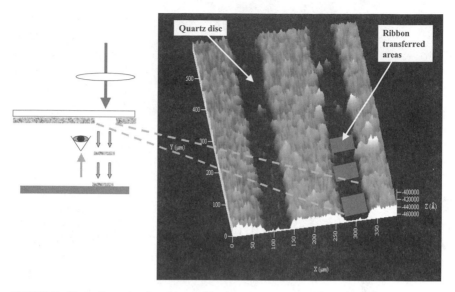

FIGURE 7   Three-dimensional surface profile of an Ag precursor ribbon after processing with a 50 μm square profile beam.

that the metal particles have a degree of "self adhesion" in addition to the adhesion at the glass/metal line interface.

Both morphology and conductivity are important parameters for direct-written metal lines. The Ag inks used in this study are designed to form a random close-packed array of Ag particles, fused together by decomposition of the Ag metalorganic precursor. Figure 10 shows the progress in optimization of conductivity of Ag metal lines, by comparing an older ink formulation (A) to a more recent, optimized composition (B). The differences between A and B are the particle size and shape, and amount of precursor and vehicle. Currently, Ag MAPLE-DW lines have been written with conductivities that are twice that of bulk Ag. These lines are uniform and reproducible on glass, alumina, Kapton, and silicon substrates.

Interdigitated capacitors were fabricated by MAPLE-DW, by depositing dielectric ink onto a prefabricated interdigitated electrode structure. Figure 11 shows an interdigitated capacitor written by MAPLE-DW. The dielectric ink was 2 μm thick, and was transferred with a 40 μm circular laser spot, as shown in Fig. 12. The density of the transferred material, after thermal treatment, is clear from SEM micrographs in Fig. 11b,c, including the metal fingers. Figure 11d shows a high-magnification SEM micrograph of a fracture cross-sectioned region. The packing density is uniform with clear evidence of the TiO_2 precursor filling the voids between the particles. Materials characterization

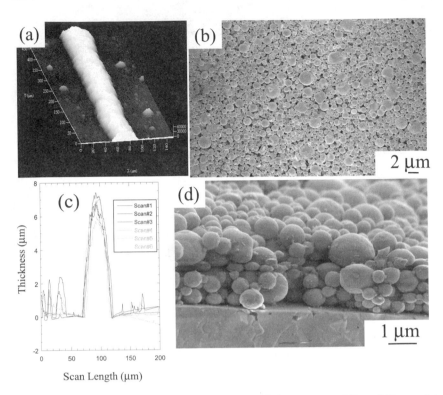

FIGURE 8   Ag line written on Kapton by MAPLE-DW, (a) shows a Tencor 3-D partial image of the 1 cm long line, (b) SEM micrograph showing the high packing density that the transferred materials exhibit, (c) Tencor 2-D line profile scans showing a 40 μm linewidth, (d) SEM micrograph of the Ag line further illustrating the packing density of the transferred material.

has shown that final dielectric properties depend strongly on the precursor choice, particle size, composition, and processing parameters. Note that parallel plate capacitors were also written in this study, but the metallic electrodes were fabricated by e-beam evaporation, not direct write. Final properties can exhibit a wide range of values as shown in Fig. 13.

The inks used in this study were designed for thermal processing below 400 °C, in order to allow deposition on polymeric substrates. However, materials properties can be improved by treatment at higher temperatures. One way to effectively increase the processing temperature, but still maintain low substrate temperatures, is to use *ex situ* or *in situ* laser surface sintering. By using *in situ* laser sintering, some of the following benefits may be realized: higher available ink temperatures, cooler substrate temperatures, faster processing, less thermal stress at the ink/substrate interface, fewer processing steps, and possibly higher yield.

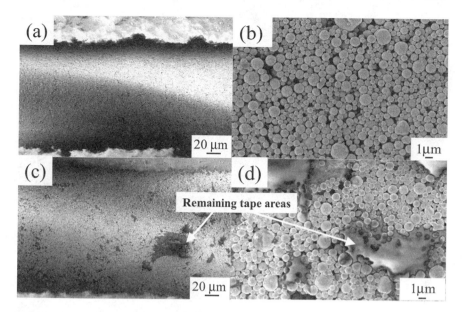

FIGURE 9 SEM micrographs of 1 cm Ag lines before and after scotch-tape adhesion testing, written by MAPLE-DW. (a) and (b) represent the Ag line prior to testing and (c) and (d) illustrate the lines after scotch-tape removal, noting the excellent adhesion and residual polymer adhering to the surface of the Ag line.

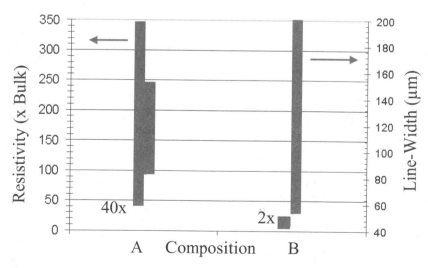

FIGURE 10 Schematic diagram showing the current progress writing metal lines with MAPLE-DW at NRL.

536

FIGURE 11   SEM micrographs of MAPLE-DW of BTO on an interdigitated capacitor structure. (a) Cross-section micrographs (b), (c), and (d) reveal that the material is uniform, dense, and exhibits low matrix porosity.

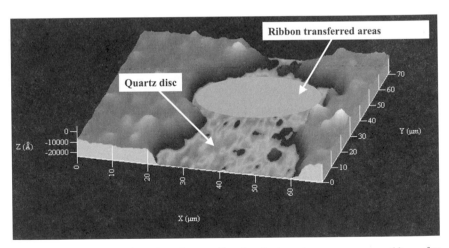

FIGURE 12   Three-dimensional surface profile of a barium titanate precursor ribbon after processing with a circular beam, 40 µm spot size. It is clear that the MAPLE-DW process efficiently removes the ribbon material with little or no debris left behind.

Capacitor Properties

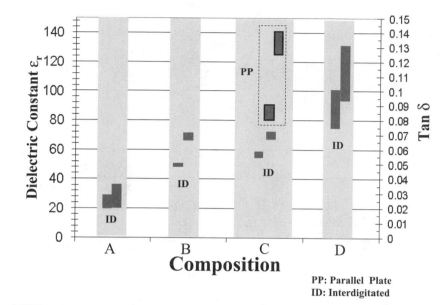

PP: Parallel Plate
ID: Interdigitated

FIGURE 13 Properties of capacitors fabricated by MAPLE-DW in both interdigitated (ID) and parallel plate (PP) geometries. A range of compositions is shown: (A) commercial BTO powder 2–3 μm diameter (irregular shaped), no precursor, (B) 150 nm BTO (spherical), 100 nm titania (spherical), no precursor, (C) 150 nm BTO (spherical), no precursor, (D) 150 nm BTO (spherical), titania precursor.

The final properties of metallic and dielectric materials written by MAPLE-DW are strongly dependent on ink composition. Aside from application specific requirements, note that inks must be designed to produce high-density (low-porosity) structures. In all cases, higher density improves the electrical performance of metals and dielectrics. Microstructure is also a critical issue in some applications. The effect of density and microstructure on different passive components and the parallel effect on device performance are given in Table 2.

4.2. RESISTORS

Thick-film resistors (22–25) are an integral part of hybrid microelectronics circuitry with a wide variety of low- and high-frequency applications in up-to-date electronics, telecommunication, and RF circuits. They are extensively used because they can satisfy certain design features like high sheet resistance,

TABLE 2   Microstructural and Device Issues for Passive Components

Passive component	Microstructure issues	Device issues
Metallic conductors	Intermediate melting point, necking, porosity	Microwave surface resistance, power loss
Dielectrics (ceramic oxides)	High melting point, difficult to neck particles, oxygen loss	Porous material has drastic effect on k, lossy
Resistors (ceramic insulator)	High melting point, most difficult to neck around insulator	Conductor/insulator composite
Resistors (polymer/insulator)	Low processing temperature, necking around insulator	Aging, electromigration
Ferrites	High melting point, difficult to neck particles, oxygen loss	Porous material has drastic effect on M_s, lossy, high H_c

breakdown field, power dissipation, and frequency response. These properties can be difficult or impossible to achieve with silicon monolithic integrated circuits. Resistors can account for over 50% of all passive components, thus making them a key target of miniaturization and direct-writing efforts. There are two major categories of thick-film resistor formulations: ceramic-metal (cermet) and polymer thick film (PTF). Cermet resistors exhibit superior device properties, but require > 800 °C heat treatment, whereas PTF resistors offer sufficiently low processing temperatures to be suitable for deposition on flexible organic substrates.

PTF resistor (23,26–29) materials are complex systems characterized by three basic constituents, a nonconducting polymer matrix, a particulate conducting phase, and an organic vehicle. Polymer thick-film resistors suffer from a number of limitations such as high TCRs (temperature coefficient of resistivity) and susceptibility to humidity which restrict their use in high-precision applications. Despite these limitations, PTF resistors are being used in flexible membrane switches, computer peripherals, telecommunications, and medical electronics products (26). The potential applications for these materials are increasing rapidly and efforts are being made to improve their properties (23).

Modi et al. has used MAPLE-DW to fabricate PTF resistor elements (30). The inks used for this study were commercially available formulations from Metech, Inc. (PCR401 series). They are based on thermosetting epoxy resin with carbon black (CB), graphite, or silver-palladium as the conducting phase. Their properties are shown in Table 3. By optimizing the laser fluence (from 0.58 J/cm^2 to 1.09 J/cm^2) and ribbon to substrate gap ($\sim 70\,\mu m$), resistors with a controlled, reproducible geometry were successfully written. These resistor elements were characterized to determine DC resistance, impedance behavior

TABLE 3 Polymer Thick-Film Resistor Inks Used for MAPLE-DW

Ink type	Nominal resistance[a]	Composition (in epoxy resin)	Viscosity (kcps at 25 °C)[b]	Solvent used
I	10 Ω/sq.	Ag/Pd	~31.0	None
II	100 Ω/sq.	CB/graphite	~32.3	None
III	1 kΩ/sq.	CB/graphite	~31.1	None
IV	10 kΩ/sq.	CB/graphite	~29.5	Butyl carbitol
V	100 kΩ/sq.	CB/graphite	~29.0	Butyl carbitol

[a] These values correspond to 16 µm print thickness.
[b] Viscosity was measured using Brookfield viscometer (model-HBT) with spindle CP-51 at 1 RPM speed.

from 1 MHz to 1.8 GHz, and temperature coefficient of resistance (TCR) from 25 to 125 °C.

The results of MAPLE-DW transfer of these resistor inks are summarized in Table 4. Note the differences in coefficient of variation (a measure of resistance reproducibility). Variation ranged from ideal, by industrial standards, for types II and III to unacceptable for type I. Characterization revealed that resistor thickness variation was the primary source of resistance variation (25), though the addition of butyl carbitol to types IV and V, necessary for laser transfer, also contributed to resistance variation.

Figure 14 shows relationships between TCRs and the sheet resistivities of various MAPLE-DW PTF resistors. Type I resistors (10 Ω/sq) show mostly positive TCR throughout the temperature range of 25 to 125 °C. Values are quite low at < 80 °C but increase substantially between 80 and 125 °C. Type II resistors (100 Ω/sq) have negative TCR which becomes less negative above ~ 100 °C. Type III (1 kΩ/sq), IV (10 kΩ/sq), and V (100 kΩ/sq) resistors show the similar trend with negative TCRs with subsequent change to less negative values as the temperature is increased. In the cases of type III and IV resistors

TABLE 4 Result Summary of Polymer Thick-Film Resistors Deposited by MAPLE-DW

Ink type	Average sheet resistance[a]	Minimum sheet resistance	Maximum sheet resistance	Standard deviation	CV (%)	Sample size
I	35.3 Ω	21.3 Ω	45.15 Ω	7.7 Ω	21.8	10
II	96.9 Ω	90.0 Ω	106.0 Ω	4.5 Ω	4.6	10
III	1.21 kΩ	1.14 kΩ	1.33 kΩ	0.06 kΩ	5.0	10
IV	22.5 kΩ	16.87 kΩ	26.16 kΩ	3.3 kΩ	14.7	10
V	15.6 kΩ	96.88 kΩ	135.10 kΩ	16.2 kΩ	14.0	4

[a] Sheet resistance is normalized to 16 µm thickness.

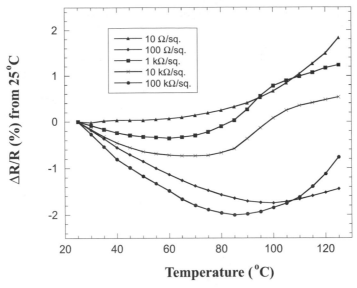

FIGURE 14 Variation of TCR with temperature of MAPLE-DW PTF resistors.

TCRs even became positive at higher temperatures. In general, with the exception of type II ($100 \, \Omega$/sq) resistors, the TCRs became more negative with increasing sheet resistivity for the other types of resistors. These observations can be explained in terms of the interplay between thermal motion of conducting phase particles and the thermal expansion of the polymer resin. The reader is referred to the published literature for further detail in this topic (25,27,28,30,31).

4.3. PHOSPHORS

In the rapidly growing information age, the current worldwide display market is $40 billion and is estimated to reach $60 billion in 2005 (Stanford Resources), and one of the fastest-growing segments in this industry is flat panel displays. Liquid crystal displays (LCD) currently dominate the flat panel display market. However, several emerging display technologies such as field emission displays (FED), electroluminescent displays (EL), and plasma displays (PD) are also being developed. These emerging technologies are all optically emissive—rather than transmissive—displays, which result in a wider viewing angle. While field emission displays, electroluminescent displays, and plasma displays have different excitation mechanisms and operating conditions, all require efficient light-emitting materials, known as *phosphors.* A significant amount of materials research is being conducted to develop efficient red, green, and blue phosphor materials.

Equally important to these industries are effective processing techniques for making the phosphor screens. Because the MAPLE-DW technique can effectively deposit particulate materials, the process is particularly useful for field emission display applications where the phosphor materials are fine powders. Current technologies to fabricate multicolor phosphor screens include multistep electrophoretic deposition, screen printing, and settling techniques (32–40). While shadow masks can be used for large pixel sizes, smaller pixel sizes require photolithographic techniques to pattern three colors (41). Photolithographic techniques require two or more process steps per color and each step can leave organic residue on phosphor particles. Under electron bombardment, it has been shown that organic residues can react and form a thin carbonaceous layer, which absorbs emitted light and degrades the phosphor performance (42). The MAPLE-DW technique offers the capability for producing clean, high-resolution (>1,000 lines per inch) phosphor screens for rapid prototyping and possibly small volume applications.

A study of MAPLE-DW of phosphor patterns has been performed by Fitz-Gerald et al. (14). Phosphor ribbons were fabricated by sputtering a thin (100 nm) gold film (explosive release layer) onto a quartz plate, followed by the deposition of a 15 µm thick particle film of phosphor powder. The phosphor powders ($ZnSi_2O_4 : Mn$—green, $Y_2O_3 : Eu$—red, and $BaMg_2Al_{16}O_{27} : Eu$—blue) were suspended in a glycerin/isopropanol solution with $LaNO_3$ and $Mg_3(NO_3)_2$ salts and electrophoretically deposited onto the gold coated quartz wafers to form. Figure 15 shows SEM micrographs of the

FIGURE 15 Scanning electron micrographs of an electrophoretically deposited $ZnSi_2O_4 : Mn$ phosphor ribbon, (a) 1 kx and (b) 15 kx images reveal a very dense powder ribbon.

$ZnSi_2O_4$: Mn ribbon after the electrophoretic deposition, and confirms a very dense deposit.

MAPLE-DW transfers of these phosphors was accomplished with both 248 nm and 355 nm laser pulses, with similar results. Figure 16a,b shows scanning electron micrographs of the $ZnSi_2O_4$: Mn ribbon after the MAPLE-DW process. Very efficient transfer of the phosphor was achieved, as very little phosphor powder remained after the single 25 ns pulse. Close inspection of the quartz ribbon after the laser transfer revealed tiny gold droplets on the quartz wafer (Fig. 16c). In addition, micrographs of the transferred material also showed gold droplets. These observations suggest that the ~ 100 nm gold layer absorbed the laser radiation and ejected the thick phosphor powders toward the receiving substrate. Calculations indicate that < 99% of the unreflected radiation is absorbed in the 100 nm of gold, which supports the idea that the gold matrix layer is absorbing and propelling the phosphor particles toward the receiving substrate. Scanning electron microscopy and 3-dimensional surface profilometry measurements of the transferred $ZnSi_2O_4$: Mn phosphor powders demonstrated that very good linewidth control and very dense thick-film

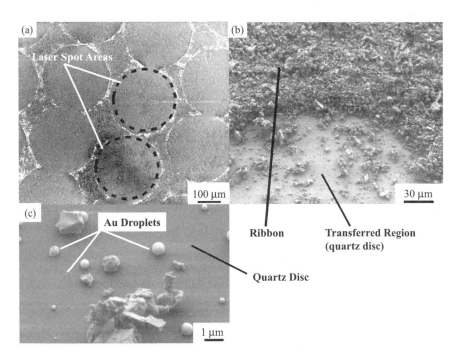

FIGURE 16   Scanning electron micrographs of a $ZnSi_2O_4$: Mn phosphor ribbon after the MAPLE-DW process at (a) 90 ×, (b) 400 ×, and (c) 10 kx resolution images reveal efficient transfer of the ~ 300 μm diameter and ~ 15 μm thick phosphor material.

transfers were achieved. Transfers of Y_2O_3 : Eu and $BaMg_2Al_{16}O_{27}$: Eu phosphors exhibited similar characteristics. Various substrate materials such as glass, alumina, Kapton, and silicon were successfully deposited using the MAPLE-DW process.

To examine the effect that the MAPLE-DW process has on the phosphor material properties, cathodoluminescence (CL) emission spectra and efficiency measurements were performed on the phosphor-coated ribbons prior to the MAPLE-DW process, and on the transferred phosphor. The emission spectra were measured from Zn_2SiO_4 : Mn and Y_2O_3 : Eu ribbons prior to the transfer and on the transferred substrates. The pre-transfer and post-transfer emission spectra were identical. However, the most significant test for the MAPLE-DW processing of phosphor materials is CL efficiency. This is particularly true for low voltage ($< 1,000$ V) excitation where the electron beam penetration is very shallow and any process-induced surface contamination can seriously degrade the performance. The CL efficiencies measured at 1 kV and 20 $\mu A/cm^2$ before and after the MAPLE-DW process were 1.3 lumens/Watt for the Zn_2SiO_4 : Mn and 1.1 lumens/Watt for the Y_2O_3 : Eu. The consistent CL spectra and efficiencies before and after the laser processing confirmed that the MAPLE-DW process did not deleteriously affect the phosphor material. This process is particularly attractive for high-resolution display applications because the linewidth control of $< 10 \mu m$ can print line densities greater than 2,500 lines/inch.

5. MAPLE-DW OF ORGANIC AND BIOMATERIALS

The fabrication of mesoscopic patterns of biological materials is a new and active field. Viable methods to generate mesoscopic-scale patterns of viable cells and active biomaterials are required to build 3-dimensional cellular structures for advanced tissue engineering (43–45), to fabricate next-generation tissue-based sensing devices (46–50), and to selectively separate and differentially culture microorganisms for a variety of basic and applied research applications (51–53). Currently, these patterns can be formed through inkjet technologies (54–56) or various soft lithographic techniques that utilize self-assembled monolayer, microcontact printing, and photolithography (57–63). Inkjet is a true direct-write approach, with the corresponding advantages, and several researchers have successfully used this technology to pattern proteins and antibodies for sensor applications (56). Soft lithography techniques create patterns of biomaterials with excellent resolution, and are capable of forming adjacent patterns of cells by using microfluidic channels to control the placement of different cells and biomaterials (61–63). Other biopatterning techniques are also laser based. Laser capture microdissection (LCM) is capable of transferring and separating specific cells from a large sample, but cannot transfer viable microorganisms (64,65). Ultraviolet photoablation can micromachine biological substrates (66–68), and laser-guidance techniques can generate patterns of limited types of biomaterials (69,70).

The transfer and patterning of biomaterials by several variations of MAPLE-DW have been demonstrated by Ringeisen et al., who successfully wrote patterns of active proteins, viable *E. coli*, and mammalian Chinese hamster ovary cells (CHO). This recent work demonstrates that MAPLE-DW is capable of not only forming mesoscopic patterns of viable bacteria and cells but also of other biological materials (tissue, proteins, enzymes, DNA, antibodies, etc.) that are active or living in aqueous solution. The capability of MAPLE-DW to generate mesoscopic patterns—not only the biomaterials of interest but also those that are needed for growth (nutrients) or for the preservation of biological activity (gel or matrix that sustains the molecular or active structure)—makes the approach unique. This is due to the capability of the MAPLE-DW technique to transfer starting materials with a wide range of rheology, from thin fluids to solid materials.

For most biological MAPLE-DW applications, the ribbon material is an active biomaterial in an aqueous buffer solution. In order to attain sufficient absorption of laser energy, emitting 193 nm laser pulses (ArF excimer) is typically used for transfer. The cells or biomaterials are stabilized on the ribbon by freezing a layer of cells and cell media onto the ribbon. Biomaterials and cells can also be transferred using room-temperature ribbons by mixing a composite of cell media (or biomolecule) and a biocompatible material such as nutrient agar, collagen gel, a polymer, or an inert ceramic (71–75). The transferred cells suffer no noticeable adverse effects from laser heating or other photon-induced processes. The absorption of the laser pulse is sufficiently localized at the ribbon interface to avoid melting in the bulk of the frozen ribbon.

The direct writing of viable *E. coli* bacteria, CHO cells, and active proteins onto various substrates with MAPLE-DW is a significant advance in biomaterial processing and shows progress in our understanding and manipulation of natural systems. Because of the gentle transfer mechanism, this approach is capable of writing mesoscopic patterns of several biomaterials classes including cells, proteins, DNA strands, and antibodies onto a variety of technologically important substrates. Because MAPLE-DW is able to micromachine substrates as well as sequentially deposit passive electronic devices adjacent to viable cells and active biomolecules, it possesses all the tools necessary to rapidly fabricate unique biosensors and bioelectronic interfaces. In the future, it will be used to produce improved microfluidic biosensor arrays, to electronically probe intercellular signaling, to control the transfer and placement of pleuripotent mammalian cells for differential culturing, and to even form 3-dimensional biological structures not found in nature (e.g., combinations of unique cells or cell arrays).

The following sections discuss the direct writing of an active hydrogen peroxide sensitive protein (horse radish peroxidase or HRP, MW = 31,000 Da), and viable patterns of bacteria and mammalian cells. The experimental results presented emphasize the universal potential that this approach has to form patterns of many active biomaterials and viable microorganisms.

5.1. ACTIVE PROTEIN PATTERNS

Patterns of the hydrogen-peroxide-sensitive protein horse radish peroxidase (HRP) were formed on silicon and glass substrates using the laser-transfer technique outlined before. Protein was transferred using both a ribbon coated with a room-temperature polyurethane/HRP composite and a frozen solution containing HRP and phosphate buffer solution. The activity of the protein pattern was tested using an optical microscope to observe the reduction of 3,3′-diaminobenzidine (DAB) by HRP in a dilute H_2O_2 solution. After several minutes, a dark brown color characteristic of decomposed DAB was observed around the transferred HRP pattern, indicating that the deposited protein retained activity after laser transfer.

5.2. VIABLE *E. COLI* PATTERNS

Patterns of viable *Escherichia coli* JM109 containing pKT230 :: gfp (green fluorescent protein reporter plasmid) (63,64,76) were written with MAPLE-DW onto Si(111), glass slides, and nutrient agar culture plates. These *E. coli* cells contained the jellyfish *Aequorea Victoria* green fluorescent protein (GFP), and have been used to assess cell viability and to positively identify the transferred microorganisms from possible contaminants (77). In order to determine whether viable *E. coli* JM109 were successfully transferred, cell patterns were first observed with an optical microscope. Figure 17a shows an optical micrograph of an *E. coli* pattern transferred using MAPLE-DW. The linewidth of the pattern is approximately 100 microns. Figure 17b,c shows micrographs of the *E. coli* pattern (portion of "R" shown) under white and black light, respectively. The characteristic fluorescence of the GFP is emitted only in the areas where *E. coli* was written. This relatively large pattern was written in order to transfer enough bacteria to observe the green fluorescence shown in Fig. 17c. When the pattern was submerged in Luria-Bertani (LB)

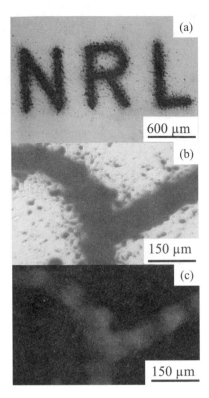

FIGURE 17   Photographs of E. coli patterns. (a) Optical micrograph of MAPLE-DW transferred E. coli pattern. Transferred E. coli under white light (b) and 365 nm UV exposure (c). Green fluorescence is observed from viable cells expressing the green fluorescent protein.

broth, fluorescence remained over a period of several days, indicating the bacteria were viable after transfer and that the composite material used as a matrix acted to immobilize the transferred cells.

The pattern shown in Fig. 17a was obtained using a quartz carrier coated with a composite mixture of E. coli, LB broth, glycerin, and a ceramic powder. Other experiments using supports coated with frozen cells and cells mixed with nutrient agar resulted in similar patterns of viable E. coli that also emitted green fluorescence upon exposure to a black light. These results demonstrate that this technique is capable of transferring patterns of viable E. coli from a variety of matrices either frozen or at room temperature. Note that these experiments were designed to demonstrate the ability of MAPLE-DW to transfer living organisms, and was not an attempt to reach the ultimate writing resolution.

Scanning electron microscopy (SEM) was used to determine if there was any laser-induced damage to the transferred *E. coli*. Figure 18a,b shows SEM micrographs of dried *E. coli* that was pipetted onto a Si(111) substrate. Figure 18a shows that the cell density is approximately 100 *E. coli*/$10^4 \cdot m^2$, and Fig. 18b demonstrates the size, shape, and structure of *E. coli* not exposed to a laser pulse. Figure 18c,d shows micrographs of dried *E. coli* after the laser-based transfer from a quartz support with frozen cell media to a Si(111) substrate. The areal density of bacteria is over 10 times smaller due to the small volume of frozen cell media transferred by the direct-write process. Figure 18d shows that transferred bacteria are undamaged by the laser energy and are identical in shape and size to the cells transferred via pipette. The external cell membrane appears intact and there is no evidence of laser heating or other destructive processes induced by the transfer. Culture experiments indicate that the transferred *E. coli* cells were viable.

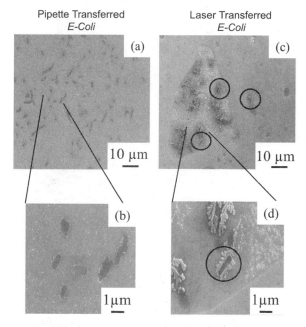

FIGURE 18 Cell observation with SEM. (a,b) SEM photos of pipette transferred *E. coli* showing undamaged cell features (not exposed to laser energy). (c,d) SEM photos of MAPLE-DW transferred *E. coli*. The crystals around the cells are due to dried LB broth that is used as the growth medium and matrix for the *E. coli*. This broth is transferred along with the cells and when dried, shows the perimeter of the *E. coli*. Comparison of panels (d) and (b) indicate that the MAPLE-DW transfer process does not alter the shape, size, or viability of *E. coli*.

5.3. Viable Chinese Hamster Ovary (CHO) Patterns

The next step in complexity for transferring living cells is to form patterns of mammalian cells. These cells are generally larger and more fragile than bacteria, and would therefore be more susceptible to the shear forces present during the laser transfer. Figure 19a is a micrograph of native (pre-transfer) CHO cells, while Fig. 19b is an optical micrograph of several CHO cells after laser transfer (the dotted circle outlines the 200 · m spot of transferred cells). Figure 19b shows the growth and reproduction of the transferred CHO cells

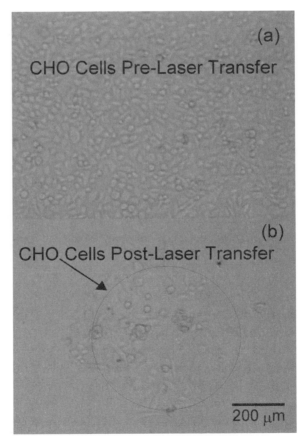

FIGURE 19   Micrograph of living Chinese hamster ovaries (CHO) on a hydrogel-coated quartz ribbon before laser transfer (a), laser-transferred 200 · m diameter pattern of living CHO cells after three days of culture (b).

after three days in growth media. We observe no damage to the plasma membrane post-transfer, and the transferred cells appear very similar to the native cells in (a). The increased size and stretched appearance (indicating attachment to the substrate) verify the viability of the transferred species.

5.4. GENERATION OF FRACTAL ANTENNAS ON LIVE HONEYBEES

There is growing interest in technologies capable of tracking flight patterns of individual insects at near-ground elevations for applications ranging from chemical and biological hazard detection to unexploded ordnance detection (UXO) (78,79). One tracking method is to place an antenna on each insect that acts to distinguish its radar signature from ground clutter. Riley et al. successfully demonstrated this concept with a harmonic radar system that uses a diode and antenna attached to a honeybee to shift the frequency of the radar waves reflected by the insect (80,81). Unfortunately, this transponder is approximately 16 mm in length, weighs 12 mg, and prohibits the insect from entering its hive.

A radar-based insect tracking system based on MAPLE-DW fractal antennas has been designed as an improvement to the Riley transponder. The planar fractal antenna designed for tracking adult worker honeybees is contained in an area of 2 mm × 2 mm and weighs less than 1 mg for a 25 GHz design. The antenna area is much smaller than the spacing between spiracles on the bee's abdomen, and the weight and size eliminate any impact the antenna will have on the bee's flight or behavior. Fractal antennas are also advantageous because they are self-loading through self-inductance and capacitance, and therefore require no matching circuitry.

Figure 20a shows a fractal antenna deposited by MAPLE-DW onto planar alumina. The antenna is conductive, resonates at 25 GHz (dielectric constant (k) = 10), and is formed from silver. The linewidth and spacing range from 20 to 40 microns. MAPLE-DW was also used to deposit the antenna on the abdomen of dead honeybees. Figure 20b shows the silver antenna deposited on the exoskeleton of the dead bee after laser ablation was used to remove insect hair over a 3 mm × 3 mm area. Even though the abdomen surface is conformal and irregular, MAPLE-DW is able to resolve the antenna linewidth and spacing successfully.

Tests were performed on live honeybees to determine if the laser processing disrupted insect viability. A total of 10 honeybees were dosed with CO_2 gas until the insects lost consciousness. The insects were then secured under the silver ribbon with a vacuum chuck. Hair from the abdomen was removed via

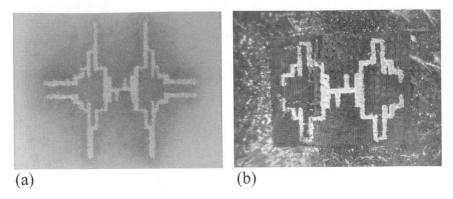

(a)   (b)

FIGURE 20   Conducting silver fractal antennae (1 × 0.8 mm) deposited by MAPLE-DW onto an alumina substrate (a), identical antennae deposited onto the abdomen of a worker honeybee (b). Hair removal (3 × 3 mm) was performed by laser micromachining prior to antennae transfer.

laser processing, and MAPLE-DW was used to coat each bee with a 2 mm patch of silver. Current laser pulse repetition rates (10 Hz) and immobilization time limited the live-bee experiments to patches instead of antenna depositions. A higher repetition rate laser would enable antennas to be deposited on live bees. All ten bees were unharmed by the laser processing and silver depositions. Each bee regained consciousness within minutes and resumed activities similar to bees not exposed to the laser treatment.

6. SUMMARY AND FUTURE WORK

MAPLE-DW is currently a new and not yet mature fabrication technique. It is expected to be adopted by industry in areas where its resolution, writing speed, flexibility, materials properties, and cost are advantageous. Despite high write speeds, it is clear that the serial nature of MAPLE-DW may prevent it from ever becoming a true mass-manufacturing technology. Fortunately, it has been demonstrated that MAPLE-DW is capable of generating mesoscopic patterns of many classes of inorganic and biological materials, which may open other doors in terms of research and commercialization. This chapter has covered metals, ceramics, polymer-based composites, phosphors, proteins, and viable cells. This is not an exhaustive list by any means, but spans a large number of potential applications. In addition, MAPLE-DW possesses the potential produce structures and devices that cannot be fabricated by current methods, particularly in the biological areas. The authors of this chapter believe that, among this large number of potential niche markets, MAPLE-DW will find acceptance as a viable research tool, a rapid prototyping tool, and/or in manufacturing technology.

Future development and improvement of MAPLE-DW will come from work in the areas of (1) direct writing of new materials; (2) improvements in inks and starting materials for MAPLE-DW; (3) improvements in instrumentation; and (4) increased basic understanding of the MAPLE-DW laser-material interactions.

ACKNOWLEDGMENTS

We gratefully acknowledge the support provided for this work from the Office of Naval Research and the DARPA MICE Program. The authors would like to thank Dr. Russel Chung, Dr. Richard Kant, Dr. Samuel Lakeou, Dr. Peter Wu, Dr. Barry Spargo, Dr. John Callahan, Dave Weir, and Matt Brooks from the Naval Research Laboratory for their discussions and assistance in this chapter. Researchers at Superior MicroPowders, Inc. (Albuquerque, NM), Potomac Photonics Inc. (Lanham, MD), and Paralec (Rocky Hill, NJ) also are acknowledged for their contributions and helpful discussions.

REFERENCES

1. D.B. Chrisey and G.K. Hubler, eds. (New York, NY: Wiley, Inc. 1994).
2. Method of Producing a Coating by Matrix Assisted Pulsed Laser Evaporation, U.S. Pat. No. 6,025,036.
3. R.A. McGill, R. Chung, D.B. Chrisey, P.C. Dorsey, P. Matthews, A. Piqué, T.E. Mlsna, and J.L. Stepnowski, *IEEE Trans. on Ultrasonics, Ferroelectrics, and Frequency Control*, 45, p. 1370 (1998).
4. A. Piqué, D.B. Chrisey, B.J. Spargo, M.A. Bucaro, R.W. Vachet, J.H. Callahan, R.A. McGill, and T.E. Mlsna, *MRS Proc.* 526, p. 421, (1998), in *Advances in Laser Ablation of Materials*.
5. J. Bohandy, B.F. Kim, and F.J. Adrian, *J. Appl. Phys.* 60, pp. 1538–1539 (1986).
6. J. Bohandy, B.F. Kim, F.J. Adrian, and A.N. Jette, *J. Appl. Phys.* 63 pp. 1158–1162 (1988).
7. I. Zergioti, S. Mailis, N.A. Vainos, C. Fotakis, S. Chen, and C.P. Grigoropoulos, *Appl. Surf. Sci.* 127–129, pp. 601–605 (1998).
8. F.J. Adrian, J. Bohandy, B.F. Kim, A.N. Jette, and P. Thompson, *J. Vac. Sci. Tech.* B5, pp. 1490–1494 (1987).
9. I. Zergioti, S. Mailis, N.A. Vainos, P. Papakonstantinou, C. Kalpouzos, C.P. Grigoropoulos, and C. Fotakis, *Appl. Phys. A* 66, pp. 579–582 (1998).
10. H. Esrom, J.-Y. Zhang, U. Kogelschatz, and A. Pedraza, *Appl. Surf. Sci.* 86, pp. 202–207 (1995).
11. S.M. Pimenov, G.A. Shafeev, A.A. Smolin, V.I. Konov, and B.K. Bodolaga, *Appl. Surf. Sci.* 86, pp. 208–212 (1995).
12. A. Piqué, D.B. Chrisey, R.C.Y. Auyeung, J.M. Fitz-Gerald, H.D. Wu, R.A. McGill, S. Lakeou, P.K. Wu, V. Nguyen, and M. Duignan, *Appl. Phys. A* 69, pp. S279–S284 (1999).
13. D.B. Chrisey, A. Piqué, J.M. Fitz-Gerald, R.C.Y. Auyeung, R.A. McGill, H.D. Wu, and M. Duignan, *Appl. Surf. Sci.* 154, pp. 593–600 (2000).
14. J.M. Fitz-Gerald, A. Piqué, D.B. Chrisey, P.D. Rack, M. Zeleznik, R.C.Y. Auyeung, and S. Lakeou, *Appl. Phys. Lett.* 76, pp. 1386–1388 (2000).

15. D. Young, R.C.Y. Auyeung, A. Piqué, and D.B. Chrisey, Time-resolved optical microscopy of a laser based forward transfer process, *Appl. Phys. Lett.* (accepted 3/2001).

16. D.E. Hare, S.T. Rhea, D.D. Dlott, R.J. D'Amato, and T.E. Lewis, Fundamental mechanisms of lithographic printing plate imaging by near-infrared lasers, *J. Imag. Sci. Technol.* **41**, no. 3, (1997).

17. D.E. Hare, S.T. Rhea, D.D. Dlott, R.J. D'Amato, and T.E. Lewis, Pulse duration dependence of lithographic printing plate imaging by near-infrared lasers, *J. Imag. Sci. Technol.* **42**, no. 2, (1998).

18. I.-Y.S. Lee, W.A. Tolbert, D.D. Dlott, M.M. Doxtader, D. Arnold, D. Foley, and E.R. Ellis, Dynamics of laser ablation transfer imaging investigated by ultrafast microscopy. *J. Imag. Sci. Technol.* **36**, 180–187 (1992).

19. W.A. Tolbert, I.-Y.S. Lee, D.D. Dlott, M.M. Doxtader, and E.W. Ellis, High speed color imaging using laser ablation transfer films with a dynamic release layer: Fundamental mechanisms. *J. Imag. Sci. Technol.* **37**, 411–421 (1993).

20. S.G. Koulikov and D.D. Dlott, Effects of energetic polymers on laser photothermal imaging materials. *J. Imag. Sci. Technol.* **44**, 111–119 (2000).

21. J.M. Sun and B.S. Gerstman, Photoacoustic generation for a spherical absorber with impedance mismatch with the surrounding media. *Phys. Rev. E.* **59**, 5772–5789 (1999).

22. R.A. West, in *Ceramic Materials for Electronics: Processing, Properties and Applications*, 2nd ed. (Marcel Dekker Inc., 1991) R.C. Buchanan, ed. pp. 435.

23. A.Y. Xiao, Q.K. Tong, and A.C. Savoca, *IEEE Components and Technology Conference*, 1999, pp. 88 (1988).

24. S. Vasudevan and A. Shaikh, *ISHM '93 Proc.*, pp. 685 (1993).

25. M.A. Jupina, C. El-Beyrouty, D.I. Amey, and A.T. Walker, *1999 International Symposium on Microelectronics*, pp. 94 (1999).

26. K. Gilleo, *Polymer Thick Film* (Van Nostrand Reinhold, 1996).

27. S.-L. Fu, M.-S. Liang, T. Shiramatsu, and T.-S. Wu, *IEEE Transactions on Components, Hybrids, and Manufacturing Technology*, **CHMT-4**(3), 283 (1981).

28. S.-L. Fu, *IEEE Transactions on Components, Hybrids, and Manufacturing Technology*, **CHMT-4**(3), 289 (1981).

29. A. Dziedzic, *Proc. of 21st International Conference on Microelectronics (MIEL '97)*, **1**, pp. 427 (1997).

30. Modi et al., unpublished data, Naval Research Laboratories.

31. E.K. Sichel, J.I. Gittleman, and P. Sheng, in *Carbon-Black Polymer Composites* E.K. Sichel, ed. (Marcel Dekker Inc., 1982) pp. 51.

32. P.D. Godbole, S.B. Deshpande, H.S. Potdar, and S.K. Date, *Mater. Lett.* **12**, 97–101 (1991).

33. H. Hu, O.G. Daza, and P.K. Nair, *J. Mat. Res.* **13**, 2453–2456 (1998).

34. D. Ivanov, M. Caron, L. Quellet, S. Blain, N. Hendricks, and J. Currie, *J. Appl. Phys.* **77**, 2666 (1995).

35. M.N. Kamalasanan and S. Chandra, *Appl. Phys. Lett.* **59**, 3547 (1991).

36. I. Koiwa and H. Sawai, *IEEE Trans. Electron Devices* **41**, 1523 (1994).

37. I. Koiwa, T. Kanehara, and J. Mita, *J. Electrochem. Soc.* **142**, 1396 (1995).

38. G. Percin, T.S. Lundgren, and B.T.K. Yakub, *Appl. Phys. Lett.* **73**, 2375 (1998).

39. U. Selvaraj, A.V. Prasadarao, S. Komarneni, and R. Roy, *Mater. Lett.* **12**, 311–315 (1991).

40. C. Surig, K.A. Hempel, and D. Bonnenberg, *Appl. Phys. Lett.* **63**, 2836 (1993).

41. S.K. Kurinec and E. Sluzky, *J. Soc. Inf. Disp.* **4**, 371 (1996).

42. C.H. Seager, D.R. Tallant, and W.L. Warren, *J. Appl. Phys.* **82**, 4515 (1997).

43. C.D. Kay, D. Puleo, and R. Bizios, *Eng. Mat. Biomed. App.* **3**, 7 (2000).

44. C.A. Heath, *Trends Biotechnol.* **18**, 391–414 (2000).

45. N. Patel et al. *J. Biomat. Sci. Polym. E.* **11**, 319 (2000).

46. L.R. Ptitsyn et al., *Appl. Environ. Microbiol.* **63**, 4377–4384 (1997).

47. S. Ramanathan, M. Ensor, and S. Daunert, *Trends Biotechnol.* **15**, 500 (1997).
48. P. Corbisier, *Res. Microb.* **148**, 534–536 (1997).
49. J.J. Pancrazio, J.P. Whelan, D.A. Borkholder, W. Ma, and D.A. Stenger, *Ann. Biomed. Eng.* **27**, 697 (1999).
50. N. Patel et al., *Langmuir* **15**, 7252 (1999).
51. S. Sattar et al., *Cell Biol. Int.* **23**, 379 (1999).
52. W.J. Pan, P.R. Haut, M. Olszewski, and M. Kletzel, *J. Hematoth. Stem Cell* **8**, 561 (1999).
53. T.E. Thomas, C.L. Miller, and C.J. Eaves, *Methods* **17**, 202 (1999).
54. J.D. Newman, A.P.F. Turner, and G. Marrazza, *Anal. Chim. Acta*, **262**, 13 (1992).
55. Roda A. et al. *BioTechniques* **28**, 492–496 (2000).
56. D.B. Wallace, W.R. Cox, and D.J. Hayes, Direct write using ink-jet techniques, in *Direct Write Technologies for Rapid Prototyping Applications*, A. Piqué, ed. (Academic Press, Boston, 2002).
57. B.J. Spargo et al., *Proc. Natl. Acad. Sci. U S A* **91**, 11070 (1994).
58. J.L. Wilbur, A. Kumar, H.A. Biebuyck, E. Kim, and G.M. Whitesides, *Nanotechnology* **7**, 452 (1996).
59. R.S. Kane, S. Takayama, E. Ostuni, D.E. Ingber, and G.M. Whitesides, *Biomaterials* **20**, 2363 (1999).
60. Y. Ito, *Biomaterials* **20**, 2333 (1999).
61. S. Takayama et al., *Proc. Nat. Acad. Sci. U S A* **96**, 5545 (1999).
62. S.N. Bhatia, U.J. Balis, M.L. Yarmush, and M. Toner. *FASEB J.*, **13**, 1883 (1999).
63. S.N. Bhatia, U.J. Balis, M.L. Yarmush, and M. Toner. *J. Biomat. Sci. Poly Ed.* **9**, 1137 (1998).
64. M.R. Emmert-Buck et al., *Science* **274**, 998 (1996).
65. R.F. Bonner et al., *Science* **278**, 1481 (1997).
66. B.A. Grzybowski, R. Haag, N. Bowden, and G.M. Whitesides, *Acc. Chem. Res.* **70**, 4645 (1998).
67. A. Schwartz, J.S. Rossier, E. Roulet, N. Mermod, M.A. Roberts, and H.A. Girault, *Langmuir* **14**, 5526 (1998).
68. R. Vaidya, L.M. Tender, G. Bradley, M.J. O'Brien, M. Cone, and G.P. Lopez, *Biotechnol. Prog.* **14**, 371 (1998).
69. D. Odde and M.J. Renn, *Biotech. Bioeng.* **67**, 312 (2000).
70. M.J. Renn and D. Odde, *Trends Biotechnol.* **17**, 383 (1999).
71. J.B. Park and S.L. Roderic, *Biomaterials: An Introduction* (Plenum Press, New York, 1992), pp. 185–222.
72. W. Bonfield, in *Advanced Series in Ceramics, Vol. 1: An Introduction to Bioceramics*, L.L. Hench and J. Wilson, eds. (World Scientific, New Jersey, 1993), pp. 281–298.
73. G.A. Skarja and K.A. Woodhouse, in *Polymers for Tissue Engineering*, M.S. Shoichet and J.A. Hubbell, eds. (VSP, Utrecht, Netherlands, 1998), pp. 73–98.
74. S.M. O'Connor et al., *Biosens. Bioelectron.* **14**, 871 (2000).
75. B.R. Ringeisen, D.B. Chrisey, A. Piqué, and R.A. McGill, Patent Disclosure (2000).
76. M. Bagdasarian et al., *Gene* **16**, 237 (1981).
77. M. Chalfie, Y. Tu, G. Euslichen, W. Ward, and D.C. Prashner, *Science* **263**, 802 (1994).
78. B. Lighthart, K. Prier, G.M. Loper, and J. Bromenshenk, *Microbial Ecology* **39**, 314 (2000).
79. J.J. Bromenshenk, R.C. Cronn, and J.J. Nugent, *J. Environ. Qual.* **25**, 868 (1996).
80. N.L. Carreck, J.L. Osborne, E.A. Capaldi, and J.R. Riley, *Bee World* **80**, 124 (1999).
81. J.R. Riley, A.D. Smith, D.R. Reynolds, A.S. Edwards, J.L. Osborne, I.H. Williams, N.L. Carreck, and G.M. Poppy, *Nature* **379**, 29 (1996).

Overview of Technologies for Pattern and Material Transfer

The great advantage of direct-write techniques is their ability to accomplish both pattern and material transfer processes simultaneously. This eliminates the need for many of the processing steps required by traditional, parallel processing techniques such as lithography. As a result, direct-write techniques fill a rapidly growing void in commercial manufacturing, as in small-lot and flexible manufacturing (including manufacturing iterative designs, rapid prototyping, and product development), and in novel areas of material deposition. Direct writing of materials is not meant to displace conventional material processing such as screen printing or lithography. Instead, direct writing is meant to augment such processing when conventional approaches take too long or are too laborious. To put these differences in perspective, this part provides a framework for relating and comparing technologies for pattern and material transfer.

CHAPTER 18

Technologies for Micrometer and Nanometer Pattern and Material Transfer

DAVID J. NAGEL

Department of Electrical and Chemical Engineering,
The George Washington University, Washington, D.C.

1. INTRODUCTION

"Energy matters" is a flippant summary statement of Einstein's famous equation, $E = mc^2$. The assertion is correct in science, engineering and other human endeavors. The same is true of another terse statement, namely "size matters." There are many reasons for this: size can be an advantage or a disadvantage. It affects performance of people, animals, plants, and systems. Also, size certainly impacts the cost of manufactured objects. This chapter is concerned with making things on micrometer and smaller scales. It is useful to begin with a review of the scale against which objects are measured, as shown in Fig. 1.

One meter characterizes the familiar macroscopic scale in our world. In the recent decades, since the development of integrated circuits in the 1960s, the micrometer, or microscopic, scale has become quite well known. More recently, the regime between the macroscopic and microscopic arenas has been termed the mesoscopic. It includes old technologies, notably watch making, under a new moniker. In the 1990s, the scale appropriate to the size of atoms and molecules came to be called nanoscopic. Much of the work today in the fields of chemistry and materials science falls under this term.

Methods for the production and assembly of parts vary widely across these regimes, as indicated in Fig. 2. In the macro- and mesoscopic areas, parts are made individually and assembled piece-wise into products by companies in already very large industries. These size scales still involve much basic and applied research, but they are characterized by commercial engineering. The micrometer scale involves parallel production of objects, using pattern transfer and other methods, with layering or bonding being primary methods of assembly. The microelectronics industry, which took off in the 1960s, is already large, and the micromechanics industry began to grow exponentially in the 1990s. The micrometer arena still requires a great deal of basic and applied research. In the nanometer regime, a radically different approach to the production of objects is emerging; while the larger regimes have something of the character of sculpture, that is, being top-down approaches to making things, the bottom-up approach of chemistry and biology will dominate nanometer technology. Molecular synthesis and self-assembly are being envis-

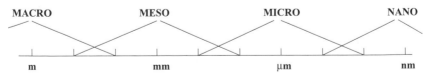

FIGURE 1   Size scale along which are shown the ranges of four major regimes of human-made objects.

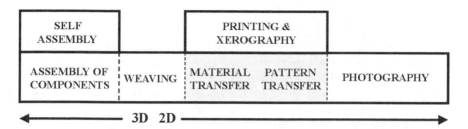

FIGURE 2 Relationships between the two (2D) and three (3D) dimensional ways to make and put parts together to make complex structures and devices, as well as two-dimensional patterns.

ioned as a primary approaches for making functional systems out of proteins and other organic materials on the nanometer scale. There are concepts for the production and use of molecular assembler devices, but they remain fraught with problems. Basic research still characterizes the state of what is called "nanotechnology." Some systems with components on the nanometer scale are already important commercially, but they have not reached the multibillion-dollar annual level that is the hallmark of industries on all the larger-size scales.

Devices and systems on the micrometer and smaller-size scales are decidedly three-dimensional. However, they are generally made by sequential use of thin-film technologies. Modern integrated circuits, for example, can have over 30 layers, which require over 200 process steps for their production. The three primary types of processes for production of familiar micrometer- and some new nanometer-scale devices are shown in Fig. 3. Pattern transfer, commonly called microlithography, is the movement of a design from an already-patterned

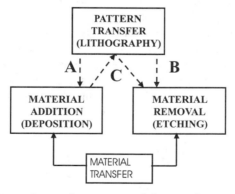

FIGURE 3 The dominant classes of processes used for manufacturing of micrometer-scale structures and devices. One class involves pattern transfer, with the other two being material transfer processes. The dashed lines indicate common process sequences, which are discussed in the text.

object, the mask, to the work piece. It is essentially the transfer of information from one medium to another. Microlithography requires the production, exposure, development, and later removal of a thin film of polymeric material, the photoresist. The deposition of usually continuous thin films of conductors, semi conductors, or insulators, and their generally partial etching, both involve the transfer of materials between the work piece and some nearby source or sink of materials.

The processes noted in Fig. 3 are used in different sequences for various purposes. These are labeled A, B, and C. In A, patterning is done prior to material deposition by ion implantation, or by the so-called lift-off process, which removes (lifts off) unwanted portions of a material layer along with the used photoresist, leaving the material that fell through the resist pattern onto the substrate. In B, patterning precedes material removal, when the substrate is etched through openings in the developed photoresist, which protects (resists) etching of the covered areas. The most common case is C, the deposition of a material that will constitute the next layer of a device, followed by production of a patterned photoresist over it and then etching away of the unwanted portions of the layer prior to removal of the no-longer-needed resist.

Often, the term "pattern transfer" is taken to mean the transfer of a pattern from the exposed and developed photoresist on the surface of a substrate into the substrate, commonly by etching. Here, we use the term "transfer" primarily to describe movements *onto* a work piece. "Pattern transfer" is putting *information* onto a substrate, whether it is by exposure of a photoresist or some other technology. "Material transfer" similarly describes the movement of some *material* onto the work piece. The preparation of unpatterned thin films is a critical commercial process, and many techniques are available for different materials. However, such processes are not of prime interest here.

Pattern transfer can be viewed as a part of information technology. The communication of the relevant information to a substrate can be done in parallel, just as most cameras capture images, or in a serial process, similar to the manner in which images are sent over the Internet or displayed on a computer monitor. Material transfer is part of the field of surface modification. Most of the activities in that arena have not directly produced patterned materials on a surface. Much of the interest in data-driven surface modification by writing patterns directly from computer files is motivated by the new opportunities for device production that the new technologies enable.

The great advantage of the direct-write material transfer processes, which are the subject of this book, is the ability to accomplish both pattern *and* material transfer processes at once. This avoids several lithographic steps, each of which has to be done many times during the production of complex micrometer-scale and smaller objects. They are, at least, spinning of a photoresist onto the substrate, its baking, exposure during pattern transfer, devel-

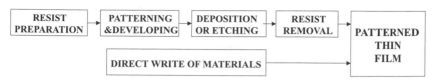

FIGURE 4   Comparison of the traditional and direct-write methods of producing of patterned thin films.

opment of either the exposed or the unexposed regions, and, finally, removal of the resist and any residues. A comparison of the process flows for traditional pattern transfer and for the new direct-write-of-materials approaches is given in Fig. 4. The relative simplicity of directly writing materials is clear. The programmability of most of the processes for direct writing of materials is also a major strength. The primary disadvantage is the sequential or serial nature of direct-write processes. A book on microfabrication (1), a recent conference proceedings (2) and this volume (3) are the primary compilations of information on processes for the direct writing of materials.

The production of micrometer- and nanometer-scale structures requires the ability to see details as well as the ability to move matter on these scales. A blind engraver might be able to produce fine-scale lines, but could not make a complex pattern involving many lines placed in precise locations relative to each other. It is interesting that several of the technologies discussed in this chapter offer the ability to both see and manipulate matter on scales approaching molecular dimensions. It is as if the engraver could feel the fine lines with the same tool used to produce them.

It must be noted that the production and the transfer of patterns are two distinct steps. The production of patterns happens in one of two primary ways: either the capture of an image by some simultaneous or sequential (scanning) process, or the creation of an arbitrary pattern using a computer program. Images are usually obtained for informational purposes, while patterns produced with computer-aided design (CAD) software are intended for replication in experiments or in commercial production. The production and uses of purposefully designed patterns are of interest here.

Pattern and material transfer technologies can have fixed or programmable geometries. Some techniques involve replication of a set pattern, such as a mask, which was made earlier by a programmable method using CAD drawings. However, many of the more recently developed methods for pattern and, especially, material transfer are flexibly programmable and do not require fixed masks. They can be controlled directly by the output of CAD programs. These are "data driven" technologies.

The objectives of this review are: (1) to provide a framework for relating and comparing technologies for pattern and material transfer; (2) to explain the basic concepts involved in each of the methods; and (3) to provide significant references to the details for each of the technologies. A few applications of pattern transfer methods are exhibited in the next section as an introduction to the reviews of such processes that follow. In a similar fashion, applications of material transfer methodologies are given in Section 10 as a prelude to the subsequent discussion of individual techniques. Many interesting patterns and structures could be displayed for each pattern and material transfer technique, but doing so would unduly lengthen this review. In some cases, a technology is best understood through its results, so some patterns and structures are shown when a technique is explained.

Two topics critical to pattern and material transfer technologies are beyond the scope of this review. First, computational simulation of the various techniques is a large and active arena. It yields understanding and promotes optimization of the various processes. Second, the actual equipment employed during the use of the transfer technologies, either in research or commercial production, is fundamentally important. In many cases, the hardware is quite remarkable for its precision and the simultaneous control required over many variables. The evaluation and comparison of such equipment, both over time and in their current and projected embodiments, are also interesting. However, a decent review of specific hardware used for pattern and material transfer would be a large study in itself.

This review has four major parts, three of which are shown in Fig. 5. The first (Sections 2 through 9) deals with technologies for only pattern transfer, including both the more traditional and the quite new lithographic methods for imprinting a pattern in a thin film of resist on a work piece. The second part (Sections 10 through 19) covers simultaneous pattern and material transfer techniques, with the focus on the technologies developed in the last decade for such transfers. The third part, Section 20, notes that, when self-assembly is possible, both the material and a pattern are effectively transferred at once. The material essentially contains its own patterning information, which may or may not be expressed, depending on conditions at the time of transfer and

| PATTERN TRANSFER ONLY | PATTERN & MATERIAL TRANSFER | MATERIAL TRANSFER ONLY |

FIGURE 5 The major classes of pattern and material transfer technologies. If self-assembly is possible, only the material needs be transferred because it intrinsically contains the information (instructions) to spontaneously form structures.

subsequently. The fourth part includes two sections. The first, Section 21, provides a brief summary of some of the major characteristics of the reviewed technologies. Then, the concluding section provides a perspective on the possible evolution of pattern and material transfer processes. Programming of DNA for control of production and self-assembly of complex structures may become very important. The appendix is a brief survey of technologies that are ancillary but critical to the transfer techniques discussed in the body of the review.

2. APPLICATIONS OF PATTERN TRANSFER TECHNOLOGIES

Micrometer-scale pattern transfer technologies grew out of the photography and printing industries. They now enable several commercially important areas, some of which have emerged only recently. Pattern transfer is fundamental to the older printed circuit industry. The best-known application of microlithography is the production of integrated circuits (ICs or "chips"). Lithography is now used to define the arrays of solder bumps used for bonding of ICs to printed circuit boards. Patterning of magnetic heads for disk drives is less known, but also very important. Manufacture of microelectromechanical systems (MEMS), and microsystems in general, also depends on the ability to produce precise patterns of the correct shape, scale, and location. Included in this arena are both static micromachined structures and dynamic micromechanisms. Microfluidics also embraces both static and dynamic components, and is becoming an important industry in itself, also dependent on pattern transfer. The industry that produces microarrays for the analysis of chemicals and biologicals similarly depends on microlithography. Production of fine patterns to guide the growth of crystals and cells on surfaces are additional applications of lithography. Examples of each of these will be given next to illustrate the wide variety of lithography applications and to motivate attention to the characteristics of pattern transfer technologies discussed in the following sections.

2.1 PRINTED CIRCUIT BOARDS

Printed circuit boards (PCBs) were once made only for the wiring together of components placed on their surfaces. Now, however, they have resistive and capacitive components embedded in them. Their complexity has increased greatly during the past decade as larger numbers of layers, connected to each other through vias, have been incorporated into PCBs. The boards in complex

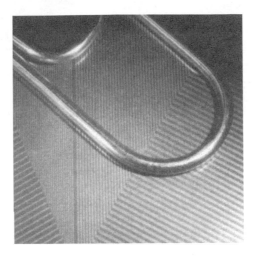

FIGURE 6   Photograph of 50 μm lines and spaces on a flexible circuit. The paper clip is 750 μm in diameter.

consumer electronics, such as the mother board in a personal computer, now have as many as 50 individual layers of conductive and insulating materials. The widths of lines and diameters of vias in PCBs continue to shrink, driven by the same performance and cost factors that force the use of ever finer lines in ICs. PCB line widths and vias are now approaching 50 micrometers, which was the width of the conductive lines on silicon in the initial ICs made in the 1960s. Figure 6 illustrates such fine lines on a flexible PCB (4).

2.2 Integrated Circuits

Plan views of ICs are familiar, including both the overall appearance of a chip and the details of the surface layer. Less commonly seen are three-dimensional views of ICs. The structure of a microprocessor and modern chip are given in Fig. 7. These show that ICs are certainly not planar (2 dimensional). In view of this character, they are sometimes termed 2.5-dimensional. The IC industry, which is completely dependent on lithography, is now about $130 billion annually (7). It is the basis for the electronics industry, operating now at the annual level of $1 trillion, which in turn is fundamental to the computer and communications industries that enable the so-called information age.

FIGURE 7 Scanning electron micrographs of IBM chips. Left: A microprocessor built with silicon-on-insulator (SOI) and copper technologies (5). Right: Seventh-generation chip technology from which the silicon dioxide insulating layers have been removed (6).

2.3 Flip Chip Bonding

Integrated circuits have usually been packaged, which requires wire bonding of the pads on the chip to the conduits that penetrate the package and are attached to a printed circuit board or other structure. The packages have to be produced, and they take up additional space on boards, which is crucial for many consumer electronics applications like cell phones. Hence, there has been growing interest in the ability to bond chips directly onto boards without packages. Then, the chip and the polymeric materials that are flowed between the bonded chip and the board, form a "package" that is no bigger than the chip. Doing such flip chip bonding requires the ability to produce solder

FIGURE 8 Part of an array of Sn-Ag-Cu solder bumps for flip chip packaging.

bumps that are very similar to each other in size and located at precise positions to match the contacts on the chip surface. Such precision mass production is now done lithographically, with results as illustrated in Fig. 8 (8). The size of the bumps and the spacings between them are about 100 μm in this example. Bumps and spacings half these dimensions are in prospect. Roughly half of the one-third of a million 300 nm wafers expected to be processed in 2003 will employ flip-chip technology (9).

2.4 THIN FILM MAGNETIC HEADS

The read and write heads in magnetic storage drives contain giant magneto-resistance materials that have to be patterned on ever finer scales with the increasing storage density of disk drives. Figure 9 shows the evolution of the feature sizes in thin-film heads (TFH) in comparison to the critical feature sizes in ICs (10). It is seen that, while the resolution required for TFH lags that of ICs, the demands of the read and write heads are not far behind those of microelectronics. In fact, the gap between finest feature sizes for the two technologies is closing. It is noted that the storage density of magnetic disk drives is doubling about every 12 months, faster than the 18-month doubling time of the transistor density in ICs (Moore's Law).

2.5 MICROMACHINED STRUCTURES

Static microstructures delineated by lithography and then etched into silicon and other substrates are the basis of passive components in many micro-

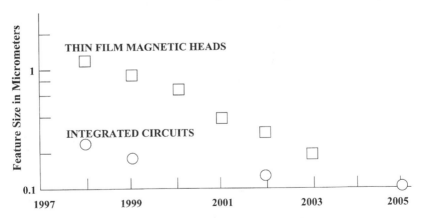

FIGURE 9   The critical feature size for thin-film heads (TFH) in disk drives and for the finest lines within integrated circuits.

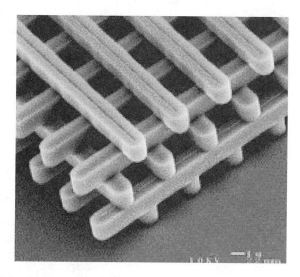

FIGURE 10   A photonic "crystal" made of silicon bars about 1 μm in cross section. The short bar in the lower right is 1 μm.

systems. They are channels for control of the flow of gases and liquids. Fine-scale structures within materials are now being produced to influence the flow of optical energy through the materials. Called "photonic materials," they serve to determine the behavior of optical waves, much like the basic structure of crystals determines the "optics" of bonding and other electrons within all solids. Figure 10 is an example of a photonic material made using lithography (11). Micromachined structures also serve as waveguides on the surfaces of chips for either optical or radio-frequency waves. Optical waveguides permit the tight integration of optics and electronics. RF waveguides above micromachined cavities have low-propagation losses because the signals couple only into air below the waveguides and not into the lossy substrate. Lithography is also used for patterning materials for the production of arrays of microlenses with diameters on the order of 100 micrometers.

2.6 MICROMECHANISMS

Dynamic (moving) micromachined structures, that is, micromechanisms, are central to the explosion of research and commercial interest in micro-sensors and -actuators. MEMS microsensors of greatest commercial importance now include pressure sensors, accelerometers, and angular rate sensors. MEMS microactuators move micromirrors that are poised to play a central role as

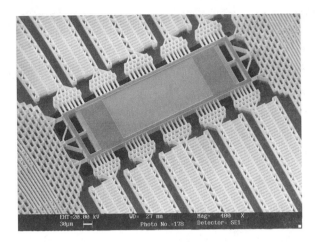

FIGURE 11 Scanning electron micrograph of a single-crystal silicon micromechanism, part of a prototype for a data storage system. The bar at the bottom of the picture is 30 μm.

switches in "all optical" fiber communication networks. Microactuators will also be important in RF systems as switches, tunable capacitors, and resonators. Micromechanisms are also at the heart of data storage devices now under development that promise terabit per square performance. Figure 11 is a micrograph of a structure made entirely of single-crystal silicon (12). Both microstructures and micromechanisms are continually being made on finer scales, like microelectronics. Nanometer-scale microsystems are now referred to as NEMS, which is short for nanoelectromechanical systems.

2.7 MICROFLUIDICS

Microsystems that manipulate fluids form the basis of the "lab on a chip" products now already on the market. All contain passive components such as channels and mixers, in addition to reservoirs for reagents and waste. Some of them also contain active components, notably microvalves and micropumps. Detection of the results of the chemical reactions performed in microfluidic devices is done in two general ways. In one, electrical sensors are integrated into the fluidic unit. In the second, optical microscopic techniques are used to detect fluorescence associated with products of reactions. Figure 12 shows one example of a lithographically patterned microfluidics system (13).

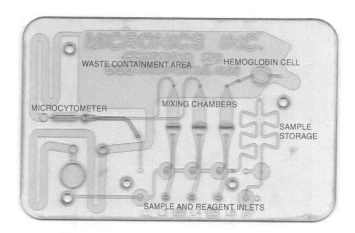

FIGURE 12   Photograph of the plastic microfluidic element from Micronics. It is the size of a credit card.

2.8 MICRO-ARRAYS

Genetic information flows from DNA first to mRNA and then to proteins during the operation of cells. Genomics is the study of the organization of DNA in genomes, the sum of all the genes in an organism. Proteomics is the study of the resultant proteins. Large numbers of experiments are needed for genomics, proteomics, and other biomedical research and for clinical analyses, such as gene expression studies and drug discovery. Performing them by conventional means, even with multiwell plates and robotic handling of samples and reagents, is both expensive and time consuming. Hence, microlithography has been used to develop large arrays of small regions that contain a variety of chemicals or biomaterials, such as oligonucleotides (short lengths of DNA). When a sample is applied to such an array, a large number of assays is done simultaneously and quickly. The results of the reactions can be detected either electrically or optically, similar to the case for microfluidic systems. Figure 13 shows two microarrays already on the market. The array from Nanogen has 99 test sites, each of which is 80 µm in size, separated by 200 µm. One of the Affymetrix Genechips™ arrays has 260,000 sites in an area 12.8 nm square, that is, the sites are approximately 25 µm on a side. Pattern transfer by lithography is as fundamental to the new micro-array industry as it is to the production of ICs.

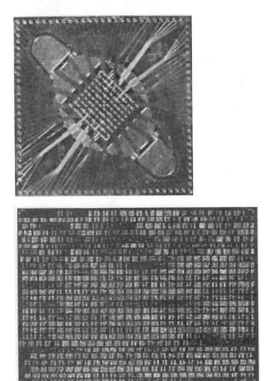

FIGURE 13 Microarrays manufactured by NanoGen (14) with electronic readout (top) and by Affymetrix (15) with optical readout (bottom).

2.9 GUIDED CRYSTAL GROWTH

Lithographically defined patterns have long been used to delineate the regions on a substrate where crystal growth would be permitted. Sometimes the growing crystal is epitaxial with a crystalline substrate. Other times, the pattern serves simply to define spots where nucleation and any kind of crystal growth can occur. Figure 14 shows an example of the latter (16). A film of gold about 1 nm thick was laid down on a sapphire substrate and patterned. Then, a vapor of ZnO lead to dissolution of that material in the gold and growth of single crystals about 100 nm in diameter and a few micrometers high. Both ends of the crystals are good mirrors, so the crystals are optical cavities. When pumped with short-wavelength laser light, the nanocrystals lase in the blue at 385 nm. Lithography has been used to align liquid as well as solid crystals.

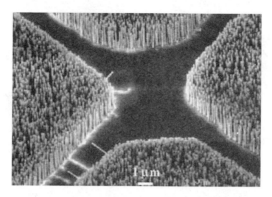

FIGURE 14   Micrograph of parts of four regions about 25 μm on a side (with rounded corners) in which nanocrystals of ZnO have grown at areal densities over $10^{10}/cm^2$.

2.10 GUIDED CELL GROWTH

There are three very large scientific and technological "revolutions" in progress now, namely information-, biological-, and nano-technology. The important role of lithography techniques in the first two of these was already noted here. To date, the biotechnology field has largely involved unstructured materials, with chemical reactions and biological growth occurring in liquid environments, especially cultures of cells. However, early work has shown that strong interactions of live cells with patterned surfaces permit the guided growth of nerve and other cells. Figure 15 shows one example (17). It is expected that

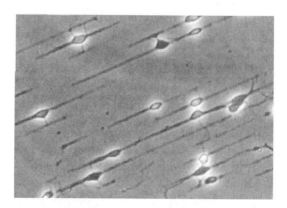

FIGURE 15   Optical micrograph of rat hippocampal (brain) cells growing along lithographically defined 5 μm wide lines of a material that attracts cells that are separated by 12.5 μm wide spaces of cell-repellant materials.

both the science and applications of structured patterns of cells will grow in importance in the coming decades. Applications of guided cell growth include research on cell immobilization, cell dynamics, cell–cell interactions, and possible clinical use of cell-based biosensors.

3. OVERVIEW OF PATTERN TRANSFER TECHNOLOGIES

3.1 HISTORICAL PERSPECTIVE

Pattern transfer is at the core of two familiar technologies: photography and xerography. In photography, the blank material (the film, which is analogous to a photoresist) is first made available, and then exposed (patterned). In xerography, the (electrostatic) pattern is produced first and then the material (toner) is supplied. Both processes can produce lines with widths approaching 10 micrometers. Both have found some uses for making prototype micrometer-scale structures and devices.

Pattern transfer is a fundamental part of printing. There are three basic approaches to printing. In letterpress printing, the surface of the tool has a "relief," like a rubber stamp, where the pattern to be duplicated is raised and inked. In the gravure or intaglio process, a metal plate is engraved or etched to have patterned grooves which contain the ink. The last method, namely lithography, involves flat or cylindrical plates without relief or grooves. The surface of a plate is treated so that only the parts forming the pattern will take up ink. The most common commercial version of lithography is offset printing, in which the ink pattern is first transferred from a cylindrical plate to an intermediate "blanket" cylinder, and thence to the paper. Avoiding plate–paper contact in this way greatly increases plate life.

Lithography is a large and complex industry with a rich history that long predates the emergence of microlithography. The word has its roots in the Greek words for stone (lithos) and writing (graphein) because the original lithographic process, developed late in the 18th century, involved pattern transfer by making some parts of a stone receptive to ink. That original process is still practiced, mainly by artists. In the 19th century and, especially, in the 20th, the graphic-communications industry developed by using many variants of lithography. A recent book on lithographic technology emphasizes the rapid changes in that industry (18). The graphic arts industries include general, newspaper, magazine, and book printing, as well as many more specialized market segments. The industry shipped about $116 billion of product in the U.S. alone in 1995 (18). It is noteworthy that the graphic-communications industry and its associated business base do not employ any of the micro-lithographic techniques of interest in this chapter. The reason is clear; ordinarily, people cannot see micrometer-sized patterns.

In micro- and nanotechnologies, the terms lithography and microlithography are often taken to be synonymous with the term "pattern transfer." Indeed, all lithographic techniques involve pattern transfer. However, not all pattern transfer methods involve lithography. In this and the following six sections, we will review lithographic techniques that have been developed during the past four decades. They vary widely in character and utility, but all offer the ability to lay down a pattern on a work piece that is needed for production of useful structures or functional devices.

The processes used for pattern transfer during the production of micro-technologies have not involved, until recently, the transfer of an "ink" by contact between the original pattern and the work piece. Rather, most of what has been described as microlithography in the past four decades is essentially an outgrowth of the processes used in classical photography. The photographic industry has historically involved the production of an image by exposure and development of a negative (the film) and the transfer of the image during printing to paper (the work piece) by "contact" or "projection" printing. This was essentially the technology used for the production of what were called printed wiring boards during much of the last century. The difference between ordinary photography and technical lithography is that the lithography "negative," called a mask, contains the man-made design needed for the required electrical connections on wiring boards.

The photographic technology used to print wiring was naturally applied to the production of the early integrated circuits in the 1960s. Figure 16 shows a photograph of equipment used in that decade for the transfer of patterns during IC manufacture (19). It worked in air and was very similar to photographic contact printers. In fact, the microlithographic technology was called contact printing because the mask and silicon wafer were in contact. This led to mask degradation and defects, so the next phase was to place the mask very near, but not in contact with the wafer—the so-called proximity method. As in contact lithography, it produced an image in the photoresist that was the same size as that in the mask. Later, projection methods were employed to produce demagnified images of the mask pattern in the resist. The equipment for projection exposures can be viewed as related to common darkroom photographic enlargers, with the optics changed to produce smaller rather than larger reproductions of pattern of interest.

By 1980, a wide variety of microlithographic technologies was demonstrated, ones that were either in development or routine use. The matrix in Fig. 17 is a rearrangement of a figure developed about 20 years ago to relate the many lithographic technologies to each other (20). It was and is useful to

FIGURE 16   Lithographic mask alignment and resist exposure tool manufactured by Electroglas, Inc. in the 1960s.

categorize them according to the quanta used to expose the photoresist, namely photons, electrons, or ions, and the means by which the pattern is produced and the transfer effected. These two salient factors are discussed next, before surveying the classes of techniques indicated in Fig. 17.

3.2 TYPES OF PATTERN TRANSFER TECHNOLOGIES

The act of exposing a resist produces chemical changes in the polymeric materials that affect the solubility during the wet or dry (that is, plasma) development stage that follows exposure and precedes the addition or removal of material from the work piece. Exposure and development of a positive resist produces a replica of the mask. With a negative resist, an inverse of the mask results. In a positive resist, the breaking, or scission, of molecular bonds leads to lower molecular weights and increased wet solubility (or lower resistance to plasma removal). In a negative resist, cross-linking by production of additional bonds between molecules within the exposed regions increases the molecular

	MASK (PROXIMITY OR CONTACT)	MASK PATTERN PROJECTION	FOCUSED BEAM (DIRECT WRITE)
PHOTONS	STEP & REPEAT	SCANNING OR STEP & REPEAT	LASER CHEMISTRY
ELECTRONS	OPEN MASK	PATTERN ON CATHODE	RASTER OR VECTOR SCAN
IONS	CHANNELING IN MASK	OPEN MASK	FIELD-ION SOURCES

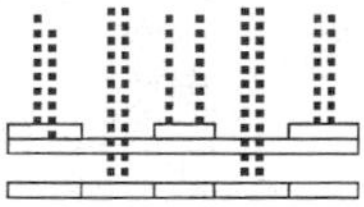

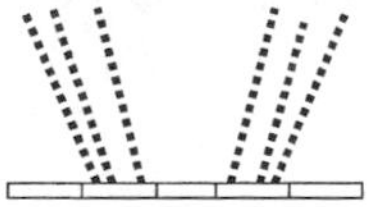

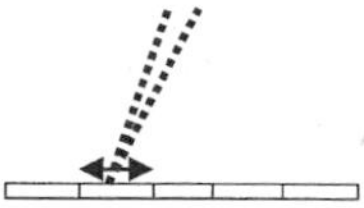

FIGURE 17   Matrix of the demonstrated lithographic technologies by 1980, relating the quanta for resist exposure to the methods for pattern transfer.

weight of the material and reduces its wet solubility (and increases its resistance to plasma removal). These changes can be caused by the absorption of energy from ultraviolet (so-called "optical") or X-ray photons, so the prefix "photo" is commonly appended to the word "resist." Similar chemical changes can be caused in the resist by electrons or ions with energies sufficient to expose the entire thickness of a resist. The terminology "resist" derives from the ability of the patterned polymeric layer to resist the subsequent processing steps involving material deposition or etching, whatever the quanta used for pattern transfer. Hence, resists have a curious duality, susceptible to exposure during pattern transfer, so that exposure times are short but resistant to change, especially removal, during the next processing steps, which often involve energetic quanta.

The characteristics of aligners for pattern production using the different quanta vary greatly. Of course, the sources are entirely different. To control transmission losses in aligners, the atmosphere in optical chip aligners has varied from clean air through inert gases to vacuum in going from Hg sources to progressively shorter photon wavelengths. Vacuum is required for all electron and ion options. Although some resists respond to any energetic quanta, it is standard practice to optimize resists for the quanta and energies.

The manner in which the pattern of interest is produced and transferred provides the other primary means of distinguishing lithographic technologies. Most techniques use some kind of mask, which is imaged onto the photoresist. The earliest masks were made by cutting out colored materials and pasting them in the desired pattern, prior to photography to produce the negative mask used for replication in the resist. In the early 1960s, the capability to move a focused electron beam over a surface in a programmed fashion grew out of technologies used in scanning electron probe analytical instruments. It was possible then to produce masks without paste-ups and photography. The method involved metallizing a glass plate, coating it with resist, exposing the resist with the programmed scanning focused electron beam, developing the resist, and etching the metal to get the desired mask pattern for its subsequent and frequent replication. The masks employed in lithography also vary greatly with the quanta employed and their energies. Chromium on quartz masks are commonly used in photolithography.

The unity-magnification contact and proximity techniques for microlithography are very similar, as discussed, so they constitute the first category of pattern transfer methods. Contact lithography tends to produce mask damage, so it is unacceptable for commercial mass production. Proximity lithography avoids most of such damage, but it requires precise control of the mask-to-wafer gap.

Projection lithographies populate the second grouping of techniques. Projection technology involves much greater separation of the mask and wafer by producing a demagnified image of the pattern in the resist, thus reducing damage to the mask. It depends on having adequate depth of focus at the resist. Masks for contact, proximity, and projection lithographies can be produced by the direct writing of a pattern from a computer file onto the resist used during production of the mask. Of course, it was similarly possible to write a pattern directly into the resist on a work piece using a focused beam of light from a laser, electrons from a hot filament, or ions from a field ionization source.

Employment of focused photon, electron, or ion beams in a programmed, scanning manner is referred to as "direct write" lithography because no mask is required. That is, the pattern is stored as a computer file, which is employed to drive a beam in two dimensions during the resist-exposure step. Direct-write techniques are the third grouping of lithographic techniques. Such methods offer fewer process steps and the flexibility that goes with re-programmability. However, the requirement to write each segment of each line or form within a complex pattern leads to long overall exposure times that are not acceptable for commodity production. Now, direct write exposure of resists is used for making masks and for developmental work. In the future, it might be the primary tool for exposure of the finest lines of ICs, as noted in Section 3.3.

Direct-write resist-exposure methods are not to be confused with the direct writing of materials, in which the need for the resist and all of its associated process steps are avoided, as shown in Fig. 4.

With this background on the quanta used during pattern transfer and the means of impressing information on those quanta, we are in a position to review each of the technologies given in Fig. 17. Not all the processes are equally important. Some that are of passing interest will be discussed only in the latter part of this section. Those that are in the mainstream of commercial device production now, or might be in the coming decade, will be treated in subsequent sections. Photon technologies are reviewed first, followed by a discussion of electron and ion techniques.

Optical lithography has been, and remains, the workhorse technique for commercial production of integrated circuits and printed circuit boards. The energies of the photons used in lithography are centrally important because the exposure of resists is energy dependent, and because of wavelength (energy)-dependent diffraction effects on resolution. Initially, visible light and, later, ultraviolet light was used for "optical lithography." Over time, there has been a steady decrease in the wavelengths of light used for lithography. This can be understood by reference to the two basic equations for optical lithography (21):

$$\text{Resolution (RES)} = k_1 \lambda / NA \qquad \text{Depth of Field (DOF)} = k_2 \lambda / (NA)^2$$

NA is the numerical aperture of the lens used to focus the pattern onto the resist. It is the ratio of the effective focal length of the lens to its open aperture (pupil). These equations show that decreasing λ or k_1, or increasing the NA, improves the resolution of optical lithography, but reduces the depth of focus for the image. The latter is critical, because it directly influences the flexibility of the resist exposure and development processes, the so-called "process latitude." A DOF of about 0.5 μm or greater is desirable, but difficult to achieve. The DOF is comparable to the 0.5 μm resist thickness commonly used for commercial chip production. When the two constants k are both equal to 0.5, the "diffraction limited" situation exists for whatever wavelength is in use at the time.

Figure 18 shows schematically the photon wavelengths that have been employed and are projected for use in optical lithography and its descendent technologies. The wavelengths λ can be converted to photon energies E using the equation $E\ (eV) = 1240/\lambda$ (nm). Line radiations from mercury lamps were used for many years. Then, the output of excimer lasers came into routine use. We are now in that era. Photons with wavelengths in the extreme ultraviolet spectrum and in the soft X-ray spectrum might be used in the future.

Contact and later proximity technologies were used with visible and ultraviolet photons to expose resists early in the IC era. Exposure of many

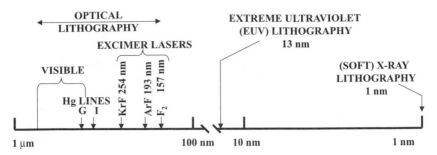

FIGURE 18   The wavelengths of radiation used in past and coming photon lithographies.

chip patterns on the same wafer was done by stepping the mask and repeating the exposure over each site. X-ray lithography, which uses unity magnification, is designed to employ the same step-and-repeat scheme. When projection lithography superceded proximity technology, step-and-repeat remained in use because it is not possible to project an image that is sufficiently sharp over a multi-chip field. Some aligners scanned an arc, in which the mask was sharply imaged, over the wafer in order to obtain the needed exposure area. These are "step and scan" technologies. Direct writing of patterns with focused laser beams cannot produce the line widths near 100-nm that are needed for current IC production. However, laser direct-write patterning methods are routinely employed to make lithography masks with features larger than 1 μm and to produce fine lines directly on some production printed circuit boards.

Electron beam contact and proximity techniques require open or "stencil" masks. The complexities of making and using such masks greatly complicates this approach, and make it very challenging for commercial chip production. Electron projection techniques include early attempts to make a cathode emit electrons in the desired pattern, which was then translated through a series of electron lenses to the resist. Difficulties with making the emitters and their limited lifetimes essentially killed this approach. Currently, two electron projection lithography (EPL) methods invented in the past two decades are being developed, SCALPEL and PREVAIL. These will be discussed in Section 7.1. Electron-beam direct writing of fine-scale patterns is too slow for complete resist exposures in commercial processes. However, it is the mainstream technique for leading-edge mask production, being the most practical way of making nanometer-scale patterns for mass replication. Whether or not electron direct-write technology will ever be used for writing lines finer than 100 nm in production will not be known for several years.

Ion-beam contact and proximity exposures require the use of channeling in a very thin single-crystal membrane that carries the absorber pattern. Such a mask is very difficult to produce and it suffers from radiation damage in use.

Hence, this approach to ion-beam lithography is moribund from a commercial perspective. Ion-projection methods are still being researched, as described in Section 7.2. Direct-write ion processes could be used to make masks, but the electron-beam approach is favored because of shorter exposure times. Focused ion beams have the nice feature of being able to sputter atoms off of a substrate, as well as initiating surface chemical changes. Direct-write ion-beam lithography can produce nanometer-scale patterns on the surfaces of a very wide variety of materials. For example, ion beams have been used to pattern printheads for microcontact printing (22). Focused ion beams are also used for the repair of defects in masks. They can repair "shorts" by removing material from regions where it is not wanted, and they can repair "opens" by initiating chemical deposition in the presence of a low-pressure atmosphere of appropriate gases.

When the matrix of quanta versus pattern forming methods shown in Fig. 17 was made 20 years ago, atoms were not employed to expose photoresists. In the interim, three approaches for lithography have been demonstrated that use neutral atoms, sometimes in a long-lived (metastable) state of excitation. One is based on flooding atoms through the openings in contact masks. The second involves the use of standing electromagnetic (EM) fields in free space to influence atomic motion. The third depends on EM fields within hollow capillaries that guide atoms even in curved structures. In the two cases involving EM fields from lasers, the wavelength has to be tuned so some transition in the atom in order to provide the coupling of the field and atom motion. None of these atom lithography techniques is likely to become viable for commercial production of ICs. However, they may have niche applications, so the technologies are briefly reviewed in the following paragraphs.

Beams of neutral atoms have energies sufficient to cause chemical reactions, but not high enough to penetrate ordinary relatively thick resists. Hence, they were initially used to expose nanometer-thick self-assembled monolayers (23) through the holes in a mask placed in contact with the resist-covered substrate (24). Impact of the atoms caused enough damage to the monolayer to render it developable compared to unexposed regions. Even without a resist, atom beams can crack hydrocarbon vapors adsorbed on the surface of a substrate. This leaves a residue of carbonaceous material that can serve as a self-developed resist. Features as fine as 70 nm were produced using this approach (25).

The first of the two atomic methods involving using EM fields to guide atoms onto a substrate is illustrated in Fig. 19. Two laser beams, obtained by splitting the output of a laser with adequate coherence, are routed by mirrors and brought together on the surface of the substrate. The spatial and temporal coherence of the laser causes standing EM waves to be set up and these affect the trajectories of atoms incident on the substrate. The atoms are influenced by

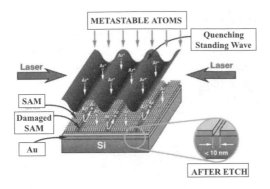

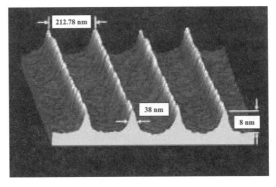

FIGURE 19   Top: Schematic showing how the coincidence of two coherent beams of laser light on a substrate can set up standing interference patterns that guide incident atoms onto a self-assembled-monolayer (SAM) resist (26). Bottom: Atomic-force microscope image of Cr atoms deposited on a Si wafer in lines that are 38 nm wide (27).

the EM field so that they fall in the regions of low field. The periodicity of the resulting pattern is determined by the wavelength and the angles of incidence of the laser light. If a photoresist is present on the surface, it will be exposed by the impinging atoms, as indicated in the schematic in Fig. 19. If not, the atoms will be deposited in rows, as shown by the data in the same figure. Use of two pairs of laser beams at right angles to each other results in a square array of high field regions and a crossed array of lines of deposited atoms. At the junctions, structures of Cr atoms 30 nm high and 90 nm wide were produced at a spacing of 213 nm in both directions (28).

In the second use of EM fields to guide atoms, capillaries with inside diameters in the range of 10–40 μm are beveled so that laser fields can be introduced into the glass (29). The evanescent EM waves that protruded into the open core of the capillary repulsed the injected metastable He atoms, and guided them through the structure. This technique provides the basis for a potential method of directly writing patterns on a thin resist using atoms as the exposing entities.

3.3 THE LITHOGRAPHY ROADMAP

Before surveying the mainstream and other lithography techniques in more detail, it is useful to summarize the industrial roadmap for commercial lithography. Roadmaps serve multiple purposes. Collaborative production and utilization of roadmaps by the semiconductor industry insure that equipment makers will produce, and equipment users will use, the right tools at the right times to meet the overall demands of electronics customers for increasing performance at the same or lower prices. The first step in the production of a roadmap for the IC industry is agreement on the critical feature sizes of what are called "nodes." These are geometric landmarks in the history of chip production that are expected to be reached at particular future times, for which equipment manufacturers must have production-capable systems on the market. The equipments, and required materials, include the most critical exposure tools, the "aligners," plus the resists and masks that are needed to make the overall process work. Consensus on the nodes, and on the tools and materials that are associated with each, is arrived at by a complex and time-consuming process. It includes consideration of available and new information generated in critical reviews and workshops, risk assessments, draft documents, peer review and, finally, decisions on the factors in the roadmap. The microprocessor, memory, and ASIC (Application Specific IC) parts of the chip industry have substantially similar, but not identical, requirements and timelines.

The primary roadmap for the microelectronics industry is the International Technology Roadmap for semiconductors, which is sponsored by the Semiconductor Industry Association in cooperation with the European Electronic Component Association, the Electronics Industries Association of Japan, the Korea Semiconductor Industry Association, and the Taiwan Semiconductor Industry Association (30). The key graphic from the 1999 ITRS roadmap for lithogrpahy is shown in Fig. 20. It displays the critical dimension (CD) nodes vertically against the first year for IC production at those CDs. Experience in the past two years indicates that the production at each future node might be a year or more sooner than shown (31). This graphic shows that production of large volumes of chips with 130 nm CDs is beginning using the 248 nm wavelength from the KrF excimer laser and phase shift masks (PSM), which will be discussed in the next section. The ArF excimer laser, with its 193 nm output, might come into production during the era of the 130 nm node. It is expected to be dominant for the 100 nm node beginning in the year 2005 or

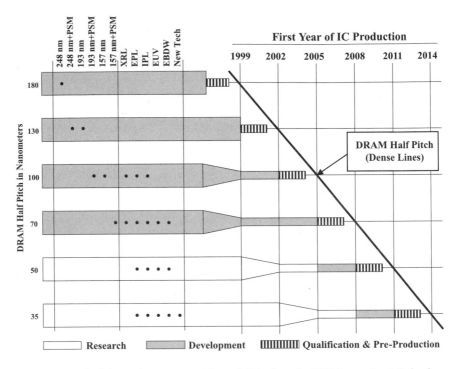

FIGURE 20 The lithography exposure tool possibilities from the 1999 International Technology Roadmap to Semiconductors. The acronyms are explained in the text. The trapezoids for the four smallest nodes indicate a narrowing of options.

sooner. The technologies that will be used for subsequent production of chips at the 70, 50, and 35 nm nodes are still uncertain. They are collectively called next-generation lithographies (NGL). Candidates are shown in Fig. 20. They include another excimer laser at 157 nm, electron and ion projection lithographies (EPL and IPL), X-ray lithography (XRL), extreme ultraviolet (EUV) lithography, and electron beam direct-write (EBDW) lithography.

The photon lithographic techniques in contention for future commercial production are clear descendants of the techniques already demonstrated twenty years ago. Optical, extreme ultraviolet and X-ray lithographies are discussed in the next three sections. The following section surveys the current status of electron and ion projection lithography techniques. Some of the new approaches to microlithography that have emerged in the past two decades offer interesting capabilities for production of research structures. One involves the use of what are called "proximal probes" for the exposure of both old and new types of resists. Another is termed lithographically induced self-construction. Several techniques for imprinting patterns deeply into a substrate have been developed as part of the interest in micromachines and for other reasons. A technology called microcontact printing or "soft lithography" is related to the letterpress method of printing in which a pattern in surface relief is inked prior to transfer of the ink to the work piece. It is both a pattern and a material transfer method. Such newer methods of pattern and materials transfer are discussed in later sections.

Many books have been written partially or exclusively on lithographic techniques. Some volumes cover semiconductor processing broadly, including lithography (32–34). Others are focussed on optical lithography (35–40). A few books include discussion of next generation lithographic technologies (41–43). Two volumes cover both next generation techniques and the micromachining processes enabled by lithography (44–45). A series of presentations in January 2000 provides a good overview of current lithography development efforts (46).

4. OPTICAL LITHOGRAPHIES

4.1 MAINSTREAM OPTICAL TECHNIQUES

The equipment and processes used for commercial optical lithography challenge the limits of physics and chemistry. The integration and simultaneous functioning of the source, mask, resist, and stage, plus critical ancillary equipment such as shutters, atmosphere and temperature controllers, and complex electronics—much of it in control loops—is a widely unappreciated aspect of a modern technological society.

The evolution of the electomechanical-optical aligners that expose the resists on wafers with precise alignment to earlier-made features can be appreciated by reference to Fig. 21 (47). The decreasing exposure wavelength; increasing numerical aperture (NA); decreasing resolution; increasing field size; decreasing depth of focus (DOF), alignment, stage accuracy, and lens distortion; and the increasing wafer size and throughput during the 20-year time span enabled the cost-effective production of denser and higher-performance chips despite an increase in aligner cost of over 25X.

Another perspective on trends in optical lithography over about the same time period can be obtained from Fig. 22 (47). Different companies offered state-of-the-art aligners at different times. The decreasing exposure wavelength

	1977 GCA 4800 DSW	1995 i-line	1997 DUV: Step Scan
M:1	10x	5x	4x
Wavelength	g-line: 436 nm	i-line: 365 nm	DUV: 248nm
Lens	0.28 NA	0.60 NA	0.60 NA
Resolution	1.25 µm	0.40 µm	0.25 µm
Field size	10 mm sq.	22 mm sq.	26 mm x 32 mm
Depth of focus	4.0 µm	1.0 µm	0.70 µm
Alignment	± 0.50 µm	± 0.06 µm	± 0.03 µm
Stage accuracy	100 nm	30 nm	15 nm
Lens distortion	250 nm	50 nm	30 nm
Wafer size	3, 4, 5 inch	5, 6, 8 inch	6, 8, 12 inch
Throughput	20 wph (4")	60 wph (6")	60 wph (8")
Cost	$300,000	$4,000,000	$8,000,000

FIGURE 21   Details on the evolution of optical lithography from 1977 to 1997. Mercury lamps emit the g and i lines. DUV, deep ultraviolet; wph, wafers per hour.

did not entirely offset the increasing NA, so that the DOF declined, and with it the process latitude became smaller and more of a problem to maintain.

There have been many advances in optical lithography that are not indicated by the trends just examined. This is especially true regarding masks and resists. Simple masks only have clear or opaque structures that are shaped like the pattern that is desired in the resist (48). Adjustments in the areal shape of the absorbers can be made to tailor the transmitted electromagnetic field. Such opical proximity correction (OPC) mask compensate for the effects of nearby features on a mask and sharpen the pattern corners. OPC masks yield improved pattern fidelity and, hence, increase process latitude. In order to get an improvement in the resolution of a mask, phase shift masks (PSM) have been developed (49). They control the amplitude of the exposing radiation at the surface of the resist. An ordinary, so-called "binary" mask either absorbs or transmits the exposure radiation. It does not selectively modify the phase of the transmitted radiation. Phase modification alters the character of the optical interference pattern produced at the focal plane, that is, at the resist. If material is removed from some of the open parts of the mask, it will introduce a shift in the phase of the transmitted light and sharpen the resulting radiation

Stepper	Resolution	Depth of focus	NA	Wavelength	Year
GCA 4800	1.25 µm	1.5 µm	0.28	g-line	1980
Nikon	1.0 µm	1.5 µm	0.35	g-line	1984
Canon	0.8 µm	1.2 µm	0.43	g-line	1987
ASML 5000/50	0.5 µm	1.0 µm	0.48	i-line	1990
Micrascan 2+	0.30 µm	0.8 µm	0.50	DUV	1995
ASML 5500/300	0.25 µm	0.7 µm	0.57	DUV	1996

FIGURE 22   The evolution in optical lithography equipment and characteristics from 1980 to 1996.

amplitude pattern. Figure 23 shows the cross section of a PSM (50). Use of a half wave shift can improve resolution by about 40% in some regions of the transferred pattern at the cost of introducing problems into other parts of the field. Using gradual transitions, rather than the sharp steps shown in Fig. 23 permits a compromise between pattern sharpening and degradation. An alternative approach to phase shifting is to thin the basic absorber pattern in places so that some radiation is transmitted through the regions that are normally-fully-absorbing. Combinations of these two approaches also are used. The enormous efforts that have gone into tailoring the chemistry and responses of resists used with each wavelength and at each node are summarized in a recent review (51).

As indicated in Fig. 18, the use of relentlessly smaller wavelengths required a shift from mercury lamps to excimer laser sources. The KrF laser with its 248 nm output is the mainstream tool for the 180 nm node in the roadmap. Figure 24 shows the overall view of a current aligner for state-of-the-art chip production at the 130 nm and possibly at the 100 nm nodes (52). The 10 W ArF laser source runs at 2 kHz. The system has a numerical aperture variable between 0.50 and 0.75. It can process over 90 mm wafers per hour using the step-and-scan approach to exposure. Such systems cost about $10 million. Comparison of the current aligners with the table-top lithography systems from 40 years ago, such as shown in Fig. 16, highlights the dramatic increase in the complexity of lithography in the past four decades.

The shift to production of 100 nm devices using the ArF laser at 193 nm with phase-shift masks will begin during the next 3 years. It is widely expected that another shift in source to an F_2 laser operating at 157 nm might be needed to meet the requirements of the 70 nm node. However, it is possible that the use of 193 nm radiation will be extended to this node. The F_2 laser is a daunting challenge to aligner manufacturers because the optics for 157 nm must be made of calcium fluoride—quartz is too absorbant at that short

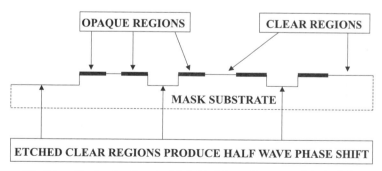

FIGURE 23   Schematic showing the principle behind the performance of phase-shifting masks for optical lithography.

FIGURE 24   The ASML ArF step-and-scan alignment and exposure tool.

wavelength. It is expected that whatever technology is used for the 70 nm node will not suffice for the 50 nm node. Hence, one of the "next generation lithography" candidates must be production worthy by about 2008.

The next three sections review the basics of EUV, X-ray, electron projection, and ion-projection lithographies, all of which are in the running for the mainstream commercial lithography tool, possibly as early as the 100 nm node. If that could be accomplished, the use of 157 nm F_2 laser sources would not be required. Before getting to the next-generation methods that might become commercially important, we pause to review some other optical techniques that have been demonstrated.

4.2 ADDITIONAL OPTICAL TECHNOLOGIES

Several optical lithography techniques are not now contenders for commercial production of advanced chips, but they offer potential for the making of patterns for research, at least. One is an extension of normal photography. The others involve manipulation of the phases and subsequent interference within the near or far fields of optical beams. Two mask techniques are now under development.

Ordinary photographic film, which contains small grains of AgBr embedded in gelatin on a plastic film, is at the heart of the photographic technique for making microstructures (53). First, the desired pattern is produced on paper with a 600-dot-per-inch computer printer. The pattern is then photographed with an 8X reduction and the film is developed. The process of development reduces the AgBr to isolated grains of metallic Ag that are distributed within the film in the desired pattern. Electroless deposition of additional Ag produces

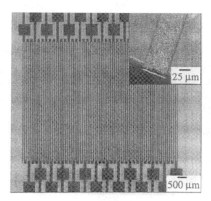

FIGURE 25   Optical micrograph of a gold wire about 50 μm wide, 2.5 μm thick and about 650 nm long.

a continuous pattern. This pattern can form the base for electrodeposition of other metals to make microstructures, such as the serpentine gold wire shown in Fig. 25 (53). Features as fine as 30 μm wide can be produced by the technique. The microstructures formed after electrodeposition are sufficiently rigid that they can be removed from the film substrate by dissolution of the gelatin and manually assembled into three-dimensional structures.

Three near-field optical lithographic techniques have been demonstrated. In one, called near-field phase-shifting photolithography, a transparent structure of organic elastomers that has features on the scale of micrometers is placed in contact with a thin resist on the surface of a substrate (54). Exposure of the resist through the elastomer results in lines within the developed resist as fine as 50 nm due to interference within the beam of the light used and the interaction of the near-field interference pattern with the resist. The second technique involves the use of a thicker resist whose surface is not flat but has significant topography (55). The surface undulations cause the resist to act as an optical element that again produces a near-field interference pattern, this time within itself. The method, termed topographically directed photolithography, is also capable of producing features as small as 50 nm.

The third near-field technology involves the use of light from a near-field scanning optical microscope (NSOM). Developed for subwavelength optical imaging of the surfaces of materials, the NSOM raster scans an illuminated aperture, usually at the end of an optical fiber, over the surface of the sample and measures the amount of light that is scattered. The aperture is typically 20% of the wavelength of the light employed. The evanescent wave that extends from the aperture and the geometry of the tip relative to the substrate determine the resolution, which is below the diffraction limit for the wavelength employed. Scanned NSOM patterns have been used to expose thin optical photoresists. In a recent study, this approach was employed for direct patterning of hydrogenated amorphous silicon (56).

Three far-field techniques for optical lithography use interference to produce the patterns of interest. The simplest employs the interference of two coherent laser beams to make a regular pattern. The second involves first making a hologram from the mask and then using that hologram to expose a resist. The last produces an interferogram at the resist to expose it.

If two parallel laser beams with sufficient coherence are brought together at an angle on a surface covered by a resist, the varying interference will produce an array of evenly spaced lines of exposed resist due to the constructive and destructive interference of the beams. Repeating the process after rotating the substrate 90 degrees will produce a square array of points of maximum exposure. Development of the resist leaves an array of fine structures. An example is shown in Fig. 26 (57). Varying either the optical wavelength or the angle of incidence on the resist enables variation of the scale of the pattern. Monochromatized beams of ultraviolet radiation from various sources can be used to produce patterns with spacings near 10 nm. A surface coated with such structures has low reflectivity, a property desirable for the surfaces of lenses and other optical elements.

In holographic lithography, the pattern to be replicated is the object that is used to make a hologram. A beam of light transmitted by the pattern is brought together with a coherent reference beam to form a hologram within a photographic film. The developed hologram can be employed to reconstruct an image of the pattern in the usual manner by using a laser beam, which is diffracted by the structure within the hologram to make the image. The reconstructed image can be viewed as is done with decorative and artistic pieces. Alternatively, the hologram can be used to produce a replica of the

FIGURE 26   Photograph of a "motheye" structure with 300 nm pitch and 500 nm deep features, as produced by optical inteference in a photoresist.

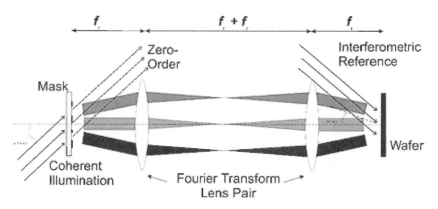

FIGURE 27   Schematic of the setup for imaging interferometric lithography.

pattern in a photoresist. A line resolution of 300 nm has been reported using holographic lithography (58). It is limited by the wavelengths of the lasers used for making and using the hologram, the resolution of the material within the hologram, and the geometry employed.

A technique called imaging interferometric lithography is shown schematically in Fig. 27 (59). It combines optical and interferometric resist exposure techniques. The optical exposure conveys information from nonrepetitive structures, while the interference pattern provides the high spatial frequencies. Three exposures are required to capture different parts of the frequency spectrum of the mask. One is the conventional optical exposure and the other two are for the orthogonal (XY) components of the high spatial frequencies. The method is comparable to a single exposure with a lens that has a numerical aperture three times larger than that of the lens actually employed.

5. EXTREME ULTRAVIOLET LITHOGRAPHY

Major changes in optical lithography include the past switch to excimer laser sources and the possible switch to calcium fluoride optics. These are really momentous changes in a very conservative industry that must have reliable tools for routine production of very large numbers of sophisticated parts in fabrication facilities whose cost now exceeds $1 billion. However, these changes pale in comparison to what is required if an order-of-magnitude decrease in wavelength to 13 nm in the extreme ultraviolet spectral region is going to be made (62). Incidentally, this approach was first termed "soft X-ray projection lithography," so some of the early literature may be confusing (63). In the early part of the last decade, the name was changed to more accurately reflect the historical nomenclature of spectroscopy, and to distinguish the technique from "X-ray lithography" (which is actually performed with what are properly called soft X-rays!). EUV lithography offers almost a 20X decrease in wavelength to 13 nm (relative to the currently used KrF laser at 248 nm). This will improve the resolution significantly, possibly to 35 nm. EUV lithography has a small numerical aperture of about 0.1, but still promises an acceptable depth of field of about 1 μm.

The basic challenges for EUV lithography are: (1) an entirely new source is required; (2) the mask must be reflective; and (3) all the optics must be reflective. Challenges (2) and (3) spring from the fact that all materials are highly absorbing in the EUV region, with mean free paths on the order of nanometers. This precludes even the use of a transmission membrane for a mask support, let alone transmission lenses. Further requirements for EUV lithography include: (4) the number of independent optical elements is significant, each with six degrees of freedom that must be maintained within micrometer tolerances; and (5) the optical elements are complex in both their shape and coatings. Work in this part of the spectrum is made difficult by the fact that simple mirrors do not work well at high angles of radiation incidence. To get the needed high reflectivities, it is necessary either to use grazing angles of incidence or to coat the optics with multiple layers of alternating high- and low-atomic-number materials. The latter coating can be viewed as an artificial crystal with the repeat spacing commonly denoted by **d**. It enables the coated optic to function like a Bragg reflector, where the radiation wavelength λ is equal to twice **d** times the sine of the incidence angle θ measured from the surface of the optic. That is, Bragg's equation applies: $\lambda = 2\mathbf{d}\sin\theta$.

An overall schematic of the contemplated EUV lithography aligner is show in Fig. 28 (64). The source of the 13-nm radiation is a xenon plasma produced by the absorption of a high-power and high-repetition-rate pulsed laser by a high pressure jet of xenon gas. This both yields radiation in the desired spectral region (to match the coatings on the optics) and avoids debris from the source (which would coat critical components). A plasma source at 113-nm radiation heated by a high-power electric discharge is being developed as an alternative to a laser-heated source (65). The radiation from the plasma is collected by grazing-incidence condenser and focused by a multilayer-covered optic onto the reflection mask. The following four optical elements are coated with multilayers. They produce an image of the mask on the resist that is demagnified four times. The mask and wafer are scanned synchronously with nanometer precision to produce the entire exposure field. Three of the four focusing optics after the mask are aspheric, as indicated by the diagram in Fig. 29 (62). The production of these mirrors with the proper form and surface finish—and their flawless coating with the multilayer reflectors, each having about 20 pairs of Mo and Si layers—is a technical tour de force. The reflective

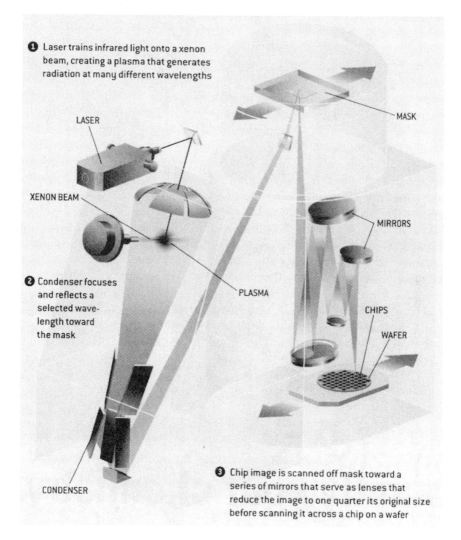

FIGURE 28 Schematic of the optical path for EUV lithography, showing the numerous components, some covered with multilayer reflectors, that must be precisely aligned to produce nanometer-scale patterns.

masks also must have multilayer coatings, which makes their repair practically impossible. The alignment of the pattern for any given layer to those already on a chip in production is also daunting. The EUV radiation is, of course, invisible, and the UV fluorescence it might produce from alignment marks is relatively difficult to employ. The use of optical fluorescence from nanometer alignment

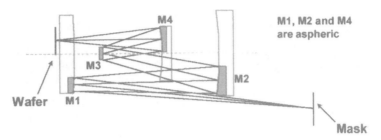

FIGURE 29 Optical path from the mask to the wafer in EUV lithography.

marks is possible in principle. The employment of visible radiation from a laser or other source for alignment requires the prior alignment of that light with the EUV radiation. In short, alignment for EUV lithography is another very difficult challenge. Metrology on the scale of a few nanometers and the production of robust resists optimized for 13 nm are also on the long list of problems, all of which must be overcome for EUV lithography to become a viable commercial tool.

The development of EUV lithography is being pursued in the U.S. by what is called a "Virtual National Laboratory," consisting of engineers from the Sandia (Livermore), Lawrence (Livermore), and Lawrence (Berkeley) National Laboratories. It is funded by EUV LLC, a group formed by Intel, which includes IBM, AMD, Motorola, Infineon, and Micron, as well as two equipment manufacturers, ASML and SVG. The prototype of the EUV lithographys system

FIGURE 30 Photograph of the Engineering Test Stand, a full-scale prototype EUV lithography system.

is shown in Fig. 30 (66). The high vacuum hardware and the scale of the system are noteworthy. The goal of the Virtual National Laboratory group is to be able to make commercial chips by 2005. Another group called EUCLIDES (Extreme UV Concept Lithography Development System) has been formed in Europe. It includes ASM, Carl Zeiss, and Oxford Instruments. Their goal, like that of the U.S. group, is to enable the production of chips with 70 nm features.

6. X-RAY LITHOGRAPHY

X-ray, or more accurately, soft X-ray lithography, is about as big a change from EUV lithography as the latter is from the excimer-laser-source technologies now in use. New sources, optics, masks, and resists are all required for X-ray lithography at 1 nm (67). The lack of Bragg coatings for mirrors with large angles of incidence at this wavelength precludes the use of projection demagnification optics. It is necessary to return to the proximity arrangement that was used in the early days of optical lithography. This is indicated schematically in Fig. 31 (68). The wavelength for X-ray lithography is determined by a balance of having photon energies that are high enough to penetrate the micrometer mask membrane, but also low enough to be stopped by the submicrometer metallic absorber pattern on the membrane and by the submicrometer polymer resist on a wafer.

The two approaches to generation of radiation at 1 nm with sufficient parallelism for X-ray lithography are very different. One is to use a synchrotron to accelerate, and a storage ring to capture, a very fine beam of circulating energetic electrons. Their centripetal acceleration in the magnetic fields around the storage ring results in the emission of a continuous band of synchrotron radiation in a wide horizontal fan that has a narrow beam spread vertically. Beam ports from the storage ring allow the X-radiation to be transported, using

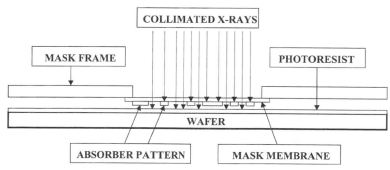

FIGURE 31 Cross-sectional schematic of the mask and nearby wafer in X-ray proximity lithography.

FIGURE 32   Overall view of the electron storage ring developed for X-ray lithography. The ports to the beam lines and exposure stations can be seen on the right.

grazing-incidence optics, to the wafer in a vertical aligner. Figure 32 shows a storage ring developed by Sumitomo Heavy Industries (69). The mutliple steppers that can be operated from one storage ring offset its high cost (tens of millions of dollars, including the alignment and exposure stations). Extraordinary reliability is required, because a great deal of production capability is lost if the storage ring has problems. This approach to X-ray lithography seems limited to large companies due to the high capital cost.

The second approach to generating 1 nm soft X-rays for lithography is to produce multimillion-degree plasmas whose natural emissions are at the desired wavelength (70). Pulsed lasers do not provide enough energy for the generation of sufficient power of 1 nm X-rays, in contrast to the case for the 13 nm radiation used for EUV lithography. It is necessary to use pulsed electrical power from energy stored in large capacitors to repetitively heat a neon plasma to X-ray emitting temperatures. The source is termed a dense plasma focus (DPF). It produces an X-ray source that is about 500 μm in diameter and 8 mm long (71). The DPF radiates into all angles, in contrast to a synchrotron X-radiation source. Hence, a great deal of effort has gone into the design, prototyping, and testing of grazing-incidence X-ray collection optics for DPF and other plasma sources (72). When run at 60 Hz, the DPF produces about 450 watts of soft X-rays (71). Such a source, integrated with a vertical aligner, is shown in Fig. 33 (73). This arrangement of one X-ray source per aligner makes X-ray lithography available to smaller companies. As with the

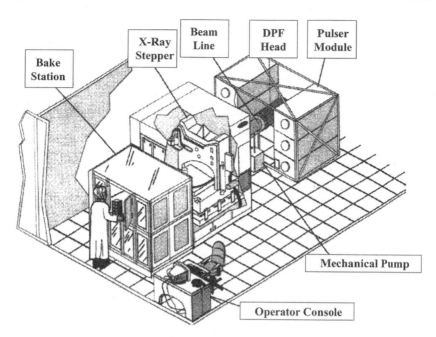

FIGURE 33   Schematic of the (vertical) aligner integrated with a pulsed plasma source of X-rays.

storage ring, reliability is key. The requirement for the system to handle high peak powers at high rates for long times is very demanding. Such a system is now in early use for production of key features in millimeter-wave microelectronics.

The history of X-ray lithography since its invention 30 years ago is an interesting case study in technology development. Over $1 billion has been spent on X-ray lithography research and development. About 10 years ago, it was thought to be the successor to optical lithography. However, the relentless advances in the capabilities of optical lithography, and problems with X-ray lithography, have delayed, and possibly, killed its chances for mainstream use. Making masks on micrometer-thick membranes, which are robust under the conditions of commercial production, and have the needed precision in both the absorber features and their placement, have been continuing problems (74). Like the development of 1 nm X-ray sources, this mask problem appears to be soluble. However, U.S. workers have essentially abandoned X-ray lithography in favor of EUV and the projection electron lithographies, which are discussed in the next section. This is not the case in Japan, where interest in X-ray lithography as a potential mass-production tool for ICs and other products remains high (69).

7. PARTICLE LITHOGRAPHIES

7.1 ELECTRON BEAM TECHNIQUES

Electron-beam direct-write (EBDW) lithography has long been shown to produce lines in resists near the limit of the resist resolution of about 10 nm (75,76). It would be the final solution to production lithography requirements if it were not a serial technique, in contrast to the parallel exposures of many pixels in the photon lithographies just reviewed. Schemes to expose entire geometrical shapes, the so-called vector scan methods, offer some relief from the slow raster-scan approach to exposing a subfield, but at the cost of more complicated systems. EBDW lithography requires bright sources of electrons, very high data rates for driving the electron deflection system in the focusing column, and proximity corrections that compensate for the overexposure of nearby features in a resist.

An immense effort has been applied to EBDW techniques because of their key role in production of the finest masks and their possible direct role in chip production. It may be that, for the 70 nm and smaller nodes in the roadmap, mixed technologies will be employed for chip production. Some other method, possibly EUV lithography, will be used for patterning most of the features on high-performance chips, with EBDW being used to delineate the finest dimensions.

One way to circumvent the speed limits of EBDW lithography is to use many beams in parallel, a possibility enabled by the emergence of micro-machining technologies. Lithography would be used to form hundreds of micrometer-scale electron-focusing columns that would be used to perform lithography. This approach is reminiscent of the use of computers to design computers. Programs to study this multi-column option are pursued in both the U.S. and Europe. Each electron microcolumn would be a few millimeters high. In one approach, an array of field emission sources would be used to produce the electrons to be focused onto the resist (77). Another technology uses an array of field emitters operated at reduced current, and a microchannel plate electron multiplier, to get adequate electron intensity (78). More than two decades of problems with making and operation field emission arrays for displays are relevant to these schemes. In a third approach, photoemission would be used to produce the electron beam. This concept is indicated schematically in Fig. 34 (79).

The multiple-column approach to electron-beam lithography is over a decade old, but it is not yet close to being a production technology. However, two technologies for electron projection lithography (EPL) are thought to be candidates for the next-generation lithography tool. These go by the names SCALPEL and PREVAIL, and will be reviewed next.

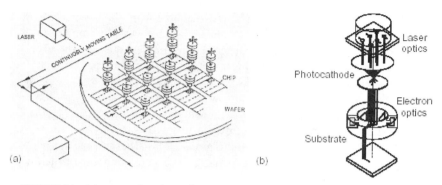

FIGURE 34   Schematic of a scheme for use of multiple microcolumns for electron-beam.

SCALPEL stands for Scattering with Angular Limitation Projection Electron Lithography. The fundamental concept is diagrammed in Fig. 35 (86). Both the open and patterned parts of the mask are transparent to the 100 keV electrons. Since electrons interact strongly with materials, those passing through the pattern are scattered more often, so they emerge in a broader cone than those electrons that scatter only a few times in the membrane alone. This permits the elimination of most of the electrons that passed through the pattern by the use of an aperture, as indicated in the diagram. At the image (resist) plane most of the incident electrons are those that went through the "open" parts of the mask. The image is a four-to-one reduction of the mask pattern. SCALPEL has two advantages: relatively little energy is absorbed in the mask and, like other

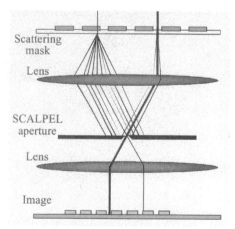

FIGURE 35   The components and electron paths in the SCALPEL technology.

projection technologies, the mask is well removed from the work piece. Both these aspects compare favorably with the situation for X-ray lithography. The subfield size in SCALPEL is one-quarter of a millimeter square at the wafer, so these areas must be stitched together to expose an entire chip. The technique has exhibited the ability to produce 80 nm lines and spacings. It is being developed by Lucent Technologies at Bell Laboratories, and is a candidate for insertion at the 100 nm node in the semiconductor roadmap.

PREVAIL, or Projection Reduction Exposure with Variable Axis Immersion Lens, technology is diagrammed in Fig. 36 (81). The central aim of this approach is to defeat the aberrations, and the associated blurring of the image, that occur off-axis in a fixed electron lens system. In PREVAIL, the axis of the magnetic lens is moved synchronously with shifting the beam laterally in order to print subfields off-axis. This synchronous shifting of the lens axis and the beam is done with a complex arrangement of deflection systems built into the illuminator and collimator lenses. PREVAIL has the same subfield size as SCALPEL, but is able to print a field size of 5 mm at 80 nm resolution without moving the wafer. A disadvantage of PREVAIL is the requirement for a stencil (partially open) mask. The technique is under development in a joint effort by IBM and Nikon, with the 100 nm node as its target insertion point.

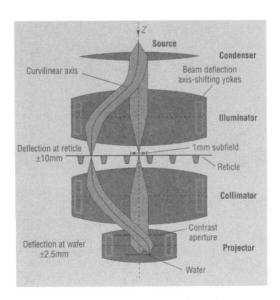

FIGURE 36   The components and electron paths in the PREVAIL technology.

7.2 Ion-Beam Technologies

Direct-write ion-beam techniques play a significant role for research (82) in the repair of masks (83), but are not candidates for production lithography. Arrays of ion microcolumns are receiving attention, similar to the case of electron microcolumns, but few expect micromachined column arrays to be commercial tools. Multiple beam methods using innovative sources are being developed (84). Ion-beam proximity technology is still being developed, but it has severe challenges regarding both the production of masks and their longevity. Projection ion-beam lithography is in the running for commercial production of leading-edge chips. Figure 37 shows both the layout of the ion projection column and the energies of the ions during the quick trip from the source to the resist (85). The system produces a four-times-demagnified image of the stencil mask on the resist. A 200 mm stencil mask with feature sizes down to 200 nm is commercially available (86). As with the two contending electron projection lithographies, ion projection lithography is vying to be the next-generation technique at the 100 nm node.

8. PROXIMAL PROBE LITHOGRAPHY

The term "proximal probe" applies to instruments in which a fine point (the probe) is placed in close proximity to the surface of a solid in a vacuum or

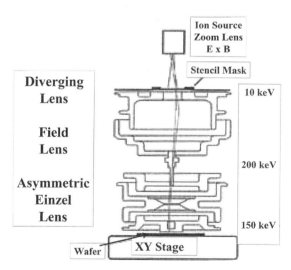

FIGURE 37   The components and the particle paths for the projection ion technology.

gaseous atmosphere or in a liquid (87). Remarkably, such probes are sensitive to the characteristics of the surface with spatial resolution on the atomic scale. The characteristics include the atomic structure and topography, electronic structure, electrical parameters, and magnetism, among others. When the probe tip on a microcantilever is scanned over the surface of a sample, maps of such properties are obtained. Several means have been devised to sense the interactions between the probe and the sample. Electrical conductivity is read out for scanning tunneling microscopy (STM). Optical measurements of the deflection of the cantilever are used for atomic force microscopy (AFM) and other proximal probes. Several table-top-scale commercial proximal probe instruments are on the market. An arrangement used for research on proximal probe lithography is shown in Fig. 38 (88).

Proximal probes, especially the STM and AFM, were initially developed for "seeing" surfaces and their properties with atomic resolution. However, it was soon found that their ability to influence a surface could be put to use as a tool to manipulate individual atoms and molecules. The situation is similar to that found two decades earlier with focused beams of heavy ions, which could be rastered over a surface to obtain an image, or moved slowly in a programmed fashion to sculpt a sample by scattering ions from its surface. The focused ion-beam methods have resolution generally on the 1-to-10 nanometer scale, while the proximal probe technologies have intrinsic resolutions on the 0.1-to-1 nanometer scale. An early demonstration of the ability to place individual atoms in desired positions is shown in Fig. 39 (89). The image of 35 xenon

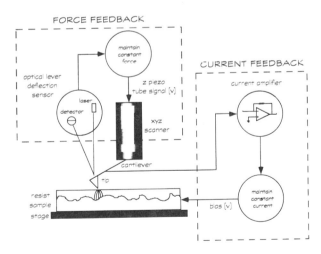

FIGURE 38 Schematic of the proximal probe lithography circuits developed at Stanford University involving both an AFM for height control and STM for current control.

atoms arranged on a nickel (110) surface to spell IBM has become one of the icons of nanotechnology. Such atomic manipulation is now used for many purposes. Figure 39 also shows an image of the result of positioning individual iron atoms on the surface of a copper (111) crystal (90). This structure, termed a "quantum corral," was made to study the effects of confining electron waves on the surface of a conductor. The manipulations of atoms on surfaces are characterized by the precise control of individual atoms with atomic positional resolution. But the processes are very slow, essentially useful only for research. The employment of tuned lasers in an Atomic Processing Microscope enables faster controllable removal and addition of atoms on surfaces, but the rates still are not commercially viable (91).

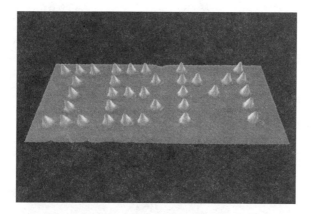

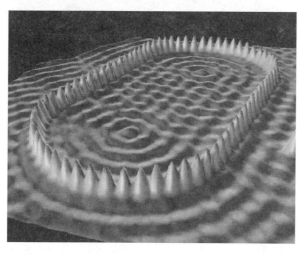

FIGURE 39   Atomic arrangements made and imaged with an AFM. The text provides details.

Proximal probes also offer the ability to modify the properties of surfaces or thin films on a surface, ignoring individual atoms and producing effects over relatively large areas at much faster pattern transfer rates. It is possible to oxidize the surface of a silicon wafer on a nanometer scale using an AFM. Figure 40 shows the result of using a negatively biased AFM to produce an oxide pattern (92). Strictly, this is an example of surface modification by simultaneous pattern and material transfer, with the oxygen coming from the atmosphere. It is included here to illustrate the chemical effects a proximal probe can produce on a nanometer scale. Such effects in photoresists are the basis of the purely lithographic use of proximal probes.

Proximal probe lithography arose from work done initially with conventional proximal probes, specifically, scanning tunneling microscopes, and unconventional resists with thickness down to monolayers (73). The electron current passing between the probe tip and the resist as it is moved over the surface in a programmed fashion modifies the chemical properties of the resist during the patterning process, similar to the oxidation of silicon shown in Fig. 40. This technique produces patterns with resolution approaching the nanometer-resolution scale, but is slow because of its sequential direct-write character.

It was realized that the use of multiple probe tips would proportionally speed up both proximal probe imaging and lithography. Figure 41 shows a pair and an array of proximal probes developed for lithography by Quate and his colleagues (93). In both cases, the probe tips are combinations of actuators (which provide the deflections), and sensors (which determine the position of the tip). The actuators are ZnO, a piezoelectric material, and the sensors exploit the piezoresistive character of silicon. A book on proximal probe lithography has been published recently (94).

FIGURE 40   AFM image of silicon dioxide lines 20 nm wide made by use of a biased AFM.

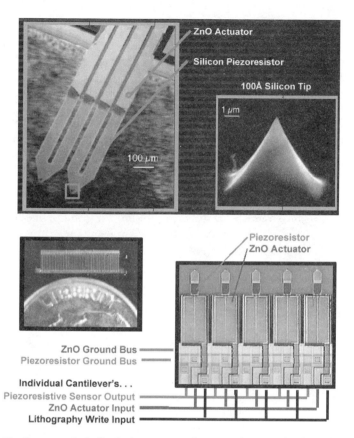

FIGURE 41 Structures, including both actuators and sensors, that were developed for research on proximal probe lithography. An array of 50 cantilevers is shown near a U.S. 10-cent coin. A magnified image of some of the cantilevers in that array is on the right.

9. OTHER PATTERN TRANSFER METHODS

The techniques for pattern transfer reviewed to this point generally result in a pattern in a thin film on the surface of a substrate. The thin film might be the disposable photoresist, or a permanent thin film that is part of the work piece. There are some other technologies for pattern transfer that are quite different from either conventional or developmental lithography techniques. These are reviewed in this section. The first is called LIGA, in which very thick resists are used and additional processing steps are taken. The next is a method termed LISC, which uses more conventional thin photoresists, but very different masks. The last technique—embossing—is old, but is now being applied to the transfer of micrometer-scale patterns.

9.1 LIGA

A three-stage process, developed in Germany, involves sequential use of lithography (**LI**thographie), electrodeposition (**G**alvanoformung), and molding (**A**bformung). In contrast to the lithography technologies surveyed so far, which use resists with thickness from nanometers to micrometers, LIGA employs resists that range from micrometers to more than millimeters in thickness. The steps in the entire LIGA process are shown schematically in Fig. 42 (95). The lithography (pattern transfer) step is characterized by the use of very thick photoresists, commonly a significant fraction of a millimeter, that are exposed through a mask by X-rays from an electron storage ring. Such radiation has the property of being very well collimated. That is, it does not diverge or converge significantly over scales comparable to the thickness of the polymer being exposed. Hence, the structure that results after development of the exposed resist has cross-sections throughout that are the same as the mask

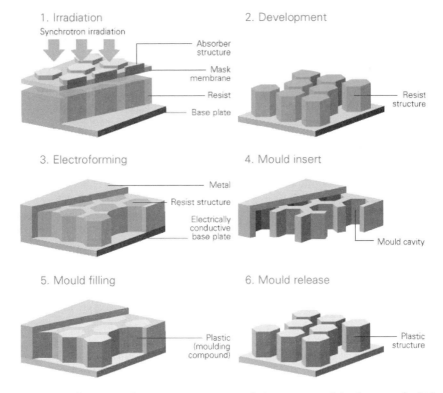

FIGURE 42   The sequential steps in LIGA, starting with the exposure and development of a thick photoresist. This can be followed by electroforming to make a mold insert, which is then used for replication of plastic parts.

pattern. The sharp delineation between developed and undeveloped parts of the polymer permits the production of structures that are 100 times deeper than they are wide (96). When harder X-rays are used for exposure, the mask pattern can be transferred even through sheets of a plastic about 1 centimeter thick. Ultraviolet (UV) radiation from an aligner can be used to expose thick epoxy and other photoresists, if they are UV transparent. The heights of UV-exposed resists are generally limited to less than 1 mm and the ratios of feature height to width are 10 or less (98).

LIGA has the interesting characteristic that useful products can result after any of the three primary stages. The patterned plastic after the initial exposure and development can itself be a product. However, this approach is not attractive for mass production because of the need for using an expensive radiation facility. The metal parts formed by electrodeposition can also be the end product. Finally, plastic parts can be made using the electroformed material as a mold. For example, a very compact dispersive grating optical spectrometer made by replication using a LIGA electroformed mold is available commercially. Electrodeposition and molding involve the transfer of material into a previously made pattern. They are discussed in Section 19.

9.2 LITHOGRAPHY INDUCED SELF-CONSTRUCTION

The next of the lithography techniques to be reviewed is both the most recently discovered and the most unconventional compared to those discussed so far. It is called lithographically induced self-construction (LISC), and it has a variant termed lithographically induced self-assembly (LISA). In both cases, a patterned mask coated with a surfactant is placed in close proximity to a substrate covered with a thin thermoplastic polymer, as indicated in Fig. 43 (97). The polymer may be chemically identical to a photoresist, for example, poly(methalmethaclylate) (PMMA), but it does not function as a normal photoresist. There is no radiative transfer between the mask and polymer, and the chemistry of the polymer is not modified during pattern transfer. Rather, when the PMMA is heated to 170 °C (which is above its softening point), electrostatic forces cause it to move laterally into shapes mediated by the nearby mask.

The techniques LISC or LISA are both hybrid contact and proximity methods, where the spacer contact controls the gap and the nearby mask dictates the pattern. The protrusions on the mask have heights near 300 nm and the PMMA has a thickness of about 100 nm. The spacer controls the gap of 100 to 400 nm between the flat polymer and the mask protrusions over areas

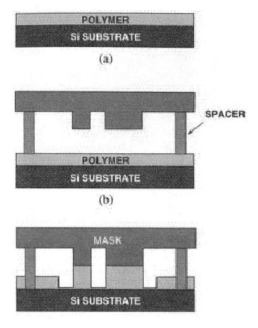

FIGURE 43   Schematic of the structures used in LISA.

that can exceed 2 cm in diameter. Figure 44 shows an atomic force micrograph of PMMA pillars formed by LISC under a mask with a pattern of dots 1.3 μm in diameter (98). The pillars have the same diameter and are 350 nm tall. It is believed that submicrometer patterns can be replicated by this technique. A pattern as large as 50 μm square was reproduced by LISC. During its formation, polymer material migrated from at least that distance away. LISC requires the employment of a mask and polymer with similar surface tensions, with the surfactant chemistry on the mask being controllable. With such similar surface characteristics, any instabilities (non-uniformities) in the initially formed pattern on the polymer surface will be filled in and the pattern in the polymer will mirror the protrusions on the mask.

    With larger differences in surface tension between the mask and polymer, instabilities produced at first in the polymer under mesas in the mask are preserved and a pattern of small dots forms even under regions of the mask that are flat. If the entire mask is flat, then an array of dots with diameters near 3 μm that are spaced a few micrometers apart forms spontaneously due to the heating cycle. The pattern shows "grains" within which the dots are in hexagonal arrays. If the mask contains patterns 10 to 100 micrometers on a side, then the self-assembled dots are confined to regions under the larger

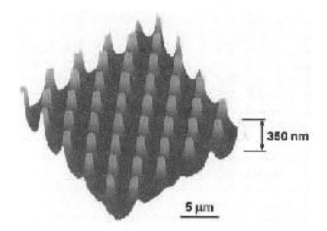

FIGURE 44   AFM of pillars formed in PMMA by LISC.

patterns. Figure 45 shows examples of dots formed by LISA under masks both without and with patterned areas (99).

A model for appearance of the patterns seen in both LISC and LISA has been developed by the discoverers, Chou and his colleagues. It invokes image charges due to the proximity of the mask protrusions. The sequential development of the patterns seen in LISA under a patterned area on the mask (which could merge to replicate the entire pattern in LISC) is shown schematically in Fig. 46 (99). The electrostatic forces dominate the polymer viscosity, surface characteristics, and gravity to produce the observed patterns.

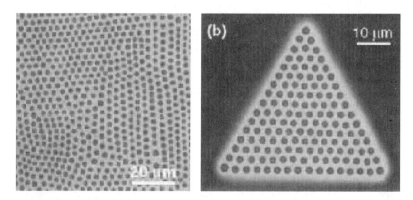

FIGURE 45   Patterns formed by LISA under a flat surface (left) and under a surface with a triangular protrusion (right).

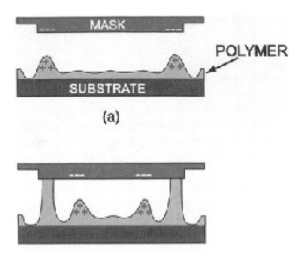

FIGURE 46   Model for the growth of structures during LISA.

It seems likely that electrical control of the processes is possible, given the role played by charge distributions.

9.3 EMBOSSING AND IMPRINTING

Embossing is one of the old technologies that has been extended to the micrometer scale in recent years. It involves bringing a patterned surface into contact with a compliant material that is to be imprinted with the pattern. When an organic material that softens with temperature, such as a wax or plastic, is heated to the softening point, impressed with a patterned structure, and allowed to cool in contact with the pattern, it will retain the pattern. An ancient example of this process, called hot embossing, is the use of wax carrying the coat of arms of the sender to seal correspondence in an envelope. Hot embossing has generally been used to produce mesoscopic structures. However, it can be employed to produce microstructures and even nanometer-scale features. Doing so requires heating of both the polymer to be shaped and the tool containing the pattern prior to forcing them together. Temperatures in the 100–200 °C range are generally sufficient. Silicon, patterned by the methods discussed later in Section 12, is an effective tool. The production of deep or narrow structures by hot embossing requires making a tool with sloping side walls and attention to the adhesive characteristics of the polymer and tool. Patterns with structures about 10 μm wide and deep can be embossed into PMMA (100). Recently, a modified wafer bonding system was used to emboss structures as fine as 400 nm entirely across a 10 cm diameter wafer (101).

A variant of embossing, called nanoimprint lithography (NIL), involves impressing a mold onto the surface of a photoresist-covered substrate (102). In this case, the pattern is transferred to the resist, commonly PMMA, by mechanical rather than chemical action. Subsequent processing of the resist to open the thinned regions to the substrate permit conventional uses of the resist for deposition onto or etching into the substrate. The molds for NIL can be prepared by a wide variety of the normal and developmental lithographic processes. For example, interferometric lithography has been used to make pillars on a mold that were 10 nm in diameter on 40 nm spacings over an area 2.5 cm square. With this mold, 10 nm holes were imprinted into PMMA (103).

A technique called Step and Flash Imprint Lithography essentially embosses a layer of liquid on a surface that is then turned into a solid (104). Hence, it avoids the elevated temperatures and pressures ordinarily required for embossing. The wafer is first coated with a transfer layer of solid organic material. Then a glass template with the desired pattern is placed near the coated wafer. The template can be micromachined by a variety of methods. A low viscosity liquid, photopolymerizable, organosilicon etch barrier material is dispensed between the template and transfer layer on the wafer, which are then brought into contact. After UV exposure to solidify the etch barrier and make it adhere to the transfer layer, the template is removed. A plasma etch transfers the pattern from the now-solid etch barrier into the transfer layer, and then the etch barrier is removed. This leaves the pattern in the transfer layer on the wafer surface ready for further processing steps. Step and Flash Imprint Lithography has produced 60 nm features.

Embossing has been employed to pattern a layer of liquid on a substrate surface (105). This process, illustrated in Fig. 47, is closely related to contact printing, except the "ink" is already on the substrate and is not brought down by the patterned "stamp." The stamp used in this process is a compliant material made by molding a polymer into a master pattern, as described in Section 14. Heating of the material after patterning is used to drive off solvents from the ink, or ultraviolet radiation can be used to cure a patterned layer of liquid prepolymer. This process can produce structures on a surface with features as fine as 200 nm over 400 mm areas (105).

10. APPLICATIONS OF MATERIAL TRANSFER TECHNOLOGIES

The applications of pattern transfer technologies, such as those discussed in Section 2, vary widely and depend largely on the characteristics of the specific

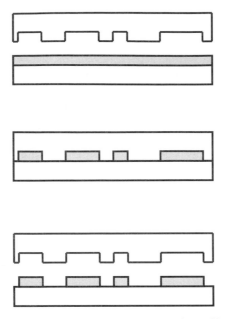

FIGURE 47 Schematic three-step process flow for patterning a layer of liquid (gray) on a surface with a soft patterned stamp.

technology employed. The primary characteristic of a material transfer technology that determines its applications is whether it uses fixed patterns, or else is a programmable (so-called direct-write, that is, direct removal or deposition) methodology. As in the case of pattern transfer technologies, fixed pattern techniques require making masks or some other tool, but can be used to make products at a high rate, while the programmable methods offer pattern flexibility at the cost of a slower speed for production of the desired product. In very general terms, fixed-pattern methods are best for large volume production and programmable methods are attractive for rapid prototyping and for small-lot production. However, this is not always true. Laser-based (programmable) methods are used for high-volume production of small vias in printed circuit boards and for trimming resistors in MEMS and other components.

The installed industrial base for material transfer technologies varies widely. Older methods, like screen printing and molding, are well honed for mass production. Newer technologies, for example, laser methods for transferring materials from a tape to a work piece, have neither the experience nor commercial base. However, some of the newer methods offer great flexibility, not only in the patterns in which material can be applied to a work piece, but also in the many types of materials that can be transferred and in the wide range of target substrates, including flexible and curved work pieces. Another big advantage of the programmable methods for direct deposition of materials is the relatively small capital investment required. A major characteristic of most techniques for simultaneous pattern and material transfer is their relatively low resolution. One micrometer is generally near or beyond the limit of patterned material transfer technologies, in contrast to most of the pattern transfer technologies, which can produce nanometer-scale lines and other designs. However, micrometer resolution is sufficient for many uses of material direct-write techniques.

Turning to the types of applications of direct deposition technologies, it is first noted that the products that result fall into the normal large classes. These are hierarchical, namely materials, components, subsystems, and systems, in order of increasing complexity. Material transfer methods can be used to produce the materials that then feed into other production technologies. Large-area patterns of materials on flexible substrates are the primary example. However, the production of components and subsystems is the main thrust of material transfer technologies (106). These include electronic, magnetic, optic, microwave, and other components. Conductors and connectors, passive components (notably resistors), reactive components (such as capacitors and inductors), antennas, and filters are among the components that have been demonstrated or commercially produced by material transfer technologies. Batteries and microsensors have also been prototyped by programmable material transfer methods (107).

The future of programmable printing of material looks bright to some workers. A recent article was entitled "Print Your Own PC" (108) envisions the ability to make microprocessors—cheaply and flexibly—that rival a current Pentium or similar microprocessor in capabilities. This is a very aggressive viewpoint, given the relatively low resolution of technologies for printing materials and the lack of a manufacturing base for many of the material transfer technologies. However, it is difficult to argue against the widespread commercial and private use of processes for printing materials, given the rate of development of both new materials and means for manipulating them.

The availability of many new technologies for simultaneous pattern and material transfer, especially tools that can make micrometer-scale features, will greatly expand the applications of such technologies. If the carpenter has only a saw and hammer, he is limited in the number and finesse of the structures that can be made. Only rough buildings are possible. The addition of new tools to his kit, especially tools that can make precision cuts, allows the carpenter to make cabinets. The same will prove to be true for the many pattern and material technologies that have been developed quite recently, especially when they are used together under computer control. Imagine the production of large areas containing micrometer-scale structures of diverse materials that can be made by some of the methods discussed here. The transfer of complex patterns of many different materials onto a substrate is a sort of micrometer equivalent of the patterns and materials that can be "ironed," that is, thermally transferred onto cloth shirts. The use of thermal techniques and others as a means to transfer patterned conductors, resistors, dielectrics, optical and many other materials in programmable sequences should enable many applications by making possible the cost-effective production of different structures and devices.

11. OVERVIEW OF MATERIAL TRANSFER TECHNOLOGIES

In the conventional and developing lithography technologies reviewed in earlier sections, there are several separate process steps, as indicated in Fig. 4. The first is material transfer with no pattern, namely the production of a thin film of the material of interest or of the photoresist employed in patterning it. The second involves pattern transfer without material transfer, that is, the exposure of a photoresist that is used to delineate patterns in a material film deposited before or after the resist, followed by resist development. Next, the thin film or substrate is etched, a thin film is deposited, or some other process like ion implantation is carried out. Finally, the resist is removed.

Technologies for the transfer of materials are diverse. There are numerous methods for producing unpatterned thin films of many materials. The spinning of layers of photoresist is a primary example. Such techniques are critical to the technologies we are reviewing, but are outside our range of discussion. The focus here is on the transfer of patterned materials, so the separate steps of resist preparation, exposure to impress a pattern, resist development, layer processing, and resist removal are not needed.

The patterns that are transferred with the material can be fixed or variable, that is, programmable. The process of simultaneous pattern and material transfer might be either subtractive, where material is removed in a pattern from the work piece, or additive, where a material is put on the work piece in a pattern. Fixed pattern subtractive technologies are reviewed in the next section. Programmable methods can be referred to as data-driven materials removal or deposition. Programmable subtractive material transfer technologies can be applied to metals and alloys, ceramics and glasses, polymers and organics, and other materials. They are reviewed in the following section. Materials that are added to a work piece can originate as solids (thin films and fine particles), liquids (homogeneous liquids, including solutions, and suspensions), or vapors. Even live cells can be put down in patterns. A discussion of additive and related technologies can be found in Sections 14–18.

Many of the methods for depositing materials on a substrate in a pattern involve the transfer of liquids. The patterns used can be fixed, as in the old technology of screen printing or in the relatively new micrometer version of rubber stamping (called microcontact printing, or soft lithography). These cases are reviewed in Section 14. Alternatively, the patterns for simultaneous transfer of a pattern and a liquid can be programmable. Here, there are two cases. In the first, some kind of dispenser, such as a fine-scale version of an ink pen, is used. Dispensing liquids in macro- and mesoscopic patterns is a very important commercial process, being used for production and assembly of many products in the automotive, electronics, pharmaceutical, and biomedical industries. Dispensing to make micrometer-sized features can be done in continuous or dropwise fashions. The alternate approach is to shoot the ink onto the work piece at the proper position using ink-jet technology. Ink jetting is used in several industries, most notably in desktop printing. A variety of techniques for liquid transfer using programmable patterns to produce micrometer-scale structures is discussed in Section 15.

Another broad class of programmable pattern and material transfer technologies involves the use of beams of photons, electrons, or ions. They serve to move material from a nearby substrate to the work piece in a fashion that is simultaneously subtractive (from the original substrate) and additive (to the work piece). Beams of quanta can also produce the material to be deposited from chemical action in the atmosphere immediately above the spot of interest on the work piece. Photons, electrons, and ions of appropriate energies can all initiate such chemical vapor deposition. Electron and ion techniques can produce structures with features on the nanometer scale. The beam-based technologies for pattern and material transfer are surveyed in Section 16. Other programmable additive techniques are the subject of Section 17.

A variety of technologies can quickly yield a three-dimensional material pattern that is often used to show how a product will appear. They have come to be called "rapid prototyping" techniques in the past decade. A wide variety of materials can be made into rapid prototypes. The rapid prototyping technologies that are capable of making small structures are reviewed in Section 18. There is another collection of techniques, some old and some new, for transferring material and producing a pattern at the same time. They include molding and a few related techniques. These are the subject of Section 19.

Several books and reviews provide coverage of many of the technologies for simultaneous pattern and material transfer. Reference (2) has many recent papers on direct writing of materials. An outstanding tabular summary of the characteristics of technologies for precision forming of materials is available in the book by Madou (109). A review by the Whitesides' group covers the techniques capable of producing structures on the micrometer and nanometer scales (110). A recent book focused on fabrication of 3D micrometer-scale structures (111). Chapters in this volume also provide broad reviews of many of the technologies discussed below (112–114).

A major feature of the programmable technologies for directly writing materials onto substrates is the ability to examine the product as it is made. This is most easily accomplished if the direct writing process involves the use of a laser beam in air. The inspection options are shown in Fig. 48 (112). This capability to do inspection during manufacture enables the correction of manufacturing errors "on the fly." Such an approach stands in contrast to the usual situation for defect detection, namely post-production examination of the characteristics and properties of a structure or device and the rejection of bad devices. The ability to quickly—and maybe automatically—repair defects could compensate somewhat for the serial nature of some technologies for simultaneous pattern and material transfer. That is, higher yields would tend to compensate for an intrinsically slower serial approach to parts manufacture.

Before proceeding to survey the panoply of methods for transferring materials onto a substrate in a patterned manner, we pause to note that both ordinary printing and xerography fall into this general category of technologies. Printing is a massive industry, in which the primary goal is to transfer thin films of materials in patterns onto a substrate, commonly paper, for ordinary viewing. The patterns have two different scales. The first is essentially limited by the resolution of the human eye to lines as fine as about 100 μm. The second is the monocolor dots that permit the appearance of diverse colors, with sizes as small as about 10 μm. Very limited experimentation on the transfer of modern nanometer-scale particles of diverse materials has been done with conventional printing technologies. This might be ascribed to the limited amounts of nanoparticle inks that are available compared to the relatively large volumes of inks needed for production presses. Also, commercial printing is

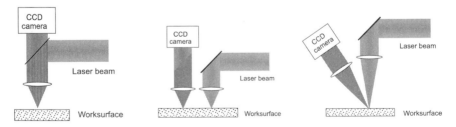

FIGURE 48   Options for inspecting a structure or device being written on a substrate as it is being manufactured: (left) through the lens with the laser beam being used for writing, (center) parallel but offset examination and (right) coincident but dependent optical paths.

geared to making large areas of patterned materials, something that has not yet been needed for quantum-scale electronic, magnetic, optical, and other new materials. Incidently, two electronic paper processes have been prototyped. In one technology, small spheres, which are white on one side and black on the other, are rotated by an applied field (115). In the other, the field is used to pull small particles into view to produce contrast on a page (116). Whether or not such technologies become commercially important, it seems likely that large-area printing of micro- and nano-materials will find some applications.

Xerography is actually a two-step process in which a pattern is first produced and then populated with the material that is to be transferred to the work piece simultaneously with the pattern transfer. The situation for xerography is similar to the extended role of photography as a means to transfer a pattern into a pre-existing material, the film, as discussed in Section 4.2. We noted that photography has been extended beyond its normal usage to include a variety of other applications. For ordinary xerography, the goal is to put a material, usually a carbonaceous powder called the toner, onto paper or plastic in the desired pattern. This is commonly done using a fixed pattern, the item that is being copied, as in ordinary office copiers. It also can be done programmably using a focused and scannable laser beam, as in office printing machines. In both cases, the transfer of the pattern and toner is fast and can be done in ordinary air. Importantly, a wide variety of other material powders, including metals, can be laid down in a pattern using the electrostatic processes that are at the base of xerography. In general, powders for such transfers range from approximately 5 to 10 μm and are relatively equi-axed. The finest lines that can be produced by commercial xerographic methods using dry toners are 1200 dots per inch, which is about 20 μm (117).

Electrostatic printing using liquid toners is also possible (118). The image is formed as usual by charging a photosensitive plate, but the toner particles are attracted from a liquid suspension. Particles of metals, ceramics, and other materials with sizes that range from 0.05 to 100 μm can be employed. The particles are transferred from the charged plate across a 50 to 150 μm gap onto the target substrate. Metal, glass, plastic, and paper substrates can be printed in this fashion with electronic interconnects and devices having features as fine as 10 μm.

12. FIXED PATTERN SUBTRACTIVE TECHNIQUES

Both new and old techniques for removing material from a work piece in a pattern with micrometer-scale features are available. Chemical and plasma processes for bulk micromachining into substrates are reviewed first. It is noted that these deep etching technologies involve pattern transfer from a developed resist on the surface of a substrate *into* the substrate. That is, they are qualitatively different than the methods discussed so far in which the pattern is transferred *onto* the work piece. The deep etching technologies are included here because of their importance for making microelectromechanical systems (MEMS) and to contrast them with the programmable subtractive processes surveyed in the following section. Next, a few technologies for patterned erosion of a substrate by chemical and mechanical means are discussed. In these processes, the tool that removes material from the substrate by chemical or mechanical means has a shape that determines the pattern that is impressed into the work piece.

Methods for producing deep and narrow structures on or in a substrate are the subject of most of this section. The aspect ratio, defined as the ratio of the thickness (of the resist or film) or depth (into the substrate) to the width of an opening (developed in resists or etched into a thin film or substrate) is commonly much less than unity. It approaches and may even somewhat exceed unity for submicrometer lithography technologies in which the line widths are comparable to the resist thicknesses. It is often desirable to have deep, narrow features in many micromachined structures and mechanisms. High ratios of the depth to the width of etched structures have many applications. Devices with such topography are described as having a "high aspect ratio." Figure 49 illustrates both low and high aspect ratio structures.

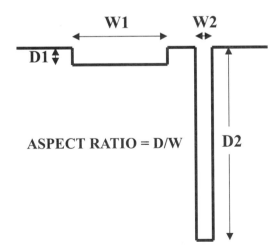

FIGURE 49   Schematic cross-sections of low and high aspect ratio structures in the surface of a substrate.

The deep etching techniques fall in a category called HARMST, which is short for high aspect ratio microsystems technologies (119,120).

12.1 Bulk Micromachining Processes

The ability to etch deeply into the bulk of a substrate material is central to the production of many structures and mechanisms that are critical to MEMS or microsystem technologies. A variety of processes are available to make high aspect ratio structures in silicon and other materials. They fall into two general classes. The first involves older methods of wet chemical etching; included are isotropic and anisotropic etching. The latter, also called orientation-dependent etching (ODE), is capable of producing very high aspect ratio structures. The second group includes plasma, or so-called "dry" processes. Deep reactive ion etching (DRIE) is the primary tool in this category. Single crystal reactive etching and metallization (SCREAM) is a process built on DRIE that also yields high aspect ratio structures.

Isotropic etching of a substrate through a patterned photoresist is characterized by the etchant attacking the substrate similarly in all directions (isotropically). The precise shape of the resulting structure depends on the chemical used, the substrate, and whether or not the system is agitated during the etching process. For example, the gaseous etchant XeF_2 will produce approximately equi-axed cavities in silicon. Anisotropic etching involves the use of chemicals that attack some planes in a single crystal faster than other planes are eaten away. Some etchants, notably a 45% solution of KOH in water and TMAH, will dissolve particular planes of silicon hundreds of times faster than others (121). By appropriate choice of the crystallographic orientation of the wafer, and of the geometry and orientation of the photoresist pattern on its surface, it is possible to produce a wide variety of high aspect ratio structures in silicon. The fundamentals of ODE are shown in Fig. 50.

DRIE is another bulk micromachining process in which a pattern earlier produced in a photoresist on the surface of a substrate is transferred into the material (122). In DRIE, the energy comes from a plasma that is generated above the substrate. The composition of the plasma is alternated every few seconds to first etch (with SF_6) and then passivate the etched structure with a polymer coating (from C_4F_8 gas). This approach is called the Bosch process after the company that developed it. Deep structures with "vertical" side walls can be produced. The defining characteristic of DRIE is its ability to produce high aspect ratio structures in silicon, as illustrated in Fig. 51 (123). Such structures have several kinds of uses, including tools for hot embossing or molding and channels for flow of gases or liquids. If structures produced by DRIE are released from the substrate, they can be micromechanical mechan-

618

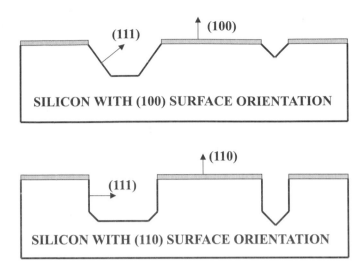

(111)   ↑ (100)

SILICON WITH (100) SURFACE ORIENTATION

▲ (110)

(111)

SILICON WITH (110) SURFACE ORIENTATION

FIGURE 50   Schematic showing the bulk micromachined structures that can be produced in the surface of silicon wafers with the indicated orientations by wet ansotropic etchings.

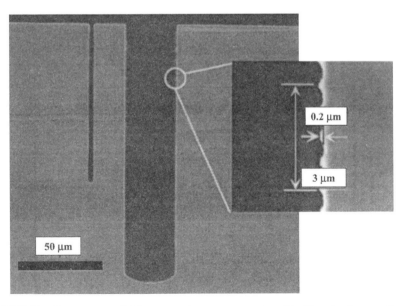

0.2 μm

3 μm

50 μm

FIGURE 51   Cross-section of high aspect ratio structures produced in silicon by deep reactive ion etching. The narrow feature etches more slowly than the wide trench. Scalloping of the sidewall due to the alternating etching and deposition processes is shown with the inset.

isms. Devices with moving parts for microsensors, such as accelerometers, and microactuators for the manipulation of optical and microwave signals are made by DRIE.

SCREAM involves sequential processes that are shown in Fig. 52 (124). It starts with DRIE to produce a high aspect ratio structure. Then the structure is coated with a passivation material, commonly silicon oxide. Further use of DRIE results in an even deeper structure of the same pattern, with the side walls protected from etching by the passivation film. Subsequent use of an isotropic etchant will undercut the structure, releasing it from the substrate. The resultant mechanism can be metallized, if desired. Complex micromechanisms made of single crystal silicon result from the SCREAM process. These have a variety of demonstrated and potential uses. For example, structures made by the SCREAM sequence of processes are under development for data storage devices with terabit-per-square-centimeter capabilities.

12.2 ELECTROCHEMICAL, ELECTRODISCHARGE AND ULTRASONIC MACHINING

A pattern can be transferred deeply into a conductive material (the work piece) by using an already patterned conductor (the tool), with both pieces immersed

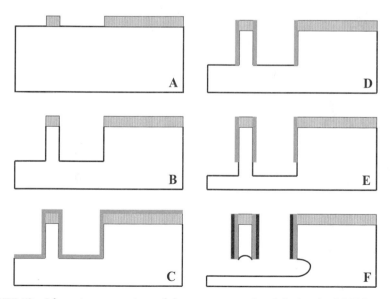

FIGURE 52   Schematic cross-sections of the structures produced during the SCREAM process sequence in steps A through F. The structure shown in Fig. 11 was made by this process.

in an electrolyte and connected to an electrical power supply. The work piece is the anode, attached to the positive side of the battery or other voltage source. As electrons are removed from the atoms of the surface of the work piece, that is, as they are "oxidized," they form soluble ions, and the surface is eroded on the atomic level. The electrons that flow into the tool reduce the hydrogen ions from water to create hydrogen gas. The metallic ions put into solution from the work piece then combine with the remaining hydroxyl (OH^-) radicals from the water to form a metal hydroxide. The process is indicated in Fig. 53 (125). The hydroxide material can be flushed out of the anode-cathode gap by a forced flow of the electrolyte. In some machines, the anode tool is oscillated during the machining process, which causes fresh electrolyte to enter the gap. Low voltages (about 10 V) and high current densities (10–100 mA/cm^2) are employed. Machining speeds in excess of 1 mm/min are possible with this technique.

When a metallic structure is brought close to another metal, and a high voltage is applied across the two pieces, arcs that jump from the cathode to the anode will erode the anode material. The high temperatures produced at the sites struck by the arcs will ablate away some of the anode material. If the tool (the cathode) is patterned, that is, if it has a structure with a complex cross-section, the pattern can be transferred by gradual erosion as the cathode is moved toward the work piece (the anode). Typically, voltages of around 100 V are employed with the two pieces immersed in oil. The very small particles removed from the work piece by the microarcs are suspended in the fluid and carried away. Electrodischarge machining, commonly known as EDM, can be used in two modes: (1) to sink a patterned die into a work piece; or (2) to cut

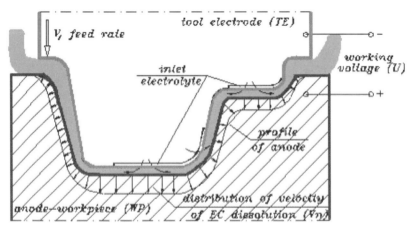

FIGURE 53   Schematic of the components and processes involved in electrochemical machining.

into the work piece with a wire. The die sinking mode can produce holes that are less than 10 μm in diameter. Holes 50 μm in diameter with aspect ratios of 10 have been made by EDM (126).

Ultrasonic machining uses essentially the same arrangement as does EDM. In this case the initially patterned piece, usually an alloy, is vibrated at frequencies in the range of 20 to 100 kHz in a slurry of hard abrasive particles. The small but frequent impacts of the abrasive on the work piece erode small particles from the work piece and produce a replica of the pattern, as illustrated in Fig. 54. The mechanical properties of the work piece, especially the hardness and elasticity, are important in ultrasonic milling. Materials such as ceramics, glass, and silicon can be patterned. A hole 9 · μm has been ultrasonically micromachined into glass (127).

13. PROGRAMMABLE SUBTRACTIVE TECHNIQUES

Historically, engraving was the first "programmable" technology for removing material from the surface of a substrate. The engraver was both the computer and the machine. Lines and spacings on the order of 100 μm can be made by engraving that is aided with only a magnifier for viewing the work piece. In a process related to engraving, a sharp point was moved on a programmable XY translation stage to scratch lines in gold. Widths about 1 μm wide could be made using this method (128).

Mechanical micromachining has advanced to the point of being able to make features finer than 10 nm (129). An extrapolation of machining capabil-

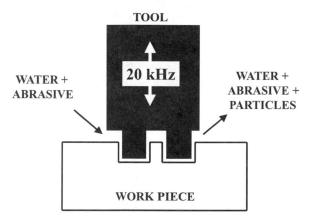

FIGURE 54   Schematic of the operation of an ultrasonic mill.

ities made in 1983 predicted that features requiring 1 nm precision will be possible in this time frame (130). However, the method lacks the general programmability of the newer technologies.

Several modern methods for removing material from a work piece in micrometer-scale patterns are now available. They involve directed beams of photons, electrons, ions, or particles. The capability to move focused beams of different quanta over the surface in a programmable manner under the control of a computer enables maskless etching of thin layers and substrates with micrometer precision. This situation parallels the use of photons, electrons, and ions to expose photoresists in direct-write lithography technologies. The employment of laser beams is most common, and very important commercially. Electron and ion beams can also be used for material removal. The physics and chemistry are very different for each type of quanta. Jets of abrasive particles work by impact-produced erosion of the work piece. The fundamentals of each programmable method for patterned material removal are briefly surveyed in the rest of this section.

13.1 PHOTON-, ELECTRON-, AND ION-BEAM METHODS

Focused lasers remove material from a target substrate by processes that range from melting to plasma formation, and include photo-induced chemical reactions. The mechanisms depend primarily on the wavelength of the laser and the power density (watts per square centimeter) focused onto the surface. Excimer lasers with wavelengths ranging from 157 to 248 nm, interact strongly with polymers. This, along with their brightness, is a major reason that they are used in current and planned aligners for production of commercial semiconductor devices. When used for resist exposure, the photon energy density on the polymeric resist is about $10 \, mJ/cm^2$ in times on the order of 100 msec. If a pulsed excimer laser beam is directed onto a polymer with energy densities exceeding about $1 \, J/cm^2$, the polymer chains are broken, the fragments are heated, and material is ablated from the affected region, which can have dimensions below 10 μm. Figure 55 is an example of excimer laser etching of an organic material, namely a hair (131). It is seen that the focused laser beam can remove material with features on the scale of the diameter of a human red blood cell, namely 8 μm.

Excimer lasers also interact strongly with ceramics and glasses, and with diamond. They are used commercially to cut and mark such materials. For example, numbers are put on the side of gem-quality diamonds for identification of individual stones. Small carbon dioxide lasers, which have a wavelength of 10.6 μm, are also able to cut thin ceramics at high linear speeds. Neo-

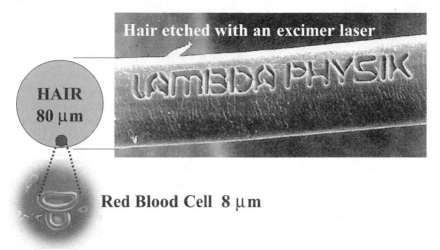

FIGURE 55   A human hair micromachined by a focused excimer laser. The diagrams on the left shown the relative sizes of a hair and a human red blood cell.

dymium-doped materials that lase at a wavelength of 1.06 μm are employed to cut and drill thin metals of interest to the electronics and other industries.

Focused laser beams can also induce chemistry near a surface to cause reactions that lead to removal of materials that are impervious to the laser radiation by itself. This technique, photochemical micromilling, can be used to both produce and modify microstructures (132). A comprehensive tabulation of chemistries and laser etching rates is available (114). Figure 56 is an example of a structured trench produced in silicon by the action of a laser

FIGURE 56   A structure produced in silicon by a focused laser in an atmosphere containing chlorine gas.

in an atmosphere of chlorine gas (133). Photo absorption produces chlorine ions that react locally with the silicon to make $SiCl_4$, which has a high vapor pressure and can be pumped off. The small steps in the structure shown in Fig. 55 are 10 μm in height.

Focused electron beams do not have the power to melt and ablate most materials. The light mass of electrons and their relatively weak interactions with the atoms of a substrate both preclude sputter removal of atoms from a target. Electrons can be used to induce local chemistry, producing species that react with the substrate and produce products with high vapor pressures, thus etching the substrate locally. The atmospheres of the requisite chemicals needed for electron-induced material subtraction are limited on the low-pressure end by removal rates and on the high-pressure side by transport of the focused electron beam to the work piece.

Focused ion beams can also induce superficial chemistry that results in the formation of volatile reaction products and selectively etches a substrate. This capability has been used to etch away materials from masks and other structures (83). As with electron beam chemical removal of materials, there is a band of pressures over which the ion-induced chemical technology works.

Ions have enough momentum per particle to sputter material from substrates. They were first demonstrated to do so over 20 years ago, when letters were written into the surfaces of substrates. This capability has been developed as a way in which to cut open microstructures and even to make relatively complex structures, an example of which is Fig. 57 (134,135). It can

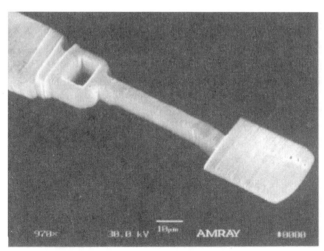

FIGURE 57   Micrograph of a shovel made from tungsten by focused ion beam milling. The bar below the image is 10 μm.

also be used to repair lithographic masks by removing material that should not be on the mask, which would produce defective resist exposures if left in place. Focused ion beams have been used to pattern micro-contact print heads, which could transfer features as fine as 100 nm over areas up to 1 mm^2 (22). Micro-contact printing is reviewed in the next section. Ions in plasmas formed by gas flow from a fine nozzle, which is biased relative to the work piece, can produce line widths as fine as 800 μm (136).

13.2 ABRASIVE JETS

When a fine hard powder, such as aluminum oxide (alumina), is propelled at a substrate in a jet of gas, it erodes the surface mechanically. The process is essentially microsandblasting. The silicon carbide nozzles used in this technology can be as small as 7 μm in diameter. Hence, patterns can be produced in the surface of a work piece with feature sizes on the order of 10 μm (137). The aspect ratios are generally on the order of 1, depending on the substrate and the rate of motion of the nozzle. Abrasive jets can be used to cut through 1 mm wafers of hard materials such as quartz.

14. FIXED PATTERN ADDITIVE MATERIAL TRANSFER METHODS

Both vapors and liquids can be put onto a work piece in a pattern by the use of some kind of a mask or other tool. Evaporation through masks is possible using a variety of sources and geometries. These are the subject of the first subsection.

Numerous methods of patterned liquid transfer, the basis of commercial printing, have been demonstrated in recent years. The "inks" employed in any of the liquid transfer methods can be homogeneous materials entirely in the liquid phase, emulsions of two or more liquid phases, or suspensions of very fine solid particles carried in one or more liquids. Many inks have classically been suspensions—for instance, the carbon particles in "India ink." However, in the recent past, several new processes for producing micrometer diameter (and smaller) particles of diverse materials have been developed (138,139). These enable the production of many more kinds of inks. Some of them still rely for their functionality on the simple optical reflectivity of the particles, but other inks containing quantum dots can have a wider variety of optical responses. For many inks, electrical properties of the resulting patterns are of interest, ranging from high conductivity to high resistivity. Suspensions of particles with desirable magnetic and other properties can also be produced and printed at ordinary temperatures to form structures and components on the surface of a wide variety of materials.

14.1 MASKED EVAPORATION

If a stencil mask is placed between a high-temperature source of vapor and a nearby substrate, the material from the source can pass through the holes in the mask and condense on the work piece in the pattern of the mask. A wide variety of heat sources have been employed for the evaporation of materials having a great range of vaporization temperatures. Resistive and electron beam heating of materials are commonly used to evaporate thin films onto a substrate. However, these sources are relatively large and cannot deal with materials having very high vaporization temperatures, such as ceramics. A focussed laser pulse can produce temperatures exceeding several thousand degrees Celsius, which is enough to vaporize any solid. Laser sources of vapors are relatively compact, having diameters on the order of tens of micrometers. Hence, they are useful for evaporation of materials through masks. The sharpness of the resulting pattern depends both on the source size and on the spacing between the mask and the substrate. Putting the mask in contact with the work piece produces the sharpest replica of the pattern in the mask. Patterns as fine as 20 µm were produced by using a pulsed excimer laser with contact between the mask and substrate (140). It is probably possible to make lines as fine as a few micrometers by masked evaporation using a pulsed laser. The mask material would have to be very thin to minimize the penumbra of the pattern. Clearly, build up of material on the mask is an issue with this technology.

14.2 SCREEN PRINTING

The use of silk or other closely woven fabrics or meshes to carry patterns—which leave some of the mesh open and the rest impervious—has a long history. In this process, the ink is a slurry of fine particles that commonly has the viscosity of a paste. It is wiped over the screen with a straight edge, forcing some of the paste through the open areas of the pattern and the underlying mesh carrier onto the nearby work piece. One of the primary advantages of screen printing is the ability to transfer both the pattern and material of interest onto a substrate at a high areal rate. Areas 40 by 60 cm can be transferred in times on the order of 1 minute. With fine mesh pattern carriers, line widths somewhat below 100 micrometers can be attained. After drying or firing, the thickness of the materials laid down by screen printing is on the order of 10 micrometers (141).

14.3 MICROCONTACT PRINTING

Rubber stamps, used in offices for over a century, require the production of patterns in thick films of an elastomer. After the film is placed on a rigid substrate, usually with a handle for convenience, the stamp can be wetted with an ink that is transferred to paper or some other substrate upon contact. Rubber stamping is both a pattern and a material transfer technology, with the material being any of many different kinds of ink.

In recent years, Whitesides and his group have extended rubber stamping to replicate patterns with features finer than micrometers by the use of poly-(dimethylsiloxane) (PDMS) and other elastomers. They term the technique "soft lithography" because of the compliant character of the stamp (142). The processes for forming the stamp and using it are indicated in Fig. 58. One of the ordinary lithography methods is used to pattern a thin film on silicon or some other substrate. After etching and silanizing the surface, the liquid PDMS precursor is cast over the pattern and polymerized by cross-linking before the elastomer is peeled off the mold. Wetting of the PDMS stamp with various liquids and suspensions is done prior to the actual stamping step. For example, a solution of hexadecanethoil in ethanol will produce a nanometer-thick self-assembled monolayer on the surface of the target substrate (23). The production of a stamp pattern that faithfully replicates the initial pattern, and permits reliable use, limits the dimensions of the patterns. In general, the depth of the pattern is in the 0.2-to-20 μm range, with the maximum limited by the stability of the structure. The width and spacing of the contact regions is 0.5 to 200 μm, with the separation limited by the tendency of the region between contacts to bulge toward the substrate being stamped (142).

The PDMS structures of interest here for their ability to do microcontact printing have alternative uses. They form the basis of a number of molding technologies that are discussed in Section 19.

Other ways of making stamps for microcontact printing have been demonstrated. In one approach to microstamp production, focused ion-beam milling was employed to pattern a silicon wafer with features as fine as 100 nm. After ion milling, PDMS was cast over the silicon mold to form the stamp, similar to the technique already described (22). In another approach, lithography was used to pattern nanometer-thick organic layers on silicon wafers or gold-coated surfaces. The regions with the monolayers of organic materials are hydrophobic and the other regions are hydrophillic. The stamp, essentially flat due to the thinness of the organic layers, is wetted with an ink by dip coating to produce a micrometer version of the original lithographic stones. Contact is used to transfer the patterned ink from the stamp to the target substrate (143). Porous hydrogel materials which can retain more ink than a flat surface, have

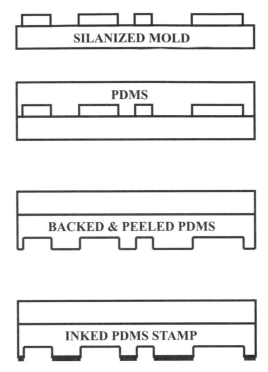

FIGURE 58 Schematic showing the steps in the preparation and use of a stamp for microcontact printing.

been employed recently as stamps for micro-contact printing (144). They have the additional advantages of being hydrophilic and having pore sizes > 50 nm that can hold a wide variety of biomolecules.

15. PROGRAMMABLE ADDITIVE LIQUID METHODS

Liquids can be applied to substrates in computer-driven patterns by a variety of methods. They fall into two major categories: dispensing and jetting. Dispensing from small apertures at the exit of a liquid reservoir is controlled by pressure applied to the liquid by mechanical means. The aperture is generally near the work piece. In some methods contact is made for part or all of the process. Dispensing can be done either in a continuous fashion and dropwise, depending on the equipment and application. Ink jetting is done at a distance from the target substrate in a pulsed fashion, driven mechanically or thermally. The various technologies for programmable liquid transfer by dispensing and jetting are discussed in the rest of this section.

15.1 Continuous Liquid Dispensing

Sealants and other liquids are dispensed in relatively large volumes at high linear velocities (greater than 10 cm/sec), commonly by robots, during the assembly of ordinary parts. Also, many liquids are dispensed under computer control during the manufacturing of microelectronics components and boards. A wide variety of volume and linear speeds is used for dispensing in the electronics industry. Companies offering dispensing equipment to the electronic industry include Asymtek, Manncorp, MRSI Group, Nordson, Ohmcraft, and Vacuum Metallurgical Corporation. These macroscopic processes are foundational to the dispensing technologies for production of micrometer-sized, and smaller, structures that are the focus of this review. We now turn to three micrometer-scale technologies for programmable dispensing of liquids that were developed in recent years.

15.1.1 MicroPen Method

The ability both to precisely control the flow rate of an ink through a fine nozzle and to position the tip of the nozzle precisely in three dimensions relative to a work piece enable what is called micropen printing (145). The ink is withdrawn slightly into the nozzle when the tip comes into light contact with the substrate. Next, the pressure is increased to squeeze some ink onto the work piece, raising the pen off the surface. Programmed motion of the substrate under the nozzle then permits dispensation of the ink onto flat and contoured sections, as illustrated in Fig. 59. At the end of a run, the ink is again withdrawn into the nozzle, which then moves away from the substrate. The micropen process has its roots in work done over 20 years ago and is aimed primarily at the production of microwave circuits and the modification of printed circuit boards. In one mode, thick films of copper conductors are laid down to form the desired pattern in the "additive" process. In a second mode, the ink acts as a directly written resist over thin films of copper on ceramic or organic substrates. In the latter "subtractive" case, an etching step then leaves the desired conducting pattern.

The micropen process can produce lines 50 μm wide with spaces one-quarter of that value. Positional control in the substrate plane is 2 μm. Deposits as thick as 250 μm can be produced in a single pass. Writing speeds range from

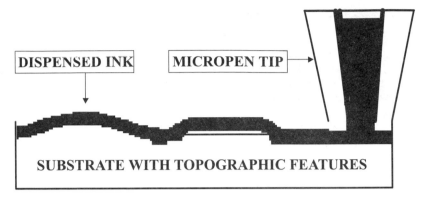

DISPENSED INK   **MICROPEN TIP**

SUBSTRATE WITH TOPOGRAPHIC FEATURES

FIGURE 59   Schematic of the MicroPen process for dispensing liquids onto a surface in a CAD-driven pattern.

about 1 mm/sec to values one hundred times as fast. Inks varying in viscosity from that of water (millipascal seconds) to that of tarry substances (nearly 1000 pascal seconds) can be accommodated. It is possible to controllably dispense volumes as small as 30 µl using the micropen technology.

15.1.2 Microwriting

Another microdispensing process, called microwriting, was developed in the last decade. It employs a pen with tip dimensions on the order of 100 µm, as shown in Fig. 60 (23,128,146). The height of the pen tip over the substrate determines the width of the liquid in contact with the surface. Self-assembled monolayers of hexadecanethiol 10 µm wide can be produced in a routine fashion.

15.1.3 Dip Pen Technology

The scanning tunneling microscope (STM) and atomic force microscope (AFM) have already been cited in Section 8 as two of the most useful members of the family of proximal probes. They were developed principally for imaging on the nanometer scale. Then, they were employed to move atoms and molecules around on substrates in a willful manner, and to induce chemical changes in molecules. It was later demonstrated that dipping the fine tip of an AFM into a liquid and subsequently moving it over a substrate permits the dispensing of the liquid with nanometer precision. The method is called dip pen nanolithography (DPN) because it has the remarkable ability to produce lines as fine as 15 nanometers with 5-nanometer control. The technique is

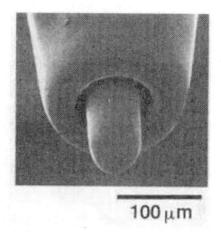

FIGURE 60 Micrograph of the pen tip used for microwriting.

shown in Fig. 61 (147). To achieve such performance, chemisorption of the "ink" by the substrate is needed. Octadecanethiol (ODT) works well on a gold surface, for example. Layers of ODT one molecule thick can be created by rastering the AFM tip over a surface.

DPN has been extended in important ways since its initial demonstration. It can now dispense additional liquids and other organic inks are under development to achieve desired characteristics, such as response to analytes of interest for applications in chemical and biological nanosensor arrays. The method has been developed into a nanoscale printer capable of producing complex patterns. By the use of registration lines laid down in the first pattern of a particular substance, later materials can be put down relative to the first pattern with nanometer precision. Because the AFM is now both an imaging

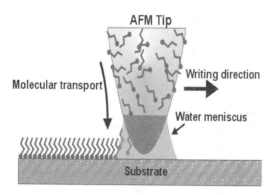

FIGURE 61 Schematic of the arrangement for dip pen nanolithography.

and patterning technique, the nanoscale patterns of single or multiple materials can be imaged using the same device, although a different tip is used to avoid contamination and smearing. It is possible that the arrays of proximal probes shown in Section 8 would be of use in DPN, with different probes dispensing different chemicals and one or more tips used for imaging. The dual character of DPN, being able to both image and modify surfaces, puts it in the same class as the focused ion-beam and proximal probe technologies already discussed. That is, DPN is essentially the nanometer analog of using micrometer and finer-scale focused ion beams to make (by sputter removal of atoms from a substrate) and to measure (by raster scanning) a contoured (patterned) surface. DPN is similar to the use of an STM or AFM to both image and move atoms and molecules, except the DPN dispenses a liquid onto a surface rather than merely moving entities around on a surface.

15.2 DROPWISE LIQUID DISPENSING

As noted, the manufacturing and electronics industries tend to use continuous dispensing techniques. The biomedical industry needs two kinds of tools for research and clinical analyses. The first is to dispense precise volumes of liquids into plates with small wells on which many experiments will be conducted simultaneously. The second is to make microarrays of chemical dots on substrates, again for parallel experimentation in drug screening, gene activation, and other topics. (Examples of microarrays are shown in Fig. 13.) Hence, a great deal of development has gone into the production of tools for the rapid and precise dropwise dispensing of liquids into microwells and onto flat substrates. Micropipette and pin arrays fall into this category. Both pipette and pin-array technologies transfer patterns and materials in one operation. However, the patterns are largely predetermined by the geometry of the arrays. That is, some variations in the resultant patterns can be obtained by programming the transfer robotic device. However, generally only square arrays are produced.

15.2.1 Micropipettes

Tandem micropipettes in robotic dispensing machines have been developed to place specific amounts of particular chemicals in the wells of microplate arrays. This technology can also be used to produce micrometer-scale arrays of dots of different substances on the surface of a substrate. Commercial micropipette arrays have up to 96 dispensers. They can elute droplets with volumes less than $(100 \ \mu m)^3$, which is 1 nl. An example of a micropipette array for robotically filling one row in a 384-well plate is shown in Fig. 62 (148).

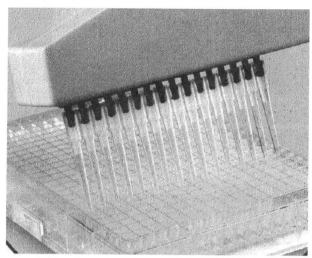

FIGURE 62   Photograph of an array of 16 micro-pipettes used for filling 384 well plates.

15.2.2 Pin Arrays

A technology for the production of arrays of chemicals for combinatorial chemistry experiments involves dipping fine pins into the chemicals of interest and then bringing them near to the surface of the substrate. The pins used in this technology have points on the micrometer scale, not nearly so finely pointed as the nanometer-scale tips employed for the dip pen technique described before. Hence, they are more robust for repeated use to produce various chemical patterns. A common microscope slide, usually coated with a polymer, is ordinarily used for the substrate in such experiments. After it has been covered with an array of dots of different chemicals, it is dipped into a chemical of interest. Optical fluorescence scanners are employed to determine whether chemicals in any of the dots have reacted.

Microarraying machines, one of which is shown in Fig. 63, can accommodate over 100 slides with X and Y dimensions in the range of 25 to 75 mm. Over 80,000 dots, typically 100 μm in diameter (with a delivery volume of about 1 nL), on center-to-center distances as small as 120 μm, can be put on the largest slides. XY position control is 1.25 μm. Slotted pins are used to control fluid flow, as in a quill or fountain pen (Fig. 64) (150). They are moved near the surface with 0.25 μm vertical position control, allowing transfer of the hanging droplet without pin contact to the substrate. Depending on their geometry, droplets with volumes ranging from 0.1 to 5 nL can be transferred. The machines for making patterns of chemicals on slides have up to 48 pins for high-speed production.

FIGURE 63 Photograph of a commercial pin-arraying machine for producing spots of chemical on substrates.

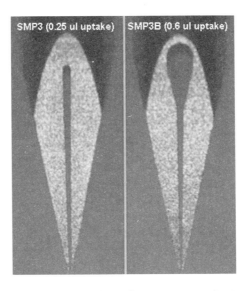

FIGURE 64 Slotted pins for microarray production.

15.3 INK (DROPLET) JETTING

The techniques already discussed in this section involve contact or near contact of the structure that dispenses the liquid onto a work piece. It is also possible to eject a liquid of interest onto a precise location on a substrate, a process called "ink" jetting. Doing so requires that a sudden pressure increase be applied to the fluid in the proximity of an orifice facing the substrate in order to form and propel a droplet to the desired location. There are two primary physical mechanisms that have been used to provide the pressure pulse. The first involves heating the fluid by an electrical pulse in order to increase its internal pressure. This approach is used in commonly available ink-jet cartridges for commercial printers. The second employs the application of an external force on the fluid, which is usually done with a piezoelectric actuator. This is used in some of the droplet dispensers to be discussed in the rest of this section. There are other physical mechanisms for droplet ejection (151). A new method, electrocapillarity, is surveyed at the end of this section.

Droplet jetting has two significant advantages. First, it shares the feature of providing software programmable patterns with the dispensing technologies. Second, it is a standoff technique and does not require actual or near contact with the substrate. That is, there is a separation between the dispenser and the substrate, which provides advantages similar to those of projection optical lithography over contact or proximity lithographies. However, this approach is another sequential method that does not have the ability to simultaneously transfer both a pattern and the desired material at high areal speeds, as do screen and microcontact printing. This fact is offset by the use of multiple print heads, similar to the case for proximal probe lithography discussed in Section 8.

Many approaches to microdroplet-pattern and material transfer have been developed for different purposes in the past decade. Some use continuous droplet streams and some employ drop-on-demand schemes. Numerous research projects have demonstrated the capabilities of droplet jetting. The companies offering microdroplet jetting equipment and services include Ink Jet Technologies, Leader Corporation, MicroDrop, MicroFab, and Packard Instruments.

MicroDrop GmbH sells both a single droplet technology, using piezoelectric actuation, and a microjet process, in which a small fast valve controls the exit of a slug of liquid (152). The droplets are in the 30–100 μm range and their volume variation is less than 1%. They travel at about 2 m/sc. At this low velocity, their surface energy exceeds their kinetic energy, so they do not splash upon impact. Spots with typical diameters of 150 μm result. Programming multiple droplets to hit a single location builds the height and width of the spot on the substrate.

A schematic for a continuous liquid jet system from MicroFab Technologies is shown in Fig. 65 (153). The system is capable of operating in a frequency range of 80 kHz to 1 MHz. Droplet diameters can be as small as 6 μm (100 fl), with 150 μm being typical and 1 mm (0.5 μl) possible. A photograph of a jet of water emerging from a 50 μm nozzle and breaking up into 100 μm droplets is included in Fig. 65. MicroFab Technologies has demonstrated the ability to produce patterns of spots on surfaces of a wide variety of materials. Included are solders and other liquid metals, ceramic powders, polymers for optical and other applications, and biomaterials (notably DNA). They have also built systems with ten nozzles in a single print head in order to speed the production of patterns of droplets.

MicroFab has demonstrated the ability to inject small amounts of liquid into a living egg, as shown in Fig. 66 (154). A pipette with a 2 μm outer diameter driven by a piezoelectric actuator was used to pump 2 pl into a 1 mm diameter frog egg. Microinjection of eggs is a routine clinical technique, and injection of

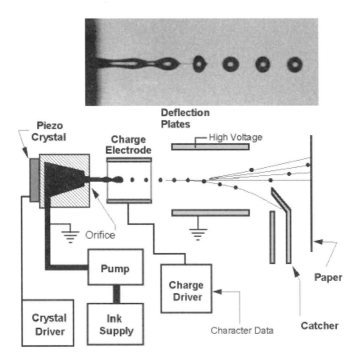

FIGURE 65   Schematic of a MicroFab Technologies piezoelectric continuous droplet jetting system, plus a micrograph of the emerging droplets of water.

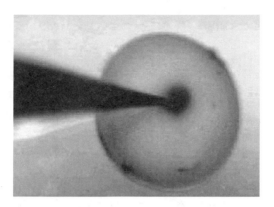

FIGURE 66 Dye injected into a live frog egg.

cells is done in research programs. A great deal of basic research remains to be done on injection of materials at significant velocities into eggs and cells. The mechanical and chemical response of living material must vary with the amount and velocity of impulsively injected materials.

Ink-jet equipment, based on piezoelectric actuators from Packard Instruments, is designed to produce programmable arrays of a wide variety of materials on nonporous substrates like glass, porous filter materials, and three-dimensional gels (155). Four piezoactuator heads on 9 mm spacings produce droplets ranging from 350 pl to 500 nl with 10 μm spatial precision in the substrate plane. Spots of materials with diameters in the range from 150 to 200 μm can be produced with densities up to 1600 per square centimeter. This equipment is targeted at the filling of microwell plates. Under appropriate control, it could be used to perform parallel printing of complex patterns, again, much as multiple proximal probes have been produced for higher-speed lithography.

The production of ink jet suspensions of nanometer scale (1–10 nm) particles of silver and gold was reported (156). The 50–70 μm droplets were ejected at velocities of 1–4 m/sec at the rate of 5 per second from a piezo-electrically actuated capillary with a 60 μm orifice. Metal traces ranging from 100 to 200 μm wide and 1 to 3 μm thick resulted after sintering.

Droplet jetting can be done with a very wide variety of matrials to make patterns of both passive and active electronic components. The formation of "plastic" transistors by a piezoelectric ink-jet method was reported recently and is shown in Fig. 67 (157). The source and drain electrodes were jet-deposited conductors, specifically the polymer poly-(ethylenedioxythiophene) (PEDOT), on a polyimide-covered glass substrate. The polyimide was patterned first to control the wettability of the surface by the PEDOT, and hence its flow. Next,

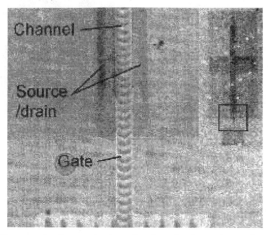

FIGURE 67   Photograph of a transistor made of conducting polymers by droplet jetting.

polymer semiconductor and insulator layers were spin coated over the source and drain. Then a PEDOT gate electrode was jet printed. The gate length is 5 μm. The transistor exhibited an on-to-off conductivity ratio of 10^5.

Electronics are used to control both the thermal and the mechanical (piezoelectric) approaches to dielectric droplet formation and, in some cases, the trajectories of the droplets. Means for the direct electronic production and control of droplets of conducting fluids have been demonstrated, and are discussed in the remainder of this subsection.

In the recent past, a new technique for rapidly moving fluid in small channels has been demonstrated. It may also be useful for jetting droplets onto substrates. The method involves the electrical control of the capillary forces in

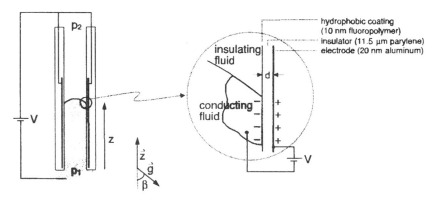

FIGURE 68   The arrangement for electrocapillarity.

small, fluid-filled channels (158). Figure 68 demonstrates the geometrical configuration. The objective is to modify the normal capillary forces between the conducting fluid of interest (colored gray) and the wall of the channel by application of a voltage to an electrode, forming a capacitor between the liquid and the electrode. The pressure that is induced in this manner is called electrocapillary pressure (ECP). Fluid velocities of several centimeters per second have been measured. It is possible to make arrays of channels driven by ECP, and control of both the individual channels and of the entire array of channels simultaneously has been demonstrated. That this technology is useful for production of fine drops remains to be seen, as does their placement at specific locations on a substrate.

Electrostatics can also be used to control the flight of liquid particles. An example of a dielectric fluid was shown in Fig. 65. Conducting liquids can also be steered electrostatically (159). Droplets of molten solder with an average diameter near 200 µm can be generated acoustically at rates of 10 to 20 thousand per second. If these are passed through a cylindrical electrode about 3 mm in diameter with a time-varying voltage, the droplets are charged to various degrees. Upon passage between plate electrodes with 3000 V, the droplets are deflected onto a substrate in a pattern determined by the charge they carry. The splats produced by the impact of individual droplets are about 375 µm in diameter.

The jetting technologies discussed in this subsection have been primarily concerned with the production of 2- or 2.5-dimensional patterns of material on substrates. They are indeed methods for rapid prototyping, that is, the fast and flexible production of usually small numbers of trial or demonstration parts. Droplet methods for making three-dimensional structures have also been commercialized. A machine with a single nozzle is sold by Solidscape (160), while a machine from 3D Systems has 352 nozzles (161). They are part of the recognized field of rapid prototyping, and are surveyed in Section 18.

16. BEAM-BASED PROGRAMMABLE ADDITIVE TECHNIQUES

The number of programmable methods for patterned liquid transfer has increased significantly in recent years. The same is true for data-driven beam-based technologies for patterned material transfer. The materials that are transferred in the beam-initiated direct-write processes can have different origins and histories. In some processes, the transferred material starts as a solid on a surface near the work piece, is vaporized by a laser pulse, and then condenses at the point of interest. The resulting patterned materials are largely two-dimensional because their lateral extent is much greater than their thickness. Such techniques are discussed in the first subsection. In methods based on beam-induced chemical reduction of vapors present over the work piece, the deposited material originates in the vapor. Photon, electron, or ion beams can be used to initiate the reactions and control the deposition. These techniques are discussed in the second subsection. Three-dimensional structures commonly result from methodologies involving beam-induced chemistry.

16.1 LASER-INDUCED MATERIAL TRANSFER

Pulsed laser evaporation of thin films from the back of transparent substrates was employed for the past two decades to propel atoms of the thin material into plasmas in Tokamaks and other machines. In the mid-1980s, the same process was first used to transfer materials from a carrier material to a nearby substrate. The basic process is illustrated in Fig. 69. Doing the transfer in a manner in which the material carrier and the target substrate can be translated relative to each other under computer control, and the laser can be similarly controlled, permits the programmable production of fine-scale patterns of many different materials and components on diverse substrates.

Laser-induced material transfer goes by a variety of names: LIFT (Laser Induced Forward Transfer), MAPLE-DW (Matrix Assisted Pulsed Laser Evaporation-Direct Write), and MELD (Microstructuring by Explosive Laser Deposition). LIFT (162) and MAPLE-DW (163) both use excimer lasers with pulse lengths of approximately a nanosecond. MELD uses picosecond Nd lasers, which decouple the heating and expansion phases of the material

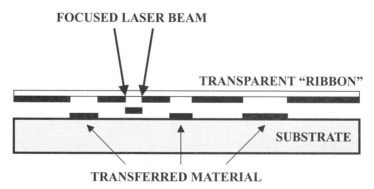

FIGURE 69   Schematic cross-section of the technique for programmable laser-induced pattern and material transfer.

transfer (164). The mechanisms and dynamics for the transfer of material from the ribbon to the substrate are delightfully complex. The type of laser (especially its wavelength, pulsed energy, pulse length, and focal spot size), the type of material, the geometry of all components, and the ambient atmosphere determine the state(s) in which transferred material arrives at the target substrate. Possibilities include all normal phases of matter, solid, liquid, gas, and plasma. The characteristics of the deposited material depend on the material as well as its phase and energy upon impact. Discussion of the details of laser-target interactions, however important to the resultant pattern on the work piece, is beyond the scope of this review. See (165).

Laser-induced material transfer can be carried out at room temperature, and in a variety of atmospheres ranging from vacuum to ordinary air. The process is reminiscent of that of old typewriters, but is actually quite different. Here the ribbon can be a solid plate or a flexible tape that can go from one roll to another, rather than being made of cloth or plastic tapes as in mechanical typewriters. The materials to be transferred include not only inks but also metals, alloys and other inorganic substances, plus organic and biomaterials, The substrates in laser-induced transfer can be virtually any relatively flat substances, not only paper. And the mechanism of transfer is the momentum imparted to the material by rapid absorption of the laser radiation, not by impact.

The ability to transfer complex organic materials—such as ordinary and chemically sensitive polymers—by laser-induced techniques, without destroying their functionality, is a testimony to the strength of covalent bonds. In the recent past, laser-induced material transfer was shown capable of putting down a pattern of live cells that remained viable after the transfer, an example of which is shown in Fig. 70 (166). The ability of cells to undergo rapid acceleration and deceleration without dying is remarkable. It is not likely to be a direct result of evolutionary experience, but rather a by-product of the general robustness of cells, and possibly their ability (albeit limited) for self-repair of damage.

Laser-induced transfer's ability to handle complex organic materials and biomaterials, combined with its programmability, make it a candidate for coating of surfaces within biochemical microfluidic sensors. A setup for doing so is shown in Fig. 71 (167). The serial character of this process might seem too slow for production of commercial parts, but biofunctionality is needed on only a small fraction of a substrate's total area. The situation is similar to that expected in the production of leading-edge integrated circuits in the future, when the highest-resolution lithographic tool might expose only very little of the entire chip. This could make electron-beam direct writing of nanometer-scale gates a viable mass-production technique in a few years.

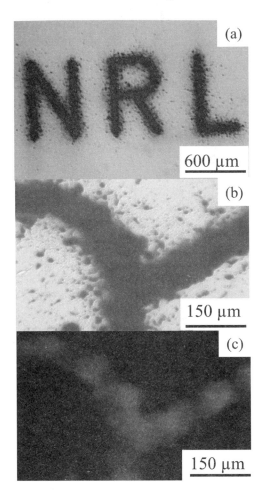

FIGURE 70   (a) and (b) are optical micrographs of patterned *E. coli* containing fluorescent tags that were transferred to a substrate by the MAPLE-DW programmable laser-induced material transfer technology at the Naval Research Laboratory (NRL); (c) is green fluorescence stimulated by 365 nm UV radiation, which shows the cells still to be viable after the transfer.

Similarly, laser-induced filling of the key parts of microfluidic devices with biochemically sensitive materials should be commercially useful.

16.2 BEAM-INDUCED CHEMISTRY

Photon, electron, and ion beams can all cause the decomposition of a vapor in the small region on a substrate where they are focused. If a product of the

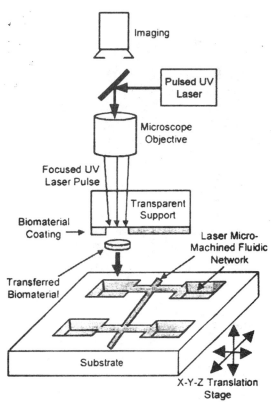

FIGURE 71 Drawing of the aparatus for coating the channels in a microfluidic device with biologically active materials.

radiation-induced reaction has a low vapor pressure, then a deposit occurs within and very close to the focal spot. The deposited material may be an oxide or other compound involving atoms of the substrate, or an entirely separate material dependent on the composition of the vapor in the chamber. The production of low-volatility products using beams permits the production of complex micro- and nanostructures on the surface of almost any solid materials. The chemical techniques involving programmable beam-induced deposition are the inverse of the methods surveyed in Section 13. There, the technologies discussed involved such energetic beams to produce reactions that lead to volatile compounds and controllable substrate erosion. In general terms, laser focal spots are on the order of 1 μm, while the focal spots of electron and ion beams can be on the order of 10 nm. Hence, three-dimensional structures can be written with particle beams.

16.2.1 Laser Beams

Laser chemical vapor deposition (LCVD) involves introducing a focused laser beam into a chamber containing the vapor of a material that can be chemically decomposed by the beam. If some of the resultant products have low vapor pressure, they will deposit on a nearby substrate. By moving the beam or the substrate, complex, three-dimensional structures of a variety of even refractory materials can be produced. Two examples are shown in Fig. 72. The helical structure was made by moving the substrate appropriately in the vapor of an organometallic compound containing tungsten (168). The complex cylindrical structure consists of sapphire (alumina). It was made by bringing two laser beams to a focus at a point in a vapor and manipulating the substrate to make the structure (169).

16.2.2 Electron and Ion Beams

The energy brought to a point on the surface of a substrate by a particle beam, in the presence of a vapor of many chemicals, can induce reactions that result in the deposition of diverse materials (83). The situation is analogous to laser-induced chemical deposition, except that laser beams are limited to focal spots on the order of a micrometer while particle beams can be focused to diameters approaching a nanometer.

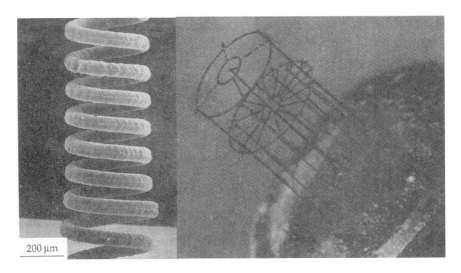

200 μm

FIGURE 72   Two laser-deposited microstructures: a helical coil of 60 μm diameter W wire that is about 350 μm in diameter, and a structure 1.2 mm in diameter made of 20 μm sapphire rods.

Electron beam-induced deposition, commonly called EBID, has been employed to make a wide variety of microstructures out of many materials. It is relatively easy to produce carbonaceous deposits because of the presence of hydrocarbon vapors in ordinary vacuum systems. Figure 73 shows the structures that can result from EBID of carbon (170). Nanotips were grown on the ends of microcantilevers to form a microscale version of a four-point electrical probe. The EBID rods are 100 nm thick, and can be grown to lengths exceeding 5 μm. Spacings between the nanotips as small as 100 nm have been achieved.

Focused ion beams have been used to make many small structures of carbon and other materials, examples of which are shown in Fig. 74 (171). In this case, gallium ions were focused to a 7 nm spot in a gaseous aromatic hydroxide. The beam collisions reduced the hydroxide and deposited the carbon structures on a silicon substrate to create the structures shown. The minimum feature that can be produced under the conditions used is 80 nm due to particle scattering and migration of the carbon from the near-surface atmosphere prior to deposition.

17. OTHER PROGRAMMABLE ADDITIVE TECHNOLOGIES

The last two sections dealt with programmable methods of writing patterns on substrates in which the transferred materials originated as uniform vapors, liquids, or thin films of solids. We now turn to a few related technologies that put down patterns using solid particles or vapors as the starting materials. The

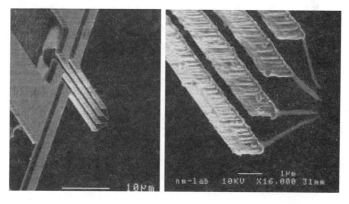

FIGURE 73   Photographs of microfabricated structures on which carbon nanotips have been grown by electron beam induced deposition. The bar in the right hand photo is 1 μm.

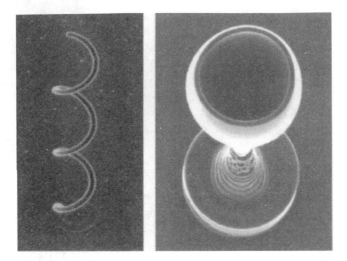

FIGURE 74   Micrographs of carbon objects made with a focused ion beam. The wire diameter in the coil is 80 nm and the goblet has a 2.75 μm diameter and submicron wall thickness.

first two, laser and flow-guided deposition, are relatively low temperature processes, so they can handle both additive and substrate materials that cannot withstand high temperatures. The next two methods, thermal spraying and jet molding, both require high temperatures, which enables them to employ metallic and ceramic powdered materials. Two other related methods, directed and jet vapor deposition, involve moving vapors of metals onto a substrate. They appear to have potential for programmable production of microstructures.

17.1 Laser and Flow Guided Deposition

Particles of a wide variety of metallic, ceramic and bio-materials, including live cells, are available in the size range from below 100 nm to over 10 μm. Methods have been developed in recent years to direct streams of such particles onto small spots on a substrate, which enables the direct writing of essentially two-dimensional patterns of a wide range of materials for many electronic and other applications (156). These technologies are surveyed in the rest of this section. The processes have the advantages of being able to produce diverse patterned materials on a wide variety of substrates at ordinary temperatures and in the ordinary atmosphere. It is noted that lines made by any of the particulate direct write processes are often subjected to post-deposition annealing to improve their density, adhesion and properties.

Two related methods begin with the production of a stream of particles that originate as suspensions, usually in water. The technology, called laser guided direct writing (LGDW), is illustrated in Fig. 75 (172). For particles smaller than about 300 nm, ultrasonic atomization is an efficient way to get the particles into the atmosphere. For larger particles and live cells, the spray process called nebulization is used to produce the same result. The focused laser beam constrains the particles and propels them at velocities of around 1 cm/sec. A hollow fiber with interior diameter of 20 µm is used to transport the particulate stream over a distance as much as centimeters toward a chosen spot on a substrate. The particle deposition rate can be varied from 1 to 10,000 per second. That rate and the size of the particles permits calculation of the volumetric deposition rates. LGDW is capable of producing line widths as fine as 2 µm as shown in Fig. 75.

If higher deposition rates are desired, another technology called flow guided direct write (FGDW) is employed (172). Figure 76 indicates the arrangement. As with LGDW, atomization and laser interactions are used to produce a stream of droplets containing particles. In the case of FGDW, the particles exit a 1-mm-diameter orifice into a second pressure-driven air stream. The flow with the particles forms the core of the combined stream that then passes through a submillimeter-diameter orifice. The sheathed stream of

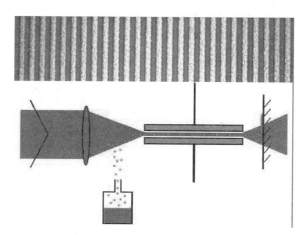

FIGURE 75 Schematic of the laser guided direct write (LGDW) process, showing the ultrasonic atomization of a suspension of fine particles, as well as their laser propulsion and capillary guidance onto a substrate. The micrograph at the top shows 2 µm-wide lines of platinum on alumina prepared by LGDW.

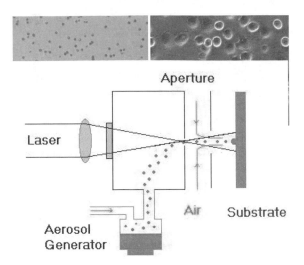

FIGURE 76 Schematic of the flow guided direct-write (FGDW) technology, with an aerosol generator to produce a stream of particle-containing droplets that are guided by laser and aerodynamic forces onto the substrate. Micrographs at the top are 3T3 mouse fibroblast cells as deposited by FGDW (left) and 55 minutes after deposition, showing normal growth.

particles in the core of the flow can be brought to a spot size on the substrate that is 5 to 10 times smaller than the second orifice to produce lines as fine as 25 μm. The FGDW method, with an aerosol generator to produce the stream of aerosol particles, is sufficiently gentle to produce patterns of live cells.

17.2 THERMAL SPRAYING AND JET MOLDING

Another of the particulate rapid prototyping technologies that is capable of producing micrometer-scale structures is thermal spraying (173). The heat that melts the 1–20 μm particles can be supplied by combustion (about 3000 °C) or a plasma created by a direct-current arc (about 25,000°C). The droplet-containing effluent from a spray nozzle naturally expands because of its high temperature relative to ambient conditions. The expansion limits the lines produced by thermal spraying to widths on the order of 1 mm. Subsequent laser trimming can produce lines from thermally sprayed material that are about 100 μm in width, as illustrated in Fig. 77. This is an example of the use of multiple material processes in sequence to achieve the desired structure and material properties.

The last particulate thermal deposition technology is called jet molding (174). In this case, metallic or ceramic particles less than 100 nm in size are

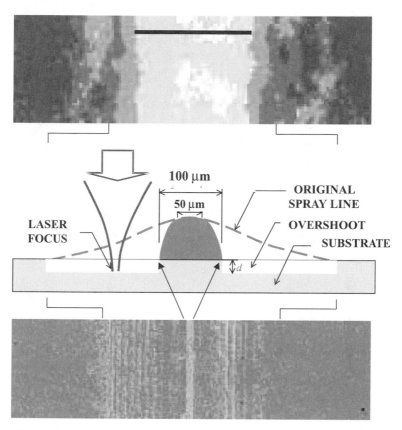

FIGURE 77 At the top is an image of an alumina line produced by thermal spraying (bar = 400 μm). The schematic in the center shows the structure produced by laser trimming of an as-deposited line. The micrograph at the bottom is a line of Ag on Ti that has been laser trimmed.

heated in a crucible. They are then carried onto a flat substrate, into a mold, or through a stencil mask by a helium jet. The work piece is under XYZ computer control. Features finer than 100 μm have been made of metals and ceramics by jet molding.

17.3 DIRECTED AND JET VAPOR DEPOSITION

As noted, both thermal spraying and jet molding transfer particulate materials onto the work piece. Two other techniques, directed vapor deposition and jet vapor deposition, use jets of inert gasses to transfer material to a substrate. In directed vapor deposition, the transferred material is melted in a crucible by an electron beam and then carried to the substrate by the gas jet (175). In jet vapor deposition, precursor gases containing the atoms to be deposited are reacted in a microwave-excited cavity through which they and the carrier helium are flowing (176). Apparently, neither of these techniques has been used to make microstructures, However, it seems that proper design of the gaseous flow system could result in the ability to make lines at least comparable to those produced by thermal spraying, that is, about 1 millimeter, which could be trimmed as shown in Fig. 77. If these techniques are made to produce micrometer-scale structures, they could be used in a programmable fashion by appropriate motions of the substrate and timely shuttering of the vapor stream.

17.4 MICRO-ELECTROCHEMICAL DEPOSITION

Electrodeposition is usually done to fill structures that are essentially molds, as discussed in Section 19.2. However, it is possible to perform electrochemical deposition of materials very locally using fine electrodes in close proximity to the work piece. The electrodes have diameters that range from 1 to 100 µm, with spacings to the conducting substrate of comparable distances. With such an arrangement, lines approximately twice the electrode diameter can be produced if the microelectrode is insulated on its sides, leaving only the end available for participation in the current flow. Plating rates of several micrometers per second can be realized, which is 100 times faster than usual plating rates with large electrodes (177). If the microelectrode is mounted on and moved by a precision XYZ stage under control, it is possible in principle to build three-dimensional metallic structures without masks or molds. Many ways to make and use microelectrodes are known (178).

18. THREE-DIMENSIONAL RAPID MICROPROTOTYPING

Rapid prototyping (RP) is the term that has been employed to describe a wide variety of technologies for making three-dimensional physical objects layer by layer directly from computer data, that is, from CAD files. For most of the three-dimensional techniques, the same steps are required to produce the desired form: (1) creation of the 3D CAD model and its conversion to a layer-wise file that can be used to drive the RP equipment, (2) production of the structure one layer at a time, and (3) cleaning and completion of the resulting piece.

The field of RP started in the early 1980s and became important during the past decade. Currently, the RP industry is valued at over $1 billion annually (107). One web site lists about 40 companies that make RP equipment (179). RP is also called solid freeform fabrication (no molds are needed), computer-aided manufacturing (CAM), layered manufacturing (because parts are built one layer at a time) and desktop manufacturing (when the equipment is small). When one of the RP methods is used to make tooling, the process it is termed rapid tooling. When directly useful products result, the process is called rapid manufacturing. In contrast to machining processes, such as milling, drilling and grinding, that subtract material to produce a part, RP adds material to build up a structure.

Many materials can be handled by RP, and they can originate from different phases. Plastics are the most important. Solid plastic parts are made from liquids consisting of the polymer precursor, which are solidified by the action of ultraviolet or other short-wavelength laser light—a significant similarity to the laser-induced photochemistry technique discussed in Section 16. The major difference is the fact that the transferred material starts as a vapor in the photochemical methods and can belong to any of several classes of materials. In alternative RP processes, the prepolymer material starts as a liquid and objects made of plastic result. In other RP techniques, waxes, paper, and other layered materials—even metals and ceramics in powder form—can be used to make objects that range in size from over 30 cm to less than a millimeter. A classification of RP technologies according to the materials they handle is available (180).

Conventional rapid prototyping includes over 30 technologies that result in three-dimensional objects directly from a computer CAD file, not all of which are on the market (111). The relatively-new programmable, maskless, data-driven technologies for adding materials to a work piece that were discussed in the past three sections can properly be termed "rapid prototyping" techniques because they produce structures quickly and flexibly. However, many of the newer methods produce structures that are two or two-and-a-half-dimensional, that is, not truly three-dimensional. This moves some people to exclude them from the RP field. It is the view of workers in the field of direct writing of materials that making primarily two-dimensional components both fast and flexibly does indeed qualify as rapid prototyping.

Another significant difference between new materials direct-write technologies and conventional RP concerns the use of the resultant structures. Much of the effort with three-dimensional RP yields static structures that serve as models of potential products that would be made by other techniques, such as injection molding. With the new methods for direct writing of materials, it is possible to make functional electronic, magnetic, optical, and other devices that can be products and not only models. Essentially, the prototype and the product can be made by the same technology.

The most important commercial three-dimensional RP techniques are: (1) stereolithography, which is the subject of the next sub-section; (2) selective laser sintering, in which a laser is used to fuse some of the particles in a thin layer of powder, as discussed below; (3) 3D printing, which is operationally similar to selective laser sintering, except that dispensed liquid adhesives are used to bond the powder particles into a solid; (4) laminated object manu- facturing, where sheets of paper, plastic, ceramic and other materials are cut with a laser to form a 3D object; (5) fused deposition modeling, in which melted plastic is dispensed onto the work piece to build up an objects; (6) Laser Engineered Net Shaping (LENS^{TM}), where a high-powered laser is used to melt powders of metals or alloys as they are fed onto the work piece; and (7) droplet jetting methods. A useful summary of these techniques (181) and a comparison of their features (182) are available. The first two of these processes can make micrometer-scale structures, so they will be discussed in more detail. The other processes have not been shown to be capable of producing structures with features on a micrometer scale.

18.1 MICROSTEREOLITHOGRAPHY OF PLASTICS

Of the older and three-dimensional RP techniques, only a few are capable of making millimeter-to-micrometer scale objects. Stereolithography is the first and best known of these processes. It is capable of making objects with feature sizes from tens of centimeters down to tens of micrometers. The technique works as indicated in Fig. 78 (183). A vertically movable platform is positioned just under the surface of liquid ultraviolet-curable polymer precursor. A laser, directed by the CAD file, photopolymerizes selected regions of the liquid thin film to form a solid. Then the elevator is lowered to create another thin liquid layer, which is again selectively polymerized by the scanning laser beam. The process is repeated many times to produce the desired solid object. In this scanning technology, the pattern transfer is from the CAD file and the material transfer is from the pool of liquid pre-polymer. In a variant called projection stereo lithography, which is also referred to as solid ground curing, a mask is made up for each layer and the exposure of the whole layer is done simultaneously and not by sequential scanning. In ordinary stereolithography, both the beam spot size and the thickness of the solidified layers are hundreds of micrometers. In microstereolithography, the spot size is about 1 μm and the layer thickness is near 10 μm.

The masks for whole-layer exposure in projection microstereolithography can be made by a variety of approaches. Actual physical masks can be used for

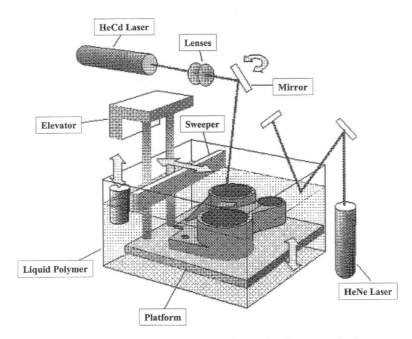

FIGURE 78   Diagram of a stereolithography system showing two lasers, one for determination of the pre-polymer liquid level (on the right) and the other for polymerization to produce the desired structure. The scanning mirror rapidly moves the laser beam to the desired spot. A computer controls the XYZ stage motion and the exposure.

each layer. Two other patterning methods have been used to produce micro-structures by projection stereolithography. In one, illustrated in Fig. 79, an electro-optic pattern generator is used to impress a pattern on a light beam (184). Using such equipment, objects are made with the number of layers varying from a few hundred to a few thousand. The rate in the Z direction is about 1.5 mm/hour, with 3 to 6 layers made each minute. Two examples of structures made by projection microstereo-lithography of polymers are shown in Fig. 80 (185). The microturbine blade was made of 110 layers., each 4.5 μm thick, and has an outer diameter of 1.3 mm. The axial hole is roughly 50% larger than the diameter of a human hair. The finest feature is about 50 μm. The small cups have an outer diameter of 200 μm and a wall thickness near 35 μm. These can be compared with the goblet less than 3 μm across that was made by ion-induced chemical vapor deposition, as shown in Fig. 74.

   In the second approach to making patterned beams for projection micro-stereolithography, a commercial array of computer-controlled micro-minors was used to determine the pattern of light striking the liquid prepolymer (186). The Digital Mirror Device made by Texas Instruments for projection display of

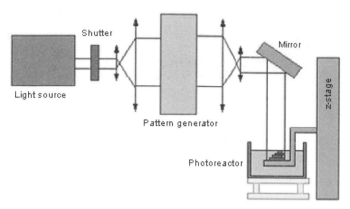

FIGURE 79   Schematic of equipment used for projection stereomicrolithography at the EPFL.

FIGURE 80   Objects made of plastic by projection microstereolithography. The white bar on the left image is 500 μm and the one on the right is 200 μm.

images contains over 750,000 micro-mirrors, each of which is 16 μm square. Putting it in the optical path and programming it appropriately enables the production of a wide variety of micrometer-scale three-dimensional plastic objects, including photonic materials.

X-ray beams will penetrate and polymerize liquid monomers in the shape of the beam. Solid plastic threads 200 μm in diameter were produced using synchrotron x-radiation (187). Rotation of the liquid container can be employed to produce complex shapes.

18.2 MICROSTEREOLITHOGRAPHY OF CERAMICS

In the RP technique of selective laser sintering, a powder of organic or other higher-temperature material is knifed across the work piece, rolled to compact it, and then exposed to laser radiation to meld the desired region onto the growing part. Microstereolithography of ceramics (188) has features in common with both microstereolithography of plastics (as the name implies) and with selective laser sintering. To make micrometer-scale parts of refractory materials, the starting material is a mixture of particles of the desired material, a dispersant (to defeat particle agglomeration), and a diluent (to control viscosity and act as a binder). The material is liquid but sufficiently viscous that a knife is used to spread it, similar to selective laser sintering of powders. Then, laser exposure solidifies the mixture, similar to the photopolymerization of the low viscosity prepolymer liquid used to make plastic parts. Thermal treatment of the part after initial fabrication burns out the polymer binder and sinters the ceramic material. Line widths of ceramics as fine as 6 μm have been produced by microstereolithography.

19. MOLDING AND RELATED TECHNOLOGIES

Casting of molten metals and of plastics into patterned structures both have long histories. Injection molding of plastics and other materials has developed into a major commercial process, capable of making parts at the rate of one every few seconds. Almost fully dense metal parts can be made by injection of metallic powder with a polymer binder that both allows for flow of the material into the mold under pressure and holds the part in the desired shape. Heating then serves both to drive off the volatile binder and to sinter the powder into the finished part. The inevitable shrinking during curing can be calibrated by tight control of the starting materials and the temperature–time profile, so that the final part has the desired dimensions. Molding by jetting materials onto a surface with topography already has been noted (174).

In this section we deal with a number of micro- and nanometer-scale technologies under the general heading of simultaneous pattern and material transfer. The pattern is in the form into which the material is placed. Molding of plastic and related materials is essentially the inverse of embossing. During embossing, as discussed in Section 9, the pattern is brought to the material. During molding, the material is put into a pattern. The next subsection discusses two injection molding technologies that operate on a micrometer scale and four recently developed methods for producing microstructures by molding. Then, chemical and electrochemical means of filling a pattern to make fine-scale structures are reviewed. Here again, there is a reciprocal relationship with the method for patterned electrochemical removal of materials that was surveyed in Section 12. Then, the HEXSIL process is briefly

656

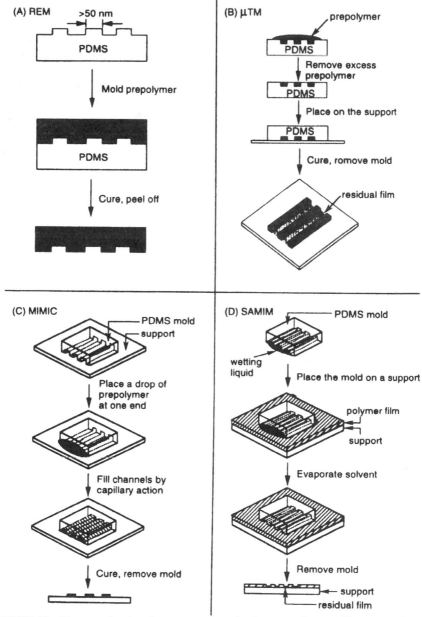

FIGURE 81 Diagrams showing the processes in technologies involving molding materials on micrometer and nanometer scales: (A) REM is replica molding; (B) μTM is micro transfer molding; (C) MIMIC is micromolding in capillaries; and (D) SAMIM is solvent-assisted micromolding.

surveyed. In it, a structure deeply etched into silicon serves as the form for production of diverse and complex microstructures. The section concludes with a discussion of how crystalline lattices are used as templates for the production of nanometer-scale structures.

19.1 MICROMOLDING

Injection molding has been extended to the micrometer and nanometer scales. Plastic parts with wall thicknesses of 20 μm, aspect ratios of 20, and features as fine as 200 nm have been made by micro-injection molding (189). Micrometer-scale particles of metals, alloys, and ceramics have also been subjected to injection molding. After molding and postprocessing to get rid of the organic binders, the parts contained features as small as about 10 μm (190). Both the pure plastic and the powder injection schemes depend on cooling to form a solid part. An alternative approach to micro-injection molding starts with a liquid polymer precursor. After injection, which is near ordinary temperatures, photo-polymerization or other methods can be used to produce solid parts (191).

Four processes developed by Whitesides and his colleagues are shown in Fig. 82 (142). These are all based on the flexible "microstamps" of poly-(dimethylsiloxane) (PDMS). Their fabrication and use for microcontact printing were outlined in Section 14.2. Replica molding (REM) involves filing a mold of PDMS made by the process described in Fig. 58. The flexibility of the PDMS mold makes release of structures with fine features easier than with rigid molds, but limits the temperature range over which REM can be practiced. PDMS is stable in air up to 186 °C (126). REM has been shown capable of replicating structures with features as fine as 10 nm. Microtransfer molding (μTM) involves filling a PDMS mold only to its surface with a pre-polymer, and then transferring the contents of the mold to another substrate such as a glass slide (142). Many polymers have been molded and transferred by μTM. Strengths of this technique include the abilities to produce submicrometer structures over areas with dimensions exceeding 1 cm in a short time (~10 min.).

Micromolding in capillaries (MIMIC) is also illustrated in Fig. 81 (142, 192,193). A PDMS mold is placed against a substrate in this technique to form a network of capillaries. When a drop of low-viscosity liquid prepolymer is touched to the openings at the edge, it naturally wicks into the network of openings. After curing the polymer, the PDMS mold is removed, leaving the pattern on the substrate. After a solvent, or a liquid carrying suspended particles, is wicked into the capillary network, evaporation leaves a layer of solute or particles with lateral dimensions defined by the capillaries, but

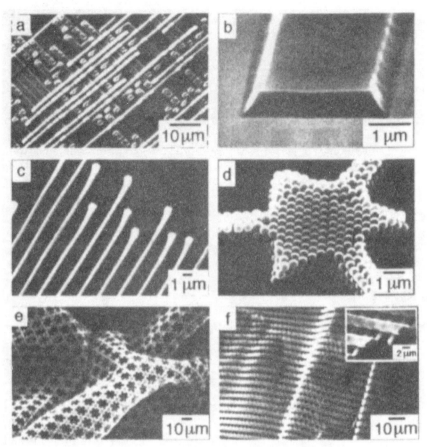

FIGURE 82   Scanning electron micrographs of structures made by the MIMIC process: (a) 2.5-dimensional structure of polyurethane on a silicon wafer; (b) end of one line of polyaniline emeraldine HCl salt from an array of such lines; (c) zirconia ceramic lines; (d) array of polystyrene beads; (e, f) free-standing patterned films of polyurethane.

thinner than the channel heights. Figure 82 illustrates the diverse patterns and materials that have been formed by MIMIC. A remarkable variety of materials has been patterned by the MIMIC technique. Included are precursors to several polymers, glassy carbon and ceramics, sol-gel materials, inorganic salts, colloidal particles, polymer beads, and biologically active macromolecules.

The last technique illustrated in Fig. 81 is much like embossing (142). In solvent-assisted micromolding (SAMIM), a PDMS mold is filled with a liquid that will soften or swell the polymer material to be patterned. When contact is made, the resultant softening of the polymer film causes it to take the form of the PDMS pattern. Lines 60 nm wide and 50 nm high have been produced in a photoresist by the SAMIM technique.

19.2 ELECTRO- AND ELECTROLESS DEPOSITION

Electrochemical deposition is the inverse process of electrochemical machining. The machining process removes material from the anode by oxidization, which is the removal of electrons from neutral atoms to form soluble ions. The deposition process plates material on a cathode by reduction, which is the supply of electrons to metallic ions from solution to form the deposited atoms. Electrodeposition can be done in a pulsed or steady (DC) manner, using aqueous or organic solvents for the electrolyte, which contains dissolved ions and contacts both electrodes. Electrodeposition has been used to fill many structures made by several micromachining processes, notably LIGA (Section 9). It has the remarkable ability to put material into even very deep and narrow (that is, high aspect-ratio) slots made by deep reactive ion etching and other methods (194).

Electroless deposition is a chemically driven process, in contrast to electrodeposition, which is driven by an applied electrical potential. In electroless deposition, differences in chemical potentials provide the impetus for the reactions. Exchange reactions are one type of electroless process. Metal ions that are initially in solution replace metal ions that are initially solid and in contact with the electrolyte. This approach has the disadvantage of consuming the initially solid material, which has to be properly placed in order to result in the desired final structure. A more practical approach to electroless deposition is to use a dissolved material in place of the sacrificial material. In this method, the metal ions, also in solution initially, are reduced and deposited while ions of an added material are oxidized, remaining in solution all of the time. Other additives to baths for electroless deposition control unwanted reactions and the temperature-dependent rates of reaction. In general, electroless deposition requires higher temperatures than does electrodeposition, but results in deposits with less stress. A useful summary of the various approaches to electroless and electrodeposition is available (195).

A technology for flexible production of microstructures based on electrodeposition called EFAB, short for electrochemical fabrication, was recently developed (196). It shares with rapid prototyping the character of building parts one layer at a time. The essential tool for EFAB is a patterned anode, which is used to lay down an elecrodeposit in the same pattern by placing it in close proximity to the cathode (substrate), with an electrolyte between the two electrodes. After deposition of a layer, the anode is removed. Then a filler material, which will be removed at the end of the entire process, is applied to

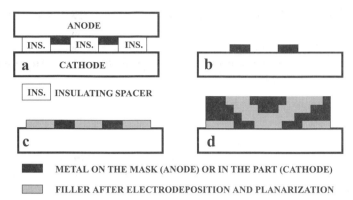

METAL ON THE MASK (ANODE) OR IN THE PART (CATHODE)

FILLER AFTER ELECTRODEPOSITION AND PLANARIZATION

FIGURE 83   Diagram showing the production of layered materials in the EFAB process: (a) the anode with the pattern for the next layer and insulators (INs.) determining the thicknenss of the next electrodeposited layer; (b) the first deposited layer of the part; (c) after electrodeposition of the filler and its planarization; and (d) four layers of the part and the sacrificial filler.

the work piece to fill in the regions between the electroplated material. (The filler material can be another material electrodeposited without a mask.) After planarization of the filler down to the surface of the just-deposited patterned material, the process is repeated, as indicated in Fig. 83. Multiple patterns are placed side-by-side on the mask structure, so that one anode tool can serve for the production of many layers. After electrodeposition of the last patterned layer, the filler layer is removed chemically. Because of the multiple steps for each layer (patterned deposition, filler deposition, and planarization, each of

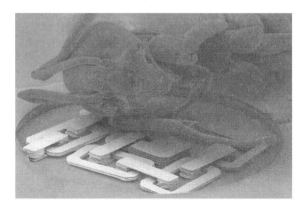

FIGURE 84   Photograph of an ordinary ant lying on a 96 µm thick, 12-layer chain with 6 links made by the EFAB process.

which requires a different experimental arrangement), EFAB is slow. It is also limited to the deposition of conductors. However, EFAB can produce complex structures under ordinary room conditions. It has been used to produce some interesting structures, with features down to about 25 μm. An example is shown in Fig. 84 (197).

19.3 HEXSIL

The sequence of steps in the HEXSIL process is shown schematically in Fig. 85 (198). A mold is first produced in a silicon wafer by a two-step process: (1) one of the conventional lithography methods is used to produce a patterned resist layer; and (2) deep reactive ion etching gives the desired mold structure. Then the entire surface is coated with SiO_2, the sacrificial layer that will be dissolved at the end of the process to free the work piece. Subsequent sequential layering of insulators and conductors (undoped and doped polycrystalline silicon) builds up the structure in the trenches of the mold. Then a conductor, commonly but not necessarily nickel, is electrodeposited, and the system is lapped to planarize the surface. Then, dissolution of the silicon dioxide frees the part. The mold in the silicon wafer can be used repeatedly.

The HEXSIL process is another example of a technique that is flexible in the parts it can yield, but requires many steps. It has two major advantages. One is the ability to make strong microstructures and the other is the ability to embed conductors in those microstructures. The conductors can serve as heating elements, so resistive heating can be used to cause the structure to deform controllably. Figure 86 shows a microgripper that is actuated by current flow (199). Other structures made by HEXSIL have been developed to serve as mechanical veneers at the end of the arms in hard disk drives. Their light weight and stiffness (high-resonance frequencies) are expected to enable tracks on the magnetic disks to be placed ten times closer, proportionally improving the storage capacity (200).

19.4 Spatial Forming

In most of the molding technologies just reviewed, the mold generally existed before its use to produce parts. In the micrometer-scale process called spatial forming, the mold is built up along with the part, both in a layer-wise fashion (201). The technique is outlined in Fig. 87. An offset printing press puts ink onto the work piece in layers about 0.5 μm at a time. The build-up of 30 layers forms the mold for the next step in the process which is cured with UV radiation. A paste with about 50% of its volume being a metal is knifed over the

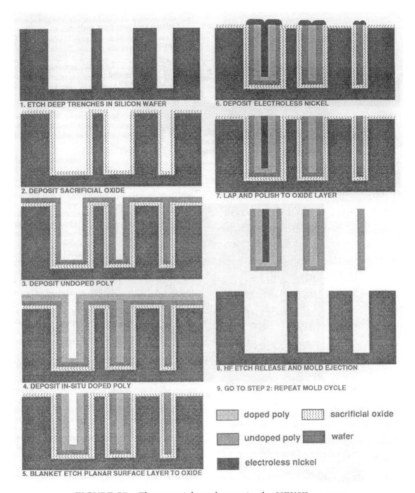

FIGURE 85   The materials and steps in the HEXSIL process.

just-made mold layer to form the uncured next layer of the part. Ultraviolet light is then used to cure that top layer of the part. Repetition of the printing, filling and curing steps results in parts up to 250 μm thick with lateral dimensions as large as 20 mm and feature sizes as small as 10 μm (184). Spatial forming has a clear similarity to the EFAB process already discussed (see Fig. 83), with printing being used instead of electrodeposition and planarization to make the mold layer, and mechanical filling of the mold and UV curing replacing patterned deposition to make a layer of the part.

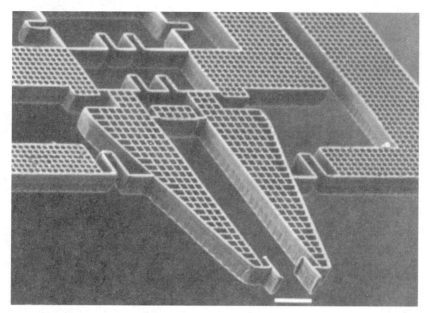

FIGURE 86 Microtweezers made by the HEXSIL process. The bar is 100 μm.

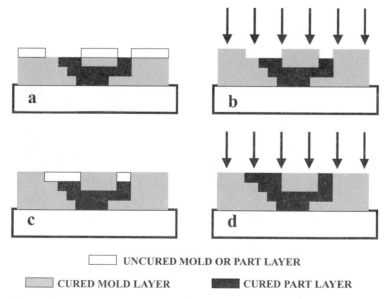

☐ UNCURED MOLD OR PART LAYER

▧ CURED MOLD LAYER　　■ CURED PART LAYER

FIGURE 87 The sequential steps in the spatial forming process: (a) printed layer of the mold, (b) UV curing of the latest mold layer, (c) mold filled with the next layer of the part and (d) UV curing of the last layer in the part.

19.5 CRYSTALLINE TEMPLATES

The surfaces of crystals include natural patterns that are increasingly being used as templates to produce nanostructures. These can be thought of as nanomolds for arrangement of atoms and molecules. The most basic structure available on a crystal surface is the pattern of the lattice itself. It is the basis of the large and important field of epitaxial crystal growth, which results in structures having different chemistries, but registered lattices from layer to layer. Many other features—including monolayer and larger steps, islands, individual atoms or molecules and a wide variety of emergent defects—are available as crystalline templates. Crystal surfaces have been used to line up individual macromolecules. Steps on such surfaces serve as templates for the production of nanoscale wires of many materials.

If a crystal is cut at an angle close to one of the dense crystal planes to produce a so-called vicinal surface, a surface with many steps is produced. If a partial monolayer of atoms is put onto a clean vicinal surface, and the energetics and temperatures are appropriate, the adatoms will migrate to the steps and form wires with nanometer cross-sections. This is also the case for high index planes in crystals that have regular and densely spaced steps, an example of which is shown in Fig. 88 (202). The wires are about 2 nm wide (which, for perspective, is also the width of the spiral DNA molecule). A review on atomic and molecular wires is avilable (203).

It was demonstrated that the arrays of dislocations that exist at the interface of two crystals can form templates for the growth of nanostructures. By controlling the angular misorientation of the two parts of a bicrystal, it is

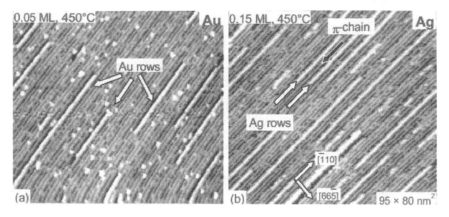

FIGURE 88 Au and Ag nanowires due to 5% and 15%, respectively, of a monolayer being evaporated onto Si (5 5 12) 2 × 1 surfaces and annealed at 450 °C. The "π chain" refers to the structure of the underlying silicon.

possible to prepare structures with spacings in the 20–100 nm range (204). The method was demonstrated with silicon crystals containing an array of structures 25 nm in size, 38 nm apart.

Another way to use nanometer-scale structures as templates is called cleaved edge overgrowth (205,206). A structure consisting of many-nanometer-thick layers of material is first grown by molecular beam epitaxy. Then, the structure is rotated 90° and cleaved to produce a surface with a pattern of parallel layers of the different chemistries with thicknesses on a molecular scale. This narrow structure can then be used directly as a template, or else further processed to introduce topography or add other chemical species. The same approach can be used with nonepitaxial multilayers grown by sputtering and a wide variety of other vacuum deposition processes. These techniques generally do not result in layer-to-layer crystalline registry, so the surface orthogonal to the layers must be exposed by techniques other than cleaving, such as cutting and polishing. This requirement is balanced by being able to make the nanometer template out of a much wider variety of materials than will grow epitaxially.

20. PATTERN AND MATERIAL TRANSFER BY SELF-ASSEMBLY

The final topic we review represents the third major class of pattern and material transfer options. Recall that we considered technologies in Sections 2 through 9, which transfer only a pattern (information) to a substrate that is usually coated with a photoresist. Sections 10 through 19 dealt with the mostly new techniques for simultaneous transfer of both information and the material constituting the patterned layer, obviating the need for the various lithography and etching steps. This section confronts the question of whether or not the transfer of only a material to a substrate can spontaneously result in the formation of a structured layer by self-assembly. The answer is yes, depending on the materials and conditions.

In self-assembly, the information needed to produce a structure is carried by the entities deposited on a surface, without appeal to externally provided information. The units may have unique shapes that fit together only in certain ways. Or, they may have a distribution of surface charges and associated fields that dictate how they will assemble. Often, the combination of the geometrical (steric) and electrostatic (chemical) forces work together to determine the structures that are possible due to self-assembly.

When studying self-assembly, it is useful to consider what entities can assemble into what structures. Figure 89 lists the input units and the output structures germane to self-assembly on scales from subnanometer to multi-

WHAT UNITS?	ASSEMBLE INTO WHAT?					
	OTHER UNITS					CRYSTALS
	MOLECULES	VIRUSES	CLUSTERS & Q DOTS	COLLOIDS & PARTICLES	3D OBJECTS	CRYSTALS
ATOMS	X		X	X		X
MOLECULES						
SIMPLE			X	X		X
COMPLEX		X		X		X
POLYMERS						X
VIRUSES						X
CLUSTERS & Q DOTS				X		X
COLLOIDS & PARTICLES					X	X
OBJECTS					X	

FIGURE 89 Matrix showing the units (on the left) that self-assemble into the larger structures (across the top), some of which themselves can serve as units for further self-assembly.

millimeter. Simple molecules without significant internal degrees of freedom behave much like atoms, except for their more complex geometries. They act as rigid entities and assemble into larger units—a behavior very different from that of more complex molecules, such as DNA, RNA, and proteins, which themselves spontaneously fold into units that can have three levels of structural organization. This folding is so complex that IBM is designing what will be the most powerful supercomputer, called the Blue Gene, to attack the protein folding problem. Proteins can then agglomerate into viruses, larger particles, and crystals, in many cases. Clusters of atoms or molecules, and closely related quantum dots of compounds, have nanometer-range dimensions. Colloids and particles are larger collections of atomic or molecular units that have sizes on the order of micrometers. Molecules and larger entities can self-assemble into diverse structures on the surfaces of substrates. For example, long chain hydrophobic alkanethiols form self-assembled monolayers on the surfaces of copper, silver, and gold. These materials can be used as nanometer-thickness photoresists (23). Further, block copolymers consisting of two or three polymers that have been reactively joined will spontaneously form complex patterns on surfaces with nanometer lateral dimensions (207).

The particular importance of crystallization in self-assembly should be noted. As indicated in Fig. 89, entities on nano- and micrometer-size scales can form crystals. The role of crystals as nanomolds—that is, templates for self-assembly—was noted in the last section. Control of crystallization to produce complex and functional nanostructures is a major scientific challenge.

The three-dimensional structures that result from self-assembly of these smaller units are sometimes themselves the entities for further assembly. For example, viruses are made from complex arrangements of proteins and DNA or RNA, such as the well-known Tobacco Mosaic virus shown in Fig. 90 (208). The self-assembled coat of protein molecules, with outer diameter of 18 nm, encloses the RNA, which has a diameter of 8 nm. These units can then form crystals (209,210). This cascade of progressively larger self-assembled structures will play an important role in the production of complex, functional nanometer and micrometer structures. It is the inverse of machines making progressively smaller machines, which is both an old idea and the basis for a new nanotechnology company (211).

In recent years, techniques have been demonstrated for the self-assembly of manufactured objects with dimensions ranging from below one micrometer to several millimeters. Figure 91 shows the structure that results from natural ordering of polystyrene latex microspheres about 1 μm in diameter, followed by settling of 20 nm gold particles into the interstices of the latex spheres (212). Calcination, chemical oxidation, or dissolution can be used to remove

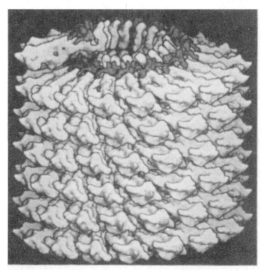

FIGURE 90   Drawing of the helical structure of the Tobacco Mosaic Virus with the interior RNA backbone and outer protein subunits.

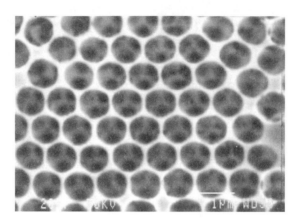

FIGURE 91   Structure made by infusing a colloidal crystal of 1 μm diameter plastic spheres with gold nanoparticles and then removing the spheres.

the latex spheres, leaving the gold structure. Such a structure, probably with less expensive nanoparticles, could form the basis for applications that require a large and controllable surface area-to-volume ratio—for example, the sorption of vapors from air or solutes from water in the front end of chemical sensors (213).

A commercial technique under development uses fluidic self-assembly (FSA) to make flat displays (214). Functional structures called "nanoblocks" on the order of 100 μm are settled into matching holes in a glass, plastic, or other substrate. After the FSA, the nanoblocks are connected electrically. Figure 92 illustrates the structures and results.

A self-assembly technique involving fluid transport is being developed to locate HgCdTe pixels for infrared imagers on silicon substrates, which might contain a 256 × 256 element readout circuit. As a step in the development of

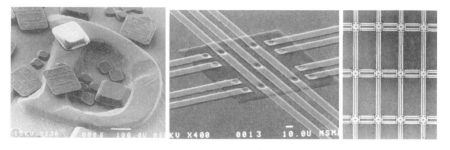

FIGURE 92   Electro-optic nanoblocks on the surface of a coin, and wired into an array after fluid self-assembly in a glass substrate.

the desired process, spheres of hydrophobic materials 150 to 200 μm in diameter were captured on 55 μm gold pads with 85% efficiency (215).

Objects that range in size from about 0.1 mm to several millimeters have been shown capable of self-assembly. The assembly process is enabled and controlled by the geometry of the mesoscale pieces and the relative energies of their surfaces and interfaces with each other and the surrounding media. The energies are controlled chemically. The process is sometimes carried out within a uniform liquid containing the pieces that will assemble into larger and three-dimensional objects. The surface tension of a drop can also serve as a template for self-assembly. Figure 93 shows a spherical structure about 1 mm in diameter that was formed of 100 μm hexagonal rings that assembled on a drop of chlorobenzene in an aqueous silver plating solution (216). Electro-deposition of Ag was then used to weld the structure into a unit.

The technique of anodizing aluminum to produce hexagonal arrays of nano-scale structures can be considered as self-assembly. In this technique, the units do not exist separately before the assembly, but rather they develop during the electrochemical process called anodization. Figure 94 shows an example of the arrays of pores that can be produced by this technique (217). Densities in excess of 10^{10} per cm^2 have been achieved. The chemical nature of the electrolyte and the applied voltages are the parameters that control the pore size and density.

The functionality of microelectronics and micromechanics certainly derives from their structured character, that is, their inhomogeneity. Crystals of smaller units are usually functionally homogeneous, because they lack the structures needed for complex behavior. It seems likely that use of a mixture of patterned

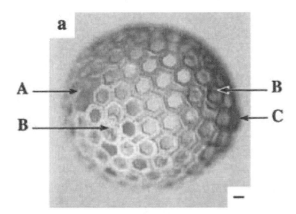

FIGURE 93 Sphere self-assembled from hexagonal pieces. (A) indicates a void in the structure, (B) exhibits double layers of hexagons and (C) is the electrode attachment point for deposition of Ag to weld the structure together.

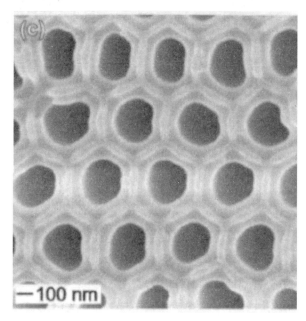

FIGURE 94   Micrograph of pores of alumina grown on the surface of aluminum by anodization.

structures and self-assembly will result in the nanometer-scale electronic and mechanical functions that are widely envisioned. Useful devices with structures having nanometer-sized features in two or three dimensions will probably be made up by a sequence of processes, some of which involve patterning while others depend on self-assembly. The situation might be somewhat similar to current manufacture of quantum well devices, for which molecular-beam epitaxy produces layers with nanometer thickness that are then patterned and etched, or even similar to the fluidic self-assembly of nanoblocks cited here.

In nature, DNA provides the recipe for the production of: (1) structures containing inorganic materials, such as teeth and bones; (2) electronic materials, notably nerves; (3) optical materials such as the retina; and (4) magnetic materials in animals ranging in size from bacteria to large marine mammals. It seems possible that DNA will be used to direct the fabrication of diverse mixed inorganic and organic materials and structures that perform electronic, optical, and other functions. The materials and structures that result from the information stored in DNA in nature now arise from self-assembly. In the future, the use of programmable DNA to direct the synthesis of artificial materials is likely to be inextricably associated with the use of DNA to also provide instructions for the self-assembly of those materials into useful structures.

21. COMPARISON OF PATTERN AND MATERIAL TRANSFER TECHNIQUES

The technologies discussed in this review are so varied that their comparison is challenging. The most basic comparison is between the methods that use patterns like masks and the programmable production techniques that do not involve such fixed patterns. Another fundamental difference in the techniques reviewed concerns strictly pattern transfer methods vis à vis simultaneous pattern and material techniques. These two distinctions are highlighted in Fig. 95.

Pattern and material transfer technologies each have a few relevant time scales. They include pretransfer activities (notably materials and equipment preparation), the times for the actual transfers (including pauses for stepping to new chip sites or spots to recommence deposition), and post-transfer times (such as etching, implantation, resist removal, pyrolysis, and sintering, among others). The times for the actual parallel transfer of all components of patterns and of patterned materials are short with fixed masks, whether they are for optical lithography or for screen printing. However, the production of the masks or screens is a separate and time-consuming step, which must precede the actual exposure of a resist or transfer of a material. The programmable technologies are inherently flexible—that is, they are reprogrammable. They avoid the need for masks, but generally require long processing times because of their serial nature, if the entire pattern is to be done by direct writing. It is possible that directly writing the key parts of chips with nanometer dimensions might be done with focused electron beams in the future. Similarly, a mix of ordinary and new direct materials writing technologies might be used for

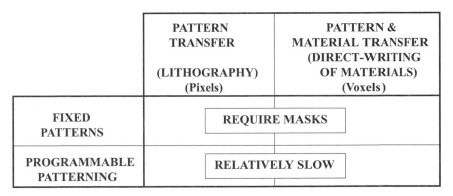

	PATTERN TRANSFER (LITHOGRAPHY) (Pixels)	PATTERN & MATERIAL TRANSFER (DIRECT-WRITING OF MATERIALS) (Voxels)
FIXED PATTERNS	REQUIRE MASKS	
PROGRAMMABLE PATTERNING	RELATIVELY SLOW	

FIGURE 95 Pattern and material transfer technologies either employ fixed patterns or are programmable, that is, driven by patterns stored in a CAD file.

commercial production of some parts, especially if there is a need for customization to meet the details of specific performance.

For manufacture of a large volume of similar products, as in the IC industry, there has been no question of the efficacy of lithographic mask production and use. The most commercially important lithographic technologies can replicate patterns with tens of nanometer features, and they have very short exposure times. Because of their use for commodity production, processes associated with masked lithography are highly advanced and precisely controlled. Uniform products are produced cheaply using lithography with masks. However, this requires several separate steps, notably earlier preparation of a thin film to be patterned and later etching, implantation, lift-off, or other processes. And, they inevitably require a resist that has to be both laid down and baked before exposure, and developed and removed after exposure. Protection, inspection, and maintenance of masks are also unavoidable aspects of the high-volume masked pattern-transfer technologies.

Some of the techniques for parallel transfer of patterns and materials use fixed masks and are relatively fast and commercially viable. The old technology of screen printing is limited to resolutions of tens of micrometers, but this is adequate for many products. The 80 µm diameter of a human hair provides a useful reference. The newer method of microcontact printing can reproduce patterns with tens-of-nanometer resolution over square centimeter areas in times on the order of minutes. It has considerable promise for commercial production.

Programmable methods for simultaneous pattern and material transfer by either subtractive or additive means are generally very flexible, but relatively slow. The resolution of such technologies varies widely. Generally, the laser-based methods are able to make patterns with micrometer and larger patterns. The techniques using electron or ion beams can achieve resolutions of tens of nanometers. Technologies involving the dispensing or jetting of liquids also have a wide range of resolutions and rates. They can generally produce features on the scale of tens to hundreds of micrometers at relatively high linear and volumetric rates. Some of the maskless direct-write methods offer the ability to make truly three-dimensional objects.

As noted in Fig. 95, an areal pattern of picture elements (pixels) is of primary interest in transferring a pattern into a photoresist or other thin film. The information-carrying capacity of an optical field can be discussed in terms of its coherence and noise (218). Smith has compared the pixel (information) rates of photo, X-ray, electron-beam and ion-beam lithographies (219). He found that optical (ultraviolet) lithography had the highest information transfer rates, with X-ray lithography using either a synchrotron or plasma source being better than either scanned electron or ion-beam lithographies.

For material transfer, the pixel transfer rate is a germane parameter only if the thickness of the material is irrelevant. This is sometimes true, for example, for films whose function is optical reflection. However, in most cases, the thickness of the deposited material is germane. For instance, the thickness of a deposited conductor determines its electrical performance; then, the volume element (voxel) is a relevant measure of the performance of a material transfer technique. A quantitative comparison of the performance characteristics for the material transfer technologies surveyed in this chapter, although yet to be made, would feature the resolution and voxel rates of each technique. Controllability, repeatability, and reliability of the direct-write and other material-transfer technologies are of great interest in that their role in commercial production of large numbers of parts could be significant.

A taxonomy of the maskless technologies for simultaneous pattern and material transfer to produce micrometer-scale structures can be made. Figure 96 shows that it is possible to array these techniques according to what is used for control of the pattern (and causes the material transfer) and the original phase of the source of the material transferred. Chemical vapor deposition (CVD) can be induced on a surface using photons, electrons, or ions. Laser

SOURCE OF MATERIAL	VAPORS	LIQUIDS & SUSPENSIONS	SOLIDS
LASER BEAMS	LASER CVD	STEREO-LITHOGRAPHY	LIFT, MAPLE-DW MELD
			LASER-GUIDED DW
ELECTRON BEAMS	ELECTRON CVD		
ION BEAMS	ION CVD		
WRITING		MICRO-PEN, MICRO-WRITING & DIP PEN	
DISPENSING		MICRO-PIPETTES & NEEDLE ARRAYS	
JETTING		CONTINUOUS & DROP-ON-DEMAND	
FLOW			FLOW-GUIDED DW THERMAL SPRAY

FIGURE 96 Matrix of the programmable techniques for directly writing materials onto a variety of substrates versus the initial state of the transferred material. CVD, chemical vapor deposition; DW, direct write. The other acronyms are explained in the text.

beams can be used with materials that originate in the vapor, liquid, or solid states. Liquid pre-polymers are solidified into objects during stereolithography. Pulsed laser transfer of solids can be transferred from a ribbon or other substrate to the work piece by LIFT (Laser-Induced Forward Transfer), MAPLE-DW (Matrix Assisted Pulsed Laser Evaporation-Direct Write) and MELD (Microstructuring by Explosive Laser Deposition). Continuous lasers are used in laser-guided direct writing with particles in the micrometer-size range. The writing and related dispensing of liquids, including suspensions (slurries) of small solid particles can be done by a variety of methods. Drying, baking, and sintering often is used after deposition to consolidate and improve the properties of solids that result from writing or dispensing particulate suspensions. Jetting of liquids is done either continuously or on demand. The placement of solid particles onto the desired region of a substrate can be accomplished by the flow of a gas that is not heated, or else is heated by chemical or electrical means.

The blanks in the matrix in Fig. 96 are understandable. Electron and ion beams generally require a good vacuum in order to protect the source and reduce atmospheric scattering. That is, they do not work with liquid sources and generally cannot transfer materials that start as solids. The use of high-energy electron beams, which can penetrate thin vacuum seals and travel significant distances (centimeters) in air, would enable an electron analog of laser stereolithography. This might allow the production of parts with nanometer-scale features, depending on electron scattering in both the thin film and air. Writing and dispensing are naturally done with liquids, with or without particles in suspension. Jetting of gasses is certainly possible, but will not generally result in deposition without another source of focused energy. It seems likely that laser CVD can be done at normal atmospheric pressure. Doing so would require a localized vapor flow, such as is now done in electron- and ion-beam CVD in vacuum systems.

A few technologies for producing regular arrays of micro- and nano-meter scale structures have been noted in this review. Figures 26, 91, and 94 are examples. A more thorough discussion of such technologies is available (220).

22. CONCLUSION

As stated in the Introduction, the first goal of this review was to provide a framework for relating and comparing technologies for the transfer of patterns and materials with micrometer and nanometer resolution. The classification discussed here is by no means the only structure for such a taxonomy. However, using what is transferred—the pattern only, both the pattern and material, or only the material with its implicit pattern—seems relatively simple and broad enough to embrace the wide range of techniques. The second goal was to explain the basic concepts for each technology, but this was done only superficially. In fact, a more self-contained and thorough review would constitute its own monograph. It is hoped that the third objective—providing some references—might help compensate for the limited depth. Nowadays, reference material resides both in conventional libraries and a large number of servers on the Internet. Printed materials are often more reliable sources of information than are those found on the Internet, but are usually not as current.

Probably the most glaring omission in this survey is the absence of material on and references to the simulation of the processes involved in the technologies that were mentioned. In this era there are two historic shifts in the conduct of research: development and commercial application of processing and other technologies. The changing availability of information was just noted. The other dramatic change is the increasing ability to usefully simulate the outcome of complex processes. This is due to ever-expanding computer power, improved algorithms and the availability of more and better values for the parameters that are needed in simulations. The computational simulation of lithographic and other processes involved in the production of integrated circuits is very advanced. This is clearly due to the economic importance of those processes, as well as to their intellectual challenges. Thorough simulation of the other techniques reviewed here much less developed in most cases and missing entirely in some. But this will gradually improve as some of the newer processes become more important commercially. Indeed, the focus here has been on information (patterns) and matter (materials) transferred by various processes. However, the energetics (efficiencies) and kinetic (rates) of the technologies are also critically important to their practice and commercialization. Simulation of all of these aspects for many of the noted techniques should happen, possibly in the near future.

There are some points, not previously highlighted in this review, that deserve attention. They include international economic competition due to the existing or potential commercial importance of the reviewed technologies. The pervasive importance of lasers is also noteworthy (221,222).

The diversity of technologies for pattern and material transfer is one of the focal points of this review. They involve physical, chemical, and biological processes, with the materials ranging from individual atoms to live cells on scales from subnanometer to beyond millimeters. The proliferation of such technologies is another focus. Almost half of the techniques surveyed here were developed in the 1990s. In fact, such is the case for most of the methods for writing patterned materials directly onto substrates. In the recent past almost all of the techniques have advanced significantly. The situation that currently exists for pattern and material transfer technologies can be likened to the status now enjoyed by tradesmen such as carpenters. They have long had a wide variety of tools and techniques that can be brought to bear on a particular project. For example, consider the production of a door frame. The carpenter saws the wood to length, planes it smooth, chisels indentations for the hinges and plate, drills holes for the lock, hammers the wood in place, checks it with a level, screws in the hardware, and finally hangs, tests, and adjusts the door. All this occurs after the lumber is iniatially sawed to prescribed dimensions and cured, and before it is sanded and finished. The point is that micromachinists, and their children who become nanomachinists, have at their disposal a significantly improved tool set for the production of complex functional devices. It can be expected that creative use of combinations of the technologies examined in this review will lead to a variety of 2.5D and 3D structures and components. They will be as diverse and useful as what is made by the carpenter using what was once "modern" technology.

It is certain that not all of the technologies discussed will become important practically. But it is also likely that some of them will turn out to be commercially significant, at least in niches. Several companies are betting on the latter. Some are referenced, and others have bought the rights to the patents generated at universities. Materials direct-write technologies may be important for rapid prototyping of fairly sophisticated devices and production of small lots, both at ordinary temperatures on rigid or flexible, flat or curved substrates. Most of the successful technologies will serve small markets, some of which might be substantial. Laser direct-write patterning of printed circuit boards has already become commercially significant, despite the serial character of direct writing. The same may prove true of the newer programmable materials-writing technologies.

Assuredly, there are techniques that have been conceived but not yet implemented—still others have yet to be imagined. An example of a method awaiting laboratory demonstration deals with a way of producing patterned arrays of nanometer scale fibers, including conductors (223). It involves bundling glass, metal, or polymer materials together to make a preform that can be heated and drawn in a manner like the production of optical fibers. The technique is illustrated in Fig. 97. The basic concept is to take glass capillaries with outer diameters of about 2 mm and inner diameters of 100 μm filled with a low-melting-point alloy or other material. The preform could be made of several such tubes, or of simply one tube in the center of a group of 2 mm diameter glass rods. Drawing produces a 100-to-1 reduction in diameter, so the 100 μm alloy diameter would be reduced to about 1 μm in the first drawing. Repeating the process after making a second preform in the desired pattern of conductors would result in a pattern of wires with diameters as small as 10 nm. Variations in the process could be used to make the final diameter larger than 10 nm if continuity and conductivity are problems. Cutting the large end of the

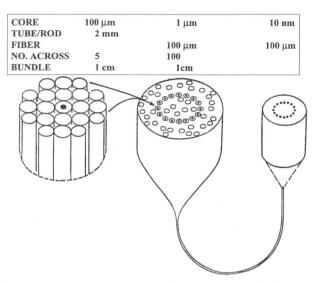

CORE	100 μm	1 μm	10 nm
TUBE/ROD	2 mm		
FIBER		100 μm	100 μm
NO. ACROSS	5	100	
BUNDLE	1 cm	1cm	

FIGURE 97   Schematics of a technology for making nanometer-scale fibers of low-melting-point alloys embedded in glass. This could also be applied to conducting polymers within other polymers.

final structure at an angle would expose the wires for normal bonding in a standard microelectronics package, as indicated in Fig. 97. The end with the nanometer-diameter conductors could be embedded in plastic and cut normally to the fiber to form a substrate for physical, (electro) chemical, and biological experimentation. Other technologies involving size reduction have already been used to make micro- and nanometer-scale structures (224).

The entire collection of demonstrated, conceptual, and eventual technologies for pattern and material transfer offers many opportunities. Prime among them is the chance to concatenate different techniques in creative ways to make new kinds of parts. We have seen examples in this review, including LIGA, EFAB, and spatial forming. Another mixed technology method, not discussed above, is called AMANDA. It involves gluing surface micromachined diaphragms to micromolded housings to make pressure sensors (225). The increasing use of the basic technologies for pattern and material transfer in new sequences to make new kinds of structures and devices is highly likely.

The existing and prospective technologies will certainly be used with new materials as they become available. For example, nanometer-scale particles such as quantum dots with valuable optical properties, will be incorporated into small structures and devices. The technologies for patterning and directly wiring materials will be applied to a wider variety of substrates. Several of the patterning techniques already have been used for flexible and curved substrates. The same is also true for some of the material transfer technologies. Printing of both electronics and optics on flexible substrates is especially attractive for some consumer devices (226). Some of the reviewed technologies can operate using rollers for transfer of patterns (for example, nanoimprinting) or materials (by microcontact printing). This facilitates production of larger areas of fine-scale structures and devices. It also tends to relieve the field-to-field alignment requirements that burden many technologies.

The trend toward ever-smaller features in microelectronics is very well known. The diffraction limit for electromagnetic radiation has been a dominant factor in the evolution of lithography. A recent paper postulates a way to avoid the diffraction limit and produce images with nanometer resolution independent of wavelength (227). Its implementation is enticing but uncertain.

The same trend toward smaller devices is now happening in micromechanics, with the emerging field of NEMS, nanoelectromechanical systems. The use of nanometer-scale lithographic methods, especially direct writing with electron beams, enables the production of devices with moving parts wherein one or more dimensions involves only tens or a few hundreds of atoms. Such components can have very high resonant frequencies that are of potential interest to the communications industry. They also enable new physics experiments because the natural frequencies of very small structures can interact with only some of the normal thermal-vibration modes of a solid from which the nanostructure was made. This effectively selects from the very large number of vibration modes, which are usually excited, only the subset that corresponds to the mechanical frequencies of the NEMS device. Those frequencies are determined by the structure and materials of NEMS. Solid-state physics and mechanical engineering are equally germane to such devices. Examples of NEMS resonators and a nanometer structure that passes only

 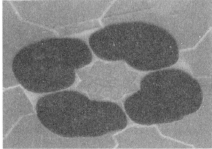

FIGURE 98   Two examples of nanoelectromechanical systems. Left: multiple nano resonators for which the highest measured frequency is 380 MHz (229). The shortest resonator has a length of 1 μm. Right: A thin silicon nitride membrane supported by phonon waveguides with which the quantum of thermal conductance was measured for the first time (230).

	MICROSYSTEMS	BIOLOGY
INFORMATION & CONTROL	MICROELECTRONICS	NERVES
SENSING	MICRO-SENSORS	SENSES
ACTUATION & MOTION	MICROMECHANICS	MUSCLES
FLUID MOTION	MICROFLUIDICS	VESSELS

FIGURE 99   Comparative aspects of microsystems and biology.

specific vibrations are shown in Fig. 98. Nanoscale fabrication and patterning of materials was the subject of a symposium in 1995 (228).

The intersection of micro- and nanostructures with live biological materials is very likely to become increasingly important. Arrays of DNA on commercial chips, guided growth of cells on a substrate, injection of live cells using microstructures, and laser direct writing with viable cells were already noted here. In general terms, such juxtapositions of human-made structures with biomaterials or living materials can be accomplished two ways. First, materials, devices, and systems can be put into live matter—from cells to an entire organism. Implantation of pacemakers (and many other electronic and mechanical devices with microstructures) into humans falls in this category. The second approach is to put bio- or living materials into or onto artificial structures. Placing DNA on chips in arrays for simultaneous experimentation and putting live cells onto substrates are in this second category. So also is the incorporation of live cells into microfluidic systems (231).

The parallels in microsystems and biology are noted in Fig. 99. Given that structures in each of these classes can perform similar functions, the possibility of mixing live material with artificial structures to produce useful microsystems is intriguing. The obvious disadvantages are having to supply life support for the living materials and its inability to function over long periods. However, insects and other animals have very small but incredibly sensitive sensory systems that might be usefully interfaced with artificial devices that have faster computational capabilities and more memory (232). Possibly, mating small-scale devices with natural organs and organisms will be a stepping stone on the much longer road to being able to replicate natural functions by means that may or may not include instructions from DNA.

Ancillary Technologies

Each of the pattern and materials transfer technologies is complex because many component parts are involved in the requisite hardware and because several process steps, each very material dependent, are needed for a single pattern transfer step. There are many challenging problems to overcome in the development and employment of these technologies. The part of this review on pattern transfer focused on the type of the quanta used to expose a photoresist and the manner in which the pattern is transferred from either a computer file or a mask. However, there are other aspects of pattern transfer that are equally critical. They include: (1) the sources of the radiation used for exposure; (2) methods for the production of masks for many of the technologies; (3) the techniques for moving the target substrate relative to the mask, and aligning the pattern for the next layer that is to be created to the earlier-patterned material already on a work piece; (4) the materials employed for the resists; (5) the processes necessarily required for the production of thin films; (6) methods used for the characterization of thin films of resist and other materials; (7) metrology, namely the set of techniques employed for quantitative measurement of position, size, and topography of any mask or work piece at any stage during and after production and for inspection of parts; and (8) packaging of components. The part of the review that dealt with material transfer technologies similarly ignored many of the same ancillary technologies that are needed for those methods. This appendix touches lightly on a few aspects of the supporting technologies.

RADIATION SOURCES

Photon, electron, and ion sources each vary greatly and have multiple roles to play. Mercury lamps were the primary light source for mass production of ICs well into the decade of the 1990s. Today, lasers with monochromatic emissions are the dominant photon sources for many of the pattern transfer technologies, including mainstream lithography. They are also used to heat plasmas to produce the 13 nm extreme ultraviolet radiation that might become the next generation lithographic tool. Very-high-power lasers may also prove viable for heating sources of 1 nm x-radiation (233). High-power electrical discharges are used to heat the plasmas to emit 1 nm radiation for use in X-ray lithography. Synchrotron radiation sources for x-ray lithography are extraordinarily complex combinations of electronic, magnetic, radio-frequency, optical, and vacuum technologies (21).

The early electron sources in transmission and scanning microscopes were simple heated filaments. Then good crystals of lanthanun hexaboride became available. They produce electron beams about an order of magnitude brighter than the hot filaments. Negative electron affinity sources are being developed for electron-beam lithography (234). At present an evolution is taking place with ion sources. Most sources for focused ion beams involve field-assisted emission of gallium ions (235); however, technologies for producing usefully bright sources of other ions have been demonstrated and are being commercialized (236).

MASKS

The masks used for projection lithography are numerous and expensive (50). Recall that modern microelectronics might include thirty layers, each of which requires a mask and several processing steps. Similarly, the leading-edge printed circuit boards now have fifty layers, many of which involve patterning. The production, protection, inspection, and repair of masks are both challenging and expensive. This is true for even "simple" masks that have only open or dark areas. It is even more the case for phase-shift masks of some use in optical lithography, and the masks to be used in extreme ultraviolet lithography, especially. The latter involve nanometer-scale patterns on optics made reflective at 13 nm by coating with 20 atomically precise layer pairs of alternating materials like silicon and molybdenum. The stencil masks with nanometer-scale features for ion projection lithography are also challenging to make and difficult to repair. Stencil masks with micrometer-scale features for evaporation and other uses are more rugged. Because of the critical importance of masks, IBM has a Mask Center of Competency (237).

STAGE MOTION AND PATTERN ALIGNMENT

The relative motion of a beam or material source and the work piece that is critical to direct writing of materials can be accomplished by beam deflection, or by motion of the material source and/or translation of the work piece on an XY stage. The motion of the stage can be effected in an open-loop system by commands sent to a precision motor from the controlling computer. The most precise stage motions are under closed-loop control, with sensors that detect stage position relative to a reference frame (238). The data from the sensors are compared with the desired stage position in the control computer, and motion commands are generated accordingly. The production of alignment marks for the sensors to observe, as well as the choice of the sensors themselves and the detailed features of the control loops, are all very challenging. A spatial phase locking technology for electron beam lithography is under development to acheive sub-5 nm alignment by use of a grid of fiducial marks on the wafer being exposed (239).

Laser radiation is used not only to expose photoresists but also for precise alignment of the wafer or other substrate to the pattern from a mask or data-driven beam. Laser interferometers also provide the sensor data for movement and positioning of a work piece. Both the mask and wafer start with six degrees of freedom. The aligners for rapid exposure of the roughly 1 cm^2 fields on wafers as large as 30 cm in diameter have to be able to position the projected pattern and the wafer relative to each other with nanometer-scale precision. The prototype for extreme ultraviolet lithography uses a magnetically levitated stage. Ultralightweight composite silicon carbide stages are being developed for next generation lithographic tools (240). Today, it is standard practice to expose 80 wafers (with diameters of 20 cm) per hour, that is, each wafer with a number of chips approaching 300 must be exposed in 45 sec, which corresponds to 150 msec per chip (241). This time has to include translation, alignment, and exposure. The translation and alignment are part of control loops involving the laser interferometers as the sensors and precision motors as the actuators. The alignment has to be done to better than 20% of the critical (smallest) dimension. That dimension is now 180 nm and it is projected to be 50 nm in a decade. That is, alignments on the scale of several nanometers, which is a few percent of the wavelength of the laser radiation used, have to be done in times near one hundredth of a second every second, minute, hour, and day over the operational lifetime of an aligner. Because of their expense, on the order of $10 million, and to keep them in top operating condition, aligners commonly make product around 60% of the time.

The computer-controlled XY stages used for direct writing of materials do not generally require nanometer precision. Because laser beams focus only to spot sizes on the order of 1 μm, and because the resolution of most dispensing and jetting technologies is roughly 10 μm, stages with XY precision near 1 μm are usually sufficient for materials direct writing. Electron- and ion-beam chemical vapor deposition can produce nanometer-scale structures, so they do require similarly precise stage control, as does nanometer lithography. Setting or control of the Z position is accomplished by a variety of mechanical, electrical and optical means for pattern and material transfer techniques. The depth of focus of optical microscopes with fast lenses (high magnification) is a common, noncontact approach to Z-axis control.

MATERIALS FOR THIN FILMS

Two classes of materials are central to the technologies for pattern and material transfer. The first group plays a supporting role, with photoresists being the primary member of the class for lithography and a wide range of sacrificial materials can be employed in the various reviewed technologies. The second group includes the materials that end up in the desired structure or component. Each of these will be noted briefly. The most fundamental point is that a range of materials which is very much greater than the collection of materials used to make ICs, comes into play in the direct write and other techniques reviewed above.

Resists are "transient" materials in most cases, and do not appear in the final product. However, the materials aspects of photoresists are varied and challenging (242–245). Resists range from thick to thin and ultrathin (in the case of self-assembled monolayers) (23,246). Some have multiple layers and are chemically amplified. Obviously, they should have both the desired resolution and short exposure times. As with photographic film, these are opposing qualities. That is, high-resolution resists and films both tend to be slow. Other critical properties of resists include their viscosity and shelf life.

Radiation-responsive materials, other than photoresists, are increasingly available for patterning with lithographic methodologies. These include photosensitive glass and photostructurable glass-ceramics (247,248). These can be used in place of photoresists, that is, as layers that are patterned and then removed, or as part of the structures produced lithographically.

The IC industry has undergone a materials revolution in recent years. Copper has supplanted aluminum for interconnects. The time-honored silicon dioxide gate material in field effect transistors will be replaced with materials that have higher dielectric constant. This serves to control a larger number of electrons for the same voltage, or maintain the number of electrons as chip voltages decrease to save power. Low dielectric constant materials are needed around interconnects in chips to reduce losses. In short, the materials employed for IC production are evolving, as usual.

The materials that are actually transferred onto a substrate in a pattern are certainly key both to the performance of the transfer technology and to the performance of the resulting structures and components. The importance of liquid transfer methods was highlighted in this review. If only pure liquids or solutions could be employed in such methods, then many useful structures could be made. However, production of metallic, magnetic, ceramic, and other

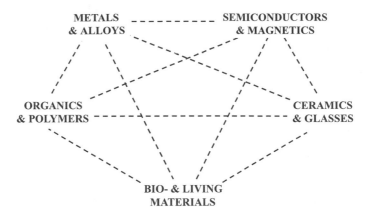

FIGURE 100   The historically separate major classes of materials can now be mixed in a wide variety of ways that are indicated by the dashed lines.

materials with useful electronic and other properties would be very limited were it not for the ability to transfer suspensions of small particles. It is clear that the diameters of the particles have to be small compared both to the diameters of orifices used to put them on a substrate and to the widths of the lines that are written or otherwise deposited. The recent past has seen a wider variety of technologies for making nanometer- and micrometer-scale particles from a wide variety of materials (139,249,250). Inks and toners for printing metallic circuit elements are now on the market under the name Parmod™ (251). Curing these materials for a few minutes at 200–300 °C after patterning by printing or photocopying yields conducting traces on a wide variety of substrates, including plastics and papers.

Thin films of materials that will be part of a component made by pattern-transfer technologies share with resists the qualities of being varied and challenging to produce with the needed properties. This is the case even if the materials of concern are simply one composition and phase. However, modern materials sciences and technologies have two major thrusts. One is the manufacture of diverse materials with nanometer-scale structures. These range from nanocrystaline metallic materials to quantum dots of semiconductors and carbon fullerenes and nanotubes. The other hallmark of current materials work is the mixing of the basic materials classes. In the middle of the last century, metallurgists, ceramicists, organic chemists, and other materials people were balkanized, each having their own journals, conferences, trade shows, and culture. Now, it is common to find glasses in plastics, metals in ceramics, semiconductors in glasses, and many other combinations of materials having extremely diverse geometries and sizes. Figure 100 is one way to convey the possibilities when materials from various traditional classes are mixed. The fundamental point here is that this burgeoning diversity of materials is available for incorporation into thin films that are patterned by lithography and into patterns of materials that are written directly with the newer programmable technologies.

PROCESSES FOR THIN FILMS

The field of thin film production and processing, which includes ion implantation, is immense. It embraces processes used to make small areas of very complex materials—for example, the production of quantum wells by molecular beam epitaxy. And, it includes roll-to-roll technologies that are used commercially on a daily basis to make thousands of square meters of decorative and other sheet products. For each material used in a pattern or material transfer technology, there are commonly multiple physical or chemical processes for the deposition of a thin film of the material. This is usually also the case for etching or removal of a material, whether it be a photoresist, a sacrificial layer in a micromachined component, or part of a layer that will remain in the work piece.

Almost any process used for the production of micro- and nanometer scale structures and devices must evolve, even if there is no change in the materials used. This certainly has been the case in the IC industry as the linewidths have decreased and the wafer sizes have increased. Consider spin coating a 300-mm-diameter wafer with a highly uniform submicrometer layer of photoresist as an example of new challenges in old processes. When new materials enter a production line, then dramatic changes in processing are required. Again citing the IC industry, the evolution from aluminum to copper conductors has had a major impact on critical processes such as chemical mechanical polishing of wafers. The need for chemical mechanical polishing was itself another, earlier revolution in wafer processing. Processing challenges for ICs have not abated (252,253).

The rates at which each of the many processes for deposition, modification, and etching of thin films can operate are critical to their uses, be it for research or making products. Rates of deposition are determined by many factors, including the source of materials, the ambient, and the sticking coefficient on the target substrate. Modification and removal technologies similarly have a wide range of parameters that determine the speed with which they can perform a desired operation.

The pattern and material transfer processes reviewed here each have their own set of enabling and limiting factors. Several of the direct-write materials technologies are limited by flow conditions—that is, by factors that normally do not limit thin-film production. An enumeration of the relevant physical and chemical processes operative in any of the transfer technologies is a first step toward their simulation, understanding, and control. Simulation of each of the processes is generally possible. Microfluidic codes are available and could be applied to the small structures used in many of the materials' direct-write technologies (254).

The recent past has seen a wide variety of new processes developed and employed for the production of both ICs and MEMS. Flip chip bonding of ICs to printed circuit boards and chip scale packaging are two large volume IC examples. Etching of sacrificial layers, avoidance of stiction and bonding by anodic, wafer fusion, and adhesive techniques are important to the growing production of commercial MEMS. Postprocessing thermal methods, such as sintering and pyrolysis, are also more important. As IC line widths continue to shrink and NEMS become more important, techniques such as molecular beam epitaxy and atomic layer deposition will be used increasingly. A recent review deals with deep reactive ion etching and wafer level bonding, two processes important for manufacture of MEMS (255).

CHARACTERIZATION OF MATERIALS AND TOOLS

The composition and structure of thin films determine their properties. Similar to the situation for processing thin films, there are many experimental techniques available for determining what is in a thin film and how the atoms or molecules are arranged. X-ray emission spectroscopy is a convenient way to determine composition nondestructively. It can yield information down to parts per thousand using either X-ray (fluorescent) or electron excitation. X-ray diffraction is the primary method for determining the structure of thin films, although electron scattering is now being used increasingly. Also nondestructive, X-ray diffractometry can detect the presence of phases at the 1% level in many cases. X-ray characterization techniques are relatively easy to apply to films that cover substantial areas of about 1 mm^2 or larger. Electron-beam methods can characterize the composition and structure of regions that are on the order of a micrometer square.

Thin patterned films put in place either by lithography (and other processes such as deposition or etching through slots in a photoresist), or by direct writing, commonly have dimensions on the order of micrometers. Also, they are often thin, so there is neither very much total material nor much area for a characterization methodology. Small spot X-ray techniques are possible with bright sources, such as synchrotron radiation. However, they require taking samples to a large facility. Electron-beam techniques for characterization of composition and structure, usually available in a scanning electron microscope, are the most convenient and, hence, widely used. Of course, such microscopes yield nice images of patterned materials over a very wide range of resolutions, as evidenced by many of the figures in this review.

Measurement of the properties of materials laid down by lithographic or material direct-write technologies is challenging, again because of the small geometric size and limited available material. Some tests, as in the measurement of adhesion, are intrinsically destructive. Others, such as the measurement of conductivity, can be done nondestructively, but require fine-scale probes. Determination of the magnetic properties of patterned materials is generally difficult because of the limited resolution of magnetic methods (due to the spatial extent to magnetic fields). Magnetic sensors in hard disc drives have submicrometer resolution, but they perform only binary sensing.

The characterization of the equipment employed to make micro- and nanometer structures and devices is no less a challenge than the characterization of the materials they produce. A wide variety of tools and associated technologies are employed to quantify the behavior of the systems used for pattern and material transfer. As one example, an interferometric plate is used in place of a mask to assay the behavior of aligners (256). Large area monolithic detectors are being developed for lithographic tool calibration (257).

METROLOGY AND INSPECTION OF PATTERNS AND STRUCTURES

Quantification of the two- and three-dimensional geometry of both masks and the various produced structures is basic to control of the materials, processes, and properties of micrometer and nanometer items made by technologies for pattern and material transfer (258). Such measurements can most naturally be carried out on images made by scanning electron and optical microscopes. However, such image analysis generally yields only two (XY) dimensions of information. The depth of focus of an optical system can be used to obtain information in the orthogonal (Z) direction. This is the case for a system new on the market, which uses stroboscopic image capture and sophisticated image analysis to determine the dynamics of micromechanisms to nanometer precision (259). Optical interferometric technologies, which are commercially available, similarly yield the three-dimensional dynamics of structures having micrometer dimensions (260). Photonic metrology of microelectronics can be done with many techniques (261), including interferometry (262).

The electron and optical metrology methods naturally give information in the XY dimensions. The stylus or probe methods of measuring features made by pattern or material transfer techniques give the height (Z direction) of a structure as a function of one of the planar directions (e.g., X). Assembly of the entire three-dimensional image is possible by raster scanning in the other (Y) direction. Proximal probe techniques, notably scanning tunneling and atomic force microscopies, give three-dimensional images with atomic resolution. They commonly scan areas of 10–100 micrometers square and can measure Z topography with sub-0.1 nm resolution over a range of several nm (263).

Defect detection and quantification are standard requirements in IC fabrication facilities. As with the case of characterization of materials and tools, the range of methods is great. Because of the complexity of modern ICs, there is a trend to integrate the tools for metrology and inspection (264).

It is emphasized that the sequential direct write technologies for making patterns of materials offer the possibility of metrology and inspection during their production. This is a qualitatively different capability compared to the use of parallel methods for patterning and material transfer. Whether or not it will prove to be practically useful remains to be determined.

PACKAGING

The packaging of microelectronics is a large field because of its criticality to the semiconductor industry (265). It is also a complex arena because of the variety of packages for ICs and the large numbers of leads for complex chips (sometimes over 400). Packaging of MEMS is similarly important, but not as advanced because the MEMS industry developed to the multi-billion dollar level only in the last decade. Packaging of MEMS has all of the challenges of IC packaging, and many more complexities. Some MEMS, notably inertial sensors like micro-accelerometers, can be sealed in packages similar to microelectronics. However, many MEMS, like pressure sensors, require contact with the ambient atmosphere. That is, their packages must have openings for communication of matter and energy to the MEMS device inside. Other MEMS packages might be sealed, but must have windows for optical radiation or leads for radio-frequency signals. Some MEMS devices get mounted on surfaces essentially without individual packages. The situation for packaging of devices made by the direct writing of materials is even less advanced than for MEMS components. It will develop significantly in the near future, as it becomes clearer which of the direct-write materials technologies are commercially significant. It is possible that some of the direct-write techniques will be used for the production of packages for diverse devices, in addition to their use for the manufacture of the devices that go into packages.

ACKNOWLEDGMENTS

The stimulus from Alberto Piqué and Douglas Chrisey to write this review is appreciated. The suggestions and assistance of Dr. Piqué during the preparation of the manuscript are recalled with pleasure. Valuable help was provided by Arturo Ayon, Alison Baski, Susan Brandow, Steven Brueck, Hans Fuller, Joseph Mangano, Christie Marrian, Andrew McGill, Harriet Oxley, Martin Peckerar, Henry Smith and David Stenger. This review is dedicated to Dr. David O. Patterson for his leadership of the Advanced Lithography Program in the Defense Advanced Research Project Agency for over a dozen years.

REFERENCES

1. M. Madou, Fundamentals of Microfabrication, Boca Raton and New York, CRC Press, 1997. Chapter 1 is an excellent overview of lithography materials, processes and technologies.
2. D. B. Chrisey, D. R. Gamota, H. Helvajian and D. P. Taylor (Editors), "Materials Development for Direct Write Technologies", Warrandale PA, Materials Research Society, 2001.
3. A. Pique and D. B. Chrisey (Editors), "Direct Write Technologies for Rapid Prototyping Applications", New York, Academic Press, 2001 (This Volume).
4. Coates Circuit Products, August 2001, http://www.coates.com/electro/circuits/advertising/pdf/flex.pdf.
5. R. C. Haavind, "Next-Generation Litho Progress, Innovative Technologies At MRS", Solid State Technology, February 2001, 32–6.
6. R. D. Miller, "In Search of Low-k Dielectrics", Science, **286**, 421–3.
7. Staff, "Electronics Industry Update", Semiconductor International, July 2001, 352.
8. D. S. Patterson, "Solder Bumping Step by Step", Advanced Packaging, July 2001, 75–80.
9. M. Ranjan, S. Kay and W. Flack, "Lithography Requirements for 300-mm WLP", Advanced Packaging, July 2001, 57–64.
10. P. Werbaneth, S. Kirshman, H. Slomowitz and J. Thomas, "Adapting Semiconductor Processing Tools to Thin Film Head Fabrications", Solid State Technology, Sept 2000, 122–8.
11. S. Y. Lin, J. G. Fleming and D. L. Hetherington, "A Three Dimensional Photonic Crysta operating at Infrared Wavelengths", Nature, 1998, **394**, pp. 251–3.
12. Kionix, Inc., August 2001, http://www.kionix.com/knx101.html.
13. B. Wiegl, "O.R.C.A. μFluidics™", August 2001, O.R.C.A. μFluidics.
14. "Introducing the NanoChip™", August 2001, http://www.nanogen.com/products/nanochip_cart.asp.
15. "The Gene Chip", August 2001, http://www.hgu.mrc.ac.uk/Users/Graham.Dellaire/hgp/genechip.html.
16. M. Huang, S. Mao, H. Feick, H. Yan, Y. Wu, H. Kind, E. Weber, R. Russo, P. Yang, "Room-Temperature Ultraviolet Nanowire Nanolasers", Science, 292, 1897, 2001 and S. K. Moore, "How to Raise UV Nanolasers", IEEE Spectrum, July 2001, 33.
17. W. Ma, Q.-Y. Liu, D. Jung, P. Manos, J. J. Pancrazio, A. E. Schaffner, J. L. Barker and D. A. Stenger, "Central Neuronal Synapse Formation on Micropatterned Surfaces", Developmental Brain Research, 1998, 111, 231–43.
18. E. A. Dennis, O. Odesina and D. G. Wilson, "Lithographic Technology in Transition", Albany, Delmar Publishers, 1997. Chapter 1 gives a survey of the components and activities within the graphic arts industries.

19. G. R. Madland, H. K. Dicken, R. D. Richardson, R. L. Pritchard, F. H. Bower and D. B. Kret, "Integrated Circuit Engineering–Basic Technology", Cambridge, Boston Technical Publishers, 1966, p. 45

20. D. J. Nagel, "Ultraviolet and X-Ray Lithography", Bellingham WA, SPIE Vol. 297, 1981, 98–110

21. D. Attwoood, "Soft X-Rays and Extreme Ultraviolet Radiation : Principles and Applications", Cambridge, Cambridge University Press, 1999

22. D. M. Longo and R. Hull, "Direct Focused Ion Beam Writing of Printheads for Pattern Transfer Utilizing Microcontact Printing", in Reference 3, 157–162.

23. A. Kumar, N. L. Abbott, E. Kim, H. A. Biebuyck and G. M. Whitesides, "Patterned Self-Assembled Monolayers and Meso-Scale Phenomena", Accounts of Chemical Research, 1995, **28**, 219–26

24. K. K. Berggren, A. Bard, J. Wilbur, J. D. Gillispy, A. G. Helg, J. J. McClelland, S. l. Rolston, W. D. Phillips, M. Prentiss and G. W. Whitesides, "Microlithography by Using Neutral Metastable Atoms and Self-Assembled Monolayers", Science, 1995, **269**, 1255–7.

25. K. S. Johnson, K. K. Berggren, A. J. Black, A. P. Chu, N. H. Dekker, D. C. Ralph, J. H. Thywissen, R. Youkin, M. Prentiss, M. Tinkham and G. W. Whitesides, "Using Neutral Metastable Argon Atoms and Contamination Lithography to Form Nanostructures in Silicon, Silicon Dioxide and Gold", Appl. Physics. Letters, 1996, **69**, 2773–5.

26. J. J. McClelland, "Lithography With Metastable Rare Gas Atoms", August 2001, http://physics.nist.gov/Divisions/Div841/Gp3/epg_files/metasam_proj.html

27. J. J. McClelland and R. J. Celotta, "Nanofabrication Of Chromium Lines By Laser-Focused Atomic Deposition", August 2001, http://physics.nist.gov/Divisions/Div841/Gp3/epg_files/atom_lines_proj.html

28. J. J. McClelland and R. J. Celotta, "Nanofabrication Of A Two-Dimensional Array Of Chromium "Dots", August 2001, http://physics.nist.gov/Divisions/Div841/Gp3/epg_files/atom_dots_proj.html

29. R. G. Dall, M. D. Hoogerland, K. G. H. Baldwin and S. J. Buckman, "Guiding Of Metastable Helium Atoms Through Hollow Optical Fibres". August 2001, http://wwwrsphysse.anu.edu.au/ampl/atommp/hollowfibres/fibre.html

30. Semiconductor Industry Association, "International Technology Roadmap for Semiconductors", August 2001, http://public.itrs.net/

31. "Industry Executive Forum", August 2001, http://www.sematech.org/public/resources/ief/ief0400/anderson.pdf

32. R. C. Jaeger, "Introduction to Microelectronic Fabrication", New York, Addison-Wesley, 1988

33. S. Wolf and R. Tauber, "Silicon Processing for the VLSI Era, Vol.1—Process Technology", Sunset Beach CA, Lattice Press, 1999 (2nd edition)

34. Y. Nishi and R. Doering (Editors), "Handbook of Semiconductor Manufacturing Technology", New York, Marcel Dekker, 2000

35. D. Elliott, "Microlithography Process Technology for IC Fabrication", New York, McGraw-Hill, 1986

36. W. M. Moreau, "Semiconductor Lithography Principles, Practices and Materials", New York, Plenum Press, 1988

37. T. Ueno, T. Ito and S. Nonogaki, "Microlithography Fundamentals in Semiconductor Devices and Fabrication Technology", New York, Marcel Dekker, 1998

38. J. N. Helbert, "Handbook of VLSI Microlithography", Norwich NY, William Andrew, 2001 (2nd Edition)

39. H. J. Levinson, "Principles of Lithography", Bellingham WA, SPIE Vol. PM97, 2001

40. C. J. Progler (Editor), "Optical Microlithogrpahy XIV", Bellingham WA, SPIE Vol. 4346m 2001

41. K. A. Valiev, "The Physics of Submicron Lithography", New York, Plenum Press, 1992
42. K. Suzuki, S. Matsui and Y. Ochiai (Editors), "Sub-Half-Micron Lithography for ULSIs, Cambridge, Cambridge Press, 2000
43. E. A. Dobisz (Editor), "Emerging Lithographic Technologies V", Bellingham WA, SPIE Vol. 4343, 2001
44. P. Rai-Choudhury (Editor), "Handbook of Microlithography, Micromachining and Micro-fabrication Vol. 1 MicroLithography", London, Institution of Electrical Engineering, 1997
45. P. Rai-Choudhury (Editor), "Handbook of Microlithography, Micromachining and Micro-fabrication Vol. 2 Micromachining and Microfabrication ", London, Institution of Electrical Engineering, 1997
46. S. Turnback and D. Etter, "Lithography, the Hidden Technology", Ausust 2001, http://www.dtic.mil/dusdst/agenda/agenda11400.html
47. H. J. Levinson and W. H. Arnold, "Optical Lithography", Chapter 1 in Reference 39, 11–138
48. "Photomask Technology (1995–2000)", Bellingham WA. SPIE Vol. CDP18, 2001
49. A. K.-K. Wong, "Resolution Enhancement Techniques in Optical Lithography", Bellingham WA, SPIE Vol. TT47, 2001
50. J. G. Skinner, T. R. Groves, A. Novembre, H. Pfeiffer and R. Singh, "Photomask Fabrication Procedures and Limitations", Chapter 5 in Reference 39, 377–474
51. R. D. Allen, W. E. Conley and R. R. Kunz, "Deep-UV Resist Technology", Chapter 4 in Reference 39, 321–375
52. PAS 5500/1100 Step & Scan Tool, August 2001, http://www.asml.com/prodtech/1100.html
53. T. Deng, F. Arias, R. F. Ismagilov, P. J. A. Kenis and G. M. Whitesides, "Fabrication of Metallic Microstructures Using Exposed, Developed Silver Halide-Based Photographic Film", Anal. Chem, 2000, 645–51
54. J. Aizenberg, J. A. Rogers, K. E. Paul and G. M. Whitesides, "Imaging Profiles of Light Intensity in the Near Field: Applications to Phase Shift Photolithography", Applied Optics, 1998, 37, 2145–52
55. K. E. Paul, T. L. Breen, J. Aizenberg and G. M. Whitesides, "Maskless Photolithography: Embossed Photoresist as Its Own Optical Element" Applied Physics Letters, 1998, 73, 2893–5
56. R. E. Hollingsworth, W. C. Bradford, M. K. Herndon, J. D. Beach and R. T. Collins, "Direct Patterning of Hydrogenated Amorphous Silicon by Near Field Scanning Optical Microscopy" in Reference 2, 87–97
57. Subwavelength Structure Anti-Reflection Surfacing, August 2001, http://www. opticalswitch.com/DataSheets/Microphotonics/Subwavelength.pdf
58. B. Omar, S. Clube, F. Hamidi, M. Struchen and S. Gray, "Advances in Holographic Litho-graphy", Solid State Technology, September 1991, 89–94
59. X. Chen and S. R. J. Brueck, "Imaging Interferometric Lithography: A Waveleigth Division Multiplex Approach to Extending Optical Lithography", J. Vacuum Science Technology B 16, 3392–7
60. D. J. D. Carter, D. Gil, R. Menon, M. K. Mondol, H. I. Smith and E. H. Anderson, "Maskless, Parallel Patterning with Zone-Plate Array Lithography", J. Vacuum Science Technology B, 1999, 17, 3449–52
61. D. M. Freeman and M. S. Mermelstein, "Synthetic Aperture Lithography", 2000, unpublished
62. J. E. Bjorkholm, " EUV Lithography–The Successor to Optical Lithography?", Intel Technical Journal, 1998, 1–8
63. J. Bokor (Editor), "Soft X-Ray Projection Lithography", Washington DC, Optical Society of America, 1991
64. G. Stix, "Getting More from Moore's", Scientific American, April 2001, 32–6
65. M. W. McGeoch, "Power Scaling of a Z-Pinch Extreme Ultraviolet Source", 2000, unpublished and M. W. McGeoch, "High Power Extreme Ultraviolet Source Based on a Z-Pinch" in Y.

Vladimirsky (Editor), "Emerging Lithographic Technologies III", Bellingham WA, 1999, SPIE Vol. 3676, 697–702

66. A. Hand, "EUV Lithography Makes Serious Progress", Semiconductor International, June 2001, 54

67. F. Cerrina, "X-Ray Lithography", Chapter 3 in Reference 39, 251–322

68. H. I. Smith and F. Cerrina, "X-Ray Lithography for ULSI Manufacturing", Microlithography World, Winter 1997,

69. H. I. Smith, "Japan Could Dominate Industry with X-Ray Lithography", Semiconductor International, February 2001, 67–72

70. D. J. Nagel and M. C. Peckerar, "Pulsed X-Ray Lithography", U. S. Patent No. 4,184,078, 1980

71. R. R. Prasad, M. Krishnan, J. Mangano, P. A. Greene, N. Qi, "Neon Dense Plasma Focus Point X-Ray Source For Sub-0.25 Um Lithography" in "Electron-Beam, X-Ray, and Ion-Beam Submicrometer Lithographies for Manufacturing IV", D. O. Patterson (Editor), 1994, Bellingham WA, SPIE Vol. 2196, 120–8 and R. Selzer, K. Mason, J. Heaton and P. Hoff, "Point Source X-Ray Manufacturing System", 2000, unpublished

72. C. A. MacDonald and A. M. Khounsary (Editors), "Advances in Laboratory-Based X-Ray Sources and Optics, Bellingham WA, SPIE Vol. 4144, 2000

73. J. Heaton, "Status of the DARPA/NAVAIR Point Source X-Ray Lithography Program", presentation at the Government Microcircuits Applications Conference, 2000 and "SAL's NanoLithography Means Faster Networks, August 2001, http://www.xraylitho.com/newhome. html

74. "Photomask X-Ray Mask Technology (Photomask Japan 1995–2000), Bellingham WA, SPIE Vol. CD18, 2001

75. M. C. Peckerar, R. Bass, K.-W. Rhee and C. R. K. Marrian, "Nanolithography with Electron Beams: Theory and Practice", Chapter 11 in Reference 3

76. M. A. McCord and M. J. Rooks, "Electron Beam Lithography, Chapter 2 in Reference 39, 139–249

77. L. R. Taylor, D. H. Lowndes, V. I. Merkulov, G. Eres, D. B. Geohagen, J. E. Hardy, D. L. Hutchinson, A. A. Puretzky, M. Simpson, C. E. Thomas, J. B. Wilgen, J. H. Whealton and E. Voelkl, "Progress in Field Emitter Sources for Digital Electrostatic E-Beam Array Lithography", 2000, unpublished

78. H. F. Lockwood, W. B. Feller, P. L. White, W. Leonard, B. White, J. E. Hefferon, R. and D. Manz, "High Throughput E.Beam Lithography", 2000, unpublished

79. M. Mankos and T. H. P. Chang, "Advances in High-Throughput Multiple Electron-Beam Lithography", presented at EUREM 2000, May 2001, http://www.etec.com/library_frames.htm

80. J. A. Liddle, L. R. Harriott, A. E. Novembre and W. K. Waskiewicz, "SCALPEL: A Projection Electron-Beam Approach to Sub-Optical Lithography", August 2001, http://www.bell-labs.-com/project/SCALPEL/9905.ngl/ngl99.pdf

81. K. Okamoto, K. Suzuki, H. C. Pfeiffer and M. Sogard, "High Throughput Electron Beam Stepper Lithography", Microlithography Worls, Winter 2000, 10–16

82. K. Edinger, "Focused Ion Beams for Direct Writing", Chapter 12 in Reference 3

83. D. K. Stewart and J. D. Casey Jr., "Focused Ion Beams for Micromachining and Micro-chemistry", Chapter 4 in Reference 40,153–195

84. K. N. Leung, T.-J. King, W. A. Barletta, R. A. Gough, Q. Li, Y. Y. Lee, V. V. Ngo, K. Scott, K. Standiford and N. Zahir, "Maskless Ion Beam Lithography at LBNL, 2000, unpublished

85. R. Mohondro, "Ion Projection Lithography: Life After Optical", Semiconductor FABTECH, 1995, Issue 3, 177–183 and "Ionenprojektionslithographie — ein Schwerpunktsprojekt der Gesellschaft für Mikroelektronik", August 2001, http://www.iaee.tuwien.ac.at/gme/ims_g.htm

86. "2000 mm Stencil Mask", August 2001, http://wwwold.ims-chips.de/products/ipl/stencilmask.htm

87. S. H. Cohen and M. L. Lightbody (Editors), "Atomic Force Microscopy/Scanning Tunneling Microscopy 3", New York, Plenum Publishing Corp, 1999

88. "Hybrid Atomic Force / Scanning Tunneling Lithography", K. Wilder, H. T. Soh, A. Atalar, and C. F. Quate, "Hybrid atomic force / scanning tunneling lithography," J. Vac. Sci. Technol. B, 1977, 15, 1811–1817 and K. Wilder, H. T. Soh, S. C. Minne, S. R. Manalis and C. F. Quate, "Cantilever Arrays for Lithography, Naval Research Reviews, 1997, Vol. XLIX, 35–48

89. D.M. Eigler and E.K. Schweizer, "Positioning Single Atoms With A Scanning Tunneling Microscope", Nature, 1990, 344, 524–526

90. M.F. Crommie, C.P. Lutz, D.M. Eigler and E.J. Heller, "Waves On A Metal Surface And Quantum Corrals". Surface Review and Letters, 1995, 2 (1), 127–137

91. R. Cassidy, "Ingenious STM Puts Atoms Right Where You Want Them", R & D Magazine, April 1993, 35, 71

92. "Local Oxidation", August 2001, http://www.research.ibm.com/nanoscience/local_oxidation.html and Ph. Avouris, R. Martel, T. Hertel, and R.L. Sandstrom, "AFM Tip Induced and Current Induced Local Oxidation of Silicon And Metal" Applied Physics A, Mater. Sci. Process., 1998, 66, Suppl., pt.1–2, p. S659

93. C. F. Quate, "Presentations", Aug 2001, http://www.stanford.edu/group/quate_group/

94. H.. T. Soh, K. W. Guarini and C. F. Quate, "Scanning Probe Lithography", Boston, Klewer Academic Publishers, 2001

95. "LIGA Technology", August 2001, http://www.imm-mainz.de/

96. W. Ehrfeld, "The LIGA Process for Microsystems", Proceedings of Conference on Micro-systems Technologies "90, Berlin, 1990, 521–8.

97. C. H. Ahn, Y. J. Kim, and M. G. Allen, "A Planar Variable Reluctance Magnetic Micromotor with a Fully Integrated Stator and Wrapped Coils", Proc. IEEE MicroElectroMechanical Systems (MEMS '92), Travemunde Germany, 1992, 93–8

98. S. Y. Chou, L. Zhuang and L. Guo, "Lithographically Induced Self-Construction Of Polymer Microstructures For Resistless Patterning", Applied Physics Letters, 1999, 75, 1004–6

99. S. Y. Chou and L. Zhuang, "Lithographically Induced Self_Assembly of Periodic Polymer Micropillar Arrays, J. Vac. Sci. Technology B., 1999, 17, 3197–3202

100. H. Becker and U. Heim, "Silicon as a Tool Material for Polymer Hot Embossing", Proc. IEEE MEMS, 1999, 228–31

101. T. Glinsner," Nanoimprinting Can Solve Pattern-Generation Problems", R & D Magazine, July 2001, 33

102. S. Y. Chou, P. R. Krause and P. J. Renstrom, "Imprint of Sub-25 nm Vias and Trenches in Polymers", Applied Physics Letters, 1995, 67, 3114–6

103. S. Y. Chou, Nanoimprint Lithography", August 2001, http://www.ee.princeton.edu/~chouweb/newproject/page3.html

104. S. C. Johnson, "Selective Compliant Orientation Stages for Imprint Lithography", MS Thesis, The University of Texas at Austin.,1999, http://sfil.org/research/papers/sjthesis.pdf

105. C. Bulthaup, E. Wilhelm, B. Hubert, B. Ridgley and J. Jacobsen, "All Printed Inorganic Logic Elements Fabricated by Liquid Embossing" in Reference 2, 225–30.

106. W. Warren, "Overview of Commercial and Military Application Areas in Passive and Active Electronic Devices", Chapter 2 in Reference 3.

107. K. H. Church, C. Fore and T. Feeley, "Commercial Applications and Review of Direct Write Technologies," in Reference 2, 3–8.

108. S. Minh, "Print your Next PC", MIT Technology Review, Nov/Dec 2000, 66–70

109. Reference 1, 358–9

110. Y. Xia, J. A. Rogers, K. E. Paul and G. M. Whitesides, "Unconventional Methods for Fabri-cating and Patterning Microstructures", Chem. Reviews, 1999, 99, 1823–48

111. V. K. Varadan, X. Jiang and V. V. Varadan, "Microstereolithography and Other Fabrication Techniques for 3D MEMS", Chichester UK, John Wiley & Sons, 2001

112. Alberto Piqué, Douglas B. Chrisey and C. Paul Christensen, "Laser Direct Write Micro-machining", Chapter 13 in Reference 3

113. J. Zhang, J. Szczech, J. Skinner, and D. Gamota, "Role of Direct Write Tools and Technologies for Microelectronic Manufacturing", Chapter 3 in Reference 3

114. H. Helvajian, "3D Microengineering via Laser Direct-Write Processing Approaches", Chapter 14 in Reference 3

115. "Electronic Reusable Paper", August 2001, http://www.parc.xerox.com/dhl/projects/gyricon/

116. "What is Electronic Ink?", August 2001, http://www.eink.com/technology/index.htm

117. See, for example, "Xerox Network Printers DocuPrint N2125 Overview", August 2001, http://www.officeprinting.xerox.com/cgi-bin/product.pl?product=N2125

118. R. H. Detig, "Electrostatic Printing: A Versatile Manufacturing Process for the Electronics Industries," in Reference 2, 71–78.

119. C. R. Friedrich, R. Warrington, W. Bacher, W. Bauer, P. J. Coane, J. Gottert, T. Hanemann, J. Hausselt, M. Heckele, R. Knitter, J. Mohr, V. Poitter, H.-J. Ritzhaupt-Kleissl and R. Ruprecht, "High Aspect Ratio Processing" in Reference 40, 299–377

120. "Fourth International Workshop on High Aspect Ratio MicroStructure Technology", August 2001, http://www.fzk.de/pmt/harmst/HarmstProgram.htm

121. K. E. Bean, "Anisotropic Etching of Silicon", IEEE Transactions on Electron Devices, 1978, ED-25, 1185–93

122. G. T. A. Kovacs, "Micromachined Transducers Handbook", Boston MA, WCB McGraw-Hill, 1998, p. 69

123. A. Ayon, unpublished, and S. D. Senturia, "Microsystem Design", Boston, Kluwer Academic Publishers, 2001, p. 70

124. Z. L. Zhang and N. C. MacDonald, "An RIE Process for Submicron Silicon Electromechanical Structures", 6th International Conference on Solid-State Sensors and Actuators (Transducers '91), San Francisco CA, 1991, 520–3

125. "General Description of Electrochemical Machining", August 2001, http://www.unl.edu/nmrc/ecm00.htm

126. "History of Micro EDM", August 2001, http://www.control.hut.fi/Kurssit/AS-74.136/materials/microfabrication2.pdf

127. K. Egashira And T. Masuzawa, "Application of Ultrasonic Machining to Micromachining", August 2001, http://www.iis.u-tokyo.ac.jp/english/publications/kenkyu/abst-199709.html#2

128. N. L. Abbott, J. P. Folkers and G. M. Whitesides, "Manipulation of the Wetability of Surfaces on the 0.1 to 1 Micrometer Scale Through Micromachining and Self Assembly", Science, 1992, 257, 1380–82

129. D. Szepesi, "Sensoren en Actuatoren in Ultraprecise Draaibanken" in Sensoren en Actuatoren in de Werktuigbouw/Machinebouw, The Hague, Centrum voor Micro-Electronica, 1993, 99–107

130. N. Taniguchi, "Current Statue In, and Future Trends of Ultraprecision Machining and Ultrafine Materials Processing", 1983, Annals of the CIRP, 32, S573–82

131. "Human Hair", August 2001, http://www.lambdaphysik.com/scientific/hair.asp

132. D. M. Allen, "The Principles and Practice of Photo-chemical Machining and Photoetching", Bristol, Adam Hilger, 1986

133. D. J. Ehrlich, R. Aucion, M. J. Burns, K. Nill and S. Silverman, "Real-World Applications of Laser Direct Writing," in Reference 2, 9–16.

134. M. J. Vasile, C. Biddick and S. A. Schwalm, "Microfabrication by Ion Milling: The Lathe Technique", J. Vac. Sci. Technology B 12, 1994, 2388–93

135. M. J. Vasile, unpublished, and Reference 1, p. 356

136. E. Pfender, "Fundamental Studies Associated with Plasma Spray Processes", Surf. Coating Technology, 1988, 34, 1–14

137. "SWAM-BLASTER™ Abrasive Nozzle Selection Chart" August 2001, http://www.crystalmark.thomasregister.com/olc/crystalmark/nozzles.htm

138. P. Atanassova, P. Atanassov, R. Bhatia, M. Hampden-Smith, T. Kodas, P. Napolitano "Direct-Write Materials and Layers For Electrochemical Power Devices", Chapter 4 in Reference 3

139. P. Atanassova, J. Caruso, H. Denham, M. Hampden-Smith, K. Kunze, T. Kodas, A. Schult, A. Stump, and K. Vanheusden, "Advanced Materials Systems for Ultra-Low-Temperature, Digital, Direct Write Technologies", Chapter 6 in Reference 3

140. D. B. Chrisey, Naval Research Laboratory, private communication

141. M. Lambrechts and W. Sansen, "Biosensors: Microelectrical Devices", Philadelphia, Institute of Physics Publishing, 1992

142. Y. Xia and G. M. Whitesides, "Soft Lithography", Annual Reviews of Materials Science, 1998, 28, 153–84

143. S. M. Miller, A. A. Darhuber, S. M. Troain and S. Wagner, "Offset Printing of Liquid Micro-structures for High Resolution Lithography," in Reference 2, 47–52.

144. B. D. Martin, S. L. Brandow, W. J. Dressick and T. L. Schull, "Fabrication and Application of Hydrogel Stampers for Physisorptive Microcontact Printing", Langmuir, 2000, 16, 9944–6

145. P. G. Clem, N. S. Bell, G. L. Brennecka, B. H. King, and D. B. Dimos , "Micropen Printing of Electronic Components", Chapter 8 in Reference 3.

146. A. Kumar, H. A. Biebuyck, N. L. Abbott and G. M. Whitesides, "The Use of Self-Assembled Monolayers and a Selective Etch to Generate Patterned Gold Features", J. Am. Chemical Society, 1992, 114, 9188–9

147. C. A. Mirkin, " Dip-Pen Nanolithography: Direct Writing Soft Structures on the Sub-100 Nanometer Length", Chapter 10 in Reference 3.

148. "Impact 2 Electronic Multichannel Pipettors", August 2001, http://www.matrixtechcorp.com/view_speed.asp?range=34

149. Advertisement for Operon Technologies, Science, 1999, 286, 649

150. "Stealth Micro-Spotting Pins", August 2001, http://www.genpakdna.com/stealthpins.shtml

151. J. Zhang, I. Shmagin, J. Skinner, J. Scczech and D. Gamota, "Material systems used by micro dispensing and ink jetting technologies," Reference 2, 41–46.

152. "Technologies", August 2001, http://www.microdrop.de/html/technologies.html

153. D. B. Wallace, W. R. Cox, D. J. Hayes, "Direct-Write Technologies for Rapid Prototyping of Sensors, Electronics, and Passivation Coatings" Chapter 7 in Reference 3.

154. H.-J. Trost, C. J. Fredrickson, D. J. Hayes and D. B. Wallace, "Ink Jet Technology for Biomedical Applications", August 2001, http://www.microfab.com/papers/papers_pdf/houston.PDF

155. "BioChip Arrayer", August 2001, http://www.packardbiochip.com/products/biochip_arrayer.htm

156. B. Szczech, C. M. Megaridis, D. R. Gamota and J. Zhang, "Manufacture of Microelectronic Circuitry by Drop-On-Demand Dispensing of Nanoparticle Liquid Suspensions," in Reference 2, 23–28.

157. P. Singer, "Plastic Transistors Formed by Ink Jet Printing", Semiconductor International, Nov 2000, 40

158. M. W. J. Prins, W. J. J. Welters and J. W. Weekamp, "Fluid Control in Multichannel Structures by Electrocapillary Pressure", Science, 2001, 291, 277–80

159. M. Orme, J. Courtier, Q. Liu, J. Zhu and R. Smith, "Charged Molten Metal Droplet Deposition as a Direct Write Technology," in Reference 2, 17–22.

160. "ModelMaker II", August 2001, http://www.solid-scape.com/mmii_techdoc.html

161. "ThermoJet Printer", August 2001, http://www.3dsystems.com/index_nav.asp?nav=products&subnav=solidobject&content=products/solidobject/thermojet/index.asp

162. I. Zergiotti, G. Koundourakis, N.A. Vainos, and C. Fotakis, "Laser Induced Forward Transfer: An Approach to Single Step Microfabricaction", Chapter 16 in Reference 3

163. J. M. Fitz-Gerald, P. D. Rack, B. Ringeisen, D. Young, R. Modi, R. Auyeung and H.-D. Wu, "Matrix Assisted Pulsed Laser Evaporation Direct Write (Maple DW): A New Method to Rapidly Prototype Organic and Inorganic Materials", Chapter 17 in Reference 3

164. A.-C. Tien, Z. S. Sacks and F. J. Mayer, "Precision Laser Metallizations", Microelectronic Engineering, 2001, in press, and F. J. Mayer, "Precision Laser Metallization", U. S. Patent No. 6,159,832, 2000

165. "Laser Books", August 2001, http://www.lasers.org.uk/books/ and J. F. Ready and D. F. Farson (Editors), "LIA Handbook of Laser Materials Processing", Orlando FL, Laser Institute of America, 2000

166. B. R. Ringeisen, D. B. Chrisey, A.Pique, H. D. Young, R. Modi, M. Bucaro, J. Jones-Mehan and B. J. Spargo, "Generation of Mesoscopic Patterns of Viable Escherichia Coli by Ambient Laser Transfer", Biomaterials, 2001, in press

167. B. R. Ringeisen, D. B. Chrisey, A.Pique, D. Krizman, M. Brooks, B. Spargo, R. Auyeng andP. Wu, "Direct Write Technology as a Tool to Form Patterns of Living Cells and Active Biomolecules", American Biotechnology Laboratory, 2001, 19, 42–4

168. M. Boman, H. Westberg and S. Johansson, "Helical Microstructures Grown by Laser Assisted Chemical Vapor Deposition, Proc. IEEE MicroElectroMechanical Systems (MEMS '92), Travemunde Germany 1992

169. O. Lehmann and M. Stuke, "Three Dimensional Laser Direct Writing of Electrically Conducting and Isolating Microstructures", Materials Letters, 1994, 21, 131–6

170. "Scanning Nano Four Point Multiprobes", August 2001, http://www.mic.dtu.dk/research/nanotech/nanotech.htm#nanohand

171. "Ultra Precise Nanometer-scale 3D Production Technology Used to Make World's Smallest Wineglass", August 2001, http://www.labs.nec.co.jp/Eng/Topics/data/r001207/ and "Ion Beam Forms Nanoscale Objects", Laser Focus World, March 2001, 56–9

172. M. J. Renn, G. Marquez, B. H. King, M. Essien and W. D. Miller, "Flow- and Laser-Guided Direct Write of Electronic and Biological Components", Chapter 15 in Reference 3

173. S. Sampath, J. Longtin, R. Gambino, H. Herman, R. Greenlaw and E.Tormey, "Direct Write Thermal Spraying of Multilayer Electronics and Sensor Structures", Chapter 9 in Reference 3

174. J. Akedo, M. Ichiki, K. Kikuchi and R. Maeda, "Fabrication of Three Dimensional Micro-structures Composed of Different Materials Using Excimer Laser Ablation and Jet Molding", MEMS '97, Proc. IEEE MEMS, 1997, 135–40

175. J. F. Groves, "Directed Vapor Deposition", University of Virginia, 1998, http://www.ipm.virginia.edu/research/PVD/Pubs/pvd_thesis5.htm

176. "Jet Vapor Deposition", August 2001, http://www.public.iastate.edu/~akshay/article/section3_2.html

177. J. D. Madden and I. W. Hunter, "Three-Dimensional Microfabrication by Localized Electro-chemical Deposition", J. MicroElectroMechanical Systems, 1996, 5, 24–32

178. M. Fleischmann, S. Pons, D. R Rolison and P. Schmidt (Editors), " Ultramicroelectrodes", Morgantown NC, DataTech Systems, 1987

179. "Commercial Rapid Prototyping System Manufacturers", Aug 2001, http://home.att.net/~castleisland/com_lks.htm

180. "The Whole RP Family Tree", August 2001, http://ltk.hut.fi/~koukka/RP/rptree.html

181. "Key US Rapid Prototyping System Vendors", August 2001, http://home.att.net/~castleisland/ind_tab.htm

182. "The Most Important Commercial Rapid Prototyping Technologies at a Glance", August 2001, http://home.att.net/~castleisland/rp_int1.htm

183. "What is Rapid Prototyping?", August 2110, http://www.ascend-modelmakers.com/over-view.htm
184. "Principle of the Microstereolithography Apparatus Developed at EPFL", August 2001, http://dmtsun.epfl.ch/~abertsch/principle.html
185. "A Few Scanning Electron Microscope Photos of Structures Made by Microstereolithography at EPFL", August 2001, http://dmtsun.epfl.ch/~abertsch/album.html
186. H. Jones-Bey, "The Big World of Little MEMS", Laser Focus World, 2001, 122–30
187. M. N. Kabler, D. J. Nagel and E. F. Skelton, "Synchrotron X-Radiation Research", Naval Research Laboratory Review, 1991, 67–84
188. X. N. Jiang, C. Sun and X. Zhang, "Microstereolithography of 3D Complex Ceramic Micro-structures and PZT Thick Films on Si Substrate", ASME MEMS, 1999, 1, 67–73
189. V. Poitter, T. Benzler, T. Hanemann, H. Wollmer, R. Ruprecht and J. Hausselt, "Innovative Molding Technologies for the Fabrication of the Components for Microsystems", 1999, Proc. SPIE, 3680, 456–63
190. T. Benzler, V. Piotter, T. Hanemann, K. Mueller, P. Norajitra, R. Ruprecht, J. H. Hausselt, "Innovations in Molding Technologies for Microfabrication", 1999, Proc. SPIE, 3874, 53–60
191. T. Hanemann, R. Ruprecht and J. H. Hausselt, "Micromolding and Photopolymerizations", Advanced Materials, 1997, 9, 927–9
192. E. Kim, Y. Xia and G. M. Whitesides, "Micromolding in Capillaries: Applications in Materials Science", J. American Chemical Society, 1996, 118, 5722–31
193. Y. Xia, E. Kim and G. M. Whitesides, "Micromolding of Polymers in Capillaries: Applications in Microfabrication", Chemistry of Materials, 1996, 8, 1558–67
194. L. T. Romankiw and E. J. M. O'Sullivan, "P;ating Techniques" in Reference 40, 197–298
195. Reference 1, p. 297
196. A. Cohen, G. Zhang, F.-G. Tseng, U. Frodis, F. Mansfield and P. Will, "EFAB: Rapid, Low-Cost Desktop Micromachining of High Aspect Ratio True 3–D MEMS", 1999, Proc IEEE MEMS, 244–51, International Microelectromechanical Systems Conference, 1999, Technical Digest, IEEE
197. "Building the Machine Tool for the Microworld", August 2001, http://www.isi.edu/efab/
198. C. Keller and M. Ferrari, " Milli-Scale Polysilicaon Structures", Tech. Digest of the 1994 Solid State Sensor and Actuator Workshop, Hilton Head SC, 1994, 132–7
199. Reference 1, p. 247
200. A. P. Pisano, University of California Berkeley, private communication
201. C. S. Taylor, XXX, "A Spatial Forming Three-Dimensional Printing Process", Proc. IEEE MEMS, 1994, 203–8
202. K. M. Jones, K. M. Saoud and A. A. Baski, "Noble Metal Row Growth on Si (5 5 12), August 2001, http://www.people.vcu.edu/~aabaski/Papers/nano99kj.PDF and A. A. Baski, K. m. Saoud and K. M. Jones, "1–D Nanostructures Grown on the Si (5 5 12) Surface, Applied Surface Science, 2001, in press
203. Joachim and S. Roth (Editors), "Atomic and Molecular Wires", Dordrecht, Kluwer Academic Publishers, 1997
204. R. A. Wind, M. J. Murtagh, F. Mei, Y. Wand, M. A. Hines and S. L. Sass, "Fabrication of Nanoperiodic Surface Structures by Controlled Etching of Dislocations in Bicrystals", August 2001, http://www.chem.cornell.edu/mah11/Nanofab.html
205. S. Tsukamoto, Y. Nagamune, M. Nishioka and Y. Arakawa, "Fabrication of GaAs Quantum Wires (~10 nm) by Metalorganic Chemical Vapor Selective Deposition Growth", Applied Physics Letters, 1993, 63, 355–7
206. J. Brunner, T. S. Rupp, H. Gossner, R. Ritter, I. Eisele and G. Abstreiter, "Exectonic Lumi-nescence from Locally Grown SiGe Wires and Dots, Applied Physics Letters, 1994, 64, 994–6

207. S. A. Jenekhe and X. L. Chen, "Self Assembly of Ordered Microporous Materials from Rod-Coil Block Copolymers", Science, 1999, **283**, 3725

208. J. Darnell, H. Lodish and D. Baltimore, "Molecular Cell Biology", New York, Scientific American Books, 1986

209. K.-B.G. Scholthof, "1898 - The Beginning of Virology... Time Marches On". August 2001, http://www.apsnet.org/education/feature/TMV/Top.html

210. "Tabamovirus Images", August 2001, http://www.virology.net/Big_Virology/BVunassignplant.html

211. "Exponential Assembly", August 2001, http://www.zyvex.com/Research/exponential.html

212. O. D. Velev, P. M. Tessier, A. M. Lenhoff and E. W. Kaler, "A Class of Porous Metallic Nanostructures", 1999, Nature, 401, 548

213. R. A. McGill, B. Ringeisen, P. K. Wu "Role of Direct Write Techniques in Chemical- and Biological-Sensor Systems: Commercial & Military Sensing Applications", Chapter 5 in Reference 3.

214. "Fluidic Self Assembly". August 2001, http://www.alientechnology.com/d/library/pdf/fsa_white_paper.pdf

215. "Self-Location of Crystalline Infrared-Detector Pixels for Fabrication of Extremely Large Arrays on Non-Crystalline and Curved Substrates", August 2001, http://www.darpa.mil/MTO/mlp/00overview/hrl.html

216. W. T. S. Huck, J. Tien and G. M. Whitesides, "Three-Dimensional Mesoscale Self-Assembly", 1998, J. American Chemical Society, **120**, 8267–8

217. A. P. Li, F. Muller, A. Birner, K. Nielsch and U. Gosele, "Polycrystalline Nanopore Arrays with Hexagonal Ordering on Aluminum", 1999, J. Vacuum Science & Technology A, 17, 1428–31

218. 218 E. Spiller, "Information Capacity of a Radiation Field", Soft X-Ray Optics, Bellingham WA, SPIE Optical Engineering Press, 1994, 59–80

219. H. I. Smith, "A Statistical Analysis of Ultraviolet, X-Ray and Charged-Particle Lithographies", J. Vacuum Science & Technology B, 1986, **4**, 148–153

220. Reference 110, p. 1839–42

221. S. M. Metev and V. P. Veiko, "Laser Assisted Microtechnology", Berlin, Springer-Verlag, 1998

222. J. Brannon, J. Greer and H. Helvajian, "Laser Processing for Microengineering Applications" in "Microengineering Space Applications", H. Helvajian (Editor), El Segundo CA, Aerospace Press, 1999, 145–199

223. D. J. Nagel, "Macro to Molecular Connectors" in F. L. Carter (Editor), Molecular Electronic Devices II, New York, Marcel Dekker, 1987, 381–403

224. Reference 110, p. 1828–9

225. W. K. Schomburg, R. Ahrens, W. Bacher, J. Martin and V. Saile, "AMANDA-Surface Micromachining, Molding and Diaphragm Transfer", Sensors and Actuators A, 1999, 76, 343–8

226. C. Marrian, "Molecular-Level, Large-Area Printing", August 2001, http://www.darpa.mil/MTO/MLP/index.html

227. J. B. Pendry, "Negative Refraction Makes a Perfect Lens", Physical Review Letters, 2000, **85**, 39669

228. F. Cerrina and C. Marrian (Editors), "Materials-Fabrication and Patterning at the Nanoscale", Pittsburgh, Materials Research Society, 1995

229. D. W. Carr, S. Evoy, L. Sekaric, J. M. Parpia, H. G. Craighead and J. M. Parpia, "Measurement of Mechanical Resonance and Losses in Nanometer Scale Silicon Wires", Applied Physics Letters, 1999, 75, 920–2

230. K. Schwab. E. A. Henriksen, J. M. Worlock and M. L. Roukes, "Measurement of the Quantum of Thermal Conductance", Nature, 2000, **404**, 974–7 and M. Roukes, "Plenty of Room Indeed", Scientific American, September 2001, 48–57

231. J. D'agnese, "Brothers with Heart", Discover Magazine, July 2001, 36–43 & 102

232. M. J. Schoning, S. Schutz, P. Schroth, B. Weissbecker, A. Steffen, P. Kordos, H. E. Hummel, H. Luth, "A BioFET on the Basis of Intact Insect Antennae", Sensors and Actuators B, 1998, 47, 235–8

233. R. Forber, H. Rieger, C. Gaeta, R. Grygier, E. Turcu, S. Campeau, G. French, B. Roberts, P. Hark, K. Cassidy, C. Kelsey, M. Powers, R. Foster, S. Lane, T. Barbee, Z. Chen, J. Burdett, D. Gibson and S. Mrowka, "Laser Plasma X-Ray Point Source and Collimators: Progress and Plans", 2000, unpublished

234. P. W. Arcuni, A. Baum, S. Presley, V. Aebi, M. Mankos, W. Owens and S. Coyle, "NEA Source for Electron Lithography, 2000, unpublished

235. "Focusing Ion Columns", August 2001, http://www.feibeamtech.com/pages/ion.html

236. "NanoFab: About the Nano 150", August 2001, http://www.nanofab.com/about.htm and "NanoLab FIB", August 2001, http://www.nanolab.uc.edu/

237. "IBM and Photronics to Collaborate on Next-Generation Lithography Mask Technologies", August 2001, http://www.chips.ibm.com/news/1999/mask/

238. B. Reuteler, "Chip Makers Move to Nanopositioning", 2110, Design News, April 2001, S37–8

239. H. I. Smith, J. G. Goodberlet, T. Hastings, F. Zhang and M. H. Lim, "Spatial Phase Locking to Achieve Nanometer Accuracy in Lithography", 2000, unpublished

240. J. Robichaud, "Ultralightweight Composite Silicon Carbide Translation Stages for Next Generation Lithography Tools, 2000, unpublished

241. J. G. Spooner, Extreme Ultraviolet: Paving the Way for Smaller, Faster Chips", August 2110, http://www.zdnet.com/eweek/stories/general/0,11011,2689597,00.html

242. L. F. Thompson, C. G. Wilson and J. m. J. Frechet (Editors), "Materials for Microlithography", Washington DC, American Chemical Society, 1984

243. E. Reichmanis, S. A. MacDonald and T. Iwayanagi (Editors), "Polymers in Microlithography: Materials and Processes", Washington DC, American Chemical Society, 1989

244. G. K. Celler and J. R. Maldanado (Editors), "Material Aspects of X-Ray Lithography", Pittsburgh PA, Materials Research Society Proceedings Vol 306, 1993

245. F. M. Houlihan (Editor), "Advances in Resist Technology and Processing XVIII, Bellingham WA, SPIE Vol. 4353, 2001

246. L. H. Dubois and R. G. Nuzzo, " Synthesis, Structure, and Properties of Model Oraganic Surfaces", Annual Reviews of Physical Chemistry", 1992, 43, 437–63

247. P. D. Fuqua, D. P. Taylor, H. Helvajian, W. W. Hansen and M. H. Abraham, "A UV Direct Write Approach for Formation of Embedded Structures in Photostructurable Glass-Ceramics," Reference 2, 79–86.

248. "Foturan", August 2001, http://www.mikroglas.com/foturane.htm

249. M. B. Ranade, Z. S. Gonen and B. W. Eichhorn, "Role of Powder Production Route in Direct Write Applications," in Reference 2, pp, 35–40.

250. S. M. Coleman, R. L. Parkhill, R. M. Taylor and E. T. Knobbe, "Sol-Gel-Derived 0–3 Composite Materials for Direct Write Electronics Applications," in Reference 2, pp. 53–58.

251. P. H. Kydd, D. L. Richard, D. B. Chrisey and K. H. Church, "Laser Processing of Parmod™ Functional Electronic Materials," in Reference 2, pp., 135–142 and "The Technology Behind Parmod", August 2001, http://www.parelecusa.com/

252. H. J. Levinson, "Lithography Process Control (Tutorial Texts in Optical Engineering)", Bellingham WA, SPIE Vol. Tt 28, 1999

253. P. Burggraaf, "A Closer Look at Some of the Most Difficult Processing Challenges", Solid State Technology, September 2000, 81–81–92

254. "CFD-ACE+ Version 6.6", August 2001, http://www.cfdrc.com/datab/software/aceplus/

255. R. Khanna, X. Zhang, J. Protz, and A. A. Ayon, "Microfabrication Protocols for Deep Reactive Ion Etching and Wafer-Level Bonding", Sensors, April 2001, 51–60.

256. K. Rebitz and A. Smith, "Characterizating Exposure Tool Optics in the Fab", Microlithography World, Summer 1999, 10–14.
257. R. R. Kunz, D. D. Rathman, S. J. Spector, T. M. Lyszczarz, B. B. Kosicki and M. Rothschild, "Large-Area Monolithic Detector Arrays for Lithographic Tool Calibration, 2000, unpublished.
258. N. T. Sullivan (Editor), "Metrology, Inspection and Process Control for Microlithography XV", Bellingham WA, SPIE Vol. 4344, 2001.
259. "Etec Product Descriptions", August 2001, http://www.etec-inc.com/products/.
260. Polytec MSV-300 Micro Scanning Vibrometer", August 2001, http://www.polytec.com/poly/lvi.htm.
261. A. C. Diebold, "Semiconductor Metrology: Will Photonics Measure Up?", Photonics Spectra, December 2000, 80–4.
262. M. L. Schattenburg, C. Chen, R. Heilmann, P. Konkola and H. I. Smith, "Interference Lithography Metrological Grids", 2000, unpublished.
263. K. Moloni and M. G. Lagally, "Advanced Probes for Surface and CD Metrology", 2000, unpublished.
264. A. E. Braun, "Defect Inspection Enters Integration Era", Semiconductor International, July 2001, 142–54.
265. C. A. Harper (Editor-in-Chief), "Electronic Packaging and Interconnection Handbook", New York, McGraw Hill, 2000 (Third Edition).

PERMISSIONS

Fig. 6. Courtesy of Coates Circuit Products; Fig. 7. Reprinted with permission from R.C. Haavind and Ref. 6. Copyright 1999 American Association for the Advancement of Science; Fig. 8. Reprinted with permission from Ref. 9; Fig. 9. Adapted from Ref. 10; Fig. 10. Reprinted with permission from Ref. 11; Fig. 11. Courtesy of Hans Fuller, Kionix. Inc; Fig. 12. Reprinted with permission from Ref. 13. Courtesy of Micronics; Fig. 13. Reprinted with permission from Refs. 14 and 15; Fig. 14. Reprinted with permission from Ref. 16. Copyright 2001 American Association for the Advancement of Science; Fig. 15. Reprinted with permission from Ref. 17; Fig. 16. Reprinted with permission from Electroglas, Inc; Fig. 19. Reprinted with permission from Refs 26 and 27; Fig. 20. Adapted from Ref. 30; Fig. 21. Adapted from Ref. 47; Fig. 22. Adapted from Ref. 47; Fig. 23. Adapted from Ref. 50; Fig. 24. Reprinted with permission from Ref. 52; Fig. 25. Reprinted with permission from Ref. 53; Fig. 26. Reprinted with permission from Ref. 57; Fig. 27. Reprinted with permission from Ref. 59; Fig. 28. Reprinted with permission from Ref. 64; Fig. 29. Reprinted with permission from Ref. 62; Fig. 30. Courtesy of Sandia National Laboratories; Fig. 31. Adapted from Ref. 68; Fig. 32. Reprinted with permission from Ref. 69; Fig. 33. Courtesy of Joseph Mangano. SRL. Inc; Fig. 34. Reprinted with permission from Ref. 79; Fig. 35. Reprinted with permission from Ref. 80; Fig. 36. Reprinted with permission from Ref. 81; Fig. 37. Reprinted with permission from Ref. 85; Fig. 39. Reprinted with permission from Refs. 89 & 90; Fig. 40.

Reprinted with permission from Ref. 92; Fig. 41. Reprinted with permission from Ref. 93; Fig. 42. Reprinted with permission from Ref. 95; Fig. 43. Reprinted with permission from Ref. 98; Fig. 44. Reprinted with permission from Ref. 98; Fig. 45. Reprinted with permission from Ref. 99; Fig. 46. Reprinted with permission from Ref. 99; Fig. 47. Adapted from Ref. 105; Fig. 50. Adapted from Ref. 122; Fig. 51. Reprinted with permission from Ref. 123; Fig. 52. Adapted from Ref. 124; Fig. 53. Reprinted with permission from Ref. 125; Fig. 55. Reprinted with permission from Ref. 131; Fig. 56. Reprinted with permission from Ref. 133; Fig. 57. Reprinted with permission from Ref. 1; Fig. 58. Adapted from Ref. 142; Fig. 60. Reprinted with permission from Ref. 23. Copyright 1995 American Chemical Society; Fig. 62. Courtesy of Ref. 148; Fig.63. Courtesy of Operon Technologies; Fig. 64. Reprinted with permission from Ref. 150; Fig. 66. Courtesy of Microfab. Inc.. Ref. 154; Fig. 67. Reprinted with permission from IEDM 2000; Fig. 68. Reprinted with permission from Ref. 158. Copyright 2001 American Association for the Advancement of Science; Fig. 70. Courtesy of A Piqu. Naval Research Laboratory; Fig. 73. Reprinted with permission from Ref. 170; Fig. 74. Reprinted with permission from Ref. 171; Fig. 78. Reprinted with permission from Ref. 183; Fig. 79. Reprinted with permission from Ref. 184; Fig. 80. Reprinted with permission from Ref. 185; Fig. 81. Reprinted with permission from Ref. 142. Copyright 1998 American Association for the Advancement of Science; Fig. 82. Reprinted with permission from Ref. 142. Copyright 1998 American Association for the Advancement of Science; Fig. 83. Adapted from Ref. 196; Fig. 84. Reprinted with permission from Ref. 197; Fig. 85. Reprinted with permission from Ref. 198; Fig. 86. Reprinted with permission from Ref. 199; Fig. 88. Courtesy of A. Baski. Virginia Commonwealth University; Fig. 90. Reprinted with permission from Ref. 208; Fig. 91. Reprinted with permission from Ref. 212; Fig. 92. Reprinted with permission from Ref. 214; Fig. 93. Reprinted with permission from Ref. 216. Copyright 1999 American Chemical Society; Fig. 94. Reprinted with permission from Ref. 217; Fig. 98. Reprinted with permission from Refs. 229 and 230.

Index

Page numbers in **Bold** refer to figures.

712

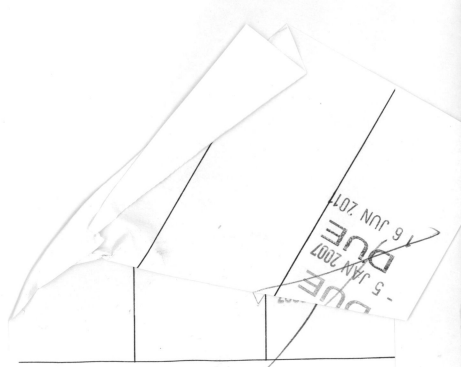

Books are to be returned on or before
the last date below.

DUE
- 5 JAN 2007

DUE
1 6 JUN 2014